MATERIALS SCIENCE - ADVANCED TOPICS

Edited by **Yitzhak Mastai**

Materials Science - Advanced Topics
http://dx.doi.org/10.5772/56700
Edited by Yitzhak Mastai

Contributors

Alexander B. Yakovlev, Yashwanth R. Padooru, George W. Hanson, Chandra S. R. Kaipa, Mihály Janóczki, Ákos Becker, László Jakab, Richárd Gróf, Tibor Takács, Stella Ezeonu, Chien-Yi Jay Huang, Hiroshi Naganuma, Hiromi Shima, Soichiro Okamura, Petr Ptáček, František Šoukal, Tomáš Opravil, Jiří Brandštetr, Vu Trung, Minh Duc Le, Tomohiro Iwasaki, Mohammad Amin Torabizadeh, Jairo Tronto, João Barros Valim, Ana Cláudia Bordonal, Zeki Naal, Chung-Wei Yang, Gaven Martin, H.P.S. Abdul Khalil, S S Suhaily, W.O. Wan Nadirah, Mohammad Jawaid, Jagannathan Thirumalai, Rathinam Chandramohan Pillai, T. A. Vijayan, R. Chandramohan, Hagay Yehuda Moshe, Zlomanov, Giovanni Franco Crosta, Michael Zinigrad, Konstantin Borodianskiy, Marios Nestoros

Published by InTech

Janeza Trdine 9, 51000 Rijeka, Croatia

Publishing Process Manager Ana Pantar
Technical Editor InTech DTP team
Cover InTech Design team

Additional hard copies can be obtained from orders@intechopen.com

Materials Science - Advanced Topics, Edited by Yitzhak Mastai
p. cm.
ISBN 978-953-51-1140-5

For EU product safety concerns:
IN TECH d.o.o., Prolaz Marije Krucifikse Kozulić 3, 51000 Rijeka, Croatia,
info@intechopen.com or visit our website at intechopen.com.

Meet the editor

Prof. Yitzhak Mastai was born in 1966 in Tel Aviv, Israel. He obtained his B.Sc in physical chemistry from Bar-Ilan University in 1989 and received his PhD from the Weizmann Institute of Science with Prof. Gary Hodes on nanomaterials synthesis (1999). He then went to the Max Planck institute of colloids and interfaces for 3 years, as postdoctoral fellow to work with Prof. M. Antonietti and Prof. H Cölfen on biomimetic chemistry and chiral polymers. In 2003 joined the staff of the chemistry department at Bar-Ilan University, where he is currently a Professor at the institute of nanotechnology at Bar-Ilan University leading the nano chirality laboratory. Prof. Mastai's earlier interests included nanomaterials synthesis and characterization. His current research is focused on the synthesis and analysis of chiral nanosurfaces, chiral self-assembled monolayers and polymeric chiral nanoparticles. Mastai has published more than 100 scientific articles and book chapters on various aspects of nanomaterials and chirality at the nanoscale and recently edited book on "Advances in Crystallization Processes".

Contents

Preface

Materials have always been important to human society, as shown by the fact that our prehistoric eras are named after the new material that defined them, e.g. the Stone Age, the Bronze Age, and the Iron Age. The present age is likely to remain known as the advanced materials or nanomaterials Age. Capitalizing on the recent advent of new instrumentation for materials synthesis and characterization there has been a nearly explosive growth in materials science research, especially in them subfield known as nanoscience, which studies the properties of materials at the nanometer length scale.

The development of novel technologies almost always relies on the use of newly available materials. Such new technologies in turn are the basis for new industries, which create jobs and better living conditions. In this framework, when studying materials that may lead to innovative applications, the scientific community has the opportunity and even the responsibility to gear its work to the objective of improving the quality of life rather than just following scientific curiosity. Today modern materials science is a vibrant, emerging scientific discipline at the forefront of physics, chemistry, engineering, biology, and medicine and is becoming increasingly international in scope as demonstrated by emerging international and intercontinental collaborations and exchanges.

The overall purpose of this book is to provide timely and in-depth coverage of selected advanced topics in materials science. This textbook goes into considerable detail concerning the many elements of knowledge needed to understand both quantitatively and qualitatively advanced subjects in materials science. The articles for this book have been contributed by respected researchers in this area and cover the frontier areas of research and developments in materials science incorporating most recent developments and applications of materials science technology. Divided into five sections this book provides the latest research developments in many aspects of materials science: It is our hope that you, as readers, will find this book useful for your work. If so, this will be the nicest reward for us.

I wish to thank all the authors for their contributions to this book and it is my pleasure to acknowledge the assistance of Ana Pantar during the write-up of this book and is preparation in final format.

Finally, I would like to acknowledge my family who has supported me through all the years of doing science, to my wife Dina von schwarze and to our children Michael, Reut and Noa for their patience and support.

Prof. Yitzhak Mastai
Department of Chemistry and the Institute of Nanotechnology,
Bar-Ilan University, Ramat-Gan, Israel

Section 1

Materials Synthesis

Chapter 1

Conducting Polymers / Layered Double Hydroxides Intercalated Nanocomposites

Jairo Tronto, Ana Cláudia Bordonal, Zeki Naal and
João Barros Valim

Additional information is available at the end of the chapter

http://dx.doi.org/10.5772/54803

1. Introduction

Layered nanocomposites represent a special class of multifunctional materials that has received a lot of attention over the last years [1-6]. The specific architecture of these composites promotes a synergistic effect between the organic and inorganic parts, generating compounds with different chemical or physical properties as compared with the isolated components. These composites not only represent a creative alternative to the search for new materials, but also allow the development of innovative industrial applications. The potential uses of layered nanocomposites include intelligent membranes and separation devices, photovoltaic devices, fuel cell components, new catalysts, photocatalysts, chemical and biochemical sensors, smart microelectronic devices, micro-optic devices, new cosmetics, sustained release of active molecules, and special materials combining ceramics and polymers, among others [7-18].

A great variety of layered nanocomposites can be prepared from the combination between polymers and layered inorganic solids [1-3]. Compared with the unmodified polymers, the resulting materials present dramatic improvement in properties such as rigidity, chemical and mechanical resistance, density, impermeability to gases, thermal stability, and electrical and thermal conductivity, as well as high degree of optical transparency.

The first successful development concerning the combination of layered inorganic solids with polymers was achieved by researchers from Toyota®, who aimed at structural applications of the nanocomposites in vehicles. These researches prepared nanocomposites by combining nylon-6 and montmorillonite (clay) using the *in situ* polymerization method [20-22]. Research conducted over the past 10 years has shown that nanocomposites containing only a small amount of inorganic silicate (2% volume), exhibit twofold larger elastic modulus and strength without sacrificing resistance to impact. Other automobile companies began to employ this type of material in their vehicles and intensified research in this area [1-3,21].

The excellent gas barrier and vapor transmission properties of these hybrid nanocomposites have led to their application mainly in food industry, more specifically in food and drink packaging. Incorporation of layered silicate nanoparticles into polymeric matrices creates a labyrinth within the structure, which physically retards the passage of gas molecules [22]. These materials can also be used to coat storage tanks in ships and lines of cryogenic fuels in aerospace systems. Compared with the unmodified polymer, nanocomposites delay fire propagation and enhance thermal stability. In contrast to the amount of additives used in traditional fireproof polymers (60%), these nanocomposites contain low layered inorganic solid loading, typically 2-5 wt%. This is due to the formation of an insulating surface layer that not only slows degradation of the polymer, but also decreases its calorific capacity [1,2]. The decomposition temperature of these nanocomposites can be increased to 100 °C, which extends the use of these materials at ambient temperatures, as in the case of automobile engines.

With respect to environmental applications, layered inorganic solids combined with biodegradable polymers have been employed as reinforcing agents. These materials, called "green" nanocomposites, are an attractive alternative for the replacement of petroleum derivatives in the production of plastics.

Depending on the nature of the components and on the preparation method, two main types of nanocomposites can be obtained from the association of layered compound with polymers, as shown in Figure 1:

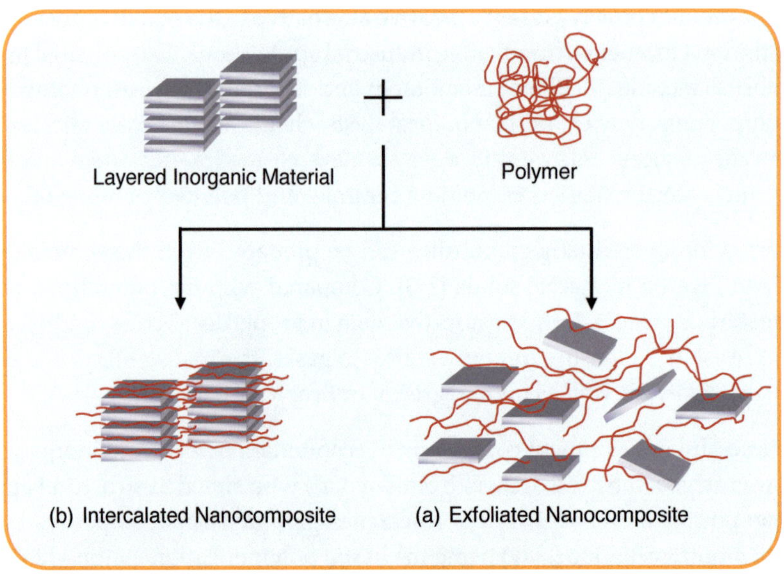

Figure 1. Schematic representation of the different types of composites produced from the interaction between layered compounds and polymers: (a) Intercalated Nanocomposite; (b) Exfoliated Nanocomposite.

- Intercalated Nanocomposite: The polymer is intercalated between the inorganic layers, producing a nanocomposite consisting of polymeric chains and alternating inorganic layers. Intercalation of the polymer often results in increased interlayer spacing; *i.e*, larger distance between two adjacent inorganic layers. (Figure 1a).
- Exfoliated Nanocomposite: The material presents no ordering along the stacking axis of the layer, or the spacing between the inorganic layers is greater than 8 nm. (Figure 1b).
- In addition to the well-defined structures cited above, a third intermediate type of structure can be found, in which the material presents characteristics of intercalation and exfoliation. In this case, there is broadening of the X ray diffraction peaks.
- Several strategies have been utilized for the preparation of organic-inorganic hybrid materials containing layered inorganic solids and polymer [1,2]:
- Exfoliation-adsorption: The layered compound is exfoliated using a solvent in which the polymer is soluble. In some layered compounds there are weak interaction forces between the layers, which can thus be easily exfoliated in appropriate solvents. The polymer may then adsorb onto the exfoliated layers which, after evaporation of the solvent, can be stacked again. As a result, the polymer is intercalated, and an ordered multilayer structure is obtained.
- *In situ* intercalative polymerization: The layered compound undergoes swelling (interlayer expansion) in a solution containing the monomer. The polymer is formed in the interlayer region. The polymerization reaction can be performed by heat or radiation treatment, using an organic initiator or a fixed catalyst.
- Melted polymer intercalation: The layered compound is mixed with the polymer matrix in the melting phase. If the layered surfaces are sufficiently compatible with the selected organic polymer under these conditions, the latter penetrates into the interlayer space, generating an intercalated or exfoliated nanocomposite. This technique does not require any solvent.
- Template Synthesis: This method can only be used for water-soluble polymers. The layered compound is formed *in situ* in an aqueous solution containing the target polymer on the basis of self-assembly forces, the polymer aids nucleation and growth of inorganic layers. As a result, the polymers are retained between the layers.

Among the inorganic solids used in the preparation of layered nanocomposites, one promising class of material is the Layered Double Hydroxides (LDHs), which have been added to polymers for the synthesis of LDH/polymers nanocomposites [23-26]. LDHs can be structurally described as the stacking of positively charged layers intercalated with hydrated anions [27]. In order to better understand the structure of the LDH, it is appropriate to start from the structure of brucite. In this $Mg(OH)_2$ structure, the magnesium cations are located in the center of octahedra, with hydroxyl groups positioned at their vertices. These octahedra share edges, forming neutral planar layers that are held together by hydrogen bonds. In this type of structure, the isomorphic replacement of bivalent cations with trivalent ones creates a positive residual charge in the layers. For charge balance to be reached in the system, anions should be present between the layers. Together with water molecules, the anions promote stacking of the layers, which culminates in the layered double hydroxide structure displaying a poorly ordered interlayer domain. Not only hydrogen bonding but also electrostatic attraction between the positively charged layers and the interlayer anions hold the layers together in LDHs. A schematic representation of the LDH structure is given in Figure 2.

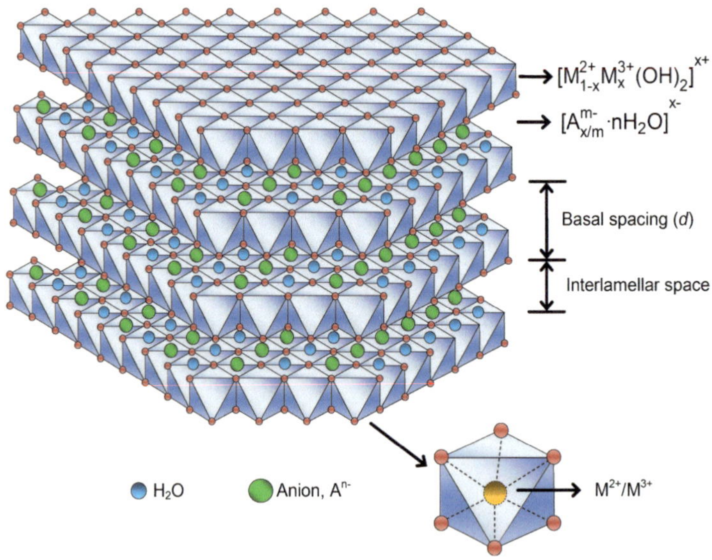

Figure 2. Schematic representation of the LDH structure.

The inorganic layers of the LDH can be stacked according to two different symmetries, resulting in rhombohedral or hexagonal unit cells of the hexagonal system. Most of the synthetic LDH belong to the hexagonal system. Only LDH with M(II)/M(III) ratio equal to 1 are orthorhombic. For the rhombohedral unit cell, the parameter c is three times the basal spacing, space group R3m. In the case of the hexagonal unit cell, the parameter *c* is twice the basal spacing, space group P63mmc. The notations 3R and 2H refer to unit cells as rhombohedral or hexagonal, respectively.

In the structure of LDH, the interlayer domain comprises the region between adjacent inorganic layers. This region is composed of randomly distributed anions and water molecules. Powder X ray diffraction (PXRD) and EXAFS studies, performed by Rousselet et al. showed the highly disordered nature of this region [28]. Besides being found in the interlayer domain, where they hydrate the intercalated anions, the water molecules can also be adsorbed between the crystallites. The water molecules onto the surface of the micro crystallites surface are called extrinsic water molecules, whereas those that are located in the interlayer domain are designated intrinsic water molecules. The global hydration status of the LDH is the addition of both terms, intrinsic hydration and extrinsic hydration. Many researchers consider the interlayer domain of LDH a quasi-liquid state, which gives high mobility to the interlayer anions.

A wide variety of anions can be intercalated into the LDH; for example, organic anions, inorganics and organic-inorganics, and polymers. The intercalation of more than one type of anion in the interlayer domain is an extremely rare phenomenon. Usually, the presence of two or more kinds of anions during the synthesis generates a competition between these anions, and the one with greater tendency to stabilize the system and/or that is present in larger amount will be intercalated. Using PXRD and *in situ* X ray energy dispersion spectroscopy techniques, Fogg et al. reported the existence of a second intermediate stage due to co-intercalation of Cl^- ions and succinate in LiAl-LDH [29]. Pisson et al. studied the exchange of Cl^- anions with succinate and tartarate anions in LDH of the system [Zn_2Al-Cl], [Zn_2Cr-Cl], and [Cu_2Cr-Cl]. The exchange reaction was monitored *in situ* by the X ray diffraction and X ray energy dispersion spectroscopy techniques. The analyses revealed the formation of a second intermediate stage in all the materials, caused by co-intercalation of organic anions and chloride ions [30]. Kaneyoshi and Jones demonstrated that terephthalate anions can adopt two different orientations in relation to the inorganic layers when they are intercalated into Mg-Al-LDH. The longer molecular axis is either perpendicular or parallel to the plane of the layers. These two orientations are known as interstratified intermediated phases. The occurrence of these two orientations of intercalated terephthalate anions was supported by the appearance of a third basal spacing, attributed to the contribution of two different orientations of anions in the interlayer domain.

A large number of natural and synthetic LDHs containing various metal cations have been studied. In order to form the LDH, the metal cations that will be part of the inorganic layer must present octahedral coordination and ion radius in the range of 0.50 to 0.74 Å. By varying the metal cations, the proportion among them, and the interlayer anion, a large variety of LDH can be prepared. Countless cations can be part of this structure: Mg^{2+}, Al^{3+}, most of the cations of the first transition period, Cd^{2+}, Ga^{3+}, and La^{3+}, among others [27]. In addition LDH displaying more than one bivalent and/or trivalent cation can be synthesized, which further expands the compositional possibilities.

The ratio between the metal cations M(II)/M(III) is also very important, because a change in this ratio between modifies the charge density in the layers, since the charge is generated from the isomorphic substitution of bivalent cations with trivalent ones in the structure of the inorganic layers [27,32]. There is controversy over the values that the x parameter in the general formula of the LDH can assume during the synthesis of these materials. According to de Roy et al., the x value should lie between 0.14 and 0.50, for the formation of an LDH where the M(II)/M(III) ratio can vary between 1 and 6 [33]. For Cavani et al., the x value must fall between 0.20 and 0.34, with the M(II)/M(III) ratio ranging between 2 and 4.37. However, some researchers have reported the synthesis of LDH with different M(II) to M(III) ratios from those mentioned above [27].

As described earlier, the interlayer domain consists of water molecules and anions, mainly. Practically, there is no limitation to the nature of anions that can compensate the residual positive charge of the LDH layers. However, obtaining pure and crystalline materials is not an easy task. Generally, simple inorganic anions with higher charge/radius ratio have greater tendency for intercalation. This is because these anions interact more strongly with the inorganic layers from an electrostatic viewpoint. For the intercalation of organic anions, especially anionic polymers, factors such as the size and geometry of the anion, the interaction between them, and the ratio between size and charge must be taken into account. Some interlayer anions are more mobile, which gives the resulting materials good exchange properties.

Several factors must be borne in mind when planning the synthesis of LDH. For instance, the degree of substitution of M(II) with M(III) cations, the nature of the cation, the nature of the interlayer anion, the pH of the synthesis and, in some cases, the controlled atmosphere. Furthermore, to obtain materials with good crystallinity, the concentrations of the solutions, the rate of the addition of the solutions, the stirring rate, the final pH of the suspension (for variable pH methods), the pH during the addition (for constant pH method), and the temperature of the mixture (typically performed at room temperature) must be controlled. There are a number of methods that can be used for the synthesis of LDH. They can be divided into two categories:

i. Direct synthesis methods: salt-base method or co-precipitation (at variable pH or at constant pH), salt-oxide method, hydrothermal synthesis, induced hydrolysis, sol-gel method, and electrochemical preparation [27,33-37].

ii. Indirect synthesis methods: simple anionic exchange method, anionic exchange by regeneration of the calcined material, and anion exchange using double phase, with formation of a salt between the surfactants [33,38,39].

Among the most extensively investigated conducting polymers are polyacetylene; poly-heterocyclic five-membered compounds like polypyrrole, polythiophene, and polyfuran; and polyaromatics such as polyaniline and poly (p-phenylene).The structures and respective electrical conductivity values of some conductive polymers are summarized in Table 1 [40].

Name	Structural Formula	Conductivity / S·cm^{-1}
Polyacetylene	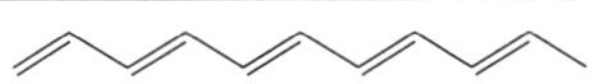	10^3 a 10^6
Polyaniline	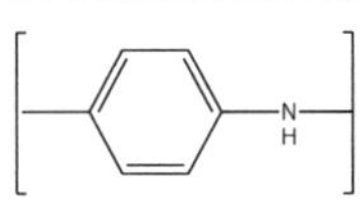	10 a 10^3
Polypyrrole	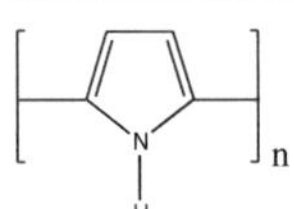	600
Poly(*p*-phenylene)	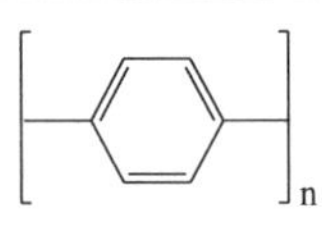	500
Polythiophene	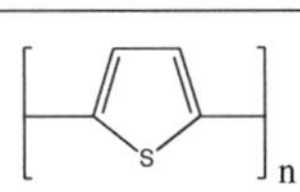	200

Table 1. Structure and electrical conductivity values of some conductive polymers [40].

http://dx.doi.org/10.5772/54803

Below, we will outline some classes of conducting polymers:

a. Polyaniline:

Polyanilines are widely studied because of their low cost, good stability in the presence of oxygen and water, and interesting redox properties. In 1835, polyaniline was first synthesized as "black aniline", a term used for the product obtained by oxidation of aniline [41]. Some years later, Fritzche analyzed the products obtained by chemical oxidation of this aromatic amine [42]. In 1862, Letheby found that the anodic oxidation of aniline in a platinum electrode, in an aqueous solution of sulfuric acid, formed a dark brown precipitate [43]. The polyaniline chain consists of units present in two main forms: (i) the fully reduced form, which contains only aromatic rings and nitrogen atoms of the amine function, shown in Figure 3a, and (ii) a completely oxidized form displaying iminic nitrogen atoms, quinonics rings, and aromatic rings, as represented in Figure 3b.

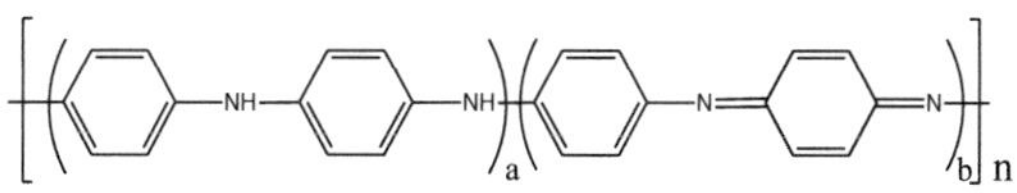

Figure 3. Representation of the general structure of the polyaniline base form: (a) reduced species (b) oxidized species.

Gospodinova and Terlemezyan examined the oxidation state of polyaniline constituents. The principal oxidation states of polyaniline are presented in Table 2 [44]. The emeraldine salt is the structural form of polyaniline that affords higher conductivity values. Polyaniline can be doped by protonation, with no change in the number of electrons in the polymer chain.

Oxidation state	Strucuture	Color	Characteristic
Leucoemeraldine	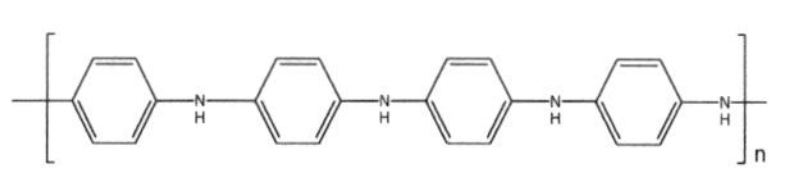	Yellow 310	Insulating, completely reduced
Emeraldine salt	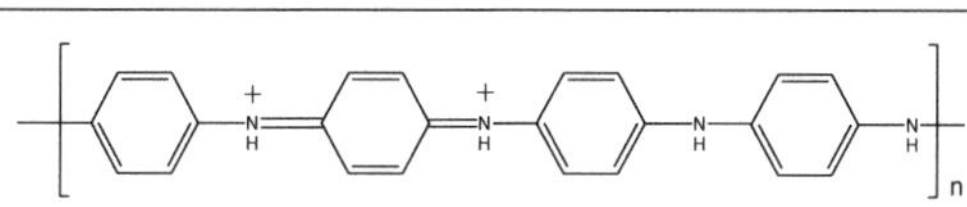	Green 320, 420, 800	Conductive, half-oxidized
Emeraldine base	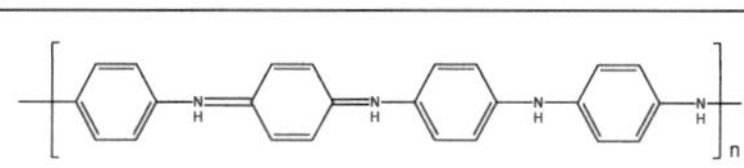	Blue 320, 620	Insulating, half-oxidized
Pernigraniline	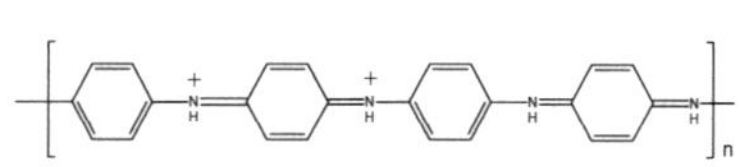	Purple 320, 530	Insulating, completely oxidized

* The numerical values refer to the wavelength (in nanometers) where absorption is maximum.

Table 2. Most important oxidation states of polyaniline [44].*

b. Polypyrrole:

The first report about pyrrole was published by Runge in 1834. This author observed a red component in coal tar and bone oil. The compound isolated and purified from this component was named of pyrrole [45]. The structural formula of pyrrole was established in 1870. In the late 19th century, the interest in pyrrole and its derivatives was aroused, following the discovery that this molecule was part of some porphyrins found in biological systems, such as chlorophyll. Approximately 100 years after the first report on the discovery of pyrrole, *i.e,* in 1970, interest in these materials increased again due to the possibility of preparing conducting polypyrroles [46].

Pyrrole is a five-membered cyclic compound (heterocyclic) containing 6 π electrons. Additionally, pyrrole has an sp^2 nitrogen, and its three σ bonds are located in the plane of the ring. The excess electron, the conjugation of the double bonds, and their ability to relocate are the structural characteristics underlying the charge conduction properties presented by the pyrrole polymer products.

c. Polythiophene:

In 1882, Meyer discovered thiophene [47]. At that time, his studies revealed that this compound, which was isolated from benzene impurities, was a new aromatic system. Thiophene is not a component of animal metabolism, but some thiophene derivatives can be found in plants. Thiophene derivatives are widely employed in many types of chemical industry, including the pharmaceutical, veterinary, polymers, and agrochemicals industries.

Thiophene is a compound analogous to pyrrole. Instead of the nitrogen heteroatom, it contains a sulfur atom with sp^2 hybridization. The sp^2 orbital, which is perpendicular to the π electron system, has an unshared electron pair. The p-orbital of sulfur donates two electrons to the π system. Polythiophene derivatives have been extensively studied, probably because most of them are soluble in organic solvents, which facilitates processing of the material.

The electrical conductivity of a solid is the result of the number of charge carriers (electrons / holes) and their mobility. Conducting polymers have a large number of charge carriers with low mobility, which is mainly caused by the large number of structural defects such as reticulation and the disordering of chains. The formation of nanocomposites by intercalation of conductive polymers into LDH can minimize the formation of reticulation defects and the disordering of polymer chains, furnishing materials with new and interesting properties. Table 3 summarizes the literature works on the synthesis and characterization of conducting polymers intercalated into LDH.

Year	Nanocomposites	Examples	Authors	Ref.
1994	LDH / polyaniline	CuCr- polyaniline-LDH CuAl- polyaniline-LDH	Challier and Slade	48
2001	LDH / aminobenzoate derivatives	LiAl-*o*-, *p*- and *m*-aminobenzoate-LDHs	Isupov et al.	49

2002	LDH / aniline sulfonate	CuCr- aniline sulfonate-LDH	Moujahid et al.	50
2003	LDH / aminobenzene sulfonate	CuCr- aminobenzene sulfonate-LDH	Moujahid et al.	51
2004	LDH / 2-thiophenecarboxylate	ZnAl-2-thiophenecarboxylate-LDH ZnCr-2-thiophenecarboxylate-LDH	Tronto et al.	52
2005	LDH / aniline sulfonate derivatives	CuCr-*o*- and *m*-aminobenzenesulfonate, 3-amino-4-methoxybenzenesulfonate, 3-aniline-1-propane sulfonate, and 4-aniline-1-butane sulfonate-LDHs.	Moujahid et al.	53
2006	LDH / pyrrol derivatives	ZnAl-4-(1H-pyrrol-1-yl)benzoate-LDH ZnCr-4-(1H-pyrrol-1-yl)benzoate-LDH ZnAl-3-(pyrrol-1-yl)-propanoate-LDH ZnAl-7-(pyrrol-1-yl)-heptanoate-LDH	Tronto et al.	54 55 56
2006	LDH / aminobenzoate derivatives	MgAl-aminobenzoate-LDH NiAl-aminobenzoate-LDH	Tian et al.	57 58
2006	LDH / aniline sulfonic	NiAl-aniline sulfonic -LDH	Wei et al.	59

Table 3. Some examples of Nanocomposites consisting of LDH/conductive polymers.

Challier and Slade reported the synthesis and characterization of layered nanocomposites of CuCr and CuAl-LDHs intercalated with polyaniline [48]. The oxidizing host matrices were prepared by the coprecipitation method, with the intercalation of terephthalate anions into CuCr-LDH and hexacyanoferrate(II) anions into CuAl-LDH. Then, the LDH precursors were submitted to an anion exchange reaction with a solution of pure aniline under reflux, for 24 h. The X ray diffractograms showed that the materials submitted to reaction with aniline exhibited basal spacings of 13.3 Å and 13.5 Å for CuCr-LDH and CuAl-LDH, respectively. This result was consistent with the intercalation of aniline molecules containing aromatic rings oriented perpendicular to the plane of the layers. FTIR analyses evidenced polymerization of the aniline molecules, since the absorption spectra displayed bands typical of the emeraldine form. According to the authors, the oxidant character of Cu^{2+}, present in layered structure of the inorganic host, helped induce oxidative polymerization of the aniline intercalated in the interlayer galleries.

Isupov et al. described the intercalation of *o*-, *p*-, and *m*-aminobenzoate anions into LiAl-LDH [49]. The incorporation of aminobenzoate anions in the host matrices was conducted by anion exchange, from an LiAl-LDH intercalated with chloride anions. The basal spacings obtained from the X ray diffractograms indicated that the anion exchange reaction was effective, with incorporation of aminobenzoate anions in the interlayer domain. To carry out the *in situ* polymerization, samples of the nanocomposites were submitted to a heat-treatment at 90 °C for 100 h, with 75% relative air humidity. For the LiAl-LDH intercalated with *m*-aminobenzoate anions, the formation of a polyconjugated system was confirmed by ESR spectra performed in vacuum at 77 K and 300 K. The spectra displayed a broad isotropic signal between 7.3 and 7.5 G, with g = 2.000 and line with Gaussian shape. Heating of the nanocomposite in air intensified the ESR signal. Formation of the polyconjugated system was also corroborated by FTIR and Raman spectroscopies. A comparison of the FTIR spectrum of LiAl-LDH intercalated with *m*-aminobenzoate and its oxidation products revealed a marked decrease in the band located at 1250 cm^{-1}, which indicates a decrease in the number of amino groups, $-NH_2$.

Moujahid et al. reported the intercalation of the *m*- and *o*-aminebenzeno sulfonate anion, 3-amine-4-methoxybenzene sulfonate, 3-aniline-1-propane sulfonate and 4-aniline-1-butane sulfonate into Cu_2Cr-LDHs [50,51,53]. These authors discussed the arrangement of intercalated molecules and their subsequent dimerization and/or *in situ* polymerization. The authors incorporated these inorganic anions between the layers was by the direct precipitation method at constant pH. After the synthesis, the resulting materials were submitted to heat treatment at different temperatures, under air atmosphere. The interlayer distances of they synthesize nanocomposites were consistent with the presence of a bilayer of guest molecules in the interlayer space. The heat treatment performed at 350 K culminated in a contraction in the basal spacing of the nanocomposites, except for the material intercalated with the *o*-amine-benzene sulfonate anions. This contraction was associated with reorientation of the intercalated molecules and/or with an *in situ* polymerization. At temperatures above 350 K up to approximately 450 K, there was no significant variation in basal spacing. ESR analysis evidenced *in situ* polymerization. The profiles of the ESR spectra changed with increasing temperature. The value of the signal g ($g = 2.0034 \pm 0.0004$) was typical of the formation of organic radicals and/or conduction electron. For the nanocomposite intercalated with the *m*-aminobenzene sulfonate anion, the ESR and CV analysis showed that the *in situ* polymerization must occur with a syndiotactic arrangement. For the nanocomposite intercalated with the *o*-aminobenzene sulfonate anion, the ESR studies indicated a very weak response of spin carriers for these materials. The electrochemical characterization did not show the presence of reversible redox processes. These results, together with the constancy of the basal spacing value obtained up to a temperature of 450 K, suggested that *in situ* polymerization was not favored when the monomer had the amino group located at the *ortho* position relative to the sulfonate group. For the 3-amine-4-methoxybenzene sulfonate anion, the presence of methoxy group in the *para* position relative to the sulfonate group made the polymerization process difficult. The ESR and CV data for this nanocomposite indicated formation of a dimer. For the nanocomposites synthesized with 3-aniline-1-propane sulfonate and 4-aniline-1-butane sulfonate, the heat-treatment at 473 K prompted an increase in g ($g = 2.0034 \pm 0.0004$), which is associated with the generation of organic radicals and/or conduction electrons.

Tronto et al. described the synthesis, characterization, and electrochemical investigation of 2-thiophenecarboxylate intercalated into ZnAl-LDH and ZnCr-LDH [52]. The materials were synthesized by the coprecipitation method at constant pH, followed by hydrothermal treatment at 65 °C for 72 h. The LDH were analyzed by PXRD, FT-IR, ^{13}C CP-MAS, TEM and CV. The basal spacing was about 15.3 Å for all the LDH which suggested the formation of bilayers of anions intercalated between the inorganic sheets. In this configuration, the 2-thiophenecarboxylate anions would be in a position in which their longer axes would lie perpendicular to the plane of the inorganic layers. Besides the phase with basal spacing of 15.3 Å, another phase with basal spacing of 7.58 Å was also detected in the diffractograms. This value was similar to some values reported for the intercalation of CO_3^{2-} anions into ZnAl-LDH and ZnCr-LDH. However, the qualitative analysis and ^{13}C CP-MAS did not confirm the presence of carbonate anions, as contaminant in the LDH. Thus, the results indicated that for this second phase, 2-thiophenecarboxylate anions were intercalated with their longer axes parallel to the plane of the inorganic layer. ^{13}C CP-MAS data further suggested that, during the synthesis, the 2-thiophenecarboxylate anions lost an acid hydrogen which led to formation of the dimer.

Tronto et al. conducted a study on the *in situ* polymerization of pyrrole derivatives, 4-(1H-pyrrol-1-yl)benzoate, 3-(Pyrrol-1-yl)-propanoate, and 7-(pyrrol-1-yl)-heptanoate, intercalated into LDH [54-56]. The materials were synthesized by co-precipitation at constant pH, followed of hydrothermal treatment for 72 h. The final LDH were characterized by X ray diffraction, ^{13}C CP-MAS NMR, TGA, and ESR. The basal spacing value coincided with the formation of bilayers of intercalated monomers. ^{13}C CP-MAS NMR and ESR analyses showed the formation of a polyconjugated system with polymerization of the monomers intercalated in the LDH during the coprecipitation and/or hydrothermal treatment processes. This result reinforced the authors assumption that the connectivity between the monomers occurred spontaneously during the synthesis, with generation of oligomers and/or syndiotactic polymers intercalated between the LDH layers. At room temperature, the ESR spectrum displayed a signal typical of the hyperfine structure (hfs). The presence of hfs suggested the existence of a proper regulatory environment for the free electrons. These electrons would be present in an organic "backbone" of small size. Thermal analysis of these materials revealed that the inorganic host matrix provided the intercalated polymers with thermal protection, because the thermal decomposition reactions happened at higher temperatures compared with the pure polymers.

Tian and cols. investigated the oxidative polymerization of m-$NH_2C_6H_4SO_3^-$ anions intercalated into NiAl-LDH using ammonium persulfate as the oxidizing agent [57]. The amount of oxidizing agent required for controlled polymerization of the intercalated monomers was systematically evaluated. The materials were characterized by PXRD, UV-Vis spectroscopy, FT-IR spectroscopy, and XPS determination. PXRD and elemental analysis data showed the co-intercalation of nitrate anions, originating from the LDH precursor, and m-$NH_2C_6H_4SO_3^-$ anions. UV-Vis results evidenced polymerization of the intercalated m-$NH_2C_6H_4SO_3^-$ anions, with the formation of small chains. The intercalated polyaniline sulfonate was present in different oxidation states and at different protonation levels, depending upon the amount of oxidizing agent that was added.

Tian and cols. also performed the *in situ* oxidative polymerization of m-$NH_2C_6H_4SO_3^-$ anions intercalated into MgAl-LDH [58]. The monomers were incorporated into the LDH via an exchange reaction using the precursor $[MgAl(OH)_6](NO_3)$ $n H_2O$. The nitrate anions remaing from the exchange reaction and co-intercalated with the m-$NH_2C_6H_4SO_3^-$ anions were utilized as oxidizing agent for the oxidative polymerization of the intercalated monomers. The resulting materials were analyzed by DTA-TG-DSC as well as UV-Vis and HT-XRD spectroscopies. In the temperature range 300-350 °C, the UV-Vis analysis confirmed the reduction of nitrate and polymerization of aniline.

Wei et al. reported the oxidative polymerization of m-$NH_2C_6H_4SO_3^-$ anions in NiAl-LDH, using intercalated nitrate anions as the oxidizing agent [59]. The LDH interlayer space was used as a "nanoreactor" for the *in situ* polymerization of the intercalated monomer. Polymerization of the monomers was acomplished by heat treatment under nitrogen atmosphere. The interlayer polymerization was monitored by thermogravimetric analysis coupled with differential thermal analysis and mass spectrometry (TGA-DTA-MS), UV-Vis spectroscopy, X ray absorption near edge (XANES), (HT-XRD) and FTIR spectroscopy. Polymerization of the monomer was observed at a temperature of 300°C.

2. Synthetic strategies for the preparation of conducting polymers / layered double hydroxides intercalated nanocomposites

The synthesis of intercalated nanocomposites of LDH/conductive polymers can be carried out using different strategies. The main ones are [23]:

1. Intercalation of monomer molecules between the LDH layers, with subsequent *in situ* polymerization. Intercalation of the monomer can occur by direct or indirect methods. The intercalation of monomer molecules with subsequent *in situ* polymerization, is widely used in the preparation of various LDH/conductive polymers. The resulting nanocomposites generally exhibit good structural organization and phase purity. This process is limited by two factors:

   i. the distance between the monomers when they are strongly linked, or grafted, to the structure of the inorganic layers. When the monomers are strongly bound to the layers, their flexibility (freedom of movement within the interlayer) is limited, so the proximity between them should be sufficient for the polymerization reaction to occur. High charge densities in the layers may shorten distance between the intercalated monomers. Functionalized monomers with long chain aliphatic groups also provide greater flexibility.

   ii. the polymerization conditions (temperature, pH, or redox reaction), which should be selected so as not to affect the layered structure of the resulting materials.

Indirect methods may also be employed for the intercalation of monomers. These methods are often utilized when the chemical nature of the interlayer space and guest species are not compatible. Such methods require the preparation of an LDH precursor intercalated with a molecule that can be easily exchanged. This LDH precursor is then placed in contact with the monomer of interest, which will replace the previously intercalated anion. To obtain the LDH/polymer, it is necessary to carry out the *in situ* polymerization reaction after the exchange with the monomer.

2. Direct intercalation of polymer molecules with low molecular weight between the LDH layers or intercalation polymers with high molecular weight by indirect methods. The incorporation of the polymer between the LDH layers, can be performed by direct method by using the direct co-precipitation reaction, nanocomposites containing polymers that have an anionic group; for example, carboxylate or sulfonate groups, can be produced

during growth of the inorganic crystal. This preparation strategy usually yields nanocomposites with low structural organization. The crystallinity of these materials can be improved by hydrothermal treatment. The indirect method requires the presence of the LDH precursor, usually containing chloride anions. This LDH precursor is placed into exchange reaction using suitable solvents in the presence of the polymer of interest.

3. Intercalation of LDH via exfoliation, when a colloidal system is formed between the LDH and an appropriate solvent, for exfoliation of the layers. Restacking of the layers in the presence of a solution containing the target monomer or polymer culminates in their intercalation by restacking of the structure of the layer. When the monomers are intercalated, a subsequent *in situ* polymerization is required for attainment of the intercalated nanocomposite LDH/polymer. This strategy is usually employed when the polymer has high molecular weight, which makes their diffusion between the LDH layers difficult. Due to its high charge density, the LDH does not have a natural tendency to exfoliation. To achieve delamination of these materials, it is necessary to reduce the electrostatic interaction between the layers. This can be done with intercalation of spacer anions, such as, dodecylsulfonate and dodecylbenzenesulfonate. Exfoliation is then obtained by placing the organically modified LDH in a solution containing a polar solvent. Addition of polymer to the solution containing the exfoliated material results in the formation of an intercalated and/or exfoliated precipitate. In some cases, the nanocomposite is only generated upon evaporation of the solvent.

In addition to the strategies described above, immobilization of the polymer between the LDH inorganic sheets can also be attained by regeneration of the layered structure using the "*memory effect*" exhibited by some LDH. In this case, a previously prepared LDH, normally MgAl-CO_3, is firstly calcined at an adequate temperature, for elimination of the interlayer anion. The calcined material, a mixed oxy-hydroxide, is then placed in contact with an aqueous solution of the polymer to be intercalated. The oxide is hydrolyzed with regeneration of the LDH structure and intercalation of the polymer. This process is accompanied by a sharp increase in the pH value. The latter can be corrected, to prevent the intercalation of hydroxyl anions. Normally, the LDH/polymers nanocomposites produced by this method do not exhibit good organization, being more suitable for the incorporation of small molecules. This method was used for the intercalation of silicates into LDH. In this case, mexinerite (an MgxAlOH-LDH, with x = 2, 3, 4) was employed as precursor for incorporation of the silicate to this end, mexinerite was previously calcined at 500 °C under air atmosphere, and then placed in contact with a solution of tetraethylorthosilicate, $Si(OC_2H_5)_4$ (TEOS). This afforded more crystalline materials than those obtained by anion exchange or direct co-precipitation, using metasilicate and ZnM-LDHs (M = Al, Cr).

An additional route for preparation of the LDH/polymer is the auxiliary solvent method. Solvents represent an important part in the swelling processes of the layered materials, since they promote separation of the layers. Schematic representation of the incorporation of polymers into layered double hydroxides is given in Figure 4.

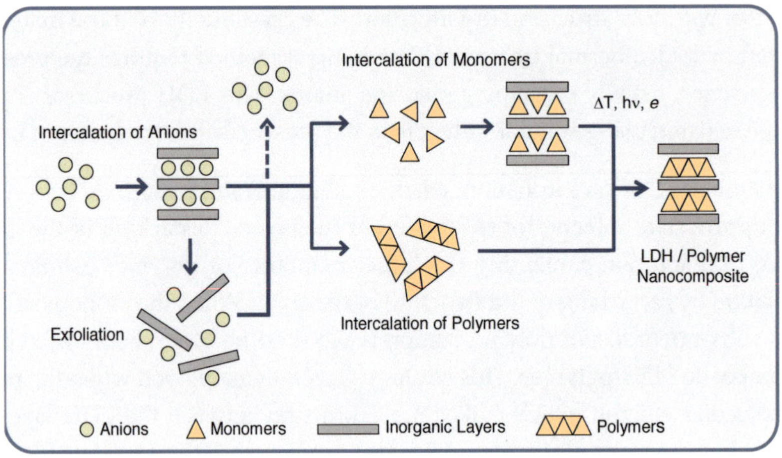

Figure 4. Schematic representation of the incorporation of polymers into layered double hydroxides. *(adapted from ref. 23)*

3. Characterization methods

This section describes the main techniques employed for the characterization of intercalated nanocomposites of conducting polymer / LDH: Powder X-ray Diffraction (PXRD), ^{13}C Cross-Polarization/Magnetic Angle Spinning (CP/MAS) NMR spectroscopy, Electron Spin Resonance (ESR) spectroscopy, Thermogravimetric Analysis (TGA), Differential Scanning Calorimetry (DSC), Fourier Transform Infrared (FTIR) spectroscopy, Ultraviolet/Visible (UV-Vis) Spectroscopy, Transmission Electron Microscopy (TEM), and Scanning Electron Microscopy (SEM).

3.1. Powder X-ray Diffraction (PXRD)

The X ray diffraction pattern (PXRD) of LDH presents basal peaks $00l$ related to the stacking sequence of the inorganic sheet. The peaks are not basal, said to non harmonics, are related to the sheet structure. For new LDHs, the indexing of the diffraction peaks can be accomplished by comparison with the PXRD of hydrotalcite, which exists in the database of diffraction equipment (JCPDS-ICDD, PDF database), or with a number of other LDHs described in the literature. Figure 5 brings a representative PXRD for an MgAl-CO_3-LDH.

The interlayer distances can be calculated from the values of 2θ, using the Bragg equation:

$$n\lambda = 2d_{hkl} \cdot sen\theta$$

where n is the diffraction order, d_{hkl} is the interlayer spacing for the peak hkl, and θ is the Bragg angle, determined by the diffraction peak. Repetition of the d value, for $n = 1, 2, 3...$, evidences the formation of a layered material. The interlayer spacing can be calculated by averaging the basal peaks according to the equation:

http://dx.doi.org/10.5772/54803

$$d=\frac{1}{n}(d_{003}+2d_{006}+...+nd_{00n})$$

The parameters *a* and *c* can be obtained according to the equation:

$$\frac{1}{(d_{hkl})^2}=\frac{4}{3}\left(\frac{h^2+hk+k^2}{a^2}\right)+\frac{l^2}{c^2}$$

where h, k, and l are the Miller indices of the corresponding peak. For a LDH with stacking sequence 3R, the *c* parameter c is equal to three times the basal spacing value.

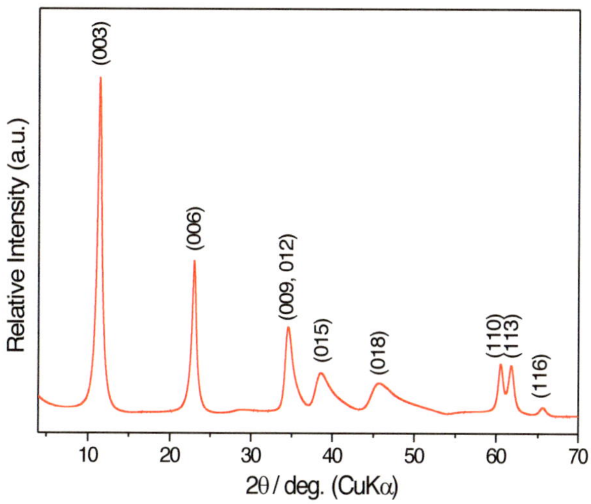

Figure 5. PXRD of synthetic Hydrotalcite.

To determine the orientation adopted by anionic species, such as monomers and polymers intercalated into LDHs, the values of interlayer spacing and/or basal spacing obtained from the PXRD data are compared with the size of anions obtained by specific computer programs, like "VASP (*Vienna Ab-initio Simulation Package*)".

When thermal treatments is performed for the *in situ* polymerization of monomers intercalated between the LDHs inorganic layers, the PXRD analysis may reveal a decrease in the value of interlayer spacing, which indicates a small contraction between adjacent layers. The presence of phases, other than the LDH can also be identified by PXRD, which is useful since thermal treatment may often generated oxides.

3.2. ^{13}C Cross-Polarization/Magnetic Angle Spinning (CP/MAS) NMR spectroscopy

In situ polymerization of monomers intercalated into LDH may be monitored by ^{13}C Cross-Polarization/Magnetic Angle Spinning (CP/MAS) NMR spectroscopy. This technique detects formation of bonds of the monomer-monomer type in polyconjugated systems. Assignment of the chemical shift values for the monomers can be carried out by computer simulation using specific computer programs, such as *"ACD/ChemSketch, version 4.04"*, provided by the company Advanced Chemistry Development Inc., and *"CS Chemdraw Ultra®"*, offered by the company Cambridgesoft Corporation. The values obtained by simulation can be compared with the values achieved experimentally.

Figure 6 contains an example of ^{13}C Cross-Polarization/Magnetic Angle Spinning (CP/MAS) NMR Spectroscopy analyses for the *in situ* polymerization of 4-(1H-pyrrol-1-yl)benzoate intercalated into ZnAl-LDH [54]. Assignment of the peaks to the carbons of the monomer is given in Table 4.

Notation	Assignment (Cn)	DMSO-D6 (ppm)	CP-MAS (75.4 MHz) (ppm)
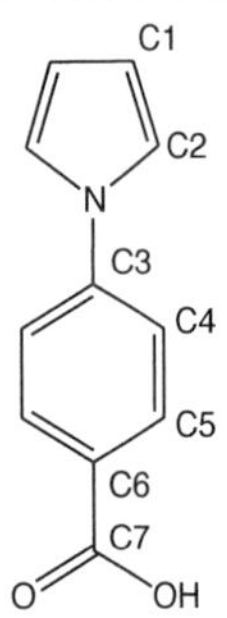	C1	111.3	111.2
	C2	118.9	120.1
	C3	143.0	142.3
	C4	118.5	115.6
	C5	130.9	132.9
	C6	127.0	122.8
	C7	166.6	174.9

Table 4. Assignment of the peaks to the carbons in the ^{13}C CP-MAS NMR analyses of 4-(1H-pyrrol-1-yl)benzoate.

The ^{13}C CP/MAS NMR spectra of all LDH were similar. In the Figure 6c and 6d, the peaks can be unambiguously assigned as carbons C7 (175.1 ppm), C5 (131.6 ppm), C3 (140.8 ppm), and C1 (113.7 ppm). The broad signal at 116.8 ppm can be attributed to the chemical shifts of the remaining carbons C4 and C6 of the six-membered ring. Several simulations of the ^{13}C NMR spectra suggest that one possible quaternary carbon, resulting from the polymerization of the monomer via condensation C2-C2, presents chemical shift in the range of 112.0 to 116.0 ppm. Therefore, the large signal at 116.8 ppm in spectrum of the polymer is ascribed to this quaternary carbon, coinciding with the chemical shifts of the remaining carbons C4 and C6. Together with the PXRD results, these data suggest that the production of oligomers and/or polymers occurs with the formation of bilayers of monomers in the interlayer space. In this arrangement, the carboxylate groups are directed to the layer, whereas the aromatic rings occupy the central region of the interlayer spacing. Therefore, the polymer obtained within the interlayer resembles a *"zig-zag"*, similar to the polymers of the syndiotactic type.

http://dx.doi.org/10.5772/54803

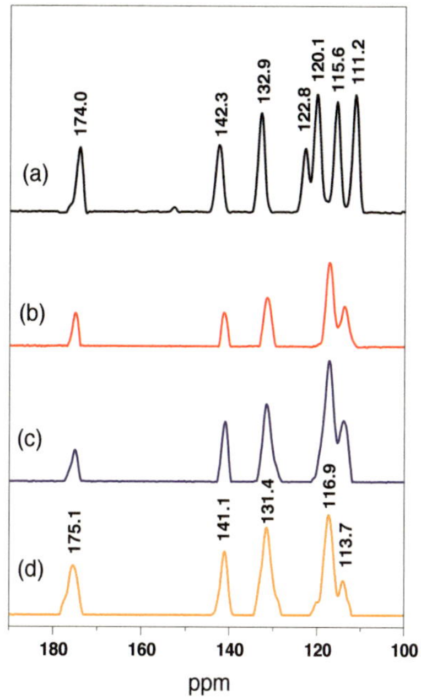

Figure 6. CP-MAS NMR spectra of: (a) pyrrolebenzoic acid; (b) ZnAl-4-(1H-pyrrol-1-yl)benzoate-LDH; (c) ZnAl-4-(1H-pyrrol-1-yl)benzoate-LDH with hydrothermal treatment; and (d) ZnAl-4-(1H-pyrrol-1-yl)benzoate-LDH after thermal treatment [54].

3.3. Electron Spin Resonance (ESR) spectroscopy

Electron Spin Resonance (ESR) spectroscopy allows for monitoring of *in situ* polymerization processes in intercalated monomers. The spectra of conducting polymers usually exhibit signs typical of the formation of polarons, with the Lorentzian profile. In these analyses, the apparatus is normally operated at 9.658 GHz, using the 1,1-diphenyl-2-picrylhydrazyl (DPPH) radical to determine the ressonance frequency ($g = 2.0036 +/- 0.0002$). The scan width can vary between 2000 and 4000 G, with a receiver gain of 100000.

Figure 7 illustrates ESR analyses for monitoring of the *in situ* polymerization of 3-(Pyrrol-1-yl)-propanoate monomers intercalated into ZnAl-LDH [55,56]. The spectra were recorded after heat-treatment at temperatures ranging from ambient to 180 °C for 2h. For material at room temperature, the ESR spectra display very weak signals. Thus, the spectrum was enlarged 16 times for comparison with those of the material treated at other temperatures. Typical signs can be noticed for the "*superhyperfine*" structure with 6 lines, and there is a sign characteristic of the formation of a polaron with $g = 2.004 \pm 0.0004$. The appearance of this "*superhyperfine*" structure suggest formation of the radical (COO·). The magnetic moment of this radical should interact with the magnetic moments of the nuclei of the aluminum atoms present in the inorganic host matrix. This hypothesis considers the nuclear spin of aluminum as $I = 5/2$ and a number of nuclei $N = 1$, which generates a spectrum of $2NI + 1 = 6$ lines. Due to the charge balance required for maintenance of the electroneutrality of hybrid systems there a regulating environment for the free electrons of the radicals (COO ·) within the interlayer spacing. These radicals are located near the aluminum cations, because the latter are responsible for the positive charge density of the layer. The signal at g = 2.004 ± 0.0004 attests to the formation of a polaron, *i.e.*, a polarized entity resulting from delocalization of the radical in structures with π conjugations. The increase in the delocalization of π orbitals favors the generation of polarons, so an increase signal in this upon heat-treatment indicates stronger connection between the monomers. The Lorentzian profile of the ESR spectrum of the material at room temperature is compatible with the formation of conjugated polymers. The ESR results agree with the NMR results and indicate that spontaneous partial polymerization and/or oligomerization of the 3-(Pyrrol-1-yl)-propanoate monomers takes place during coprecipitation of the nanocomposites.

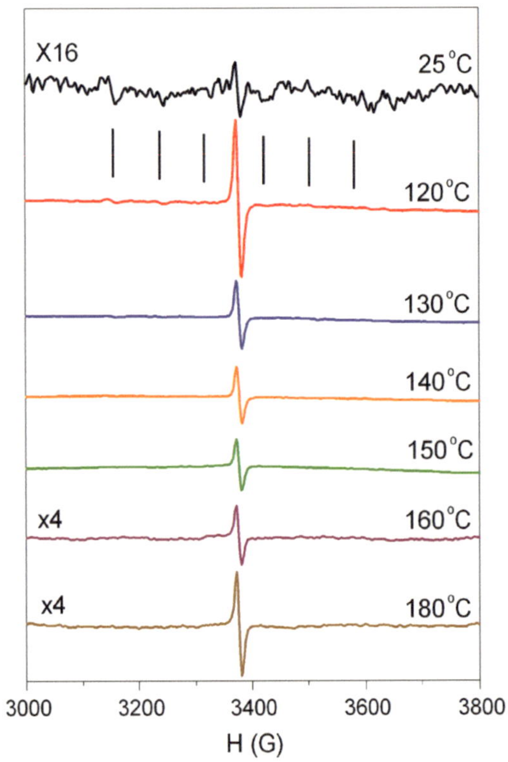

Figure 7. ESR spectra of Zn-Al-3-(Pyrrol-1-yl)-propanoate-LDH as a function of the heat-treatment temperature [55,56].

3.4. Thermogravimetric Analysis (TGA) and Differential Scanning Calorimetry (DSC)

The thermal stability of LDHs intercalated with conductive polymers as well as the amount of water, intercalated and adsorbed, in the nanocomposites can be determined by thermogravimetric analysis. The results are obtained as a curve mass decrease (%) *versus* temperature. For the LDHs, the thermal decomposition steps generally overlap, especially in the case of LDHs intercalated with organic molecules.

TGA is important in the thermal *in situ* polymerization of nanocomposites, since it is necessary to determine the temperature that should be used for polymerization of the intercalated monomers. The thermal decomposition of intercalated organic compounds takes place at higher temperatures, so it is possible to achieve greater thermal stability for conducting polymers intercalated into an inorganic host matrix (LDH).

Figure 8 displays an example of TGA/DSC analysis for the ZnAl-LDH intercalated with 3-aminobenzoate monomers and for the pure monomer.

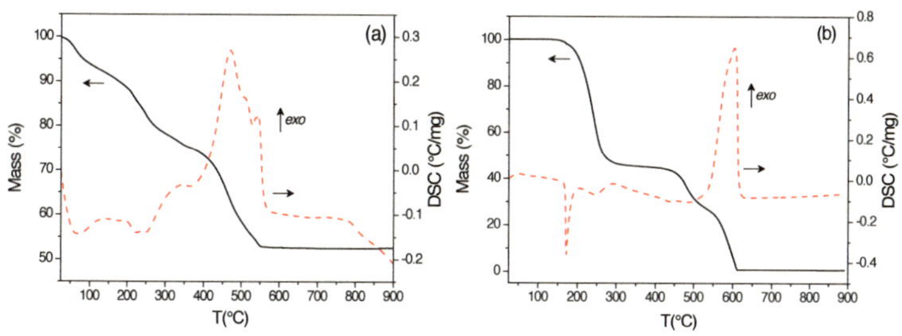

Figure 8. TGA-DSC of (a) ZnAl-3-aminobenzoate-LDH; (b) sodium 3-aminobenzoate [60].

For the nanocomposite material, Figure 8a, the early stages of thermal decomposition are associated with loss of adsorbed and intercalated water. In this temperature range, the DSC curve reveals the occurrence of endothermic processes. Dehydroxylation of the inorganic sheets and decomposition of the anion intercalated species happen concomitantly. The DSC curve also indicates the occurrence of an exothermic process during decomposition of the intercalated organic species.

3.5. Fourier Transform Infrared (FTIR) spectroscopy

FTIR analysis is carried out in KBr pellets, pressed from a mixture of 2% of the LDH samples in previously dried KBr. The spectra are recorded over a wavelength range going from 4000 to 400 cm^{-1}. FTIR spectroscopy data provide information about the functional groups and possible interactions between the organic and inorganic parts of the nanocomposites. Identification of the *in situ* polymerization of monomers intercalated into LDH by this technique is difficult because of several overlapping spectral bands.

Figure 9 contains the FTIR spectra of (a) sodium 3-aminobenzoate, (b) pure sodium poly-3-aminobenzoate, and LDH intercalated with sodium 3-aminobenzoate submitted to different treatments [61]. For the pure monomer, the bands (not shown in the figure) at 3408, 3349, and 3223 cm^{-1} are related to $\nu(NH_2)$ symmetric and anti-symmetric stretching, whereas the band at 1628 cm^{-1} is characteristic of $\delta(NH_2)$ symmetric deformation. The bands at 1560 and 1411 cm^{-1} are typical of ν(-COO$^-$) symmetric and anti-symmetric stretching, respectively. The band at 1312 cm^{-1} is due to ν(C-N) stretching. The bands at 1266 and 1115 cm^{-1} are attributed to δ(NH_2) symmetric and asymmetric deformations. The bands at 776 and 676 cm^{-1} are ascribed to δ(COO$^-$) out of the plane symmetric and asymmetric deformation. Concerning the polymer, the FT-IR spectrum of pure poly-3-aminobenzoate undergoes significant changes, especially in the area relative to the vibrations of the aromatic ring and the functional group NH_2. The bands due to (-COO$^-$) in the plane stretching at the 1700 and 1400 cm^{-1}, and (NH_2) out of the plane deformation at 1698, 1634, 1566, 1509 and 1441 cm^{-1} are fairly broad. The broad overlapping bands in the region between 1300 and 1110 cm^{-1}, refer to δ(NH_2) symmetric and asymmetric. Alterations in the spectrum of the polymer are expected because the amine and *p*-methylenes groups of the 3-aminobenzoate molecules interact during polymerization. As for the heat-treated nanocomposites, there is virtually no changes in the profile of the spectra. The bands in the regions between 1700 and 1360 cm^{-1}, related to stretching of the carboxylate group and the aromatic ring are displaced and broader. The bands relative to ν(-COO$^-$) symmetric and anti-symmetric stretching can be observed in the regions near 1554 and 1384 cm^{-1}. In the region between 1300 and 1110 cm^{-1} there is a shoulder around 1303 cm^{-1} and weak at 1266 cm^{-1}, corresponding to δ(NH_2) symmetric and asymmetric deformation. Analysis of the bands in the regions below 1200 cm^{-1} is highly compromised because of the large overlap of bands with medium and weak intensity. The bands in the regions below 700 cm^{-1} are due to metal-oxygen-metal vibrations occurring in the inorganic host matrix.

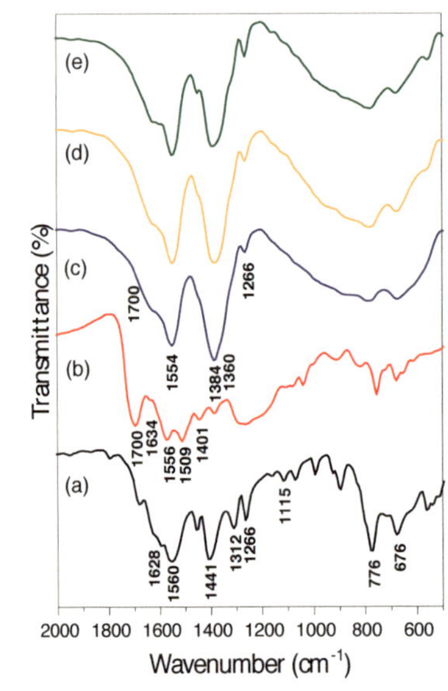

Figure 9. FTIR spectra of (a) sodium 3-aminobenzoate; (b) pure sodium poly-3-aminobenzoate; (c) MgAl-3-aminobenzoate-LDH; (d) MgAl-3-aminobenzoate-LDH with hydrothermal treatment; and (e) MgAl-3-aminobenzoate-LDH heat-treated at 160 °C [61].

3.6. Ultraviolet/Visible (UV-Vis) spectroscopy

The UV-Vis spectra are collected between 200 and 800 nm. Samples are prepared by dissolution of the material in concentrated HCl and subsequent dilution in water.

Figure 10 depicts the UV-Vis spectra pure sodium Poly-3-aminobenzoate and ZnAl-AMB-LDH with different Zn:Al molar ratios(2:1, 3:1 and 6:1) [60].

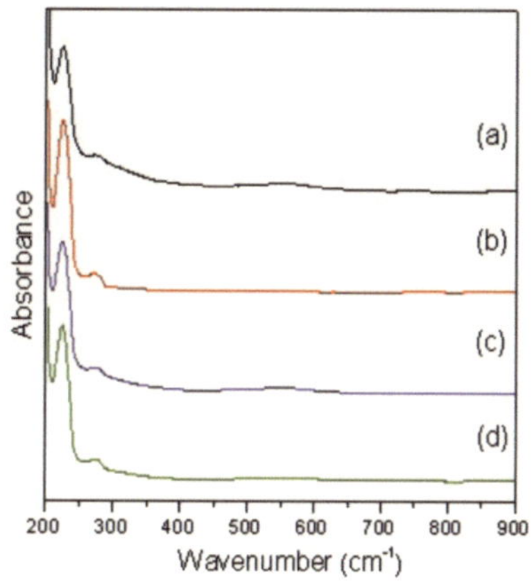

Figure 10. UV-Vis absorption spectra of the materials prepared with (a) pure sodium poly-3-aminobenzoate; (b) Zn_2Al-3-aminobenzoate-LDH; (c) Zn_3Al-3-aminobenzoate-LDH; and (d) Zn_6Al-3-aminobenzoate-LDH.

All the LDH display a band at about 225 nm, after polymerization, the band verified for the monomers is dislocated to lower wavelengths ~215 nm, and a band at ~ 275 nm appears. The latter band is less pronounced for Zn_6Al-3-AMB-LDH prepared by anion exchange in double phase, which is attributed to the n-π* transition due to the presence of non-shared electrons in the COO^- group. After polymerization a peak at ~ 315 nm ascribed to π-π* transition related to conjugation of rings in the polymeric chain is detected. As for the LDH, the first absorption peak intensifies ongoing from the compounds prepared with Zn/Al ratios of 2:1 and 3:1 to 6:1. In the case of the materials prepared by exchange in double phase only for the compounds with Zn/Al ratios of 2:1 and 3:1 the band intensifies. The compound with Zn/Al ratio of 6:1 has the least intense peak.

3.7. Transmission Electron Microscopy (TEM)

The best TEM images are generally achieved when LDHs are dispersed in an epoxy resin, centrifuged, and kept at 70 °C for the 72 h, for drying. After drying, the materials are cut in an ultra-microtome and transferred to hexagonal copper bars appropriated for TEM image acquisition. An alternative approach is to prepare a suspension containing ethanol and LDH. The copper grid is then immersed into the suspension and dried at ambient temperature.

Figure 11 reveals very orderly particles in which the darkest lines represent the inorganic layers and the clearest lines refer to the intercalated conductive polymers [54]. There is good pillaring of the sheets, with a large sequence of darker lines. The basal spacing value estimated from the TEM images can be compared with the one obtained PXRD analysis.

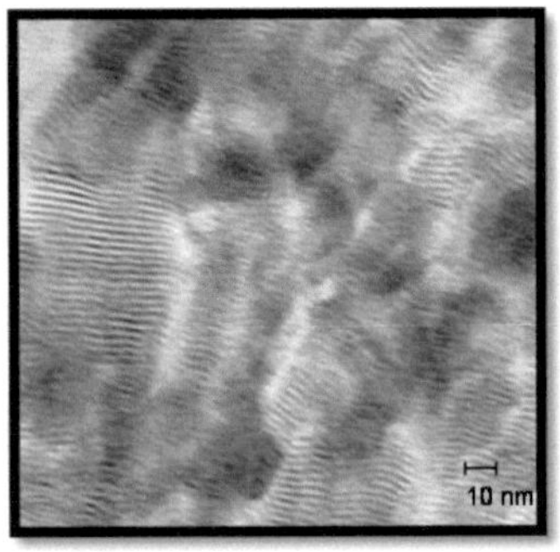

Figure 11. TEM micrographs for ZnAl-4-(1H-pyrrol-1-yl)benzoate-LDH with hydrothermal treatment.

3.8. Scanning Electron Microscopy (SEM)

The morphology of the crystallites and nanocomposite particles can also be analyzed by SEM.

For these analyses, the samples are usually supported on the sample port by powder dispersion on double-sided conductive adhesive tape. Because LDHs do not present enough conductivity for generation of good images it is necessary to cover the samples with gold before the measurements, using a sputter equipment.

Figure 12 shows the SEM images of LDH intercalated with 3-aminobenzoate. There is superposition of the sheets, with formation of aggregates on the surface of the cristallyte [60].

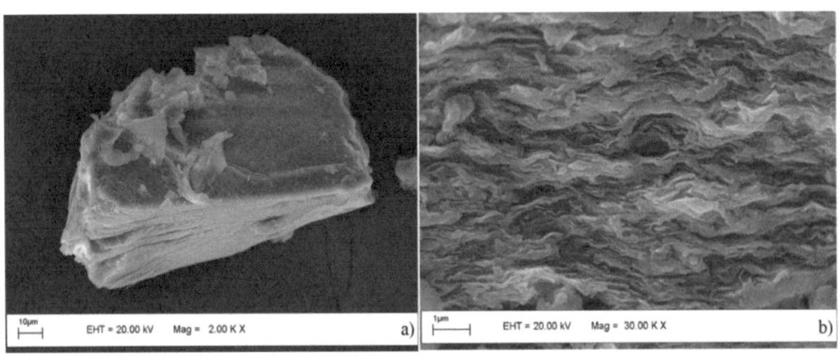

Figure 12. SEM images of ZnAl-3-aminobenzoate-LDH [60].

3.9. Cyclic Voltammetry (CV)

Cyclic Voltammetry (CV) experiments are conducted on potentiostats. The supporting electrolyte is 0.1 mol/dm^3 $LiClO_4$ solution, and a conventional electrochemical cell arrangement involving three electrodes is utilized: Platinum wire as the counter electrode, as the reference electrode (Ag/AgCl/KCl(sat)), and glassy carbon, prepared by dip-coating in an aqueous suspension of monomers intercalated into LDH as the working electrode. The potential of the liquid junction is disregarded. CV experiments enable evaluation of the oxidation and reduction processes of the intercalated monomers. A typical voltammogram of ZnAl-LDH intercalated with 3-aminobenzoate anions is presented in Figure 13. The oxidation process involved in the polymerization of 3-aminobenzoate intercalated into LDH can be noticed. Moreover, Zn^{2+} oxidation can be verified.

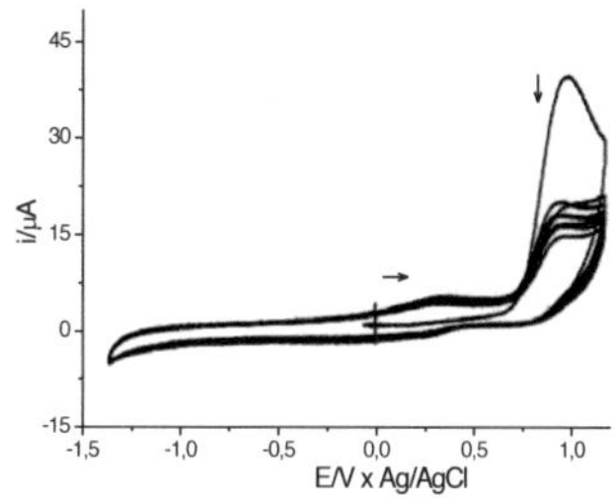

Figure 13. CV for ZnAl-3-aminobenzoate-LDH [60].

There is an irreversible oxidation peak at about 0.960V, and the amplitude of this peak diminishes upon consecutive scanning. This peak is ascribed to 3-aminobenzoate oxidation. A similar behavior has been previously observed for 2-thiophenecarboxylate anions intercalated into ZnAl-LDH.

4. Conclusion

Layered Double Hydroxides (LDHs) are materials whose layered architecture enables separation of the inorganic part (double hydroxide), and the organic portion (conductive polymer), thus culminating in a hybrid composite. The "*growth*" of conductive polymers in limited spaces, like the interlayer region of the LDHs, has been shown to be a very promising method for the improvement of the properties of conductive polymers.

On the basis of literature works, it is possible to deduce that, the guests species (monomers) are generally intercalated in a bilayer arrangement within the LDH layers. In this arrangement, the functional groups of the monomers are directed to the inorganic layer, and the aromatic rings occupy the central region of the interlayer spacing. The nature of the substituent group (aliphatic or aromatic) influences the structural organization and the *in situ* polymerization of the resulting hybrid materials.

During the synthesis, some nanocomposites undergo spontaneous polymerization, while others have to be submitted to thermal or electrochemical treatments to reach polymerization. Monomers containing substituents with aliphatic chains, tend to undergo polymerization in milder conditions, because the aliphatic chains provide small mobility of the intercalated monomers, thereby faciliting formation of polyconjugated systems. In some cases, thermal treatment may cause collapse of the layered structure, with consequent formation of oxide.

The thermogravimetric analysis data show that, compared with the pure polymer, the LDH-intercalated conducting polymer is more thermally stable. This stability is provided by the inorganic coverage offered by the LDH layers.

In the case of materials intercalated with conducting polymers, there is initial removal of one electron from the polymeric chain, *e.g*, through *p* doping. This results in the formation of an electronic state denominated polaron. Generation of the polaron can also be interpreted as π electron redistribution. Moreover, the formation of this entity is associated with distortion of the polymeric chain, which transforms the aromatic form into the quinoid form. The production of polaron may be also due the presence of electronic state located in the energy region found in the middle of gap. The quinoid structure presents smaller ionization energy and larger electronic affinity than the aromatic form. Polaron is chemically defined as a radical ion of spin = 1/2. As the concentration of polarons increases, they tend to recombine, stabilizing the structure and forming a "bipolaron". "Bipolaron" is defined as a pair of equal diamagnetics dication with spin equal to 0 and equal charges. The formation of "Bipolaron" is associated with strong distortion to the LDH net work.

Acknowledgements

This work was supported by the Brazilian agencies: Fundação de Amparo à Pesquisa do Estado de Minas Gerais (FAPEMIG), Fundação de Amparo à Pesquisa do Estado de São Paulo (FAPESP), and Conselho Nacional de Desenvolvimento Científico e Tecnológico (CNPq).

Author details

Jairo Tronto[1*], Ana Cláudia Bordonal[2], Zeki Naal[3] and João Barros Valim[2]

*Address all correspondence to: jairotronto@ufv.br

1 Universidade Federal de Viçosa - Instituto de Ciências Exatas e Tecnológicas - Campus de Rio Paranaíba - Rio Paranaíba - MG, Brazil

2 Universidade de São Paulo - Faculdade de Filosofia Ciências e Letras de Ribeirão Preto - Departamento de Química - Ribeirão Preto – SP, Brazil

3 Universidade de São Paulo - Faculdade de Ciências Farmacêuticas de Ribeirão Preto - Departamento de Física e Química - Ribeirão Preto – SP, Brazil

References

[1] Pinnavaia TJ, Beall GW. Eds. Polymer-Clay Nanocomposites; Jonh Wiley & Sons Ltd.: New York; 2000.

[2] Alexandre M, Dubois P. Polymer-Layered Silicate Nanocomposites: Preparation, Properties and Uses of a New Class of Materials. Materials Science & Engineering R-Reports 2000;289(1-2) 1-63.

[3] LeBaron PC, Pinnavaia TJ. Clay Nanolayer Reinforcement of a Silicone Elastomer. Chemistry of Materials 2001;13(10) 3760-3765.

[4] Lagaly G. Introduction: From Clay Mineral-Polymer Interactions to Clay Mineral-Polymer Nanocomposites. Applied Clay Science 1999;15(1-2) 1-9.

[5] Darder M, Aranda P, Ruiz AI, Fernandes FM, Ruiz-Hitzky E. Design and Preparation of Bionanocomposites Based on Layered Solids with Functional and Structural Properties. Materials Science and Technology 2008;24(9) 1100-1110.

[6] Podsiadlo P, Kaushik AK, Arruda EM, Waas AM, Shim BS, Xu J, Nandivada H, Pumplin BG, Lahann J, Ramamoorthy A, Kotov NA. Ultrastrong and Stiff Layered Polymer Nanocomposites. Science 2007;318(5847) 80-83.

[7] Merkel TC, Freeman BD, Spontak, RJ, He Z, Pinnau I, Meakin P, Hill AJ. Ultrapermeable, Reverse-Selective Nanocomposite Membranes. Science 2002;296(5567) 519-522.

[8] Ogara JE, Ding J, Walsh D; Waters investments ltd (WATE-Non-standard) Waters technologies corp (wate-non-standard), assignee. Hybrid inorganic-organic material used for separation devices, comprises polymerized scaffolding nanocomposite containing scaffolding functionality capable of chemically interacting with surface of another material patent WO2004105910-A2; GB2419886-A; DE112004000906-T5; US2007141325-A1; JP2007515503-W; GB2419886-B; WO2004105910-A3; JP2012042477-A WO2004105910-A2 09 Dec 2004 B01D-000/00 200505.

[9] Chen H, Yu H, Hsiao W, Chen X, Xiao W, You X; Du Pont Apollo Ltd (Dupo), assignee. Backsheet for a photovoltaic module, comprises a nanocomposite layer comprising a polymeric matrix including a polymer consisting of polyethylene terephthalate and silicate nanoparticles, polymeric layers, and adhesive layers patent US2011259415-A1; CN102280505-A US2011259415-A1 27 Oct 2011 H01L-031/0216 201172.

[10] Unnikrishnan L, Mohanty S, Nayak SK, Singh N. Synthesis and Characterization of Polysulfone/Clay Nanocomposite Membranes for Fuel Cell Application. Journal of Applied Polymer Science 2012;124(SI) E309-E318.

[11] Yuan S, Li Y, Zhang Q, Wang H. ZnO/Mg-Al Layered Double Hydroxides as Strongly Adsorptive Photocatalysts. Research on Chemical Intermediates 2009;35(6-7) 685-692.

[12] Cavani F, Clause O, Trifiro F, Vaccari A. Anionic Clays with Hydrotalcite-Like Structure as Precursors of Hydrogenation Catalysts. Advances in Catalyst Design, 1991; 186-190.

[13] Vial S, Prevot V, Leroux F, Forano C. Immobilization of Urease in ZnAl Layered Double Hydroxides by Soft Chemistry Routes. Microporous and Mesoporous Materials. 2008;107(1-2) 190-201.

[14] Hsu SL, Chang K, Shiu L, Jang G; HSU S L (HSUS-Individual) CHANG K (CHAN-Individual) UNIV NAT CHENG KUNG (UNCK), assignee. Fabricating Polybenzoxazole Clay Nanocomposite for Microelectronic Industry, by Ion Exchange Reaction of Modifying Agent and Layered Clay, and Polycondensation Reaction of Diacid Chloride and Bis(o-aminophenol) Monomer Patent US2003139513-A1; TW576855-A; US7081491-B2 US2003139513-A1 24 Jul 2003 B29C-039/14 200373.

[15] Bonifacio LS, Gordijo CR, Constantino VRL, Silva DO, Kiyohara PK, Araki K, Toma HE. Optical Changes and Writing on Hydrotalcite Supported Gold Nanoparticles. Journal of Nanoscience and Nanotechnology. 2008;8(1) 274-279.

[16] Carretero MI, Pozo M. Clay and Non-Clay Minerals in the Pharmaceutical and Cosmetic Industries Part II. Active ingredients. Applied Clay Science. 2010;47(3-4) 171-181.

[17] Cunha VRR, Ferreira AMD, Constantino VRL, Tronto J, Valim JB. Layered Double Hydroxides: Inorganic Nanoparticles for Storage and Release of Species of Biological and Therapeutic Interest. Quimica Nova. 2010;33(1) 159-171.

[18] Bernardo E, Colombo P, Hampshire S. Advanced Ceramics from a Preceramic Polymer and Nano-Fillers. Journal of the European Ceramic Society. 2009;29(5) 843-849.

[19] Okada A, Fukushima Y, Kawasumi M, Inagaki S, Usuki A, Kurauchi T, Kamigaito O, Sugiyama S. Toyota Chuo Kenkyushok, assignee. Composite with High Strength and Excellent High Temp. Properties has Layers of Silicate Mineral Homogeneously Dispersed in a Poly-Amide Matrix. patent DE3632865-A1; JP62074957-A; JP62252425-A; US4739007-A; JP7309942-A; JP7310012-A; JP96022946-B2; DE3632865-C2; JP2663113-B2; JP2724547-B2; US37385-E.

[20] Usuki A, Kojima Y, Kawasumi M, Okada A, Fukushima Y, Kurauchi T, Kamigaito O. Synthesis of Nylon 6-clay Hybrid. Journal of Materials Research. 1993;8(5) 1179-1184.

[21] Kojima Y, Usuki A, Kawasumi M, Okada A, Fukushima Y, Kurauchi T, Kamigaito O. Mechanical-Properties of Nylon 6-clay Hybrid. Journal of Materials Research. 1993;8(5) 1185-1189.

[22] Yano K, Usuki A, Okada A. Synthesis and Properties of Polyimide-Clay Hybrid Films. Journal of Polymer Science Part A-Polymer Chemistry. 1997;35(11) 2289-2294.

[23] Leroux F, Besse JP. Polymer Interleaved Layered Double Hydroxide: A New Emerging Class of Nanocomposites. Chemistry of Materials. 2001;13(10) 3507-3515.

[24] Wang G, Cai FL, Si LC, Wang ZQ, Duan X. An Approach Towards Nano-Size Crystals of Poly(acrylic acid): Polymerization Using Layered Double Hydroxides as Template. Chemistry Letters. 2005;34(1) 94-95.

[25] Wang GA, Wang CC, Chen CY. The Disorderly Exfoliated LDHs/PMMA Nanocomposite Synthesized by In Situ Bulk Polymerization. Polymer. 2005;46(14) 5065-5074.

[26] Darder M, Lopez-Blanco M, Aranda P, Leroux F, Ruiz-Hitzky E. Bio-nanocomposites based on layered double hydroxides. Chemistry of Materials. 2005;17(8) 1969-1977.

[27] Cavani F, Trifiro F, Vaccari A. Hydrotalcite-Type Anionic Clays: Preparation, Properties and Applications. Catalysis Today. 1991;11(2) 173-301.

[28] Roussel H, Briois V, Elkaim E, de Roy A, Besse JP. Cationic Order and Structure of [Zn-Cr-Cl] and [Cu-Cr-Cl] Layered Double Hydroxides: A XRD and EXAPS Study. Journal of Physical Chemistry B. 2000;104(25) 5915-5923.

[29] Fogg AM, Dunn JS, O'Hare D. Formation of Second-Stage Intermediates in Anion-Exchange Intercalation Reactions of the Layered Double Hydroxide [LiAl2(OH)(6)]Cl Center Dot H2O as Observed by Time-Resolved, In Situ X-ray Diffraction. Chemistry of Materials. 1998;10(1) 356-360.

[30] Pisson J, Taviot-Gueho C, Israeli Y, Leroux F, Munsch P, Itie JP, Briois V, Morel-Desrosiers N, Besse, JP. Staging of Organic and Inorganic Anions in Layered Double Hydroxides. Journal of Physical Chemistry B. 2003;107(35) 9243-9248.

[31] Kaneyoshi M, Jones W. Exchange of Interlayer Terephthalate Anions from a Mg-Al Layered Double Hydroxide: Formation of Intermediate Interstratified Phases. Chemical Physics Letters. 1998;296(1-2) 183-187.

[32] Morpurgo S, LoJacono M, Porta P. Copper-Zinc-Cobalt-Aluminium-Chromium Hydroxycarbonates and Mixed Oxides. Journal of Solid State Chemistry.1996;122(2) 324-332.

[33] de Roy A, Forano C, El Malki K, Besse JP. In Synthesis of Microporous Materials; Ocelli, M. L.; Robson, M. E. (ed.) Van Nostrand Reinhold: New York; 1992. p. 108-169.

[34] Lopez T, Bosch P, Ramos E, Gomez R, Novaro O, Acosta D, Figueras F. Synthesis and Characterization of Sol-Gel Hydrotalcites. Structure and Texture. Langmuir. 1996;12(1) 189-192.

[35] Reichle WT. Synthesis of Anionic Clay-Minerals (Mixed Metal-Hydroxides, Hydrotalcite). Solid State Ionics. 1986;22(1) 135-141.

[36] Taylor RM. The Rapid Formation of Crystalline Double Hydroxy Salts and Other Compounds by Controlled Hydrolysis. Clay Minerals. 1984;19(4) 591-603.

[37] Indira L, Dixit M, Kamata PV. Electrosynthesis of Layered Double Hydroxides of Nickel with Trivalent Cations. Journal of Power Sources. 1994;52(1) 93-97.

[38] Kooli F, Depege C, Ennaqadi A, de Roy A, Besse JP. Rehydration of Zn-Al Layered Double Hydroxides. Clays and Clay Minerals. 1997;45(1) 92-98.

[39] Crepaldi EL, Pavan PC, Valim JB. A New Method of Intercalation by Anion Exchange in Layered Double Hydroxides. Chemical Communications. 1999;2155-156.

[40] Kumar D, Sharma RC. Advances in Conductive Polymers. European Polymer Journal. 1998;34(8) 1053-1060.

[41] Syed AA, Dinesan MK. Polyaniline – A Novel Polymeric Material - Review. Talanta. 1991;38(8) 815-837.

[42] Fritsche, J. Ueber das Anilin, Ein Neues Zersetzungsproduct des Indigo. Journal für Praktische Chemie, 1840; 20 453-459.

[43] Letheby HJ. On the Production of a Blue Substance by the Electrolysis of Sulphate of Aniline. Journal of the Chemical Society, 1862;15 161-163.

[44] Gospodinova N, Terlemezyan L. Conducting Polymers Prepared by Oxidative Polymerization: Polyaniline. Progress in Polymer Science. 1998;23(8) 1443-1484.

[45] Runge F.F. Ueber Einige Produkte der Steinkohlendestillation. Annalen der Physik. 1834; 31 65-78.

[46] Skotheim TA, editor. Handbook of Conducting Polymers. M. Dekker: New York, 1998.

[47] Meyer V. Ueber Benzole Verschiedenen Ursprungs. Berichte der DeutschenChemischenGesellschaft.1882;15 2893-2894.

[48] Challier T, Slade RCT. Nanocomposite Materials - Polyaniline-Intercalated Layered Double Hydroxides. Journal of Materials Chemistry. 1994;4(3) 367-371.

[49] Isupov VP, Chupakhina LE, Ozerova MA, Kostrovsky VG, Poluboyarov VA. Polymerization of m-NH2C6H4COO Anions in the Intercalation Compounds of Aluminium Hydroxide [LiAl2(OH)(6)][m-NH2C6H4COO] center dot nH(2)O. Solid State Ionics. 2001;141(SI) 231-236.

[50] Moujahid EM, Dubois M, Besse JP, Leroux F. Role of Atmospheric Oxygen for the Polymerization of Interleaved Aniline Sulfonic Acid in LDH. Chemistry of Materials. 2002;14(9) 3799-3807.

[51] Moujahid EM, Leroux F, Dubois M, Besse JP. In Situ Polymerization of Monomers in Layered Double Hydroxides. Comptes Rendus Chimie. 2003;6(2) 259-264.

[52] Tronto J, Sanchez KC, Crepaldi EL, Naal Z, Klein SI, Valim JB. Synthesis, Characterization and Electrochemical Study of Layered Double Hydroxides Intercalated with 2-Thiophenecarboxylate Anions. Journal of Physics and Chemistry of Solids. 2004;65(2-3) 493-498.

[53] Moujahid EM, Dubois M, Besse JP, Leroux F. In situ Polymerization of Aniline Sulfonic Acid Derivatives into LDH Interlamellar Space Probed by ESR and Electrochemical Studies. Chemistry of Materials. 2005;17(2) 373-382.

[54] Tronto J, Leroux F, Crepaldi EL, Naal Z, Klein SI, Valim JB. New Layered Double Hydroxides Intercalated with Substituted Pyrroles. 1. In Situ Polymerization of 4-(1H-pyrrol-1-yl)benzoate. Journal of Physics and Chemistry of Solids. 2006;67(5-6) 968-972.

[55] Tronto J, Leroux F, Dubois M, Taviot-Gueho C, Naal Z, Klein SI, Valim, JB. New Layered Double Hydroxides Intercalated with Substituted Pyrroles. 2. 3-(Pyrrol-1-yl)-Propanoate and 7-(pyrrol-1-yl)-Heptanoate LDHs. Journal of Physics and Chemistry of Solids. 2006;67(5-6) 973-977.

[56] Tronto J, Leroux F, Dubois M, Borin, JF, Graeff, CFD, Valim JB, Hyperfine Interaction in Zn-Al Layered Double Hydroxides Intercalated with Conducting Polymers. Journal of Physics and Chemistry of Solids. 2008;69(5-6) 1079-1083.

[57] Tian X, Wei M, Evans DG, Rao G, Yang H. Controlled Polymerization of Metanilic Anion within the Interlayer of NiA1 Layered Double Hydroxide. Clays and Clay Minerals. 2006;54(4) 418-425.

[58] Tian X, Wei M, Evans DG, Rao G, Duan X. Tentative mechanisms for In Situ Polymerization of Metanilic Acid Intercalated in MgAl Layered Double Hydroxide Under Nitrogen Atmosphere. Advanced Materials Research.2006;11-12295-298.

[59] Wei M, Tian X, He J, et al. Study of the In Situ Postintercalative Polymerization of Metanilic Anions Intercalated in NiAl-Layered Double Hydroxides under a Nitrogen Atmosphere. European Journal of Inorganic Chemistry. 2006;17 3442-3450.

[60] Bordonal AC. Materiais Híbridos Orgânico-Inorgânicos: Polímeros Condutores Intercalados em Compostos Lamelares. MS thesis. Universidade de São Paulo; 2012.

[61] Tronto J. Síntese, Caracterização e Estudo das Propriedades de Hidróxidos Duplos Lamelares Intercalados com Polímeros Condutores. PhD thesis. Universidade de São Paulo; 2006.

Chapter 2

Optical Properties of Multiferroic $BiFeO_3$ Films

Hiromi Shima, Hiroshi Naganuma and
Soichiro Okamura

Additional information is available at the end of the chapter

http://dx.doi.org/10.5772/54908

1. Introduction

1.1. Background

Lightwave communication systems are predominantly used for handling high-speed data traffic. Long-distance ground-based systems particularly depend on optical fibers. Several business and research facilities employ direct fiber connections, and fiber to the home (FTTH) technology is foreseeable in the near future. These developments are driven particulary by the high demand for bandwidth necessary for many computers contributing to internet traffic. Lightwave communication systems are one of the fastest growing industrial fields because of a few important inventions and extensive research and development by physicists and engineers. The key components of a long-distance lightwave communication system are semiconductor lasers, low-loss glass fibers, optical amplifiers, and photodetectors. Apart from these key elements, several additional functions are required to enable modulating, switching, and combining the optical signals. In addition, network traffic management and switching, routing, and distribution systems are essential. Therefore, we focused on the development of optical components such as modulators and switches.

Here, some of the novel optical devices for modulating or switching light signals are introduced. First, the operating principle of an electro-optic spatial light modulator (EOSLM) is described. In general, a spatial light modulator is an optical device that achieves spatial modulation of incident light. Figure 1(a) shows the basic structure of an optical switching cell in an EOSLM. [1, 2] An electro-optic thin film is fabricated on a large-scale integration (LSI) circuit together with top and bottom electrodes. The top electrode is made of transparent conductive material such as indium tin oxide (ITO). A dielectric multi-layer mirror (DMM) is deposited onto the ITO electrode, and a Fabry-Perot resonator is formed between the upper mirror and the bottom platinum electrode. In this case, the minimum reflectance becomes zero because the reflectance of the DMM and the bottom Pt layer is the same. If the refractive index can be controlled by applying an electric field, the cell can switch incident light at a specific wavelength as shown in Fig. 1(b).

Next, the operating principle of a Mach-Zehnder modulator (MZM) is described. Figure 2 shows a schematic diagram of an electrooptic-type MZM. [3-5] Incident light is split into two waveguides. The output amplitude depends on the phase difference at recombination. As shown in the top right part of the figure, in-phase recombination produced a "1" bit output while anti-phase recombination produces a "0" bit output owing to half-wave phase shifting caused by an applyied voltage. A half-wave phase shift can be caused by either the electro-optic effect or the thermo-optic effect. When using the thermo-optic effect, the optical switch structure includes a heater on one side of the waveguide instead of a top electrode. [6,7]

To realize these novel thin-film optical devices, many researchers have intensively studied materials that exhibit the electro-optic effect. Traditional electro-optic materials include (Pb,La)(Zr,Ti)O_3 (PLZT) and $LiNbO_3$. PLZT is the most promising candidate for such applications because it has a high transparency in its polycrystalline form. [7] It is well known that bulk PLZT with a 65:35 Zr:Ti ratio shows large electro-optic e ▪ ect coe ▪ cients: a Pockels coe ▪ cient (linear electro-optic coe ▪ cient) of 6.12×10^{-10} m/V and a Kerr coe ▪ cient (quadratic electro-optic coe ▪ cient) of 9.12×10^{-16} m^2/V^2 with La contents of 8 and 9 at.%, respectively. [8] The next generation of new multiferroic materials, such as $BiFeO_3$ (BFO), which exhibits a giant remanent polarization of 100 $\mu C/cm^2$ in the thin-film form, [9,10] is now reaching maturity and has recently attracted considerable attention because of its potential applications in novel multifunctional devices. In recent years, the electric and magnetic properties of BFO films have been a topic of intense research, with regard to their magnetoelectric (ME) effect [11-13], while few reports on their optical properties have been published. [14-16] Therefore, it is necessary to know the basic optical properties of BFO films, such as the optical constant and thermooptic property, for the development of various optical applications. Understanding the optical potential in multiferroic materials leads us to additional noble material selections and thus degrees of freedom.

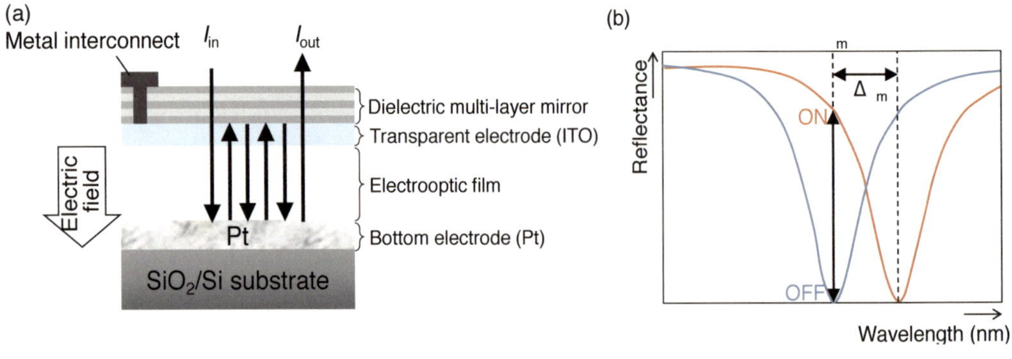

Figure 1. (a) Cross-sectional view of optical switching cell structure of the EOSLM, and 1(b) simulated optical switching properties of unit cells.

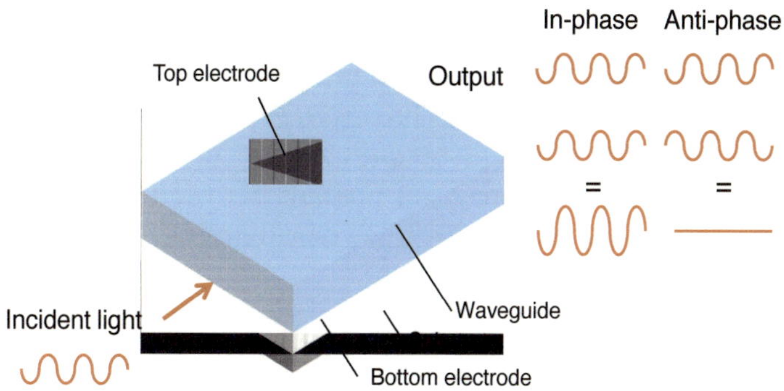

Figure 2. Schematic illustration of the electrooptic-type Mach-Zehnder modulator.

1.2. Objectives

To realize smaller and faster optical modulators or switches, it is necessary to embed electro-optic or thermo-optic materials in semiconductor integrated circuits. Accordingly, we have investigated the electro-optic and thermo-optic coefficients of materials in polycrystalline film form. We explain in detail the optical property evaluation method used in this study to exclude extrinsic effects, and then, we carefully determine the basic optical and thermo-optical properties of multiferroic $BiFeO_3$ polycrystalline films. Finally, we would like to understand the potential application of these optical properties to noble multifunctional devices using multiferroic $BiFeO_3$ polycrystalline films. This chapter describes the basic film preparation method and the basic method for evaluation of optical properties to increase understandings for beginner in this research field.

1.3. Outline of this chapter

This chapter is divided into 4 sections. Section 1 describes the background and objective of this study. In Section 2, we summarize the fundamentals of multiferroic $BiFeO_3$ used in this study, the method for fabricating multiferroic thin films, the basic optics principles related to this study, and the detailed method for evaluating optical properties. In Section 3, we discuss the thermooptic effect of multiferroic BFO films. The polycrystalline BFO films are fabricated on $Pt/Ti/SiO_2/Si$ substrates, respectively. First, their fundamental properties are evaluated. Next, the optical constants of these films are evaluated as a basic optical property. In Section 4, we summarize the conclusions of this chapter.

2. Experimental procedures

2.1. Fundamentals

Ferroelectricity is a property of certain materials that allows for spontaneous reversible electric polarization by applying an external electric field. The ability of a crystal to exhibit spontaneous polarization is related to its symmetry. Of the 32 point groups, which describe all crystalline systems, 11 are centrosymmetric and contain an inversion center. The remaining 21 point groups without an inversion center can exhibit piezoelectricity, except for the point group *432*. Among the 21 groups without an inversion center, 10 polar groups possess a unique polar axis. Such asymmetric crystals can show unique electrical as well as optical characteristics, *e.g.*, the electro-optic effect, acousto-optic effect, photorefractive effect, nonlinear optical effect, etc. It is possible to control light dynamically using these effects to change the refractive index of materials.

In this chapter, the fundamentals of materials, fabrication, and evaluation are mentioned. In particular, the basic characteristics of $BiFeO_3$ potential ferroelectric films are mentioned together with their optical properties.

2.2. $BiFeO_3$ multiferroics

Perovskite-oxide has a structural formula of ABO_3, in which A is a large cation such as Bi^{3+}, Ba^{2+}, or Pb^{2+}, and B is a medium-sized cation such as Fe^{3+}, Ti^{4+}, or Zr^{4+}. These cations are located in cages formed by a network of oxygen anions, as shown in Fig. 3(a). Ferroelectric perovskites are a subgroup of the perovskite family. They are cubic at high temperatures and become polar non-cubic, *i.e.*, tetragonal, rhombohedral, *etc.*, below their Curie temperature. In the cubic phase, the cations are located at the center of an oxygen octahedron, while in the polar phases, they are shifted off center. The direction of the displacement of the cations in oxygen can be switched by applying an electric field, as shown in Fig. 3(b).

Multiferroic materials have more than one primary ferroic order parameter such as ferroelectricity, ferromagnetism, and ferroelasticity in the same phase. Multiferroic materials have attracted considerable attention, not only in terms of scientific interest but also because of their potential applications in novel functional devices. Bismuth ferrite ($BiFeO_3$, BFO) has long been known to be a multiferroic material that exhibits antiferromagnetism ($T_N \approx 643$ K) [17] and ferroelectricity ($T_C \approx 1103$ K) [18] when in bulk form. The structure and properties of thesingle-crystal form of BFO have been extensively studied. It has a rhombohedrally distorted perovskite structure with space group *R3c* [19] at room temperature (RT), as shown in Fig. 4. A perovskite-type unit cell with a rhombohedral structure has a lattice parameter of $a = b = c =$ 0.3965 nm and $\alpha = 89.3^{\circ}$ at RT. [19,20]

The Fe magnetic moments are coupled ferromagnetically within the pseudocubic (111) planes and antiferromagnetically between the near planes; this is called the *G*-type antiferromagnetic order. If the magnetic moments are oriented perpendicular to the [111] direction, the symmetry also permits a canting of the antiferromagnetic sublattices resulting in a macroscopic magnetization called weak magnetism. [21,22]

According to a first-principles calculation, the spontaneous polarization of BFO changes depending on whether the crystal structure is rhombohedral or tetragonal. The tetragonal structure of the BFO (SG: *P4mm*) possesses P_s of around 150 μC/cm² along the [001] direction, and the rhombohedral structure (SG: *R3c*) possesses P_s of around 100 μC/cm² along the [111] direction without strain. [23,24] At the beginning of this research, in the case of bulk form, the

http://dx.doi.org/10.5772/54908

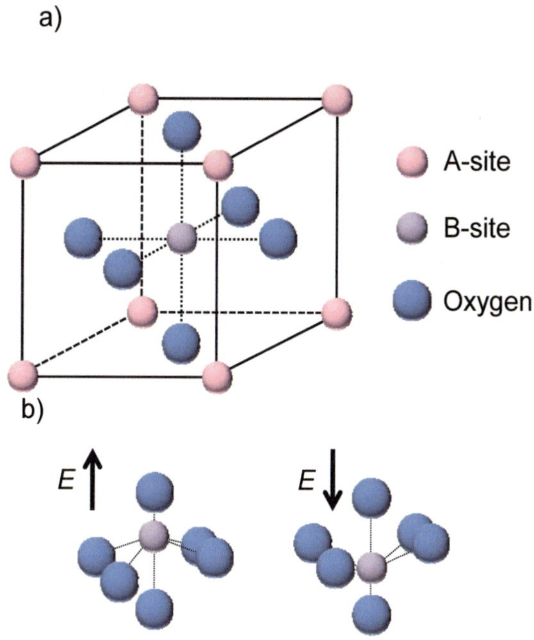

Figure 3. The crystal structure of typical perovskite-oxide: (a) The cubic phase, and (b) Schematic illustration of dipole switching with applying electric field $E>E_c$.

spontaneous polarization was 3.5 μC/cm² along the [001] direction, indicating a value of 6.1 μC/cm² along the [111] direction at 77 K. [25] Recently, it was reported that the bulk form of BFO showed P_s 60 μC/cm² along the [111] direction 9er, in thin-film BFO, large remanent polarizations ranging from 100 to 150 μC/cm² have been reported. [9,10] The difference between the thin-film and bulk values, initially attributed to epitaxial strain, could also result from mechanic constraints in granular bulk ceramics and from leakage effects in crystals caused by defect chemistry or the existence of second phases.

2.3. Fabrication method

2.3.1. Preparation method using chemical solution deposition

Chemical solution deposition (CSD) [26] is the method for fabrication of thin films using a precursor solution; several types of metal-organic compounds such as metal alkoxide and metal carboxylate compounds can be used as the precursor solution. The fabrication of thin films by this approach involves four basic steps:

i. Synthesis of the precursor solution;

ii. Deposition by spin-casting or dip-coating, where the drying processes usually depends on the solvent;

iii. Low-temperature heat treatment for drying, pyrolysis of organic species (typically 300–400°C), and formation of an amorphous film;

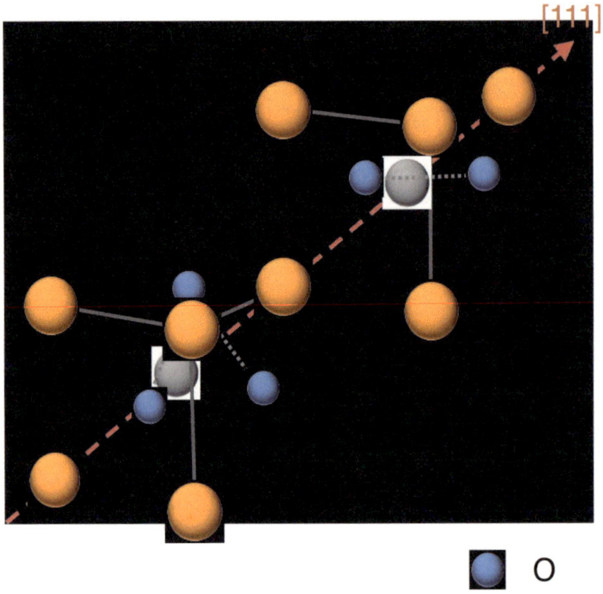

Figure 4. Schematic view of the *R*3*c* structure built up from two cubic perovskite $BiFeO_3$ unit cells. The cations are displaced along the [111] direction relative to the anions, and the oxygen octahedral rotate with alternating sense around the [111] axis.

iv. Higher-temperature heat treatment for densification and crystallization of the coating into the desired oxide phase (600–1100°C).

For most solution deposition approaches, the final three steps are similar despite differences in the characteristics of the precursor solution, and for electronic devices, spin-casting has been used almost exclusively. Depending on the solution route employed, different deposition and thermal processing conditions may be used for controlling film densification and crystallization in order to prepare materials with optimized properties.

For the fabrication of perovskite thin films, the most frequently used CSD approaches may be grouped into three categories:

i. Sol-gel processes that use 2-methoxyethanol as a reactant and solvent.

ii. Chelate processes that use modifying ligands such as acetic acid.

iii. Metal-organic decomposition (MOD) routes that use water-insensitive metal carboxylate compounds.

Other approaches that have also been used, although less extensively, include the nitrate method, citrate route, and Pechini process. In this study, a sol-gel solution and an enhanced-MOD (EMOD) solution (symmetric) were used because of their manageability.

http://dx.doi.org/10.5772/54908

2.3.2. *Synthesis of the precursor solution for CSD*

Processes based on 2-methoxyethanol are most appropriately considered sol-gel processes, and the key reactions leading to the formation of the precursor solutions are hydrolysis and condensation of the alkoxide reagents, in which metal-oxygen-metal (M-O-M) bonds are formed:

Hydrolysis:

$$M(OR)_n H_2O \rightarrow M(OR)_{n-1}(OH)ROH.$$

Condensation:

$$M(OR)_n M(OR)_{n-1}(OH) \rightarrow M_2O(OR)_{2n-2}ROH.$$

In some cases, an alcohol exchange reaction occurs in a practical synthesis process.

Alcohol exchange:

$$M(OR)_n xR'OH \rightarrow M(OR)_{n-x}(OR')_x xROH.$$

where OR is a reactive alkoxy group and OR′ is the less reactive methoxyethoxy group.

In addition, to prepare a compound oxide material using two or more kinds of metal compounds in order to increase the homogeneity of the precursor solution, a double alkoxide with M-O-M′-O bonds may be synthesized for refluxing in an inactive gas atmosphere.

Synthesis of double alkoxide:

$$xM(OR)_m M'(OR')_n \rightarrow xMM'(OR)_m(OR')_n$$

In addition, if a metal alkoxide and carboxylate compounds are used, the synthesis is occasionally accompanied by an esterification reaction.

Esterification reaction:

$$M(OR)_n M'(O_2CR')_m \rightarrow (OR)_{n-1}MOM'RO_2CR'(O_2CR')_{m-1}RO_2CR'$$

For reproducible thin films, byproducts such as esters produced during the synthesis should be removed from the precursor solution through fraction distillation.

2.3.3. *Preparation conditions of BFO polycrystalline films*

The BFO films were formed through CSD. A precursor solution for CSD was prepared from bismuth acetate (99.99%, Aldrich), iron acetylacetonate (99.9%, Wako) in a solvent of 2-methoxyethanol (99.7%, Aldrich), and acetic acid (99.5%, Wako). The solution was heated to 80–100°C while stirring for 30 min to promote the dissolution of the precursors, followed by stirring at RT for 1 day before film deposition. The solution was synthesized on the basis of a stoichiometric composition. The concentration of the precursor solution was adjusted to be 0.05 M.

Figure 5 shows the process flow for the fabrication of the polycrystalline BFO film through CSD. A (111)Pt/Ti/SiO_2/(100)Si substrate was spin coated with the precursor solution at 3000 rpm for 30 s. The spin-coated film was dried at 150°C for 1 min and pyrolized at 400°C for 2 min in air. After the processes from spin coating to pyrolysis were repeated 10 times, the film was fired at 550°C for 5 min in air through rapid thermal annealing (RTA). This sequence was repeated 10 times. The film was polycrystalline with a random orientation and was approximately 650 nm thick.

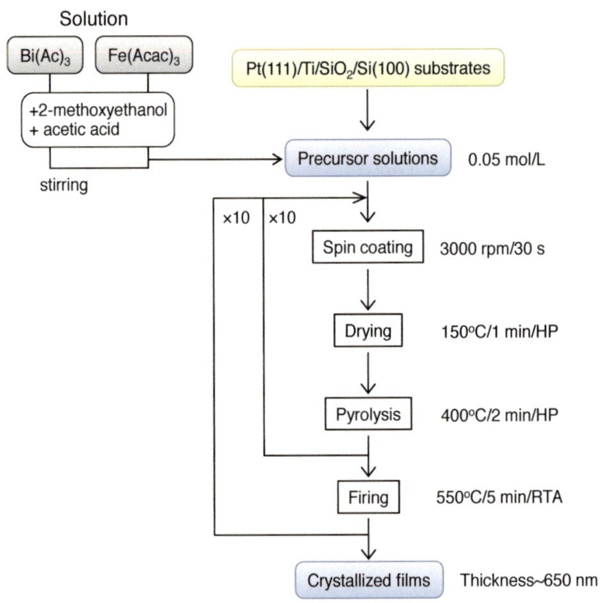

Figure 5. The process flow for fabrication of BFO films by CSD method.

2.4. Optical properties of ferroelectrics

2.4.1. Refractive index

The general definition of refractive index n is:

$$n = \frac{c_0}{v} \tag{1}$$

where c_0 is the speed of light in vacuum, and v, the speed of light in a material. The refractive index is related to the dielectric constant ε_r through the following equation:

$$\varepsilon_r = n^2 \tag{2}$$

This relationship is only valid when the interacting electric field has a frequency on the order of THz or higher and when the material is isotropic. The general behavior of condensed matter in an alternating electric field is that moving charges cause a frequency-dependent phase shift between the applied field and the electric displacement. Mathematically, this is expressed by writing the permittivity ε as a complex function:

$$\varepsilon(\varpi) = \varepsilon_1(\varpi) - i\varepsilon_2(\varpi) \tag{3}$$

The real part ε_1 characterizes the electric displacement, and the imaginary part ε_2 denotes the dielectric losses. The loss tangent is defined as

$$\tan\delta = \frac{\varepsilon_2}{\varepsilon_1} \tag{4}$$

Since light is an alternating electromagnetic wave with the electric and magnetic field vibration directions perpendicular to one another, the electric field induces an electric polarization in a dielectric crystal and the light itself is influenced by the crystal. The alternating frequency of light is so high (λ= 500 nm corresponds to a frequency of approximately 600 THz) that only electronic polarization can follow the electric field change. Therefore, the relative permittivity of an optically transparent crystal is small, typically less than 10. It is known that a dielectric material shows wavelength dispersion of its refractive index at optical frequencies.

2.4.2. Optical indicatrix and anisotropy of refractive index

In a microscopically anisotropic medium, the refractive index is generally different for different crystal directions. Ferroelectric materials, particularly in film form, can be both optically isotropic and optically anisotropic. Ferroelectric ceramics or polycrystalline films are an example of the former type; their isotropic behavior is due to the random orientation of their constituent grains. Ferroelectric single crystals or epitaxial films are an example of the latter type, and they can be divided into optically uniaxial and optically biaxial crystals. If a coordinate system is chosen to coincide with the three principal axes of a crystal, we have the following relations:

$$\varepsilon_x = n_x^2, \quad \varepsilon_y = n_y^2, \quad \varepsilon_z = n_z^2. \tag{5}$$

The optical anisotropy of a crystal is characterized by an optical indicatrix (or index ellipsoid) defined as

$$\frac{x^2}{n_x^2} + \frac{y^2}{n_y^2} + \frac{z^2}{n_z^2} = 1, \tag{6}$$

where n_x, n_y, and n_z are the principal refractive indexes, as shown in Fig. 2.6. The optical indicatrix is mainly used to find the two refractive indexes associated with the two independent plane waves that can propagate along an arbitrary direction *k* in a crystal. The optical indicatrix is used as follows: The intersection ellipse between the optical indicatrix and a plane through the origin point normal to the propagation direction *k* is found. The two axes of the intersection ellipse are equal in length to $2n_1$ and $2n_2$, where n_1 and n_2 are the two refractive indexes.

In the case of a biaxial crystal, there are two optical axes, and the refractive indexes are different in all three principal directions, $n_x \neq n_y \neq n_z$. In the common case of a uniaxial crystal, we have $n_x = n_y = n_o$ and $n_z = n_e$, where n_o and n_e are the ordinary and extraordinary refractive indexes, respectively. The refractive index along the optical axis corresponds to the extraordinary index n_e, and the refractive index perpendicular to the optical axis corresponds to the ordinary index n_o, as shown in Fig. 6. The existence of two rays with different refractive indexes is called birefringence. The birefringence Δn is usually defined as:

$$\Delta n = n_e - n_o \tag{7}$$

Since the value of n_e may be either higher or lower than n_o, birefringence may take on positive or negative values. If $\Delta n > 0$, the crystal is said to be optically positive, whereas if $\Delta n < 0$, it is said to be optically negative. For light propagating in a different direction from the principal axis in a uniaxial crystal, the situation becomes more complicated. A light wave with the wave vector κ, as shown in Fig. 6, has a constant ordinary index, whereas the extraordinary refractive index is dependent on the angle θ as:

$$\frac{1}{n_e(\theta)^2} = \frac{\cos^2\theta}{n_o^2} + \frac{\sin^2\theta}{n_e^2}. \tag{8}$$

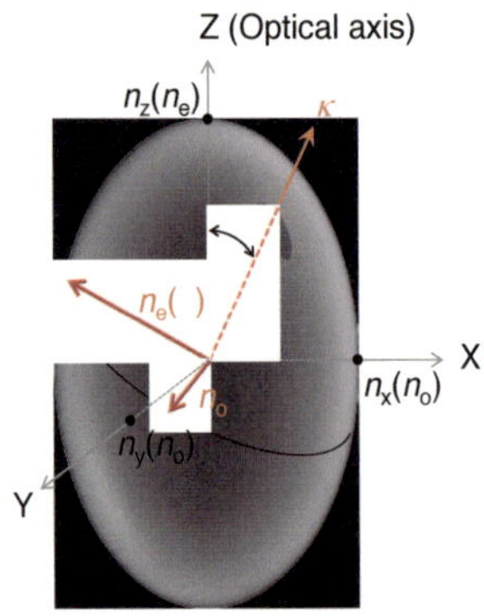

Figure 6. Schematic view of the *R*3*c* structure built up from two cubic perovskite $BiFeO_3$ unit cells. The cations are displaced along the [111] direction relative to the anions, and the oxygen octahedral rotate with alternating sense around the [111] axis.

2.4.3. *Thermooptic effect*

The thermo-optic effect refers to thermal modulation of the refractive index of a material. The refractive index of a material can be modulated as a function of its thermooptic coefficient α as

$$n(T) = n_0 - \alpha \cdot T, \tag{9}$$

where T is the temperature and $n(T)$ and n_0 are the refractive indexes at an arbitrary temperature and at 0°C, respectively. The thermo-optic coefficient relates the changes in the optical indicatrix ΔB_{ij} with the changes in temperature ΔT. Since temperature is a scalar, the thermo-optic effect is a symmetric second-rank tensor similar to the dielectric constant. The temperature dependence of the refractive index is generally small, except near phase transformations. The situation is analogous to low-frequency dielectrics. For silica and alumina, the permittivity is nearly independent of temperature, but ferroelectrics exhibit enormous changes near T_c. The refractive index of common oxides increases with density. Because of thermal expansion, density decreases with increasing temperature, thus decreasing the refractive index as well. Thermal expansion makes a small negative contribution to the temperature coefficient of refractive index dn/dT. This effect is often influenced by changes in the electronic band gap or by phase changes. These effects can be either positive or negative, depending on the nature of the energy levels or on the location of the phase transformation.

2.5. Evaluation method of optical properties

Spectroscopic ellipsometry

Ellipsometry determines the optical constants and thickness of materials in layered samples by fitting a parameterized model to the measured data for simultaneously analyzing data from multiple samples. Figure 7 shows a schematic illustration of ellipsometry. The linearly polarized incident light is reflected after interacting with the sample. The polarization of the light changes from linear to ellipsoidal from this interaction. We measure the polarization state using the ratio of the reflection coefficients for the light polarized parallel (p) and perpendicular (s) to the plane of incidence. This ratio, called the ellipsometric parameter, is defined as [27-30]

$$\rho = \frac{R_p}{R_s} = \tan(\psi)\exp(i\Delta), \tag{10}$$

where R_p and R_s are the Fresnel reflection coefficients of polarized light parallel and perpendicular to the incident plane, respectively. Here, tan (ψ) and Δ are the amplitude and phase of ρ, respectively. The Fresnel reflection coefficients are represented as follows:

For p-polarized light,

$$R_p = \frac{\tilde{n}^2\cos\varphi - \sqrt{\tilde{n}^2 - \sin^2\varphi}}{\tilde{n}^2\cos\varphi + \sqrt{\tilde{n}^2 - \sin^2\varphi}}, \tag{11}$$

and for *s*-polarized light,

$$R_s = \frac{\cos\varphi - \sqrt{\tilde{n}^2 - \sin^2\varphi}}{\cos\varphi + \sqrt{\tilde{n}^2 - \sin^2\varphi}}. \tag{12}$$

Here, ñ is the complex refractive index, and Φ, the incident angle.

In this study, a Gaussian oscillator was used to model a dielectric function to represent film properties. Gaussian oscillators represent the normal distribution for the ε_2 spectrum; the ε_1 spectrum is determined by the ε_2 values because the Kramers-Kronig (KK) relation couples the real and imaginary parts of the complex dielectric constant. When a Gaussian oscillator is used as a dielectric function, complex dielectric constants (ε_1, ε_2) are calculated as [31,32]

$$\varepsilon_2 = A_n \exp\left[-\left(\frac{E - E_n}{B_n}\right)\right] + A_n \exp\left[-\left(\frac{E + E_n}{B_n}\right)\right], \tag{13}$$

$$\varepsilon_1 = 1 + \frac{2}{\pi} P \int_0^\infty \frac{\xi\varepsilon_2(\xi)}{\xi^2 - E^2} d\xi, \tag{14}$$

where A_n is the amplitude of the oscillator; E_n, the central energy of the oscillator; B_n, the broadening of the oscillator; and P, Cauchy's principal value. Optical constants (n, k) are equivalent to the complex dielectric constants (ε_1, ε_2). Therefore, optical constants (n, k) can be determined from the Gaussian oscillator parameters, which show the best fitting for experimental values of Φ and Δ. Model fitting was carried out by minimizing the mean square error (MSE) function defined as [27,29]

$$\mathrm{MSE} = \frac{1}{2N - M} \sum_{i=1}^{N} \left[\left(\frac{\psi_i^{\mathrm{mod}} - \psi_i^{\mathrm{exp}}}{\sigma_{\psi_i}^{\mathrm{exp}}}\right)^2 + \left(\frac{\Delta_i^{\mathrm{mod}} - \Delta_i^{\mathrm{exp}}}{\sigma_{\Delta_i}^{\mathrm{exp}}}\right)^2\right], \tag{15}$$

where N is the number of (ψ, Δ) pairs, M is the number of variable parameters in the model, σ is the standard deviation on the experimental points, and the superscripts "mod" and "exp" represent the calculated and experimental values, respectively.

http://dx.doi.org/10.5772/54908

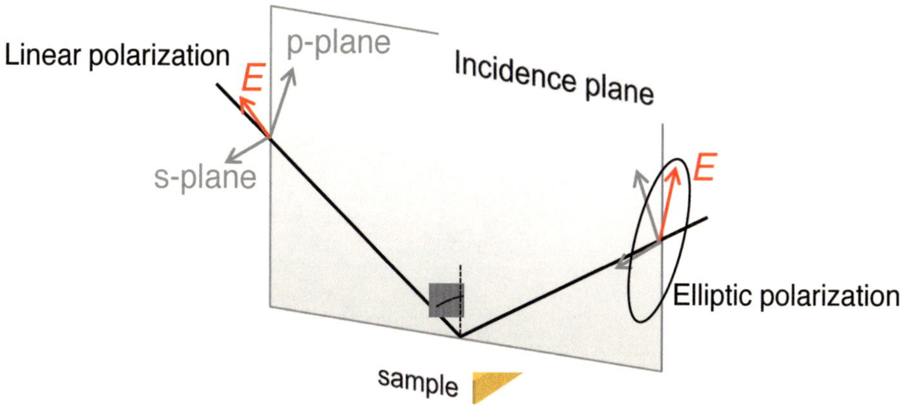

Figure 7. Schematic illustration of the ellipsometry.

3. Optical properties of multiferroic $BiFeO_3$ films

3.1. Fundamentals

Multiferroic materials, which simultaneously exhibit ferroelectricity and magnetic ordering, have attracted considerable attention, not only in terms of scientific interest but also because of their potential applications in novel functional devices. BFO is one of the few materials that exhibit dielectric and magnetic ordering at room temperature. [17,18] BFO also exhibits a large remanent polarization of 100 $\mu C/cm^2$ in thin film form. [9,10] Therefore, the electric and magnetic properties of BFO films have been the subject of intense research in recent years. [11-13] However, few reports on their optical properties [14,15] or their applications [16] have been published. It is important to know the exact optical properties of BFO films in order to develop various optical applications. To apply BFO films to optical devices, the electro-optic, magneto-optic, and thermo-optic effects of the films can be controlled by modulating their refractive indices. Recently, a Mach-Zehnder-type optical switch, which employs the thermo-optic effect, is a topic of immense interest in the photonics field. [33-35] In this section, we examine the optical constant and the temperature dependence of the refractive index of polycrystalline BFO films.

3.2. Experimental machines

- Crystal structure and orientation: X-ray diffractometer (PANalytical, X'pert PRO MPD)
- Morphology: field emission scanning electron microscope (JEOL, JIB-4500FE), transmission electron microscope (Hitachi, HF-2000), atomic force microscope (SII, SPI3800N)
- Electrical property: ferroelectric test system (Toyo, FCE)
- Optical property: spectroscopic ellipsometer (J. A. Woollam, M-2000) with a heating stage, as shown in Fig. 8(a) and 8(b).

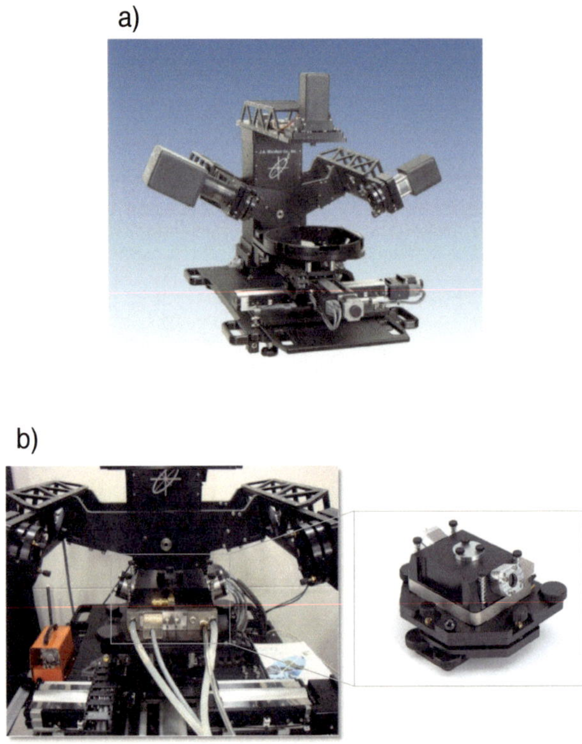

Figure 8. (a) Spectroscopic ellipsometer (J. A. Woollam, M-2000) used in this study, and (b) the heating stage used for annealing sample from RT to 600°C.

3.3. Fundamental properties of polycrystalline BFO film

Polycrystalline BFO films were successfully formed on the Pt/Ti/SiO_2/Si substrates through CSD. Figure 9 shows the XRD θ-2θ pattern of the polycrystalline BFO film. The XRD analysis, it was confirmed that the BFO film was crystallized into a single perovskite phase with a random orientation.

Figure 10 shows the *P-E* hysteresis loop of the polycrystalline BFO film, measured with a single triangular pulse of 100 kHz at room temperature. From this figure, the ferroelectricity of the film can be confirmed. By a positive-up-negative-down (PUND) measurement technique [36] using a pulse train with an amplitude of 1.53 MV/cm and a width of 5 μs, a remanent polarization of 30 $\mu C/cm^2$, relative permittivity of 280, and leakage current density of 7.6 A/cm^2 were estimated at room temperature.

Figure 11 shows a) the surface morphology and the cross-sectional images of (b) the cleavage face and (c) the worked surface using a focused ion beam. From Fig. 11(a) and (b), it can be seen that the polycrystalline BFO film consists of small randomly grown grains, whereas the bottom Pt layer has a columnar grain growth. From Fig. 11(c), voids in the film were confirmed. Film thickness was estimated to be approximately 650 nm from cross-sectional images.

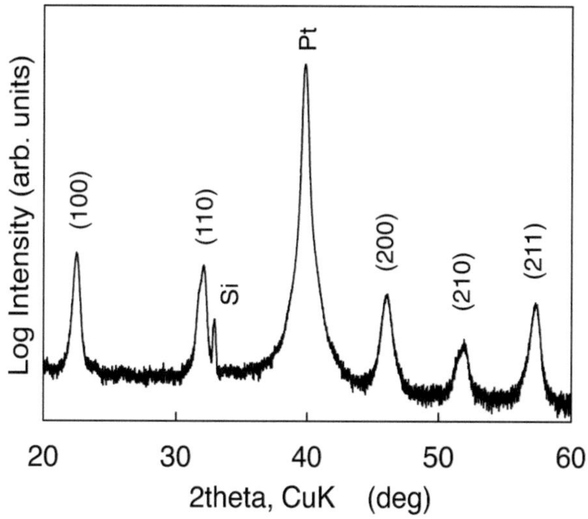

Figure 9. The XRD θ-2θ pattern of the polycrystalline BFO film.

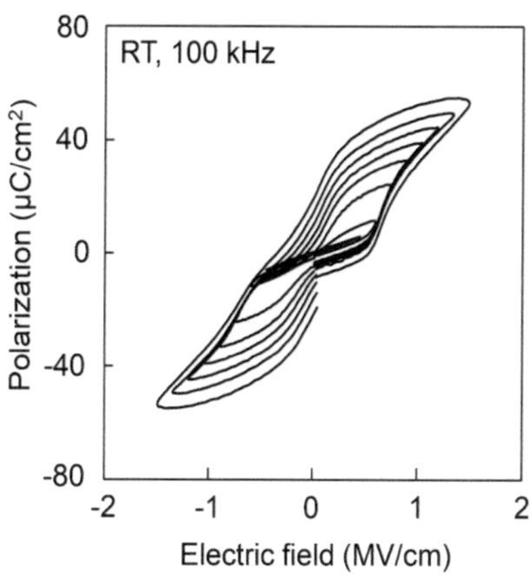

Figure 10. The *P-E* hysteresis loop of the polycrystalline BFO film measured with a single triangular pulse of 100 kHz at room temperature.

3.4. Optical constants of polycrystalline BFO film

Ellipsometric spectra in (Δ, ψ) were recorded at tincident angles of $\theta_i = 50^o$, 60°, and 70° in a spectral range of 245–1670 nm. Figure 12 shows the multilayer model used in this study. It was assumed that the model consisted of ambient (air), a surface layer, a bulk layer, and a substrate (Pt). The optical constants of the surface layer were represented by the Bruggeman effective medium approximation (EMA) [37] consisting of a 0.50 bulk film/0.50 void mixture. In the polycrystalline BFO film, 4 Gaussian oscillators were assumed to represent the film properties.

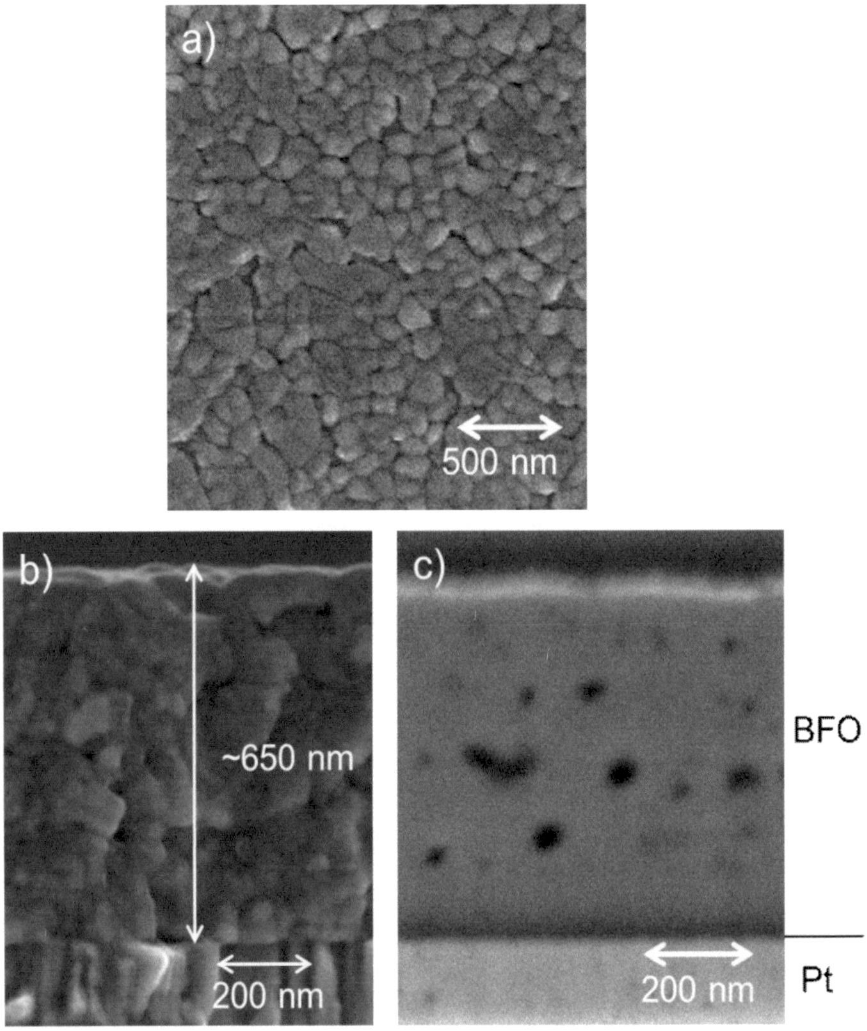

Figure 11. (a) The surface morphology and the cross-sectional images of (b) cleavage face and (c) worked surface by focused ion beam.

Furthermore, the density gradient along the film thickness, which affects the refractive index gradient, was also introduced by applying the EMA. Figure 13(a)–(d) shows the ψ and Δ spectra of the polycrystalline BFO film measured at various incident angles and fitting curves. The fitting parameters are summarized in Table 1. Figure 13(a) and (b) shows the results without the assumption of the refractive index gradient. In this case, the fitting curve did not represent the experimental result well and the MSE was relatively large. Figure 13(c) and (d) shows that when the refractive index gradient was considered, the MSE decreased from 106.6 to 66.6, and the fitting curves represented the experimental results well. This refractive index gradient seems to be caused by the distribution of voids in the film, as shown in Fig. 11(c). The total thickness of the polycrystalline BFO film was estimated to be approximately 650 nm. This value coincided with that observed in the cross-sectional SEM image shown in Fig. 11(a).

Figure 14 shows the optical constant of the polycrystalline BFO film in the wavelength range of 245 - 1670 nm calculated from the best-fitting results. In Fig. 14, the solid and broken lines represent the maximum and minimum values, respectively, of each optical constant in the graded layer. A large refractive index of 3.22 was estimated for the polycrystalline BFO film; this value is higher than that of rutile-type TiO_2 film, which has the highest refractive index among oxides reported thus far. At a wavelength of 600 nm, the refractive index of our BFO film was 3.22 and that of the TiO_2 film was reported to be approximately 2.6. [38] Figure 15 shows the depth profile of the refractive index and the extinction coefficient of the BFO film at a wavelength of 500 nm. This profile shows that the refractive index near the substrate is larger than that at the surface area. The reason for this gradient is not yet clear, although one possible explanation is the existence of voids in the film.

To determine the optical band gap, we plot $(\alpha E)^2$ vs. E for the polycrystalline BFO film, as shown in Fig. 16(a), where α and E are the absorption coefficient and photon energy, respectively. The absorption coefficient α is given by

$$\alpha = \frac{4\pi k}{\lambda}, \tag{16}$$

where k is the extinction coefficient and λ, is the wavelength. A good linear fit above the band gap indicates that the BFO film has a direct gap. [15] The linear extrapolation of $(\alpha E)^2$ to 0 induces band gaps of 2.79 eV for the polycrystalline BFO film. This means that the absorption edge of the BFO film is 445 nm. This value for the polycrystalline BFO film was similar to that in recent reports. [15,39,40] Next, we plot $(\alpha E)^{1/2}$ vs. E in Fig. 16(b) for the BFO film. The $(\alpha E)^{1/2}$ vs. E plot did not show 2 clear slopes as was expected for an indirect gap material. [41] Finally, the polycrystalline BFO film was found to show a sufficiently low light loss at a wavelength greater than 600 nm so that its large reflective index at visible wavelengths is useful for electric-optic devices such as an electro-optic spatial light modulator.

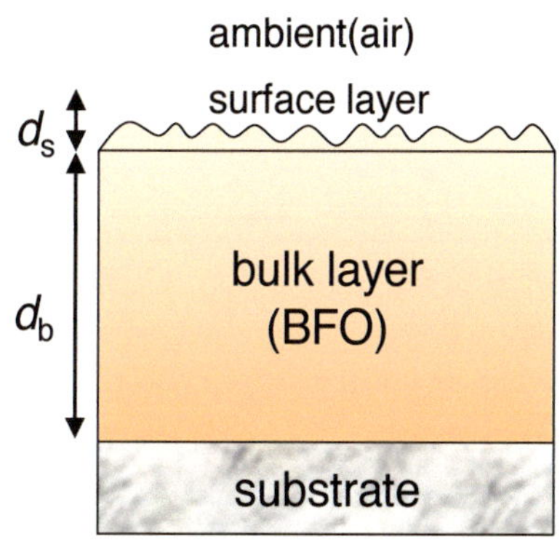

Figure 12. A multilayer model used in this study.

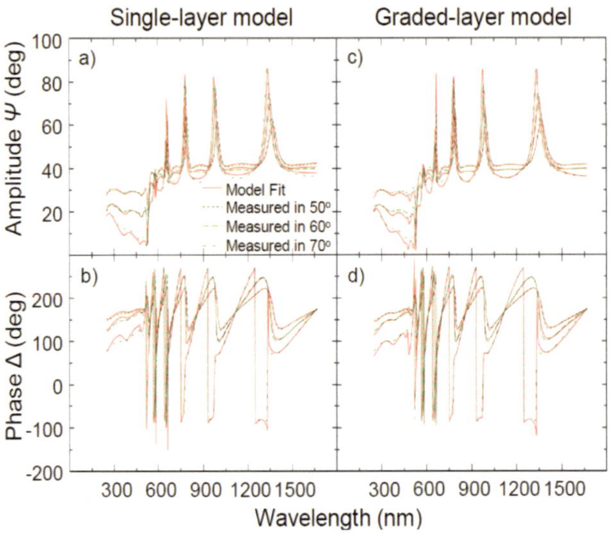

Figure 13. Ψ and Δ spectra of the polycrystalline BFO film measured at various incident angles and fitting curves obtained using the model (a) - (b) without and (c) - (d) with refractive index gradient.

$$\varepsilon_1 = 1 + \frac{2}{\pi} P \int_0^{\infty} \frac{\xi \varepsilon_2(\xi)}{\xi^2 - E^2} d\xi, \quad \varepsilon_2 = A_n \exp\left[-\left(\frac{E - E_n}{B_n} \right) \right] + A_n \exp\left[-\left(\frac{E + E_n}{B_n} \right) \right]$$

Fitting parameters		Sample-Model	
		BFO-single	BFO-graded
Thickness (nm)	d_s	3.566±0.635	4.597±0.224
	d_b	667.423±3.31	644.839±1.92
Amplitude A_n	A_{n1}	6.771	9.834
	A_{n2}	8.981	10.973
	A_{n3}	3.638	5.008
	A_{n4}	0.735	0.716
Central energy E_n (eV)	E_{n1}	4.175	4.119
	E_{n2}	7.460	7.348
	E_{n3}	3.024	3.052
	E_{n4}	2.491	2.476
Broadening B_n (eV)	B_{n1}	1.201	1.205
	B_{n2}	4.254	4.254
	B_{n3}	0.527	0.527
	B_{n4}	0.215	0.215
MSE		106.6	66.66

Table 1. The fitting parameters in single layer model and graded layer model of the polycrystalline BFO films. Gaussian oscillator defined as following equation.

http://dx.doi.org/10.5772/54908

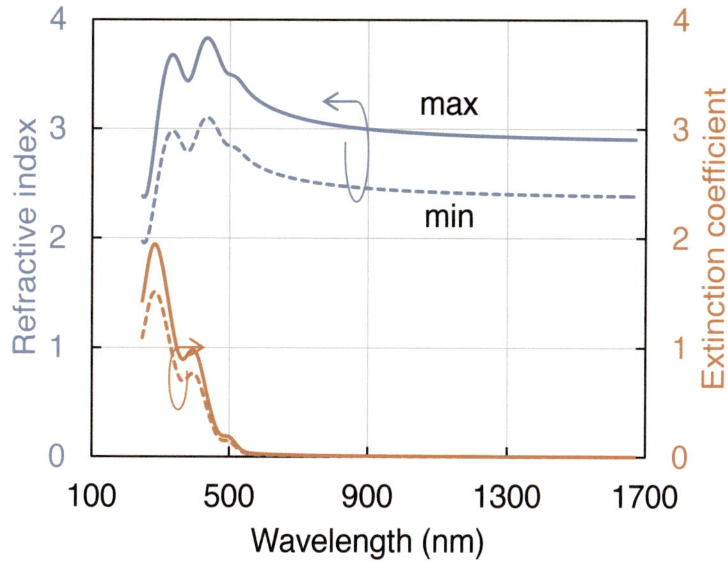

Figure 14. Refractive index and extinction coefficient of the polycrystalline BFO film; the solid and broken lines show maximum and minimum values, respectively.

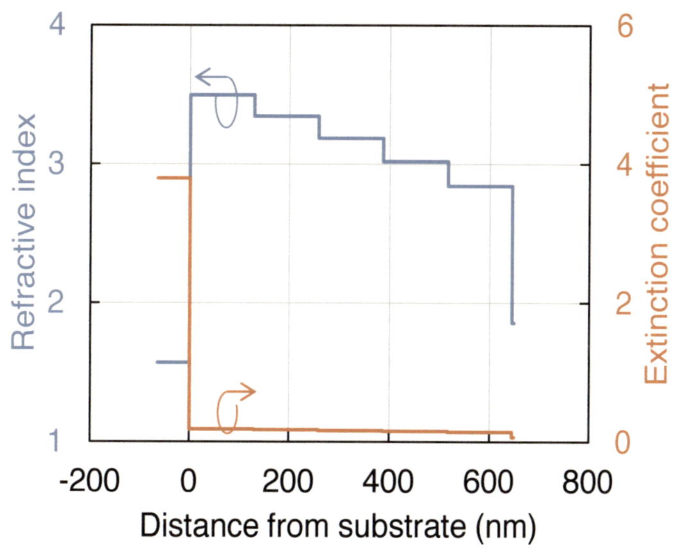

Figure 15. Depth profile of refractive index and extinction coefficient of the BFO film at a wavelength of 500 nm.

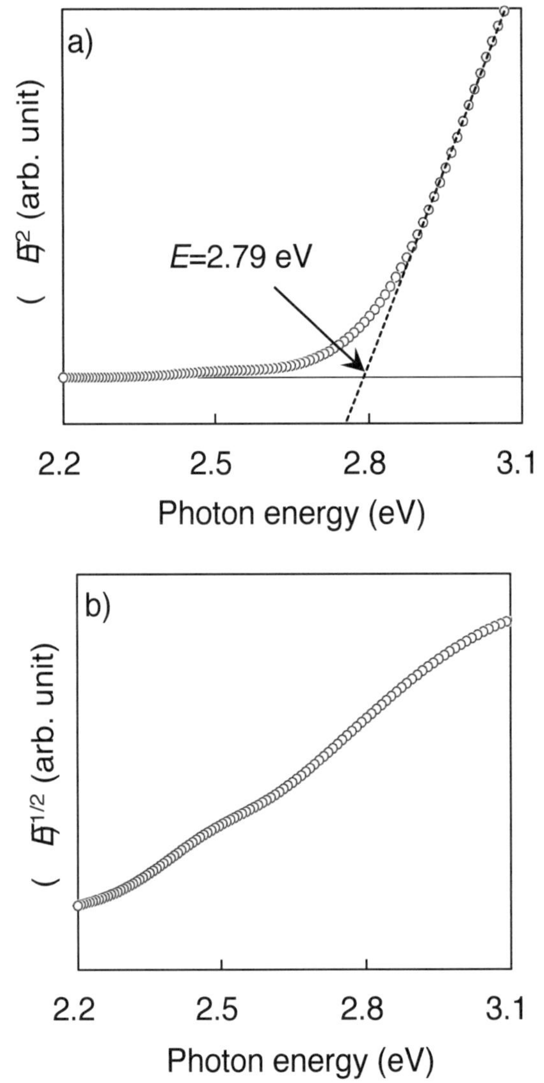

Figure 16. Plots of (a) $(\alpha E)^2$ and (b) $(\alpha E)^{1/2}$ vs. photon energy E for the polycrystalline BFO film. The linear extrapolation of $(\alpha E)^2$ to 0 gives band gaps of 2.79 eV.

3.5. Thermooptic property of polycrystalline BFO film

3.5.1. Temperature dependence of lattice space

Figure 17 shows the lattice spacing of Si (400), BFO (100), and Pt (111) as a function of temperature, estimated from XRD patterns. All the *d*-spaces monotonically increased with increasing temperature because of thermal expansion. Thermal expansion coefficients, estimated from Fig. 17, are shown in Table 2, which includes reference data [42-44] for comparison. From this table, we can see that our experimental values are larger than those in reference data, except for the Si substrate because of the effect of in-plane compressive stress.

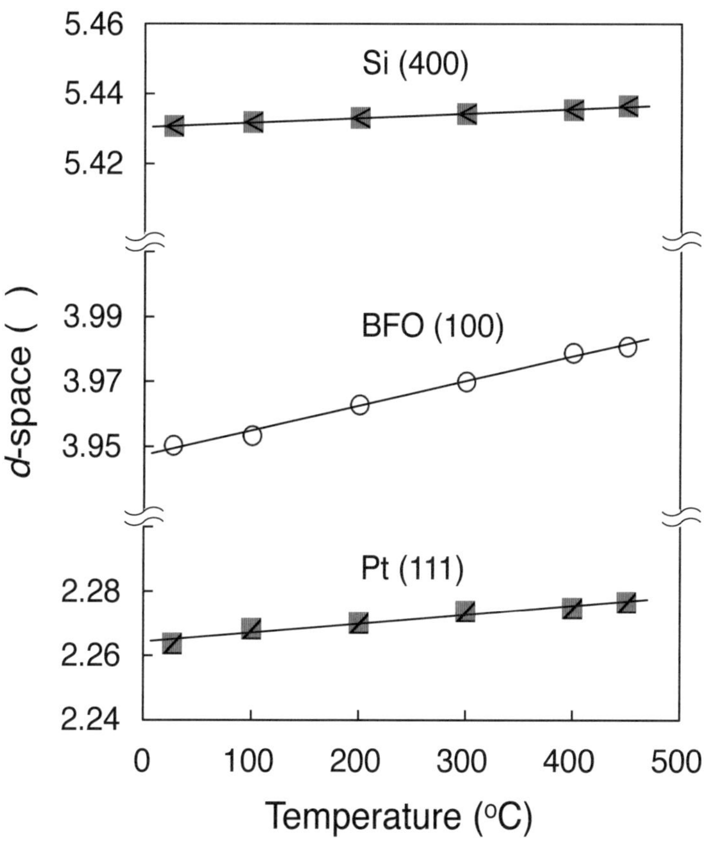

Figure 17. The lattice spacing of Si (400), BFO (012), and Pt (111) as a function of temperature estimated from XRD patterns.

Material	Thermal expansion coefficient	
	Experimental data ($\times10^{-6}$ K^{-1})	Reference data ($\times10^{-6}$ K^{-1})
BFO	19	10-14
Pt	12	9
Si	2.4	2

Table 2. The thermal expansion coefficients of the BFO, Pt, and Si.

3.5.2. Temperature dependence of refractive index

Ellipsometric spectra in (Δ, Ψ) were recorded at incident angles of $\theta_i = 70^o$ in a spectral range of 245–1670 nm. The sample was loaded onto the customized heating stage and heated from RT to 530°C. In these measurements, one Gaussian oscillator was assumed to represent film properties, and the central energy E_n of the oscillator was fixed at 5 eV for each temperature. Figure 18(a) – b) shows the wavelength dispersion of the refractive index and the extinction coefficient measured at 50 and 530°C, respectively. From this figure, we observe that the refractive index decreases with increasing temperature for all wavelengths. The refractive index variation with increasing temperature is large for shorter wavelengths. Based on these results, we investigated the origin of the temperature dependence of refractive index dispersion, as shown in Fig. 18. In the short-wavelength region, the lattice vibration becomes more intense with increasing temperature; therefore, in a high-temperature region, the amplitude of the oscillator decreased, and the broadening of the oscillator expanded. At the same time, density is decreased with increasing temperature owing to thermal expansion; therefore, the refractive index decreased in all wavelength regions. This combination seems to be responsible for the refractive index curve shown in Fig. 18(a).

Figure 19 shows the temperature dependence of the refractive index in a long- wavelength region. It is found that refractive index decreases with increasing temperature at each wavelength, although there was some variability. The thermo-optic coefficient was estimated from the slopes of linear approximation. A thermo-optic coefficient of 0.8×10^{-4} K^{-1} was obtained at a wavelength of 1550 nm for the BFO film. The thermo-optic coefficients of polymers and glasses, which are known to be typical thermo-optic materials, have been reported to be approximately 2×10^{-4} K^{-1} and 0.1×10^{-4} K^{-1}, respectively. [45,46] It is found that the BFO film shows a large refractive index and a thermo-optic coefficient comparable to that of these traditional materials. Finally, it can be concluded that the BFO also has potential for use in an electro-optic-type Mach-Zehnder modulator.

http://dx.doi.org/10.5772/54908

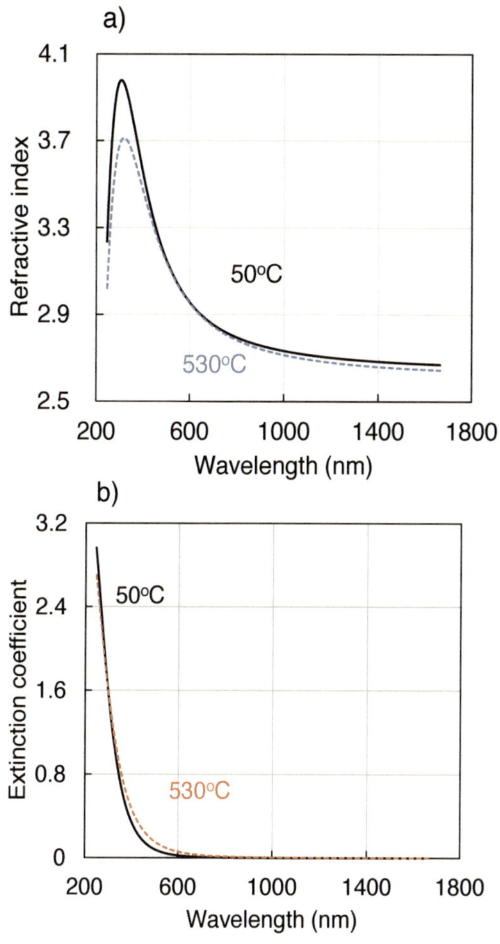

Figure 18. The wavelength dispersion of a) refractive index and b) extinction coefficient measured at 50 and 600°C, respectively.

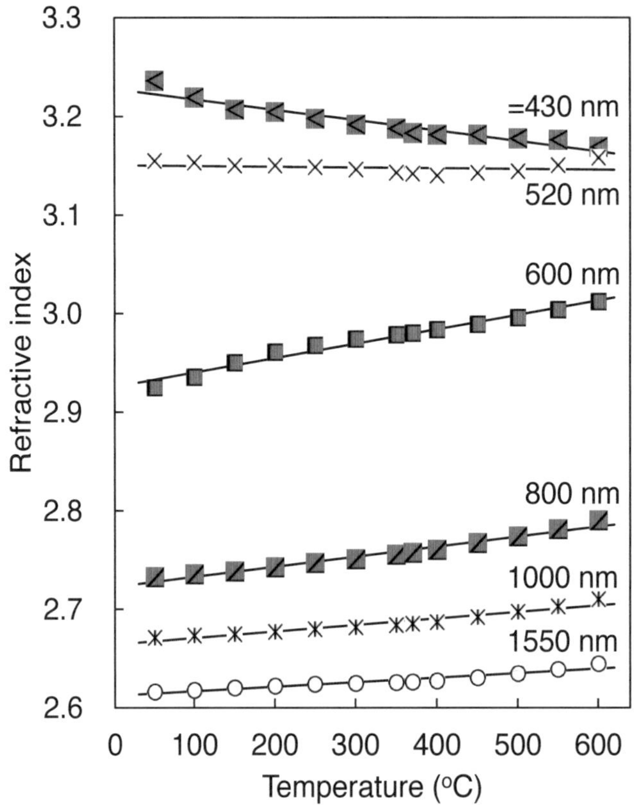

Figure 19. The refractive indexes at various wavelengths as a function of temperature.

4. Conclusions

The optical and thermo-optic properties of $BiFeO_3$ (BFO) films were studied. Polycrystalline BFO films were formed on $Pt/Ti/SiO_2/Si$ substrates, and their basic optical and thermo-optic properties were systematically evaluated. The new findings are summarized as follows:

i. The polycrystalline BFO films were evaluated using a spectroscopic ellipsometer. Gaussian oscillators were assumed as a dielectric function to represent film properties, and the graded model was assumed by introducing a refractive index gradient. As a result, large refractive indexes of 3.22 and 2.91 were estimated for the polycrystalline BFO film at wavelengths of 600 and 1550 nm, respectively, these refractive indexes are higher than that of the rutile-type TiO_2 film, which is known to have a high refractive index. The optical band gap of the BFO film at RT was estimated as a direct transition to be 2.79 eV, which corresponds to the absorption edge of 445 nm. It was found that the BFO film shows sufficiently low light loss at wavelengths larger than 600 nm.

ii. The thermo-optic properties of the BFO films were evaluated using a spectroscopic ellipsometer with a heating stage. The refractive index of the polycrystalline film decreased with increasing temperature. We considered that this change in refractive index is caused by the balance between the increase in refractive index due to the enhancement of the oscillator dispersion and the decrease in refractive index due to the decreased density of the film. In addition, we obtained thermo-optic coefficient of 0.8×10^{-4} K^{-1} at a wavelength of 1550 nm for the BFO film, which is larger than those of typical thermo-optic polymers (2×10^{-4} K^{-1}) and glasses (0.1×10^{-4} K^{-1}).

These results suggested that the BFO films have a high potential for application as an optical material with a high refractive index, and that the effectual refractive index change can be controlled by the balance of two factors, the activation of the oscillator and the thermal expansion coefficient, even in the same material.

Acknowledgements

The author thanks Assistant Prof. Takashi Nakajima, Tokyo University of Science, Prof. Akiharu Morimoto, Asso. Prof. Takeshi Kawae, Kanazawa University, Prof. Takashi Yamamoto, Ass. Prof. Ken Nishida, National Defense Academy, Dr. Takashi Iijima, AIST, Dr. Koichi Tsutsumi, Dr. Masahiro Matsuda, Dr. Michio Suzuki, J. A. Woollam Japan, and Dr. Toshiyasu Tadokoro Techno-Synergy, Inc. This study was partially supported by a Grant-in-Aid for JSPS Research Fellows (No. 217990) from the Japan Society for the Promotion of Science.

Author details

Hiromi Shima[1*], Hiroshi Naganuma[2*] and Soichiro Okamura[1]

*Address all correspondence to: shima@nda.ac.jp, naganuma@mlab.apph.tohoku.ac.jp

1 Department of Applied Physics, Tokyo University of Science, 1-3, kagurazaka, Shinjuku-ku, Tokyo, Japan

2 Department of Applied Physics, Graduate school of Engineering, Tohoku University, Japan

References

[1] Fujimori, Y.; Fujii, T.; Suzuki, T.; Kimura, H.; Fuchikami, T.; Nakamura, T.; and Takasu, H. (2005) *IEDM Tech. Dig.*, 2005 p.p. 935.

[2] Bitou, Y.; and Minemoto, T. (1998) High-contrast spatial light modulator by use of the electroabsorption and the electro-optic effects in a GaAs single crystal, *Appl. Opt.*, Vol. 37, August 1998, pp. 4347-4356.

[3] Thapliya, R.; Okano, Y.; and Nakamura, S. (2003) Electrooptic Characteristics of Thin-Film PLZT Waveguide Using Ridge-Type Mach–Zehnder Modulator, *J. Lightwave Tech.*, Vol. 21, January 2003 pp. 1820-1827.

[4] Jacobsen, R. S.; Andersen, K. N.; Borel, P. I.; Pedersen, J. F.; Frandsen, L. H.; Hansen, O.; Kristensen, M.; Lavrinenko, A. V.; Moulin, G.; Ou, H.; Peucheret, C.; Zsigri, B.; and Bjarklev, A. (2006) Strained silicon as a new electro-optic material, *Nature*, Vol. 441, May 2006 pp. 199-202.

[5] Shimizu, T.; Nakada, M.; Tsuda, H.; Miyazaki, H.; Akedo, J.; and Ohashi, K. (2009) Gigahertz-rate optical modulation on Mach-Zehnder PLZT electro-optic modulators formed on silicon substrates by aerosol deposition, *IEICE Electro. Exp.*, Vol. 6, December 2009 pp. 1669-1675.

[6] Xie, N.; Hashimoto.; T and Utaka, K. (2009) Very Low-Power, Polarization-Independent, and High-Speed Polymer Thermooptic Switch *IEEE Photonics Tech. Lett.*, Vol. 21, December 2009 pp. 1861-1863.

[7] Reed, G. T.; Mashanovich, G.; Gardes, F. Y.; and Thomson, D. J.; (2010) Silicon optical modulators, *Nature Photonics*, Vol. 4, July 2010 pp. 518-526.

[8] Haertling, G. H,; and Land, C. E.; (1971) Hot-Pressed (Pb, La)(Zr,Ti)O_3, Ferroelectric Ceramics for Electrooptic Applications*J. Am. Ceram. Soc.*, Vol. 54, October 1971, pp. 1-11.

[9] Wang, J.; Neaton, J. B.; Zheng, H.; Nagarajan, V.; S. Ogale, B.; Liu, B.; Viehland, D.; Vaithyanathan, V.; Schlom, D. G.; Waghmare, U. V.; Spaldin, N. A.; Rabe, K. M.; Wuttig, M. & Ramesh, R. (2003) Epitaxial $BiFeO_3$ Multiferroic Thin Film Heterostructures, *Science* Vol. 299, February 2003, pp. 1719-1722.

[10] Yun, K. Y.; Rincinschi, D.; Kanashima, T.; Noda, M.; and Okuyama, M.; (2004) Giant Ferroelectric Polarization Beyond 150 $\mu C/cm^2$ in $BiFeO_3$ Thin Film*Jpn. J. Appl. Phys.* Vol. 43, 2004 pp. L647-L648.

[11] Bibes, M.; and Barthelemy, A.; (2008) Towards a magnetoelectric memory*Nature materials*, Vol. 7, 2008 pp. 425-426.

[12] Ramesh, R.; and Spaldin, N. A. (2007) Multiferroics: progress and prospects in thin films *Nature materials*, Vol. 6, 2007 pp. 21-29.

[13] Chu, Y. H.; Martin, L. W.; Holcomb, M. B.; Gajek, M.; Han, S. –J.; He, Q.; Balke, N.; Yang, C. –H.; Lee, D.; Hu, W.; Zhan, Q.; Yang, P. –L.; Rodriguez, A. F.; Scholl, A.; Wang, S. X.; and Ramesh, R.; (2008) Electric-field control of local ferromagnetism using a magnetoelectric multiferroic*Nature materials*, Vol. 7, 2008, pp. 478-482.

[14] Iakovlev, S.; Solterbeck, C. –H.; Kuhnke, M.; and Es-Souni, M.; (2005) Multiferroic $BiFeO_3$ thin films processed via chemical solution deposition: Structural and electrical characterization *J. Appl. Phys.*, Vol. 97, 2005 pp. 094901-1-094901-6.

[15] Kumar, A.; Rai, R. C.; Podraza, N. J.; Denev, S.; Ramiez, M.; Chu, Y. –H.; Martin, L. W.; Ihlefeld, J.; Heeg, T.; Schubert, J.; Schlom, D. G.; Orenstein, J.; Ramesh, R.; Collins, R. W.; Musfeldt, J. L.; and Gopalan, V.; (2008) Linear and nonlinear optical properties of $BiFeO_3$ *Appl. Phys. Lett.*, Vol. 92, 2008, pp. 121915-1-121915-3.

[16] Choi, T.; Lee, S.; Choi, Y. J.; Kiryukhin, V.; and Cheong, S. –W. (2009) Switchable Ferroelectric Diode and Photovoltaic Effect in $BiFeO_3$ *Science* Vol. 324, 2009, pp. 63-66.

[17] Kiselev, S. V.; Ozerov, R. P.; and Zhdanov, G. S.; (1963) Sov. Phys., Vol. 7 1963, pp. 742.

[18] Venevtsev, Yu. N.; Zhadanow, G.; and Solov'ev, S.; (1960) Sov. Phys. Crystallogr., Vol. 4 1960, pp. 538.

[19] Kubel, F.; and Schmid, H.; (1990) Structure of a Ferroelectric and Ferroelastic Monodomain Crystal of the Perovskite $BiFeO_3$ *Acta Crystallogr. B*, Vol. 46 1990, pp. 698-702.

[20] Filip'ev, V. S.; Smol'yaninov, I. P.; Fesenko. E. G.; and Belyaev, I. I.; Kristallografiya, Vol. 5 1960 pp. 958.

[21] Dzyaloshinskii, I. E.; S(1957) Sov. Phys. JETP, Vol. 5 1957 pp. 1259.

[22] Moriya, T. (1960) Anisotropic superexchange interaction and weak ferromagnetism *Phys. Rev.*, Vol. 120, 1960 pp. 91-98.

[23] Neaton, J. B.; Ederer, C.; Waghmare, U. V.; Spaldin, N. A.; and Rabe, K. M.; (2005) First-principles study of spontaneous polarization in multiferroic $BiFeO_3$ *Rhys. Rev. B*, Vol. 71 2005 pp. 014113-1-014113-8.

[24] Ederer, C.; and Spaldin N. A. (2005) Effect of Epitaxial Strain on the Spontaneous Polarization of Thin Film Ferroelectrics, *Physical Review Letters.*, Vol. 95, December 2005, pp. 257601-1-4.

[25] Teague, J. R.; Gerson, R.; and James, W. J.; (1970) Dielectric Hysteresis in single crystal $BiFeO_3$ *Solid State Commun.*, Vol. 8 1970 pp. 1073-1074.

[26] Schwartz, R. W. (1997) Chemical Solution Deposition of Perovskite Thin Films *Chem. Mater.*, Vol. 9 1997 pp. 2325-2340.

[27] Aspnes, D. E. (1974) Optimizing precision of rotating-analyzer ellipsometers *J. Opt. Soc. Am.*, Vol. 64 1974 pp. 639-646.

[28] Jayatissa, A. H.; Suzuki, M.; Nakanishi, Y.; and Hatanaka, Y. (1995) Microcrystalline structure of poly-Si films prepared by cathode-type r.f. glow discharge *Thin Solid Films*, Vol. 256 1995 pp. 234-239.

[29] Yao, H.; Snyder, P. G.; and Woollam, J. A.; (1991) Temperature dependence of optical properties of GaAs *J. Appl. Phys.*, Vol. 70 1991 pp. 3261-3267.

[30] Yao, H.; Woollam, J. A.; and Alterovitz, S. A.; (1993) Spectroscopic ellipsometry studies of I-IF treated Si (100) surfaces *Appl. Phys. Lett.*, Vol. 62 1993 pp.3324-3326.

[31] Brendel, R.; and Bormann, D.; (1992) An infrared dielectric function model for amorphous solids *J. Appl. Phys.*, Vol. 71 1992 pp. 1-6.

[32] Kim, C. C.; Garland, J. W.; Abad, H.; and Raccah, P. M.; (1992) Modeling the optical dielectric function of semiconductors: extension of the critical0point parabolic-band approximation *Phys. Rev. B*, Vol. 45 1992 pp. 11749-11767.

[33] Khan, M. H.; Shen, H.; Xuan, Y.; Zhao, L.; Xiao, S.; Leaird, D. E.; Weiner, A. M.; and Qi, M. (2010) Ultrabroad-bandwidth arbitrary radiofrequency waveform generation with a silicon photonic chip-based spectral shaper *Nature Photonics*, Vol. 4 2010 pp. 117-122.

[34] Yamada, H.; Chu, T.; Ishida, S.; and Arakawa, Y. (2006) Si Photonic Wire Waveguide Devices *IEEE Journal of selected topics in quantum electronics*, Vol. 12 2006 pp. 1371-1378.

[35] Wang, X.; Xu, L.; Li, D.; Liu, L.; and Wang, W.; (2003) Thermo-optic properties of sol-gel-fabricated organic–inorganic hybrid waveguides *J. Appl. Phys.*, Vol. 94 2003 pp. 4228-4230.

[36] Naganuma. H.; Inoue, Y.; and Okamura, S.; (2008) Evaluation of Electrical Properties of Leaky $BiFeO_3$ Films in High Electric Field Region by High-Speed Positive-Up–Negative-Down Measurement *Appl. Phys. Exp.*, Vol. 1 2008 pp. 061601-1-061601-3.

[37] Landauer, R.; in Electrical Transport and Optical Properties of Inhomogeneous Media, ed. J. C. Garland and D. B. Tanner (AIP, New York, 1979) p. 1.

[38] Ting, C. –C.; Chen, S. –Y.; and Liu, D. –M.; (2000) Structural evolution and optical properties of TiO_2 thin films prepared by thermal oxidation of sputtered Ti films *J. Appl. Phys.*, Vol. 88 2000 pp. 4628-4633.

[39] Ihlefeld, J. F.; Podraza, N. J.; Liu, Z. K.; Rai, R. C.; Xu, X.; Heeg, T.; Chen, Y. B.; Li, L.; Collins, R. W.; Musfeldt, J. L.; Pan, X. Q.; Schubert, J.; Ramesh, R.; and Schlom, D. G. (2008) Optical band gap of $BiFeO_3$ grown by molecular-beam epitaxy *Appl. Phys. Lett.*, Vol. 92 2008 pp. 908-1-908-3.

[40] Hauser, A. J.; Zhang, J.; Mier, L.; Ricciardo, R. A.; Woodward, P. M.; Brillson, L. J.; and Yang, F. Y.; (2008) Characterization of electronic structure and defect states of thin epitaxial $BiFeO_3$ films by UV-visible absorption and cathodoluminescence spectroscopies *Appl. Phys. Lett.*, Vol. 92 2008 pp. 222901-1-222901-3.

[41] Pankove, J. I.; Optical Process in Semiconductors (Prentice-Hall, Englewood Cliffs, NJ, 1971) p. 34.

[42] Amirov, A. A.; Batdalov, A. B.; Kallaev, S. N.; Omarov, Z. M.; Verbenko, I. A.; Razumovskaya, O. N.; Reznichenko, L. A.; and Shilkina, L. A.; (2009) Specific Features of the Thermal, Magnetic, and Dielectric Properties of Multiferroics $BiFeO_3$ and $Bi_{0.95}La_{0.05}FeO_3$ *Phys. Solid State*, Vol. 51 2009 pp. 1189-1192.

[43] Watanabe, H.; Yamada, N.; and Okaji, M. (2004) Linear Thermal Expansion Coefficient of Silicon from 293 to 1000 K *Inter. J. Thermophys.*, Vol. 25 2004 pp. 221-236.

[44] Okada, Y.; and Tokumaru, Y.; (1984) Precise determination of lattice parameter and thermal expansion coefficient of silicon between 300 and 1500 K *J. Appl. Phys.*, Vol. 56 1984 pp. 314-320.

[45] Zhang, Z.; Zhao, P.; Lin, P.; and Sun, F.; (2006) Thermo-optic coefficients of polymers for optical waveguide applications *Polymer*, Vol. 47 2006 pp. 4893-4896.

[46] Jewell, J. M.; Askins, C.; and Aggarwal, D. (1991) Interferometric method for concurrent measurement of thermo-optic and thermal expansion coefficients *Appl. Opt.*, Vol. 30 1991 pp. 3656-3660.

Chapter 3

Atomic Layer Deposition on Self-Assembled-Monolayers

Hagay Moshe and Yitzhak Mastai

Additional information is available at the end of the chapter

http://dx.doi.org/10.5772/54814

1. Introduction

Atomic layer deposition (ALD) is an advanced technique for growing thin film structures. ALD was developed by Tuomo Suntola and co workers in 1974. At first, the method was called Atomic layer epitaxy (ALE). However, today the name "ALD" is more common. The motivation behind developing ALD was the desire to achieve a technique for creating thin film electroluminescent (TFEL) flat panel displays. [1]- [7]

Several types of materials including metals [8], metal oxides [9], metal nitrides [7] and metal sulfides [10] can be deposited into ALD thin films, depending on the precursors used. ALD advantages are: precise and easy thickness control, superior conformality, the ability to produce sharp interfaces, the substrate size is limited by the batch size and straightforward scale up and repetition of the process.[2-4,6] ALD is appropriate for deposition processes which require angstrom or monolayer level control over coating thickness and/or are maintained on complex topographies of the substrate. No other method for thin film creation can get close to the conformality obtained by ALD.[4] ALD also has several limitations. The Achilles' heel of the method is that ALD is a slow process and therefore is not economic for many industrial processes.[2]

Atomic layer deposition controlled film growth is a significant technology for surface chemistry. In the last four decades, ALD has developed into a system used for depositing thin films in a variety of products. For example, ALD is used in microelectronic production, construction of optical and magnetic devices, flat panel displays, catalysts, and energy conversion including solar cells, utilizing fuel cells, storage batteries or supercapacitors, nanostructures as AFM tips, biomedical purpose and more. [11]

Self Assembled Monolayers (SAMs) are ordered molecular (organic molecules in most cases) assemblies formed by adsorption of molecules on a solid substrate. The surface properties of the surfaces formed are determined by the nature of the adsorbed molecules. [12] A typical surfactant molecule for SAMs is built from three main parts. The first part has a high affinity to the solid surface and is called the "headgroup". The headgroup forms a chemical interaction with the substrate. While adsorbing, the molecules make an effort to adsorb at all surface sites, resulting in a close-packed monolayer. The second molecular part is the alkyl chain. The Van der Waals interactions between these chains cause the SAMs to be ordered. The third part which is exposed at the surface is called the "terminal group". The chain can be terminated with several different groups e.g. CH_3, OH, COOH or NH_2, allowing the SAMs to be applied for the modification of surface properties. Thus, SAMs can modify the surface free energies of the substrates, ranging from reactive, high energies, to passive, low energies. [12],[13]

This book chapter will focus on a new application of ALD as a novel method for thin film deposition on SAMs. Since ALD is very sensitive to surface conditions, it is an ideal method for film deposition on SAMs. Examples for the application of ALD on SAMs are, for instance, surface patterning and selective deposition of thin films. ALD is a very suitable method for the deposition of thin films with three dimensional structures.[8] This book chapter will cover the most recent and novel applications of ALD used for the preparation of chiral nanosized metal oxide films using chiral SAMs.

1.1. Principle of technique

ALD is a Chemical vapor deposition (CVD) process with self-limiting growth and is controlled by the distribution of a chemical reaction into two separate half reactions; the film is done in a growth cycle. Throughout the process, the precursor materials have to be separate. A growth cycle includes four stages: 1) Exposure of the first precursor, 2) purge of the reaction chamber, 3) exposure of the second precursor, and 4) a further purge of the reaction chamber [2,5] (Figure 1). In the first stage, the first precursor reacts with all the sites on the substrate receiving a single molecular layer of the first precursor. The second stage consists of Argon flowing and pumping of the residue of the first precursor to avoid unwanted gas phase reactions between precursors, a reaction which will prevent acceptance of a single molecular layer. In the third stage, the second precursor reacts with one molecular layer of the first precursor to get a single molecular layer of the target material. The fourth stage consists of pumping the residuals of the second precursor [2,5,6]. The cycle ends after four stages. The film thickness is determined by the number of cycles because one cycle deposits one molecular layer (Figure 2). [9] Every stage in the process has to be fully completed before the next stage starts. This means that all the sites on the substrate must react with the precursor and the extra precursor molecules must be removed. The molecular size of the precursor determines the film thickness per cycle. The film density obtained depends on the molecular volume of the precursor- that is to say, a molecule with steric hindrance will probably prevent the formation of a monolayer while small molecules without steric hindrance will allow the formation of a full monolayer. The density of the reactive sites on the substrate is also significant for the nature of the film obtained. One cycle can take from half a second to a few seconds depending on the reactivity between the gas precursors and the solid substrate. In ALD, spontaneous reactions are desired. [2], [5], [6]

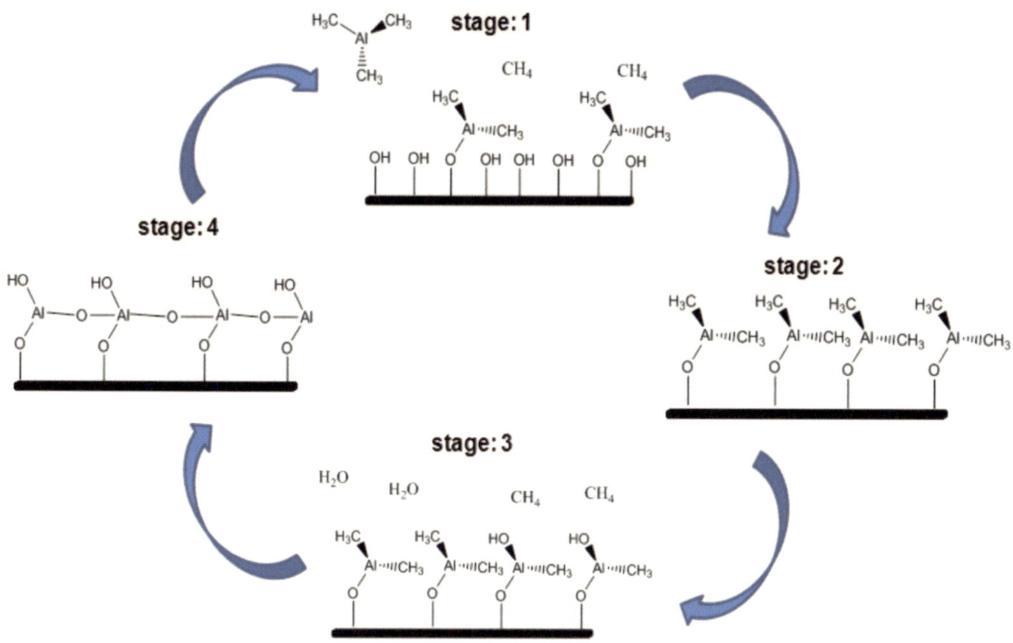

Figure 1. ALD growth cycle includes four stages: 1) Exposure of the first precursor, 2) purge of the reaction chamber, 3) exposure of the second precursor, and 4) a further purge of the reaction chamber.

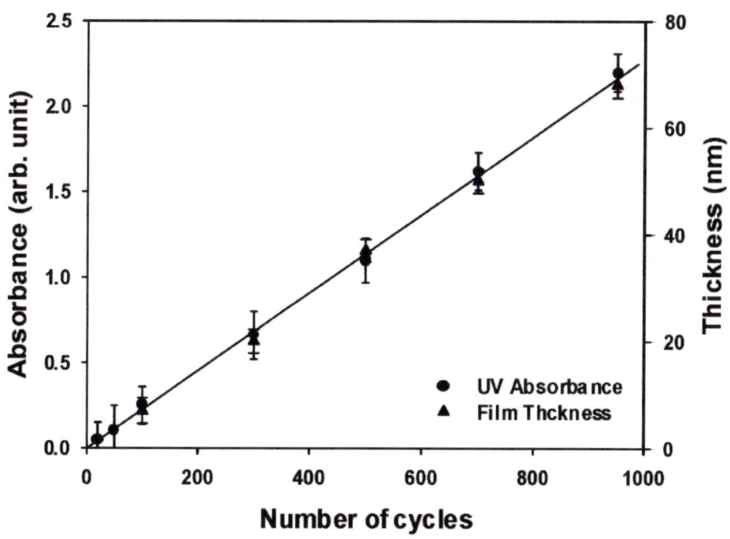

Figure 2. UV absorbance and thickness of the TiO_2 thin films versus the number of ALD cycles. [9]

Depending on the precursors used, ALD can deposit several types of materials, including metals as Pt[14], Ru[15] and Ir[16]; metal oxides as ZnO[17], TiO_2[9], ZrO_2[18] and HfO_2[19]; metal nitrides as Hf_3N_4[20] and Zr_3N_4[20]; metal sulfides as PbS[10] and Polymers as Polyimide [16]. Using the right precursor is one of the keys for a successful ALD process. Good ALD precursors have to include a number of properties during deposition conditions. First, they have to be stable, evaporable and react with the substrate to completeness. In addition, they must be safe, non toxic and inexpensive. Finally, there should be no etching of the substrate or the growing film and inert volatile byproducts. [2,5-7]

The material type of ALD thin films depends on the precursor type. Normally halides, alkyl compounds, and alkoxides are used as metal precursors. Nonmetal precursors include water, hydrogen peroxide, and ozone for oxygen; hydrides for chalcogens; ammonia, hydrazine, and amines for nitrogen; hydrides for the fifth group in the periodic table. [2], [5]

ALD precursors can be in any state of matter- gas state, liquid state, or solid state. In order to have an effective feeding of precursor molecules to the system, vapor pressure should be high enough. The precursor is also heated sometimes. There must be enough precursors to cover all sites on the substrate surface. [2], [5]- [7]

1.2. Advantages and disadvantages

The ALD technique has a number of advantages: ALD has angstrom or monolayer level control on thickness, the film thickness depends only on the number of reaction cycles. ALD has large area deposition ability, the area size depends only on the ALD chamber size. ALD is a very suitable method for the deposition of thin films with three dimensional structures. [8] As a result, ALD has excellent conformality to substrate surfaces. ALD is a reproducible process, can work on low temperatures and uses highly reactive precursors. The ALD method allows processing of different materials in a continuous process. [2], [5], [7]

ALD's weak point is its slow growth rate; one monolayer is deposited per cycle. The monolayer thickness is a few angstroms; if a cycle takes a few seconds, micron thickness deposition will take a few hours. Consequently, ALD is not a useful method for many applications. The growth of films in micrometer size takes too long for ALD to be an economic industrial process. This problem is sometimes overcome by using a big chamber to be able to contain many substrates per batch, but a single wafer process is still more ideal. ALD is an unselective process. Generally, precursor molecules react with all surfaces. In order to achieve selectivity control, pretreatment is necessary. The places which should not be deposited on have to be passivated. As a chemical process, ALD has a risk of impurities. The impurities can come from gas precursors and/or a carrier gas, the process requires material with a high degree of cleanliness. Impure chemicals can lead to the incorporation of impurities and to the growth of poor quality films. [2], [5], [7]

1.3. ALD process at low temperature

The ability to perform ALD at low temperatures (ALD-LT) is very important. It is critical for ALD on SAMs and it is the subject of this chapter. SAMs as well as polymers or biological samples are thermally sensitive materials. At high temperatures, they decompose. [3] In the case of SAMs, there is also disabsorption from the surface. Inter-diffusions of materials occur at high temperature processes, it has a devastating effect on nano-structured devices. ALD at low temperatures avoids these effects. To carry out ALD-LT, a catalyst is sometimes used [3], [21], [22] although there are reactions that occur without catalysts. [3], [23] Nanostructures of biological structures have very interesting effects. For example, a lotus leaf shows highly hydrophobic behavior due to its nanostructures. The coat of the lotus leaf can be copied by ALD-LT, achieving similar effects. ALD-LT was also used on a tobacco mosaic virus (TMV) on protein spheres [24] and on cellulose fibers from filter paper.[3], [25]

2. ALD on self-assembled-monolayer

The use of ALD for depositing thin films onto different SAMs has great potential applications. SAMs are thin organic films which form spontaneously on solid surfaces. The SAM head group has to connect to the substrate strongly enough for stable monolayers to form. Typical SAM head groups are alkanethiols [X-(CH_2)n–SH] which are formed on metal surfaces such as Ag, Au, and Cu, and alkyltrichlorosilanes [X-$(CH_2)_n$–$SiCl_3$] formed on SiO_2, Al_2O_3, and other oxide surfaces. [26]- [28]

In general, SAMs are formed by immersing the substrates into a solution comprising the precursor molecules or by bringing the SAM precursors to the substrate surface as vapors. [16] SAMs are well known to modify the physical and chemical properties of surfaces. The surface features can be controlled by using the appropriate SAM. Potential applications include control of wetting and friction behaviors, passivating layers, protection of metals against corrosion, preparation of chiral surfaces, molecular electronics, chemical sensing, soft lithography and more. [10], [26], [29] Figure 3 shows water droplet angle measurements demonstrating the formation of hydrophobicity by ODTS (octadecyltrichlorosilane) SAM on originally hydrophilic SiO_2. [26]

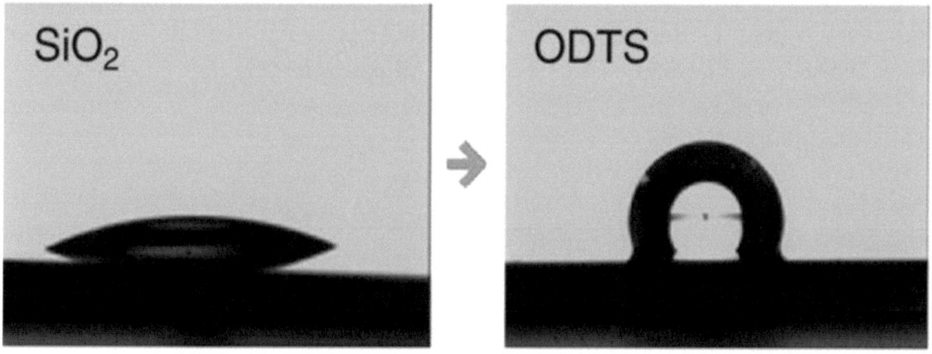

Figure 3. Contact angle measurements showing the control of surface energy by ODTS SAM before and after treatments on SiO_2 substrate. [26]

ALD onto SAMs is interesting because the ordered structure of the monolayer can act as a template for the growth of structured thin films. The SAM can be used to engineer the properties of the interface to the original substrate, when the ALD coatings are protecting the SAM. [30]

The thermal stability of the SAMs under the ALD process conditions is very important because of the potential loss of ordering at elevated temperatures. In order to maintain the SAM order during the growth of the ALD film, the ALD process must be done under conditions which are compatible with the thermal budget of the underlying SAM film. [30]

2.1. Area-Selective ALD on SAM

Patterned SAMs are commonly used as growth-preventing masks for selective-area ALD. Selective-area ALD is the growth of thin films on the substrate surface on designated sites only. Selective-area ALD requires that the chosen regions of the surface are inert to ALD precursors. In this case the function of the SAM is to protect the surface against deposition. ALD grows only on areas without a SAM, on the desired sites of the surface [16]. High-resolution patterns can be created by printing SAMs by means of soft lithography, [31]- [33] or by removing SAMs using electron beams, [34] ion beams, photolithography, [35] or scanning probe microscopy. [36] Generating patterned SAMs in the most economical way is a critical necessity for using patterned SAMs in advanced applications. Photolithography can transfer an entire pattern on a photomask to a SAM at a given time. Therefore, it is the most practical among various patterning methods. [18]

During the past decade, several groups have used SAMs as a chemical resist to block various ALD precursors, including ZnO [17] [37], TiO_2[9], [38]- [40], ZrO_2 [18], HfO_2 [19], [41]- [45], Ru [15], Ir [16], [40], [46], Pt [14], [42], [45], [47]- [49], PbS [10] and Polyimide [16]. In 2001 Yan et al. reported on selective area ALD growth of ZnO, they used a microcontact printing (or μCP) with poly(dimethylsiloxane) (PDMS) stamp as a soft lithographic technique. PDMS creates a hydrophobic surface in the stamped area, leaving the ink-free hydrophilic surface unmodified. The pattern consists of arrays of cylinders having cross sectional diameters ranging from 1.0–40 μm with center-center distances of 100 μm. As ALD precursors they used diethylzinc (DEZ) and deionized water. The deposition process for ZnO consists of two self-limiting chemical reactions, repeated in alternation (ABAB...). Each AB reaction cycle deposits a single mono-layer of ZnO, [50] as shown in Eqs. (1) and (2), where asterisks indicate the outermost surface functional groups.

$$A: Zn-OH^{*} + Zn(CH_2CH_3)_2 \rightarrow Zn-O-Zn-CH_2CH_3^{*} + C_2H_6 \uparrow \quad (1)$$

$$B: Zn-CH_2CH_3 + H_2O \rightarrow Zn-OH^{*} + C_2H_6 \uparrow \quad (2)$$

The deposition was carried out at a substrate temperature of 125 °C. The exposure times for DEZ and water vapor were 0.7 sec and 0.5 sec, respectively. ZnO nucleation and growth do not occur on the; 2 nm thick SAM-patterned areas, but only on the bare, hydrophilic unpatterned areas as illustrated in Figure 4. [17]

Park et al. reported on a patterning method of TiO_2 thin films using microcontact printing of alkylsiloxane SAMs, followed by selective atomic layer deposition of the TiO_2. Park et al. approach consists of two key steps. First, the patterned alkylsiloxane SAMs were formed by using microcontact printing. Second, the TiO_2 thin films were selectively deposited onto the SAM-patterned Si substrate by atomic layer deposition. [9]

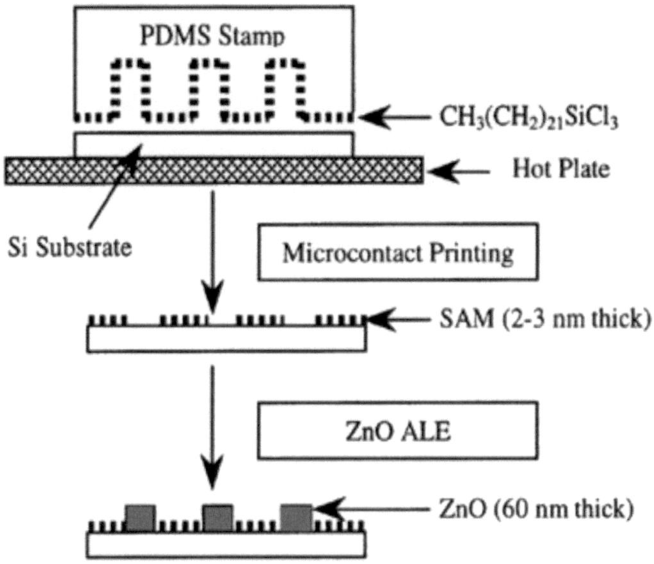

Figure 4. Schematic outline of the patterning and deposition procedures used for selective area ZnO thin film growth. [17]

Seo et al. reported on a patterning method for TiO_2 thin films using microcontact printing of alkanethiolate SAMs on gold, followed by selective atomic layer deposition of the TiO_2. Seo et al. approach consists of three key steps. First, patterned CH_3-terminated alkanethiolate SAMs on gold were formed by using microcontact printing. Second, the remaining regions of gold were coated with OH-terminated alkanethiolate SAMs. Third, the TiO_2 thin films were selectively deposited onto the SAM-patterned gold substrate by atomic layer deposition. [38] Both groups used PDMS stamp as a soft lithographic technique and same conditions for the preparation of TiO_2 thin films. As ALD precursors they used Titanium isopropoxide ($Ti(OPri)_4$) and deionized water. The $Ti(OPri)_4$ and water were evaporated at 80 and 20 °C, respectively. The cycle consisted of 2 sec exposure to $Ti(OPri)_4$, 5 sec Ar purge, 2 sec exposure to water, and 5 sec Ar purge. The total flow rate of the Ar was 20 sccm. The TiO_2 thin films were grown at 100 °C under 2 Torr. The deposition process for TiO_2 also consists of two self-limiting chemical reactions, repeated in alternation (ABAB...). Each AB reaction cycle deposits a single monolayer of TiO_2 [9], [38]. Figure 5 illustrates AFM images and cross sections of micropatterned TiO_2 thin films, which were selectively deposited onto the monolayer-patterned gold substrate by ALD. The patterned SAMs showed high selectivity for TiO_2 ALD;

hence, the patterns of the TiO_2 thin films were defined and directed by the patterned SAMs generated with microcontact printing. The TiO_2 thin films are selectively deposited only on the regions exposing the OH groups of the MUO-coated gold substrates, because the regions covered with the ODT monolayers do not have any functional group to react with ALD precursors. These AFM images clearly show that the patterned TiO_2 thin films retain the dimensions of the patterned SAMs used as templates with no noticeable line spreading. [38]

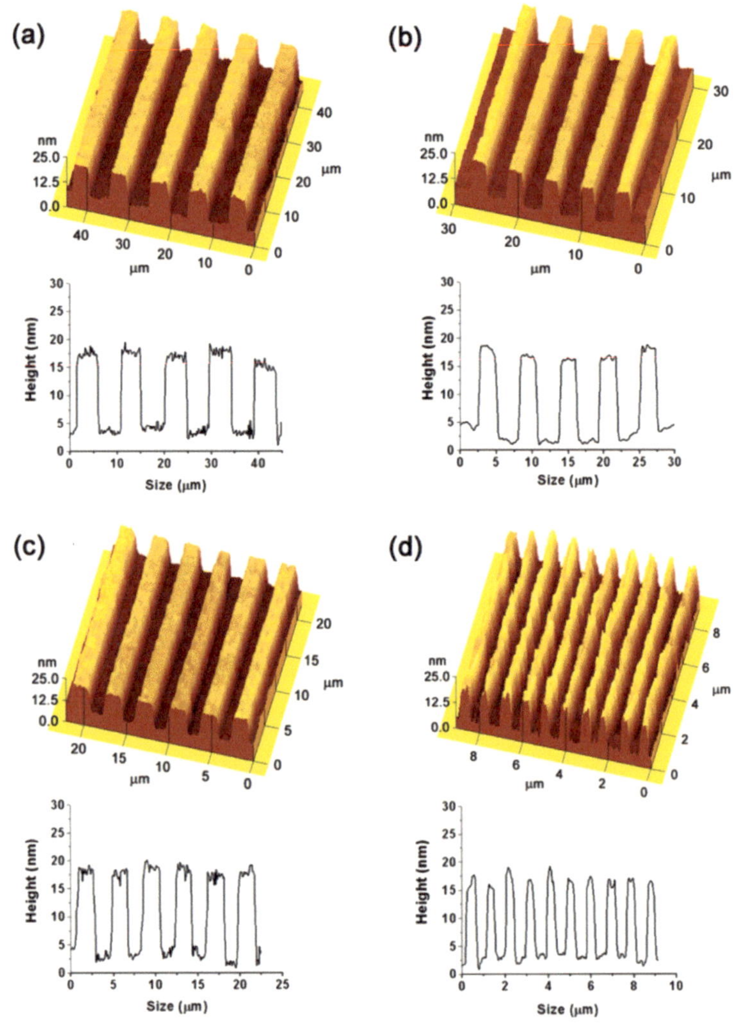

Figure 5. AFM images and cross sections of the patterned TiO_2 thin films generated by using selective ALD on the SAM patterned gold substrates: (a) 3.7 µm lines with 5.6 µm spaces, (b) 1.9 µm lines with 3.7 µm spaces, (c) 1.8 µm lines with 1.9 µm spaces, (d) 0.5 µm lines with 0.4 µm spaces [38].

The successful use of poly(methyl methacrylate) (PMMA) [39] and octadecyltrichlorosilane (OTS) SAMs [40] as a mask layer to obtain the direct patterned deposition of TiO_2 films has been reported.

Lee and Sung reported on a fabrication method using photocatalytic lithography of octadecylsiloxane SAMs, followed by selective deposition of ZrO_2 thin films using ALD. Lee et al. approach consists of three key steps. First, the alkylsiloxane SAMs were formed by immersing Si substrate in alkyltrichlorosilane solution. Second, photocatalytic lithography using a quartz plate coated with patterned TiO_2 thin films was done to prepare patterned SAMs of alkylsiloxane on the Si substrate. The patterned SAMs of the octade octadecylsiloxane on the Si substrate were made by using the quartz plate coated with the patterned TiO_2 thin films under UV irradiation in air. The photocatalytic lithography is based on the fact that the decomposition rate of the alkylsiloxane monolayers in contact with the TiO_2 is much faster than that with the SiO_2 under UV irradiation in air. These patterned SAMs define and direct the selective deposition of the ZrO_2 thin films. Third, ZrO_2 thin films were selectively deposited onto the SAMs-patterned Si substrate by ALD. A ZrO_2 thin film was selectively deposited using $Zr(OC(CH_3)_3)_4$ and water as ALD precursors. [18]

Chen et al. investigated a series of self assembled molecules as monolayer resists for HfO_2 atomic layer deposition. A series of n-alkyltrichlorosilanes of chain lengths ranging from 1 to 18 carbon atoms was used to form self-assembled monolayers on the oxide-covered silicon substrates. The ALD precursors for HfO_2 deposition were hafnium tetrachloride ($HfCl_4$) and water. The HfO_2 ALD process includes two self-limiting chemical reactions, repeated in alternating ABAB sequences shown in Eqs. (3) and (4), where asterisks indicate the outermost surface functional groups.

$$A: Hf-OH^* + HfCl_4 \rightarrow Hf-O-HfCl_3^* + HCl\uparrow \tag{3}$$

$$B: Hf-Cl^* + H_2O \rightarrow Hf-OH^* + HCl\uparrow \tag{4}$$

Each AB reaction cycle produces an HfO_2 layer terminated by hydroxyl groups, with the hydrochloride byproduct pumped away. After each exposure, the reaction chamber and the gas manifold were purged with nitrogen to avoid possible gas-phase reactions and to eliminate the possible physisorption of the precursors on the substrates. Deposition was carried out at a substrate temperature of 300 °C. The exposure times for $HfCl_4$ and water vapor were both 2 sec, followed by 3 min of nitrogen purging after each precursor was introduced into the chamber. Chen et al. explained the blocking mechanism by three important factors that influence the blocking efficiency of the monolayer organic films: chain length, tailgroup structure, and headgroup reactivity. This investigation shows that to achieve satisfactory deactivation toward the ALD process, it is crucial to form densely packed, highly hydrophobic organic monolayers. This in turn requires deactivating agents with high reactivity, low steric effect tail groups, and minimum chain length. [19]

Park et al. presented a method that combines SAM passivation and high-k dielectric deposition. Tetradecyl-SAM is formed on a Ge (100) surface via a thermal method. Part of the SAM is then removed by proper annealing, and a HfO_2 film is deposited by ALD. The system development was based on a previous research on the electrical properties of SAMs on Ge surfaces showing that SAM/Ge interfaces are electrically stable compared to Ge surfaces covered with native oxide. [51] Therefore, a combination of SAM passivation on a Ge surface with high-k gate dielectric deposition is suggested for scaling down gate oxides. [43]

Liu et al. created sub- 10 nm patterns of high dielectric constant (high-k) HfO_2 on Si substrate, by combining the use of the reassembled S-layer proteins as nanotemplates and an area-selective ALD process. To realize area-selective ALD of metal oxide-based high-k material nanopatterns into the nanotemplates composed of protein architectures, it is necessary to modify the S-layer proteins to introduce different surface functional groups upon them and the silicon substrate surface used in Liu et al. study. As a result, ALD only happens on the Si substrate and cannot take place on the modified surface of the S-layer proteins. ODTS used as an effective monolayer resist on a hydrophilic SiO_2 surface toward ALD of HfO_2, was chosen to modify the surface of the S-layer proteins but not the Si surface. Specifically, the ODTS-modified S-layer proteins are terminated with aliphatic chains ($R=(CH_2)_{17}CH_3$), while the Si surface exposed through the pores defined by the protein units is terminated with -OH or -H functional groups. Since atomic layer deposition has been achieved ideally on surfaces with -OH groups and with an incubation time on surfaces with -H groups, it is therefore feasible to achieve area-selective ALD on a surface with a contrast between aliphatic groups and -OH/-H terminations. [52], [53] ODTS-modified S-layer protein nanotemplate was selectively removed by thermal annealing. Therefore, S-layer proteins reassembled on Si substrate acted as a promising nanotemplate for the sub-10-nm nanopatterning of high-k oxides for future Metal Oxide Semiconductor Field Effect Transistor (MOSFET) applications. [44]

Park et al. demonstrated selective deposition of Ruthenium using contact printed self-assembled monolayer resists by selective area atomic layer deposition. Ruthenium is of interest for advanced metal/oxide/semiconductor (MOS) transistor gate electrodes to reduce polysilicon depletion effects and as nucleation layer for copper interconnect layers. [54] Ruthenium is considered as a viable candidate for p-type MOS devices because it has a vacuum work function near the conduction band edge of silicon, good thermal stability, and low resistivity of the oxidation phase. [55] Selective deposition enables direct formation of $Ru/HfO_2(SiO_2)/Si$ capacitor stacks, and the effective work function of ALD Ru is characterized on HfO_2 and SiO_2 dielectrics. They used PDMS stamps and OTS SAMs to prepare the patterned organic monolayer. ALD Ru was carried out using bis-(cyclopentadienyl)rutheniumg ($RuCp_2$) as a precursor and dry oxygen. $RuCp_2$ is solid at room temperature with vapor pressure of ~10 mTorr at the bubbler temperature of 80 °C. The ALD chamber was evacuated to $5*10^{-6}$ Torr, and the precursor and oxidant gases were introduced into the reactor in separate pulses (3 and 6 sec, respectively) with a 20 sec Ar purge between each reactant. Argon was also used as a carrier gas for the $RuCp_2$ pulse. [15]

Färm et al. reported on selective deposition of Iridium by using octadecyltrimethoxysilane (ODS), SAMs prepared from gas phase using a process where water-vapor pulses were given alternately with ODS. SAMs were patterned by a simple lift-off process. [46] In another work, narrow lines of OTS was printed by PDMS stamp which had 1.5 μm wide print lines and 1.5 μm wide spaces between. They also presented the passivation of copper surfaces using 1-dodecanethiol ($CH_3(CH_2)_{11}SH$) SAMs against iridium ALD growth. 1-dodecanethiol was chosen as a SAM precursor because it has relatively long carbon chain, it is liquid and volatile enough so that SAM were prepared from the vapor phase using moderate heating. [16] Iridium was grown only on non-SAM areas at 225 °C from Ir (pentanedione)$_3$ and O_2. [16], [40], [46]

Chen and Bent reported on deposition of Pt for the positive patterning area-selective ALD. Pt is a promising electrode material for dynamic random-access memories because of its high chemical stability in an oxidizing atmosphere and its excellent electrical properties. [56], [57] It is also a promising gate metal candidate owing to its high work function (5.6 eV) and compatibility with high-k dielectrics. [58] In the Chen and Bent paper, a deposition of Pt occurs on a SiO_2 film, providing a model process for the deposition of a gate metal on a dielectric. They used 1-octadecene as a monolayer, which undergoes a hydrosilylation reaction selectively on the hydride surface. Following monolayer attachment onto oxide patterned silicon wafers, Pt thin films were selectively deposited onto the substrates. ALD of a Pt thin film was carried out using methylcyclopentadienyl(platinum)trimethyl ($CH_3C_5H_4Pt(CH_3)_3$) and dry air. Exposure times for the Pt precursor and air were 3 and 2 sec, respectively, followed by a 60 and 45 sec N_2 purge after each precursor was introduced into the chamber. The Pt ALD process includes two self-limiting chemical reactions, repeated in the alternating ABAB sequences shown in Eqs. (5) and (6), where asterisks indicate the outermost surface functional groups and OBP, H_2O, CO_2 are reaction byproducts. [14], [42]

$$A: Pt + O_2 \rightarrow Pt - O^* \tag{5}$$

$$B: CH_3C_5H_4Pt(CH_3)_3 + Pt - O^* \rightarrow Pt + CO_2 \uparrow + H_2O \uparrow + OBP \uparrow \tag{6}$$

Figure 6 illustrates the Auger electron spectroscopy (AES) analysis of the patterned lines at higher spatial resolution. Figure 6a shows a SEM image of the patterned lines used for the study. In the SEM image, the oxide and the deactivated hydride regions (areas 1 and 2 in Fig. 6.a, respectively) were chosen for AES compositional analysis. The Auger survey scans shown in Figure 6.b reveal that, in the deactivated lines (area 2), the Pt signal is below the AES detection limit (0.5 %), whereas significant Pt is seen in area 1. AES line-scan images that compare the amounts of C and Pt as a function of position are displayed in Figure 6.c. A cross-sectional line (similar to the dashed line shown in Fig. 3a) was obtained perpendicular to the patterned lines. The C and Pt spectra clearly show the alternation as expected, and the edges of the Pt lines are sharp. [14]

Jiang and Bent reported on area selective atomic layer deposition of Platinum on Yttria stabilized zirconia (YSZ) substrates using microcontact printed SAMs. Jianga and Bentb technique can be used to deposit Pt on an YSZ solid oxide electrolyte for the catalyst in solid oxide fuel cell (SOFC). Pt is the catalyst used for a number of reactions, including the O_2 reduction reaction at the cathode of a SOFC, and is especially useful at the lower operating temperatures below (600 °C) that are desired for integratable fuel cell systems. [48], [49]

Lee et al. reported on capability of SAMs to block the deposition of PbS thin films by ALD. ODTS SAMs were chosen to modify the surface termination because of their ability to

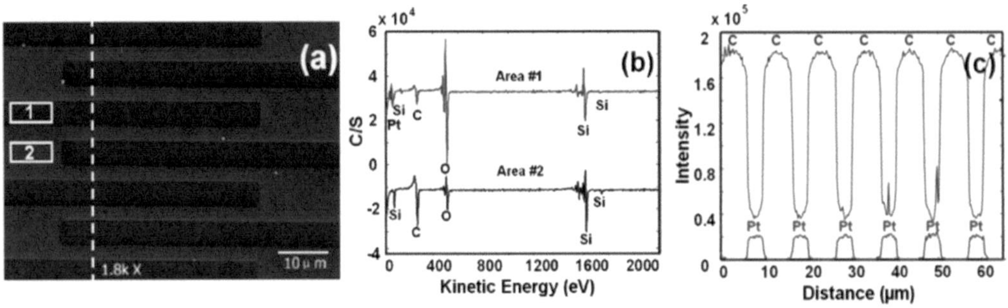

Figure 6. AES analysis on a patterned structure after the area-selective Pt ALD process: a) SEM image of the patterned area, b) AES selected-area survey composition scan, and c) AES defined line scan. [14]

deactivate ALD reactions as well as their good chemical and thermal stability. Microscale patterns of ALD PbS with high spatial and chemical selectivity were fabricated on ODTS patterned Si/SiO_2 substrates. ODTS was selectively grown only on the oxide patterns defined by the photolithography and deactivated PbS deposition during the ALD process. Hence, materials were selectively deposited by the ALD process only where ODTS was not present. The ALD precursors used were bis(2,2,6,6-tetramethyl-3,5-heptanedionato)lead(II) (Pb(tmhd)2) and H_2S. The base pressure of the ALD chamber was 50 mTorr. The substrate temperature was maintained at 160 °C, and the precursor was sublimated at 140 °C. [59] The PbS ALD process includes two self-limiting chemical reactions, repeated in an alternating ABAB sequence. Lee et al. postulated the following ligand-exchange reactions, which are typical of ALD half-reaction chemistry. In both reactions, a gas-phase precursor molecule reacts with the surface functional species and saturates the entire surface in a self-limiting manner. Each AB reaction cycle produces a PbS layer terminated by sulfhydryl groups, with the corresponding byproducts pumped away. [10]

Färm et al. reported on passivation of copper surfaces for selective-area ALD using 1-dodecanethiol SAMs against polyimide ALD growth. Polyimide is a new material for selective-area ALD and has potential applications as an insulating material in copper interconnects. As test substrates, silicon with evaporated copper dots was used. SAMs were prepared on the copper surfaces from the vapor phase. Polyimide was deposited from 1,2,3,5-benzenetetracarboxylic anhydride (pyromellitic dianhydride) and 4,4-oxydianiline at 160 °C. [16]

SAMs have typically been created by dipping the solid substrates into a solution containing the precursor molecules. The vapor process in preferred and involves preparing SAMs by bringing the precursors to the substrate surface as vapors. The vapor process has some advantages over the liquid process, e.g., when SAMs have to be formed on three-dimensional structures. The vapor process can prevent problems related to the absorption of liquids into the porous structures. [16] The vapor-phase process also requires fewer precursors than the liquid-phase processes. Moreover, the aggregation of the precursor molecules prior to deposition on the substrate surface, which can cause defects in the arrangement of the SAMs in the liquid-phase process, is significantly reduced using the vapor-phase process. Aggregated precursors had lower vapor pressures than the single-molecule precursors and thus were rarely vaporized. [45] The vapor-phase SAM formation can be carried out in a vacuum system allowing easier combination with the ALD reactor. In principle, SAM formation can be performed in the ALD reactor itself. [45] When SAMs are prepared as an initial stage of the ALD process, the patterning of the SAMs has to be done by relying on the chemical selectivity of the SAM formation. [16] Silane [40], [45], [46] and thiol [16] SAMs has been formed from the vapor phase for selective-area ALD of TiO_2 [36], HfO_2 [45], Ir [16], [40], [46], Pt [45] and Polyimide. [16]

The common way to block ALD by using SAMs is limited when the height of the deposited inorganic film exceeds the height of the self-assembled monolayer (~2 nm). In that case the growth will not be area-selective anymore near the interface where the already deposited inorganic film meets the end of the alkyl tails. Near that interface, the ALD reactants are able to adsorb on the inorganic film.

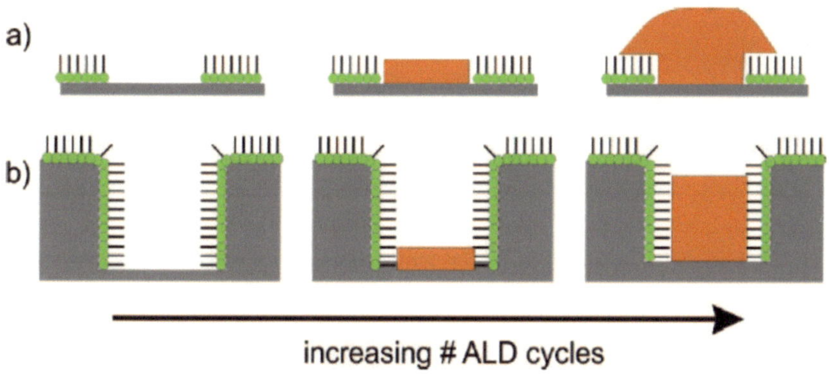

Figure 7. (a) Conventional area-selective ALD in which the substrate is planar and contains patterns of self-assembled monolayers. With increasing number of deposition cycles there occurs also sideways film growth originating from adsorption of ALD reactants on the previously deposited ALD film. b) Blocking the lateral ALD growth independent of deposited film thickness by combining surface modification and topographical features. [37]

The inorganic film is not confined anymore to the original pattern of the SAM and the lateral dimension of the film will increase when more ALD cycles are carried out (see Figure 7.a). Robin at el. have shown a new concept to enable construction of nanoscale lateral structures by area-selective ALD. The concept is based on providing chemical inertness by surface modification combined by nanoscale topographical structures (Figure 7.b). Whereas surface modification, as traditionally used in area-selective ALD, is only a chemical barrier for film growth, Robin et al. shows that the topographical structures are also a physical barrier for film growth. Their concept allows ALD synthesis of constructs that have lateral dimensions many times smaller than the film thickness. Robin et al. used cicada wings as a prototypical example from nature; however, their concept can be also applicable on other types of designed substrates that combine surface modification (including SAMs) with nanoscale topographies. [37]

2.2. Surface study by ALD on SAM

Lee et al. studied a surface free energy by atomic layer deposition of TiO_2 on mixed SAMs. They studied ALD growth modes as a function of surface free energy. [60] Mixed SAMs have been used to modify the surface free energy of the Si substrates. By using solutions containing two different silanes, it formed SAMs containing mixtures of them. The influence of the surface free energy of the Si substrates on the growth modes of TiO_2 thin films has been studied with AFM, XPS and contact-angle analysis. Mixed SAMs with several surface compositions of H_3C-Si and HO-Si were formed on the Si substrates. The surface free energy of the SAM-contact samples was derived from the contact-angle data by using water and diiodomethane, as shown in Table 1. [60]

HO–Si : CH_3–Si	contact angle θ [°]		surface free energy [mN/m]		
	water	diiodomethane	γ_s	γ_s^d	γ_s^p
1 : 0	40	25	64	41	22
4 : 1	51	34	56	38	18
2 : 1	62	40	49	36	13
1 : 1	70	45	43	34	9
1 : 2	78	50	38	32	6
1 : 3	89	55	33	31	3
0 : 1	108	60	29	29	0

Table 1. Contact angle (θ) and surface free energy (γ_s) of Si substrates coated with SAMs (d: dispersive part, p: polar part). [60]

The surface free energy of the mixed SAM-contact samples, ranging from 64 to 29 mN/m, appears to be determined primarily by the surface composition of the of H_3C-Si and HO-Si, which means that the surface free energy of solid substrate can be controlled by mixed SAMs. The TiO_2 thin films were grown on the mixed SAM-coated Si substrates by atomic layer deposition from titanium isopropoxide and water. The ALD growth mode of the TiO_2 film changes as function of the surface free energy of the Si substrates, and the surface free energy can be modified by changing the ratio of the components of the mixed SAMs. A two-dimensional growth mode is observed on the SAM-coated substrates with high surface free energies. As the surface free energy decreases, a three-dimensional growth mode begins to dominate. From the results, Lee et al. have found that the mixed SAMs can be used to control the growth modes of the atomic layer deposition by modifying the surface free energy of the substrates. [60]

Xu and Musgrave used density functional theory (DFT) for investigated surface reactions between trimethylaluminum (TMA) as precursor for alumina and SAMs terminated with different functional groups. [30] They show that the reaction of TMA and the -OH-terminated SAM is favored both thermodynamically and kinetically over the reaction with -NH_2- and -CH_3-terminated SAMs. Reactions on the -NH_2-terminated SAM form more stable complex intermediates; however, because the ligand exchange barrier is large, the precursors are trapped in the adsorbed complex state. Furthermore, although there is a thermodynamic driving force for this reaction, the reaction is relatively slow compared to the -OH-terminated case and desorption of the precursor is favored over ligand exchange. In the case of the -CH_3-terminated SAM, there is no thermodynamic driving force for the reaction and the reaction barrier is large. The reaction path and predicted energetics for reactions of TMA and -OH/NH_2/CH_3-terminated SAMs as shown in Figure 8. [30]

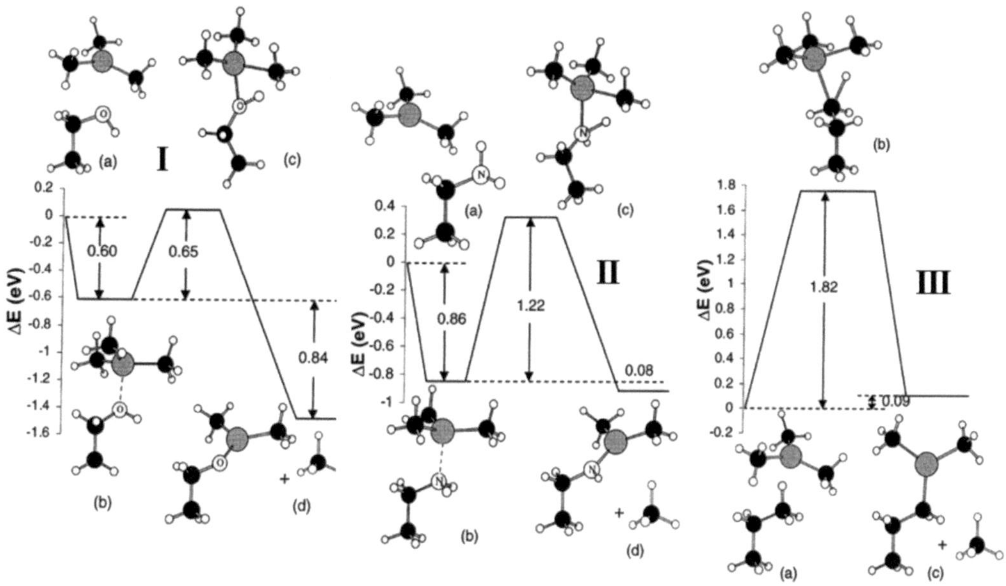

Figure 8. I) Reaction path and predicted energetics for reactions of TMA and -OH-terminated SAM. The stationary points correspond to (a) CH_3CH_2OH + TMA, (b) complex TMA•$OHCH_2CH_3$, (c) transition state, and (d) CH_3CH_2O-$Al(CH_3)_2$+ CH_4. II) Reaction path and predicted energetics for reactions of TMA and -NH_2-terminated SAM. The stationary points correspond to (a) $CH_3CH_2NH_2$ + TMA, (b) complex TMA•NH_2-CH_2CH_3, (c) transition state, and (d) CH_3CH_2NH-$Al(CH_3)_2$ +CH_4. III) Reaction path and predicted energetics for reactions of TMA and -CH3-terminated SAM. The stationary points correspond to (a) $CH_3CH_2CH_3$ + TMA, (b) transition state, and (c) $CH_3CH_2CH_2$-$Al(CH_3)_2$ + CH_4. [30]

The energetics of the reactions does not depend on the length of the SAM using ethyl and pentyl groups as models. [30] After the initial TMA adsorption on the -OH-terminated SAM, the second half-reaction of ALD growth of Al_2O_3 (Al-CH_3*→Al-OH*) is calculated and the mechanism and energetics are consistent with their previous results for ALD of Al_2O_3 using TMA and water. [61] Because these adsorption reactions are highly localized, the conclusions are not only limited to the effect of surface functionalization on ALD reactions on SAMs; they can also be extended to reactions on other substrates and to ALD reactions involving other precursors which form dative-bonded complexes. [30]

Lee et al. used DFT simulation for study reactions between $Pb(tmhd)_2$ precursor to ODTS SAMs and SiO_2, the results showed an increased activation barrier and a higher overall reaction energy for the $Pb(tmhd)_2$ precursor on an ODTS-terminated substrate than on a SiO_2 surface. [10]

2.3. ALD on chiral SAMs

H. Moshe et al. proposed a new innovative type of stable chiral nanosized metal oxide surfaces. [62] The structure and chirality of this type of chiral surface is based on chiral self-assembled monolayers (SAMs) coated with nanosized films of metal oxide materials deposition by ALD. The idea underlying this new design of nano-chiral surfaces is that the ceramic nanolayers coating the chiral SAMs protects the chiral SAMs that would otherwise be destroyed under the reactions conditions, thereby preserving their enantioselective nature. In Figure 9, the overall structure of the new nanoscale hybrid chiral surfaces based on chiral SAM and ceramic nanolayers is shown.

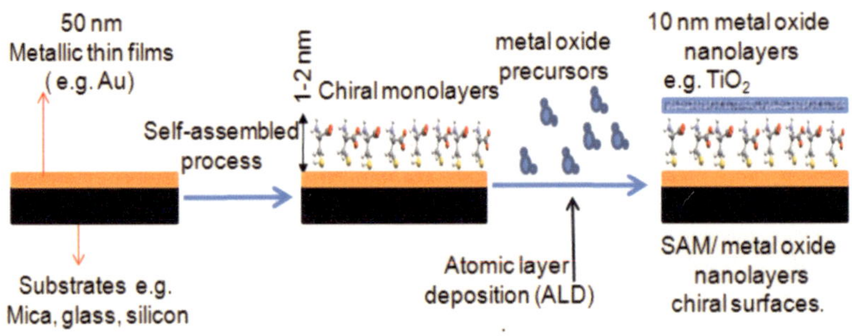

Figure 9. Illustration of the design and the synthesis paths for the new chiral SAM/ceramic nanolayers surfaces.

In their study, H. Moshe et al. used TiO_2 to form the protective nanolayers for the chiral SAM since its synthesis does not demand high temperatures that may harm the chiral SAMs. In the research, they utilize the atomic layer deposition (ALD) technique since it provides excellent thickness control and produces very dense and uniform layers. The first step in the synthesis of this type of nano-chiral surface requires the preparation of chiral SAMs. For the chiral SAMs preparation, they used enantiomers of cysteine and glutathione. TiO_2 films were grown by ALD using $Ti(N(CH_3)_2)_4$ and water as the precursors. Their research focuses on evidence of the chirality of the SAM/metal oxide nanosurfaces. Generally, several techniques [63] can be used to study the chiral nature of nano-sized surfaces such as chiral AFM, STM, second-harmonic generation (SHG) and isothermal titration calorimetry. [64] However, due to the unique structure of their nanosized chiral surfaces, they are rather limited in the techniques that can be used to prove the chirality of these surfaces. In their work, H. Moshe et al. have selected several techniques namely quartz microbalance (QMB), second-harmonic generation circular-dichroism (SHG CD) spectroscopy, enantioselective crystallization and chiral adsorption measurements as the methods to study the chirality of the SAM/TiO_2 nanolayer surfaces.

3. Conclusions

In this book chapter the preparation, properties and applications of ALD as a novel method for thin film deposition on Self Assembled Monolayers have been briefly reviewed. We have reviewed a selective-area atomic layer deposition of a variety of materials such as metals, metal oxides, and polymers. First, we presented a brief introduction reviewing the ALD method and principle of operation. Second, we discussed the ability of SAMs to shape the surface of the substrate before the ALD deposition stage. ALD is very sensitive to surface conditions and therefore offers an ideal method for film deposition. Third, we reviewed procedures for properties and applications of ALD on SAMs. We included a variety of molecules and materials and different conditions used for atomic layer deposition. Fourth, we discussed studies which used ALD on SAMs in order to learn surface properties. Finally, a novel application of ALD for the preparation of chiral nanosized metal oxide films using chiral SAMs was discussed. Future work is aimed at the modification and functionalization of surfaces by SAMs used as templates for ALD.

ALD is a technique with high control capabilities. SAMs are a simple and versatile method used for surface design. Integration of the ALD technique and the SAM method can increase the ability to study and engineer substrate surfaces.

Acknowledgements

H. Moshe would like to acknowledge the Department of Chemistry, Bar Ilan University for funding.

Author details

Hagay Moshe and Yitzhak Mastai

Department of Chemistry and the Institute of Nanotechnology, Bar Ilan University, Israel

References

[1] Suntola, T, & Antson, J. (1977). Method for producing compound thin films. U.S. Patent #4,058,430, Issued Nov. 25

[2] Leskela, M, & Ritala, M. (2003). Atomic layer deposition chemistry: recent developments and future challenges. *Angewandte Chemie International Edition*, , 42

[3] Knez, M, Niesch, K, & Niinisto, L. (2007). Synthesis and surface engineering of complex anostructures by atomic layer deposition. *Advanced Materials*, , 19

[4] George, S. M. (2010). Atomic layer deposition: an overview. *Chemical Reviews*, , 110

[5] Ritala, M, & Leskelä, M. (2002). Atomic layer deposition. *Handbook of Thin Film Materials*, Ed.: Nalwa, H. S.) Academic Press, San Diego, , 1, 103-159.

[6] Niinistö, L, Nieminen, M, Päiväsaari, J, Niinistö, J, Putkonen, M, & Nieminen, M. (2004). Advanced electronic and optoelectronic materials by Atomic Layer Deposition: An overview with special emphasis on recent progress in processing of high-*k* dielectrics and other oxide materials. *Physica Status Solidi (a)*, , 201

[7] Becker, J. S. (2002). Atomic layer deposition of metal oxide and nitride thin films. Ph.D. dissertation,Harvard University.

[8] Aaltonen, T. (2005). Atomic layer deposition of noble metal thin films. Ph.D. dissertation, University of Helsinki.

[9] Park, M. H, Jang, Y. J, Sung-suh, H. M, & Sung, M. M. (2004). Selective atomic layer deposition of titanium oxide on patterned self-assembled monolayers formed by microcontact printing. *Langmuir*, , 20

[10] Lee, W, Dasgupta, N. P, Trejo, O, Lee, J, Hwang, R, Usui, J, Prinz, T, & Area-selective, F. B. Atomic Layer Deposition of Lead Sulfide: Nanoscale Patterning and DFT Simulations. *Langmuir*, , 26

[11] Tanskanen, J. T, Bakke, J. R, Bent, S. F, & Pakkanen, T. A. (2010). ALD growth characteristics of ZnS films deposited from organozinc and hydrogen sulfide precursors. *Langmuir*, , 26

[12] Zaera, F. (2012). The surface chemistry of atomic layer depositions of solid thin films. *The Journal of Physical Chemistry Letters*, , 3

[13] Ulman, A. (1996). Formation and structure of self-assembled monolayers. *Chemical Reviews*, , 96

[14] Chen, R, & Bent, S. F. (2006). Chemistry for positive pattern transfer using area-selective atomic layer deposition. *Advanced Materials*, , 18

[15] Park, K. J, Doub, J. M, Gougousi, T, & Parsons, G. N. (2005). Microcontact patterning of ruthenium gate electrodes by selective area atomic layer deposition. *Applied Physics Letters*, , 86

[16] Färm, E, Vehkamäki, M, Ritala, M, & Leskelä, M. (2012). Passivation of copper surfaces for selective-area ALD using a thiol self-assembled monolayer. *Semiconductor Science and Technology*, , 27

[17] Yan, M, Koide, Y, Babcock, J. R, Markworth, P. R, Belot, J. A, Marks, T. J, & Chang, R. P. H. (2001). Selective-area atomic layer epitaxy growth of ZnO features on soft lithography-patterned substrates. *Applied Physics Letters*, , 27

[18] Lee, J. P, & Sung, M. M. (2004). A new patterning method using photocatalytic lithography and selective atomic layer deposition. *Journal of the American Chemical Society*, , 126

[19] Chen, R, Kim, H, Mcintyre, P. C, & Bent, S. F. (2005). Investigation of self-assembled monolayer resists for hafnium dioxide atomic layer deposition. *Chemistry of materials*, , 17

[20] Becker, J. S, Kim, E, & Gordon, R. G. (2004). Atomic layer deposition of insulating hafnium and zirconium nitrides. Chemistry of materials, , 16

[21] Gasser, W, Uchida, Y, & Matsumura, M. (1994). Quasi-monolayer deposition of silicon dioxide. Thin Solid Films , 250

[22] Klaus, J. W, Sneh, O, & George, S. M. (1997). Growth of SiO_2 at room temperature with the use of catalyzed sequential half-reactions. *Science,* , 278

[23] Luo, Y, Slater, D, Han, M, Moryl, J, & Osgood, R. M. (1997). Low-temperature, chemically driven atomic-layer epitaxy: *In situ* monitored growth of CdS/ZnSe(100). Applied Physics Letters, , 71

[24] Knez, M, Kadri, A, Wege, C, Gösele, U, Jeske, H, & Nielsch, K. (2006). Atomic layer deposition on biological macromolecules: metal oxide coating of tobacco mosaic virus and ferritin. Nano Letters, , 6

[25] Kemell, M, Pore, V, Ritala, M, Leskelä, M, & Linden, M. (2005). Atomic layer deposition in nanometer-level replication of cellulosic substances and preparation of photocatalytic TiO_2/cellulose composites. *Journal of the American Chemical Society,* , 127

[26] Kim, H, Lee, H. B. R, & Maeng, W. G. (2009). Applications of atomic layer deposition to nanofabrication and emerging nanodevices. *Thin Solid Films,* , 517

[27] Laibinis, P. E, Whitesides, G. M, Allara, D. L, Tao, Y. T, Parikh, A. N, & Nuzzo, R. G. (1991). Comparison of the structures and wetting properties of self-assembled monolayers of n-alkanethiols on the coinage metal surfaces, copper, silver, and gold. *Journal of the American Chemical Society,* , 113

[28] Blum, A. S, Kushmerick, J. G, Long, D. P, Patterson, C. H, Yang, J. C, Henderson, Y. C, Yao, Y, Tour, J. M, Shashidhar, R, & Ratna, B. R. (2005). Molecularly inherent voltage-controlled conductance switching. *Nature Materials,* , 4

[29] Akkerman, H. B, Blom, P. W. M, De Leeuw, D. M, & De Boer, B. (2006). Towards molecular electronics with large-area molecular junctions. *Nature,* , 441

[30] Xu, y, & Musgrave, C. B. (2004). A DFT study of the Al_2O_3 atomic layer deposition on SAMs: effect of SAM termination. *Chemistry of materials,* , 16

[31] Kumar, A, Biebuyck, H. A, Abbott, N. L, & Whitesides, G. M. (1992). The use of self-assembled monolayers and a selective etch to generate patterned gold features. *Journal of the American Chemical Society,* , 114

[32] Kumar, A, & Whitesides, G. M. (1993). Features of gold having micrometer to centimeter dimensions can be formed through a combination of stamping with an elastomeric stamp and an alkanethiol "ink" followed by chemical etching. *Applied Physics Letters,* , 63

[33] Kumar, A, Biebuyck, H. A, & Whitesides, G. M. (1994). Patterning self-assembled monolayers: applications in materials science. *Langmuir,* , 10

[34] Carr, D. W, Lercel, M. J, Whelan, C. S, Craighead, H. G, Seshadri, K, & Allara, D. L. (1997). High-selectivity pattern transfer processes for self-assembled monolayer electron beam resists *Journal of Vacuum Science & Technology A*, , 15

[35] Huang, J. Y, & Hemminger, D. A. (1994). Photopatterning of self-assembled alkanethiolate monolayers on gold: a simple monolayer photoresist utilizing aqueous chemistry. *Langmuir*, , 10

[36] Xu, S, & Liu, G. (1997). Nanometer-scale fabrication by simultaneous nanoshaving and molecular self-assembly *Langmuir*, , 13

[37] Ras, R. H. A, Sahramo, E, Malm, J, Raula, J, & Karppinen, M. (2008). Blocking the lateral film growth at the nanoscale in area-selective atomic layer deposition. *Journal of the American Chemical Society*, , 130

[38] Seo, E. K, Lee, J. W, Sung-suh, H. M, & Sung, M. M. (2004). Atomic layer deposition of titanium oxide on self-assembled-monolayer-coated gold. *Chemistry of materials*, , 16

[39] Sinha, A, Hess, D. W, & Henderson, C. L. (2006). Area-selective ALD of titanium dioxide using lithographically defined poly(methyl methacrylate) films. *Journal of The Electrochemical Society*, G465-G469, 153

[40] Färm, E, Kemell, M, Ritala, M, & Leskelä, M. (2008). Selective-area atomic layer deposition with microcontact printed self-assembled octadecyltrichlorosilane monolayers as mask layers. *Thin Solid Films* , 517

[41] Chen, R, & Bent, S. F. (2006). Chemistry for positive pattern transfer using area- selective atomic layer deposition. *Advanced Materials*, , 18

[42] Chen, R, & Bent, S. F. (2006). Highly stable monolayer resists for atomic layer deposition on Germanium and Silicon. *Chemistry of Materials* , 18

[43] Park, K, Lee, Y, Im, K. T, Lee, J. Y, & Lim, S. (2010). Atomic layer deposition of HfO_2 on self-assembled monolayer-passivated Ge surfaces. *Thin Solid Films* , 518

[44] Liu, J, Mao, Y, Lan, E, Banatao, D. R, Forse, G. J, Lu, J, Blom, H. O, Yeates, T. O, Dunn, B, & Chang, J. P. (2008). Generation of oxide nanopatterns by combining self- assembly of S-layer proteins and area-selective atomic layer deposition. *Journal of the American Chemical Society*,, 130

[45] Hong, J, Porter, D. W, Sreenivasan, R, Mcintyre, P. C, & Bent, S. F. (2008). ALD resist formed by vapor-deposited self-assembled monolayers. *Langmuir*, , 23

[46] Färm, E, Kemell, M, Ritala, M, & Leskelä, M. (2006). Self-assembled octadecyltrimethoxysilane monolayers enabling selective-area atomic layer deposition of iridium. *Thin Solid Films*, , 12

[47] Jiang, X, Huang, H, Prinz, F. B, & Bent, S. F. (2008). Application of atomic layer deposition of platinum to solid oxide fuel cells *Chemistry of materials*, , 20

[48] Jiang, X, & Bent, S. F. (2007). Area-selective atomic layer deposition of platinum on YSZ substrates using microcontact printed SAMs. *Journal of The Electrochemical Society*, 154, DD656, 648.

[49] Jiang, X, Chen, R, & Bent, S. F. (2007). Spatial control over atomic layer deposition using microcontact-printed resists. *Surface & Coatings Technology*, , 201

[50] Ott, A. W, & Chang, R. P. H. (1999). Atomic layer-controlled growth of transparent conducting ZnO on plastic substrates. *Materials Chemistry and Physics*, , 58

[51] Schoell, I. D, Sharp, S. J, Hoeb, M, Brandt, M. S, & Stutzmann, M. (2008). Electronic properties of self-assembled alkyl monolayers on Ge surfaces. *Applied Physics Letters*, , 92

[52] Chen, R, Kim, H, Mcintyre, P. C, & Bent, S. F. (2004). Self-assembled monolayer resist for atomic layer deposition of HfO_2 and ZrO_2 high-κ gate dielectrics. *Applied Physics Letters*, , 84

[53] Lao, S. X, Martin, R. M, & Chang, J. P. (2005). Plasma enhanced atomic layer deposition of HfO_2 and ZrO_2 high-k thin films. *Journal of Vacuum Science & Technology A*, , 23

[54] Kim, H. (2003). Atomic layer deposition of metal and nitride thin films: Current research efforts and applications for semiconductor device processing. *Journal of Vacuum Science & Technology B*, , 21

[55] Zhong, H, Heuss, G, & Misra, V. (2000). Electrical properties of RuO_2 gate electrodes for dual metal gate Si-CMOS. *Electron Device Letters, IEEE*, , 21

[56] Hiratani, M, Nabatame, T, Matsui, Y, Imagawa, K, & Kimura, S. (2001). Platinum film growth by chemical vapor deposition based on autocatalytic oxidative decomposition. *Journal of The Electrochemical Society, C524-C527*, 148

[57] Nayak, M, Ezhilvalavan, S, & Tseng, T. Y. (2001). High-Permittivity (Ba, Sr)TiO_3 thin films. Handbook of Thin Film Materials, Ed.: Nalwa, H. S.) Academic Press, San Diego, , 3, 99-167.

[58] Wilk, G. D, Wallace, R. M, & Anthony, J. M. (2001). High-κ gate dielectrics: Current status and materials properties considerations. Journal of Applied Physics, , 89

[59] Dasgupta, N. P, Lee, W, & Prinz, F. B. (2009). Atomic layer deposition of lead sulfide thin films for quantum confinement. *Chemistry of Materials*, , 21

[60] Lee, J. P, Jang, Y. J, & Sung, M. M. (2003). Atomic layer deposition of TiO_2 thin films on mixed self-assembled monolayers studied as a function of surface free energy. *Advanced Functional Materials*, , 13

[61] Choy, K. L. (2003). Chemical vapour deposition of coatings. *Progress in Materials Science*, , 48

[62] Moshe, H, Vanbel, M, Valev, V, Verbiest, T, Dressler, D, & Mastai, Y. Chiral nanosized metal oxide surfaces. unpublished yet.

[63] Chena, Q, & Richardsonb, N. V. (2004). Physical studies of chiral surfaces. *Annual Reports Section "C"*, , 100

[64] Shval, A, & Mastai, Y. (2011). Isothermal titration calorimetry as a new tool to investigate chiral interactions at crystal surfaces. *Chemical Communications*, , 47

Chapter 4

Plasma Electrolytic Oxidation of Valve Metals

Alex Lugovskoy and Michael Zinigrad

Additional information is available at the end of the chapter

http://dx.doi.org/10.5772/54827

1. Introduction

Plasma electrolytic oxidation (PEO) is also known as micro-arc oxidation and spark anodizing is often regarded as a version of anodizing of valve metals (Mg, Al, Ti, and several others) and their alloys. Indeed, the essence of both anodizing and PEO is the production of oxide layers on a metal surface by the action of electricity in a convenient electrolyte. An oxide layer has a complex composition and includes various oxides of a base metal, alloy additives and species coming from the electrolyte. For both anodizing and PEO, an oxide layer forms due to electrochemical oxidation of the metal constituents and inclusion of some components of the electrolyte with possible further interactions in the vicinity of the electrode.

However, some features of PEO are clearly different than those of the anodizing. Normally, low-voltage direct currents are used for anodizing and the formation of the oxide layer occurs under a quiescent regime. The produced layer has relatively homogeneous structure with more or less evenly distributed blind pores. The thickness of an oxide layer is limited by ~ 20-50 μm for most cases, because the electrical conductivity of the oxide layer is low and the formation of the layer effectively breaks the electric circuit in the cell. It can be said that the formed layer "passivates" the metal surface in the course of anodizing and thus prevents its own further growth (Fig. 1, stages I - II). It deserves to be notes that sparking is considered undesirable for anodizing, because it is an indication of cracks and inhomogeneities in the formed layer.

For PEO, significantly higher voltages and (normally) alternating currents cause intense sparking due to micro-arc discharges that break down the oxide layer (Fig.1, stages I – III). Extreme temperatures and pressures [1] develop in the discharge channels and cause complex phase-transformation processes that result in the production of a compact, thick hard layer, which often has attractive abrasion and corrosion resistances. Although PEO layers have a relatively high porosity [2], they can effectively protect the base metal against corrosion because the pores formed by a discharge can subsequently "heal" by molten oxides due to high local temperatures in the vicinity of plasma discharge channels [3] and are therefore impermeable to corrosion media. The improved corrosion stability of PEO-treated metals as compared to bare metals has been reported for aluminum [4-8] and magnesium alloys (see, for instance, [9, 10]).

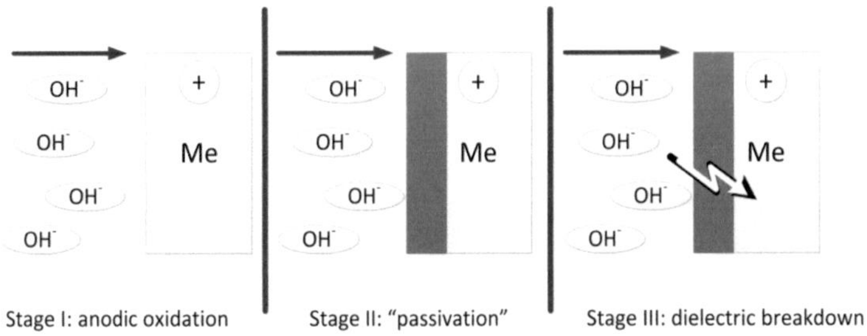

Figure 1. Main stages of an oxide layer formation in the processes of anodizing (stages I - II) and PEO (stages I – III)

Both direct and alternating current can be used for PEO. However, AC regime is preferable, because pores formed during a cathodic breakdown "heal" by molten oxides during the next anodic pulse [3], the electrolyte in the metal vicinity is refreshed and the produced oxide layers are more uniform. Industrial 50–60 Hz sine-wave AC voltages of 100–600V are most frequently used for the PEO processing. Due to the partial rectifying effect of the valve metal oxide, complex sew-like waves are observed in practice (Fig. 2).

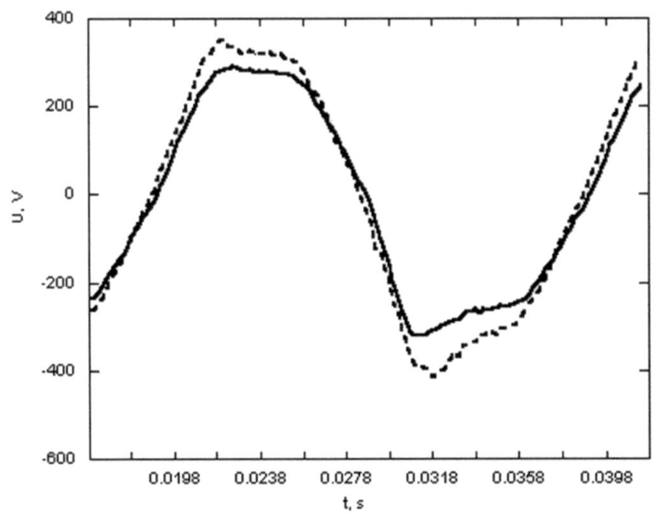

Figure 2. Observed voltage profile of a typical PEO processing of an aluminium alloy in alkaline silicate electrolytes containing 10 g/L of $Na_2O\ SiO_2$ (solid line) or of $Na_2O\ 3SiO_2$ (dotted line). Industrial 50 Hz sine-wave voltage of 200V was supplied for both cases.

The detailed mechanism of the PEO process has not yet been revealed; however, most investigators (see, for example, [3, 11]) agree that during each AC period several principal steps occur: (1) a barrier oxide layer forms on the boundary between the metal and the electrolyte during the initial anodic semi-period; (2) the potential difference between the two sides of the dielectric oxide layer increases as the anodic semi-period advances until (3) dielectric breakdown takes place. The breakdowns through the oxide layer are accompanied by sparks, so that the process actually occurs in a mode of micro-arc discharges. Fresh portions of the electrolyte are injected into the bare metal surface during the breakdowns, and the process continues as long as the voltage is sufficient for new breakdowns which perforate the growing oxide layer. Relaxation of the metal and oxide and partial reduction of the oxidized species occur during cathodic semi-periods. Gas micro-phase formation (nucleation) and annihilation (cavitation) processes apparently contribute to formation of the oxide layer, but these processes have been scarcely studied because of obvious experimental difficulties.

The production of oxide layers by PEO was most often studied for aluminum and its alloys (for example, in [5-8], [12-17]), magnesium and its alloys ([9, 10] and others), titanium and its alloys (for example, in [18-19]). Studies of PEO on other metals (zirconium [20], zirconium alloys [21] and steel [22]) are scarce.

The various versions of PEO differ from each other with respect to the profile of the applied voltage and the composition of electrolyte. The oxide layer produced always consists of two sub-layers: an outer brittle sub-layer, which typically has a hardness of 500–1000 HV and a porosity of more than 15%, and an inner functional sub-layer, whose typical hardness is 900–2000 HV and whose typical porosity is 2–10%. The outer brittle sub-layer can be easily removed by polishing, and the inner harder sub-layer can be finished to a smooth marble surface.

Oxide layers can be produced in several types of electrolytes, whose action can differ (see the survey in [11]). Acidic and alkaline electrolytes dissolve moderately the base metal, phosphate and polymer electrolytes passivate it, and fluoride electrolytes interact with it in more complex and less understood ways.

The most frequently used electrolytes for the PEO processing of aluminum and aluminum alloys are aluminate [23, 24], phosphate [24, 25] and, most often, alkaline silicate solutions (for example, [7, 11]). Magnesium and its alloys are normally PEO processed in alkaline phosphate [26-28] or alkaline silicate electrolytes [10, 29, 30] often containing fluorides [27-30]. Aluminate, phosphate and silicate electrolytes are used for titanium and its alloys (cf. a comparative study in [31]).

Since silicate electrolytes are frequently used for the PEO processing of the valve metals, many aspects of their influence on the properties of produced oxide layers have been intensively studied. For example, it was established that the addition of silicates to the electrolytes stabilizes the oxide layer toward alkaline attacks [12], causes some increase in the thickness of the oxide layers, but reduces their hardness and wear resistance as compared to alkaline electrolytes without silicates [13].

Less information is available about the effect of various forms of "water glasses," i.e., polymer silicates of various composition, on the structure and properties of PEO layers. Little or nothing is known about the difference between oxide layers obtained in silicate electrolytes having identical or close element composition, but containing silicates of different SiO_2-to-Na_2O ratios (silicate indexes).

Another point of interest is the influence of the fluoride additives on the structure and properties of the PEO oxide layers produced not on aluminum alloys only, but also on other base metals.

Here we try to summarize these two effects (the role of the silicate index and the influence of fluorides) in a comparative study of the PEO processing of a magnesium alloy and of an aluminum alloy.

2. Experimental

Rectangular flat (3 x 15 x 30 mm) specimens of aluminum A5052 alloy (Al as the base and approximately 2.5% of Mg) and magnesium AZ9110D alloy (Mg as the base and 8.3-9.7% Al, 0.15% Mn min., 0.35-1.0% Zn, 0.10% Si max., 0.005% Fe max., 0.030% Cu max., 0.002% Ni max., 0.02% max. others) were cut, polished with #1200 grit SiC abrasive paper and rinsed in tap water prior to be PEO processed. The oxidation was performed in AC mode by the industrial 50 Hz sine voltage at the end current density 6.6 ± 0.2 A / dm^2 for 30-60 minutes on a home-made 40 kVA PEO station with a water-cooled bath made of stainless steel, which served as the counter electrode. Potassium hydroxide KOH (Finkelman Chemicals, technical grade), KF (Merck, 99%), sodium silicate $Na_2O\ SiO_2\ 5H_2O$ (pentahydrate, Spectrum, practical grade), and water glass $Na_2O\ 3SiO_2$ (Spectrum, practical grade) having the silicate indexes n = 1 and n=3, respectively were used for the preparation of the electrolytes.

Conductivities and pH of the electrolytes were measured by a YK-2005WA pH/CD meter, the thickness of oxide layers was first roughly measured by a coating thickness gauge CM-8825 and then more exactly by SEM. The surface morphology, structure and composition were inspected on SEM JEOL JSM6510LV equipped with an NSS7 EDS analyzer (Correction Method Proza – Phi-Pho-Z was used for the quantitative analysis). Cross-section samples prepared according to standard metallographic protocols [32] were used for SEM, EDS, XRD and microhardness measurements. Microhardness was measured on Buehler Micromet 2100, HV_{25}. X-ray Diffractometer (XRD) Panalytical X'Pert Pro with Cu Kα radiation (λ=0.154 nm) was used with the full pattern identification made by X'Pert HighScore Plus software package, version 2.2e (2.2.5) by PANalytical B.V. Materials identification and analysis made by the PDF-2 Release 2009 (Powder Diffraction File). Phase analysis identification made by XRD, 40kV, 40mA. The XRD patterns were recorded in the GIXD geometry at a=1°and 5° in the range of 20-80º (step size 0.05º and time per step 2s).

Autolab12 Potentiostat with a standard corrosion cell was used for corrosion tests. Potentials were measured against Ag|AgCl reference electrode and then related to SHE.

http://dx.doi.org/10.5772/54827

3. Results and discussion

3.1. PEO of aluminum A5052 alloy in different alkaline silicate electrolytes

Two sodium silicates were taken for the comparison. The first, Na_2O SiO_2 will be hereafter referred to as the "n=1 silicate" and the second, Na_2O $3SiO_2$ will be referred to as the "n=3 silicate," in accordance to their silicate index, that their SiO_2-to-Na_2O ratios. PEO processing was performed in the electrolytes containing 1 gr/L (17.9 mmol/L) KOH and various amounts of the silicates as specified in Table 1. Conductivities of the electrolytes were at least 4-5 mS/m and all the electrolytes had pH = 11-13 (see Table 1).

$Na_2O{\bullet}nSiO_2$	5 g/L	10 g/L	15 g/L
$Na_2O{\cdot}SiO_2$ (n = 1)	12.68 / 10.27	12.74 / 15.5	12.80 / 22.7
molarity	0.021	0.041	0.062
$Na_2O{\cdot}3SiO_2$ (n = 3)	11.08 / 4.53	11.18 / 5.47	11.24 / 6.52
molarity	0.024	0.047	0.071

Table 1. Typical electrolyte parameters (pH / Conductivity, mS/m)

As seen from Table 1, both the basicity and the conductivity are strongly affected by the silicate index, which is not surprising because the molar fraction of sodium oxide is 0.5 for Na_2O SiO_2 and only 0.25 for Na_2O $3SiO_2$. The values of pH of the electrolytes only weakly depend on the concentration of a given silicate, while their conductivities are roughly proportional to the concentration of Na_2O SiO_2 or Na_2O $3SiO_2$. As one could expect, better conductivities of the "n=1 electrolytes" must facilitate the PEO process.

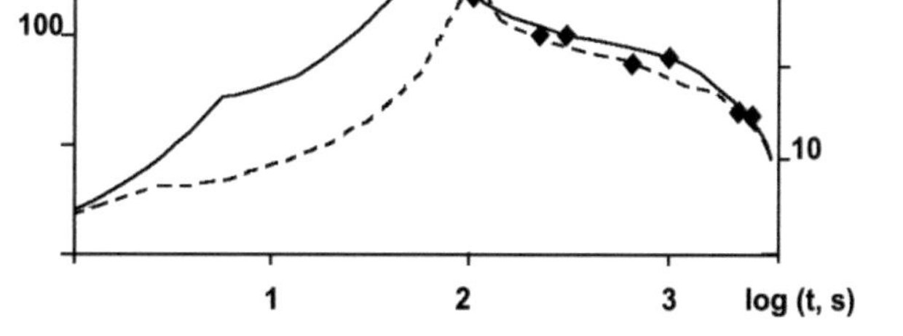

Figure 3. Typical voltage (U) and current density (j) amplitudes in the PEO process of an aluminium alloy in n=1 and n=3 electrolytes.

Indeed, lower current densities are needed for the plasma process initiation when $n = 1$ (Fig. 3). As seen from Fig. 3, not only the initial current densities, but also the process voltages are higher for "n = 3 electrolytes". Visual changes both in voltage and in current density are observed after 100 - 200 seconds. These changes can indicate that a steady state has been achieved, when only significantly fewer discharges occur and the oxide layer has mainly been formed.

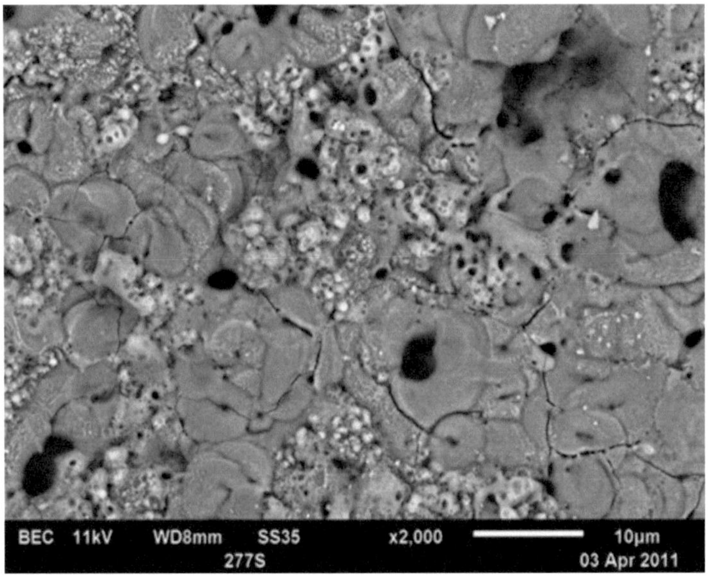

Figure 4. A SEM image (Secondary electron detector) of the morphology of an oxide layer produced by PEO of an aluminium alloy in an alkaline silicate electrolyte.

Oxide layers produced after 30-60 min of PEO have porous morphology with blind "crater-like" pores, which are the results of plasma discharges through the oxide (Fig. 4). No difference in morphology was observed for the two types of electrolytes.

The morphology and elemental composition of a pore obtained by the EDS are presented in Fig. 5 and Table 2. As follows from the data of the elemental analysis, the interior of a pore contains much less silicon and much more aluminum than the exterior close to the surface. This is not surprising, because aluminum comes from inside (from the metal substrate), while silicon is provided by the electrolyte and only with difficulties can penetrate to the depths of the oxide layer.

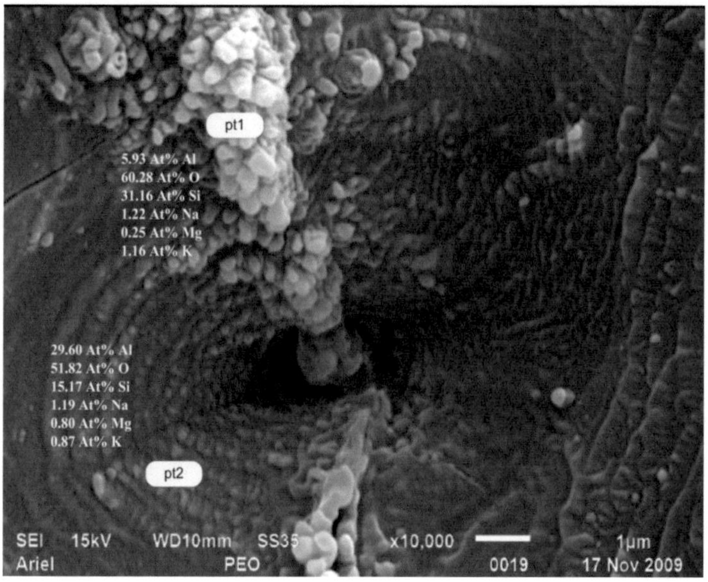

Figure 5. Morphology and elemental composition of a pore at two characteristic points: point 1 (pt1) on the surface of the specimen; point 2 (pt2) inside the pore.

At.%	Al	Si	O	Na	Mg	K
Point 1	5.9	31.2	60.3	1.2	0.3	1.2
Point 2	29.6	15.2	51.8	1.2	0.8	0.9

Table 2. Element composition of the oxide coating on the surface (point 1) and inside a pore (point 2)

The oxide layers, formed after 30 minutes of PEO, are 20-60 μm thick and consist of two clearly pronounced sublayers: a denser inner sublayer and a loose porous outer sublayer (Fig. 6).

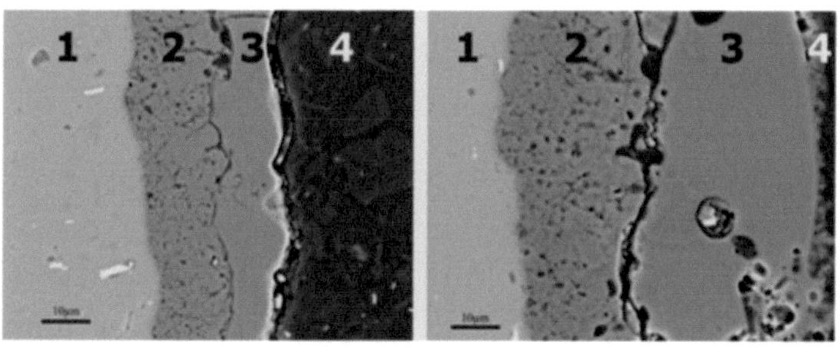

Figure 6. The structure of oxide layers on cross-sections of specimens obtained by PEO in (left image) 0.05 mol/L n=1 and (right image) n=3 electrolytes: (1) non-oxidized base alloy, (2) inner denser oxidized sublayer, (3) outer loose sub-layer, (4) resin wrapping. Back-Scattered Electron SEM image, x1,000.

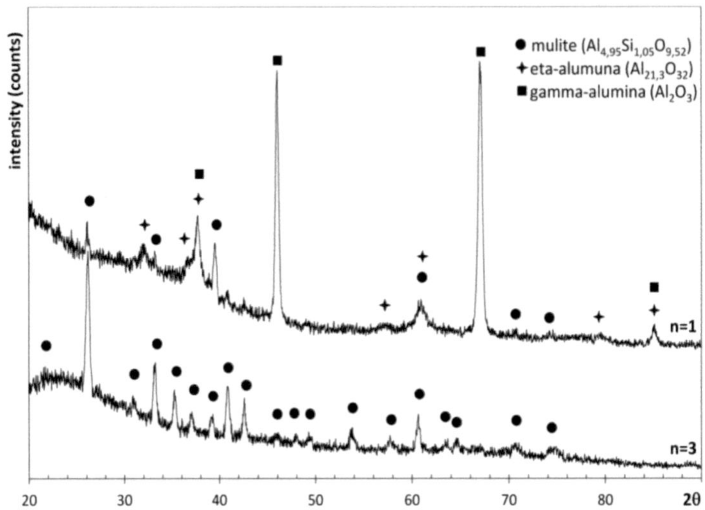

Figure 7. XRD patterns for oxide layers on specimens obtained by PEO in 0.05 mol/L n=1 and n=3 electrolytes. Minor phases are not shown.

While the outer sublayer contains fewer pores, the pores in it are much larger than in the inner sublayer. The mean density of the outer sublayer is lower than of the inner sublayer. Both sublayers are considerably thicker for the "n=3 electrolytes", but they contain 2-5 times as much silicon as for the "n=1 electrolytes". XRD phase analysis (Fig. 7) shows that the oxide layer contains 60-70% of γ-alumina, 20-30% of η-alumina and about 1% of quartz for the "n=1 electrolytes." For "n=3 electrolytes" the oxide layer consists mainly of mullite $3Al_2O_3 2SiO_2$ and varied (for different concentrations of the n=3 silicate in the electrolyte) amounts of amorphous silica, quartz and various types of alumina.

Silicate concentration in electrolyte, mol·L^{-1}	Microhardness, Vickers (HV)			
	n=1 electrolyte, outer sublayer	n=1 electrolyte, inner sublayer	n=3 electrolyte, outer sublayer	n=3 electrolyte, inner sublayer
0.019	840	1100	770	1060
0.025	1130	1380	1280	1570
0.050	890	1050	710	980
0.075	920	1630	700	910
Mean hardness	945	1290	865	1130

Table 3. Microhardness of oxidized sublayers produced by PEO in electrolytes containing different concentrations of "n=1" and "n=3" silicates measured on cross-sectioned specimens perpendicularly to the section planes.)

Results of hardness measurements of oxide layers are presented in Table 3. Obviously, alumina-quartz layer formed in the "n=1 electrolytes" is harder than mullite layer formed in the "n=3 electrolytes."

Corrosion tests were made after a specimen was masked by resin except for a square window having the area of 1 cm^2 on the oxidized surface. Thus prepared specimen was held for 1 hour in 1% NaCl for the achievement of steady-state corrosion and then its voltammetric curve was measured using Linear Sweep Voltammetry (25 mV/sec). Broader potential range (normally, OCP ± 500 mV) was first studied for the determination of the corrosion potential. Narrower potential range of ± 50-70 mV relatively to the previously roughly determined corrosion potential was then measured and used for Tafel slope analysis. All potentials were measured against the Ag|AgCl reference electrode and then recalculated to the standard hydrogen electrode potentials.

The results of thus measured corrosion characteristics of "bare" Al5052 alloy and different PEO oxidized specimens are given in Table 4. As follows from Table 4, corrosion current densities measured on oxidized samples are at least 3-4 times lower than for the untreated alloy. Corrosion potentials for all the oxidized samples are considerably more positive than for the untreated alloy, which evidences the increase of anodic stability in the test solution. The most noble corrosion potentials are observed for lower concentrations of both n=1 and n=3 silicates and correlate with higher microhardness of oxide layers (Table 3) observed for these concentrations. We could carefully assume that the content of γ-alumina in an oxide layer plays the key role in the shift of corrosion potentials to the positive direction.

Silicate concentration in electrolyte, mol·L^{-1}	n=1 electrolyte, E_{corr}, V vs. SHE	n=1 electrolyte, i_{corr}, A/cm^2 x 10^6	n=3 electrolyte, E_{corr}, V vs. SHE	n=3 electrolyte, i_{corr}, A/cm^2 x 10^6
"bare" Al5052	-1.126	15.99		
0.013			-0.525	0.08
0.025	-0.497	3.60	-0.815	2.66
0.050	-0.796	4.30	-0.998	0.98
0.075			-0.972	1.68
0.100	-0.942	1.93		
0.150	-0.995	3.77		

Table 4. Corrosion current densities and potentials of Al5052 alloy oxidized in different electrolytes.

The results summarized in Table 4 are better than those obtained for anodizing [14], similar to those obtained for much more expensive protection methods and similar or batter than those obtained by PEO in other silicate electrolytes [6, 15-17]. All the measurements evidence that specimens treated in "n=3 electrolytes" have better corrosion protection than those treated in "n=1 electrolytes." The microscopic inspection of cross-sections evidences (Fig. 4) that even though the outer sublayer produced in "n=3 electrolytes" contains large caverns and the inner sublayer looks more porosive than for the "n=1 electrolytes", the larger thickness of the layer produced in the "n=3 electrolytes" presents a more difficult barrier for the diffusion of corrosive media and therefore forms better protection of the metal substrate. It deserves to be reminded that the oxide layers produced in the "n=1 electrolytes" only contain oxide phases (alumina and quartz), in contrast to the mullite layer produced in the "n=3 electrolytes." Obviously, the milder mullite better fills pores in the oxide layer than harder oxides do. Somewhat similar results were obtained by another research group [7] for another Al alloy (2219) and also demonstrated that higher silicate contents in silicate-alkaline electrolytes increase the corrosion resistance of PEO coatings.

3.2. Fluoride influence on the properties of oxide layer produced by PEO

The surface of PEO layers produced on both the magnesium and the aluminum alloys is normal for PEO coatings and consists of "volcanic" pores chaotically distributed on a fused surface (Fig. 4).

Long PEO processing times (30-90 min) result in thick coatings for both alloys, but the initial stages demonstrate a clear difference between the two metal alloys. For the PEO treatment in the alkaline silicate electrolyte (0.08 mol /L KOH + 0.08 mol / L Na_2SiO_3) without the fluoride addition, the 15 minute process produces approximately 20 μm thick non-continuous oxide layer on the magnesium alloy and only about 5 μm thick non-continuous layer on the aluminum alloy (Fig. 8).

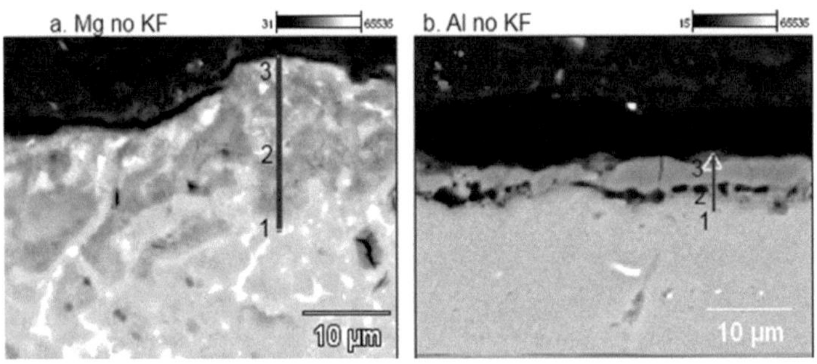

Figure 8. SEM images (x3000) with EDS linear scans of cross-sectioned PEO layers on (a) the magnesium and (b) aluminum alloys obtained after 15 minute oxidation. The black zone in the topmost parts is polymer tar fixing the specimen. The base metals are in the bottom part of each image. The element composition at points 1, 2, 3 of each image is specified in Table 5.

Both on the magnesium and on the aluminum alloys the coating is porous, but the porous are blind. Element compositions at points 1, 2 and 3 along lines drawn from the pure base metal to the outer border of the coatings (Fig. 8) are given in Table 5.

Mg alloy		Al alloy	
Point 1	Mg 82 at%, Al 8 at%, O 9 at%, Si 1 at%	Point 1	Al 94 at%, Mg 2 at%, O 4 at%
Point 2	Mg 67 at%, Al 13 at%, O 17 at%, Si 3 at%	Point 2	Al 61 at%, Mg 3.5 at%, O 34 at%, Si 1.5 at%
Point 3	Mg 60 at%, Al 9 at%, O 27 at%, Si 4 at%	Point 3	Al 70 at%, Mg 2.5 at%, O 15 at%, Si 9.5 at%, K 3 at%

Table 5. Element compositions at representative points 1, 2 and 3 as shown in Fig. 8.

As follows from Fig. 8 and Table 5, the oxidized layer on the magnesium alloy is not only thicker, but also more uniform than that on the aluminum alloy, for which large voids containing relatively high amounts of oxygen are formed between the base metal and the oxide layer. It deserves to be noted that the percentage of oxygen is lower than what should be expected according to the stoichiometry of magnesium and aluminum oxides (Mg : O = 1:1 and Al : O = 2 : 3). Therefore, the oxide layer never consists of the oxides only, but contains some excess metal atoms.

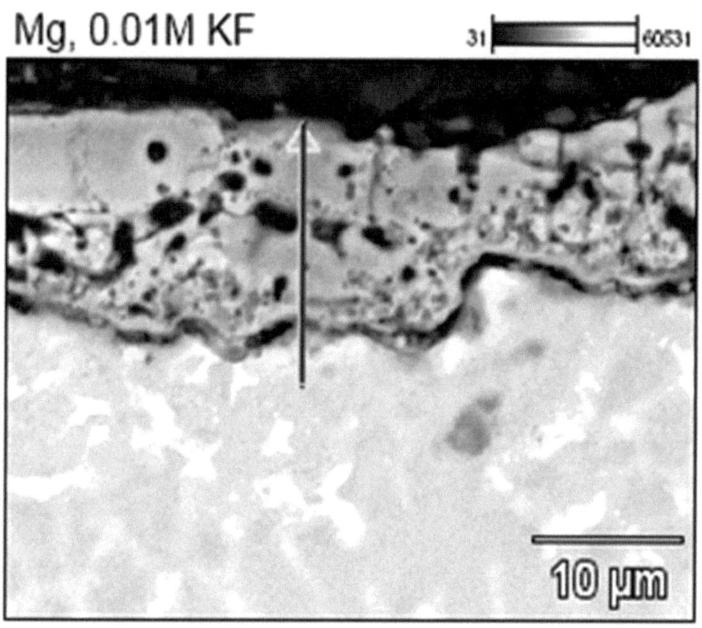

Figure 9. BSE SEM images (x3000) with EDS linear scans of cross-sectioned PEO coating on the magnesium alloy obtained after 15 minute oxidation in the electrolyte containing 0.01 mol/ L KF.

As KF has been added to the electrolytes, the situation with the oxidation of magnesium alloy becomes different. While the total thickness of the coating remains relatively the same (~ 20 µm after 15 minutes), its structure and composition are clearly distinguished from what was observed without the fluoride. Already for the lowest studied KF concentration (0.01 mol / L) the coating is continuous and consists of two very different sublayers (Fig. 9). The inner sublayer is only 2-3 μm thick and contains about 70 at% of Mg, 25 at% of O and 5 at% of Al. The outer sublayer is 15-18 μm thick, porous and contains large voids filled by light elements (darker sites in Fig. 9). The typical composition of the outer layer is 50-60 at% O, 35-40 at% Mg, 10 at% Si and 2-4 at% Al. As the content of KF in the electrolyte increases, the structure of the layers does not change, but significant amounts of fluorine are detected in the coating.

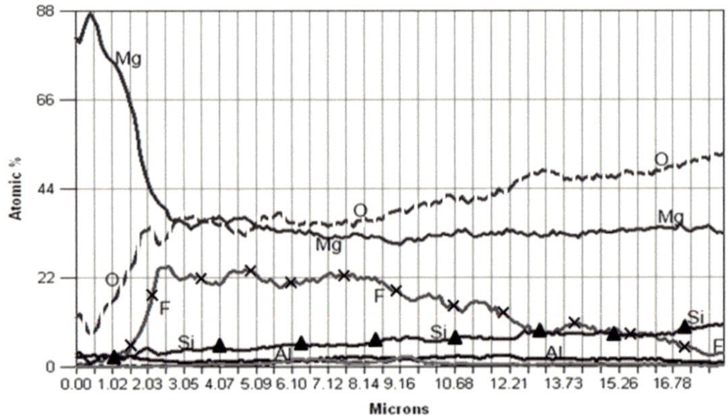

Figure 10. Element composition profiles (EDS linear scan) across the PEO coating on the magnesium alloy obtained after 15 minute oxidation in the electrolyte containing 0.2 mol/ L KF.

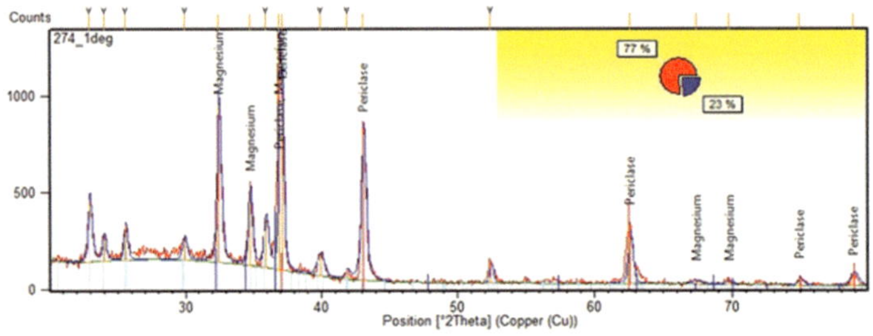

Figure 11. Small angle (1°) XRD pattern for the surface of the PEO coating on the magnesium alloy obtained after 15 minute oxidation in the electrolyte containing 0.02 mol/ L KF.

Interestingly, the maximal amount of fluoride was detected in the most inward part of the outer sub-layer (Fig. 10). According to small angles (1º-5º) XRD measurements, the surface consists of Periclase MgO (77%) and metal Mg (23%). Deeper layers of the coating demonstrate the increase of Mg at the expense of MgO. No fluorine-containing phases could be identified with confidence (Fig. 11).

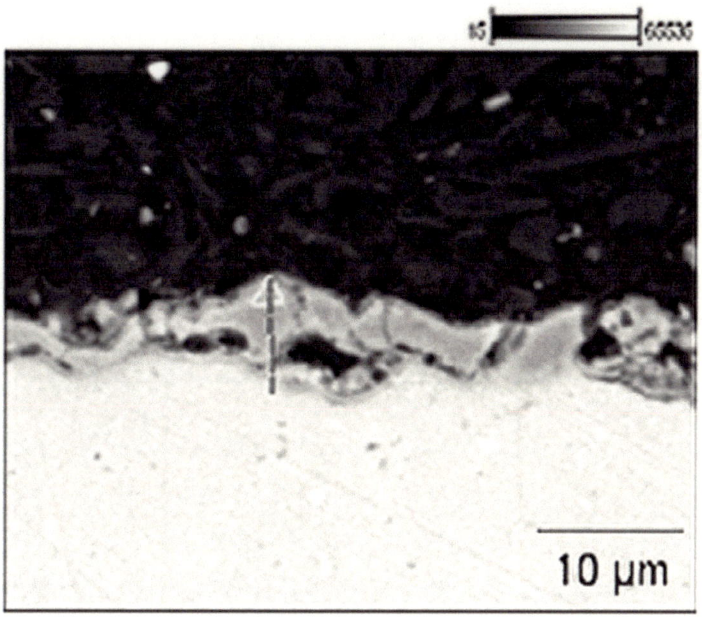

Figure 12. BSE SEM images (x3000) with EDS linear scans of cross-sectioned PEO coating on the aluminum alloy obtained after 15 minute oxidation in the electrolyte containing 0.2 mol/ L KF.

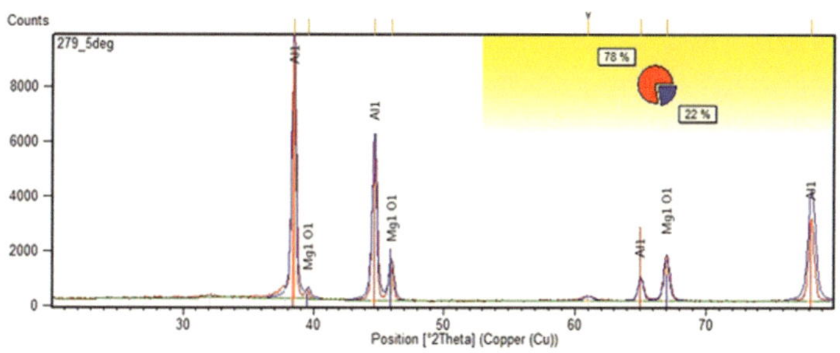

Figure 13. Small angle (5º) XRD pattern for the surface of the PEO coating on the aluminum alloy obtained after 15 minute oxidation in the electrolyte containing 0.05 mol/ L KF.

The addition of KF to the electrolytes for the PEO processing of the aluminum alloy has another effect. While for lower concentrations of KF (<0.05 mol / L) only a very thin porous coating layer is formed, higher KF contents result in the formation of 5-10 μm oxide layer. As the thickness of the coating increases, two sublayers are revealed. As for the magnesium alloy, the thin inner sublayer is denser and the thicker outer one is more porous (Fig. 12). As evidenced by EDS, the outer sublayer contains 60 at% of O and 40 at% of Al. As much as 9% fluorine is found in the inward part of the outer sublayer. XRD measurements show that the surface (1º incident beam) consists of two crystal phases, whose composition is Al_2MgO_4 (56%) and MgO

(44%). Deeper layers (5º incident beam) are formed by metal Al (78%) and MgO (22%). A sample of an XRD pattern is given in Fig. 13.

The most obvious observation, which can be made from the comparison of the PEO of the magnesium and of the aluminum alloys, is that magnesium is oxidized much more easily than aluminum. While for all the studied magnesium systems 20 μm coating was produced after 15 minutes of the PEO, only 5-10 μm coating on the aluminum alloy could be obtained for the same process time. An interesting issue is the ratio "Mg:Al" at different depth of the coatings (Table 6).

	C_{KF}, mol / L	Base metal	Inner sublayer	Outer sublayer
	0	10.3	6.7*	
Mg alloy	0.01	10.3	14	8.2
	0.1	10.3	15	6.4
	0	0.02	0.06	0.04
Al alloy	0.01	0.02	0.04	0.06
	0.1	0.02	0.04	0.05

*No sublayers were observed for this PEO layer.

Table 6. "at% Mg : at% Al" ratio at different depths.

As follows from Table 6, magnesium content at the inner sublayer is always higher than in the base metal, except for the oxidation of the magnesium alloy in the electrolyte containing no fluoride, when no sublayers can be seen. The trend is less straightforward for the outer sublayer, which can be explained by the fact that the latter is thicker, more porous and much less uniform. XRD phase analysis confirms than the key role at the initial stage of the oxidation is played by magnesium oxide and only on the surface aluminum starts to be oxidized to form Al_2MgO_4. These can be explained by two facts: (1) that the amphoteric aluminum is readily dissolved by the alkaline electrolyte while magnesium is not; (2) that according to Ellingham diagrams [33] the oxidation of magnesium is thermodynamically more favorable than that for aluminum in the entire range of temperatures below 1500K.

Many authors report that they could not identify a fluoride containing phase in PEO coatings on aluminum or magnesium alloys obtained in alkaline fluoride-containing electrolytes (see, for instance [34-37]). This is consistent with our XRD observation; however we could clearly see the presence of fluoride on EDS cross-sectional profiles of the coatings (Fig. 4). This means that considerable amounts of amorphous fluorides are found in the coatings very close to the base metal. Summarizing, the action of fluoride additives can be assumed as follows: (1) fluoride anions are first chemisorbed to the metal surface and create on it barrier layer; (2) as the dielectric breakdown occurs, oxide layer is formed due to the exchange of fluorine by oxygen atoms from the electrolyte; (3) fluoride remains in the vicinity of the metal and takes part in the formation of an amorphous phase. This process is much more favorable for the magnesium alloy due to the easier breakdown of magnesium fluoride as compared to aluminum fluoride.

4. Conclusions

Plasma Electrolytic Oxidation in alkaline silicate electrolytes containing 0.013-0.150 mol/L of sodium silicates having silicate index n=1 or n=3, was performed on Al5052 aluminum alloy. For all the electrolytes studied, 20-90 μm thick oxide layer was obtained and its composition, structure and properties were studied. For each sample, the oxidized layer consists of a denser inner and looser outer sublayer. While for "n=1 electrolytes" the oxidized layer is mainly formed by several kinds of alumina, the principal constituent of the oxidized layer for "n=3 electrolytes" is mullite.

Measurements of microhardness evidenced that it is apparently not influenced by the kind of silicate (n=1 or n=3) and by its concentration in the electrolyte.

Electrolytes with silicate index n=3 ensure better corrosion protection than those with n=1. This might be caused by the milder and more plastic nature of the oxide layer produced in the "n=3 electrolutes" as compared to those produced in the "n=1 electrolytes."

Corrosion protection parameters are significantly better for all PEO oxidized samples than for the untreated Al5052 alloy.

The formation of PEO coating on magnesium and aluminum alloys in the presence of fluoride starts with the fluorination of the metal surface and formation of a dielectric metal fluoride layer. Electric breakdowns destruct this layer and form oxide layers containing also amorphous fluorides.

The fluoride-supported PEO process proceeds more easily for the magnesium than for the aluminum alloys. This difference might be caused by the easier breakdown of the dielectric layer containing magnesium fluoride as compared to that containing aluminum fluoride. This is consistent with the values of dielectric permeability of magnesium fluoride (4.87) [38] and of aluminum fluoride (2.2) [39].

The external surface of the coating is enriched by magnesium as compared to the base metal.

Author details

Alex Lugovskoy* and Michael Zinigrad

*Address all correspondence to: lugovsa@ariel.ac.il

Chemical Engineering Department, Ariel University Center of Samaria, Ariel, Israel

References

[1] Hussein, R. O, Nie, X, Northwood, D. O, Yerokhin, A, & Matthews, A. Spectroscopic study of electrolytic plasma and discharging behaviour during the plasma electrolytic oxidation (PEO) process. J. Phys. D: Appl. Phys. (2010).

[2] Curran, J. A, & Clyne, T. W. Porosity in plasma electrolytic oxide coatings. Acta Materialia (2006)., 54-1985.

[3] Sah, S. P, Tsuji, E, Aoki, Y, & Habazaki, H. Cathodic pulse breakdown of anodic films on aluminium in alkaline silicate electrolyte- Understanding the role of cathodic half-cycle in AC plasma electrolytic oxidation. Corrosion Science (2012)., 55-90.

[4] Barik, R. C, Wharton, J. A, Wood, R. J. K, Stokes, K. R, & Jones, R. L. Corrosion, erosion and erosion-corrosion performance of plasma electrolytic oxidation (PEO) deposited Al_2O_3 coatings. Surface & Coatings Technology (2005)., 199-158.

[5] Xue, W, Shi, X, Hua, M, & Li, Y. Preparation of anti-corrosion films by microarc oxidation on an Al-Si alloy. Applied Surface Science (2007)., 253-6118.

[6] Nie, X, Meletis, E. I, Jiang, J. C, Leyland, A, Yerokhin, A. L, & Matthews, A. Abrasive weary corrosion properties and TEM analysis of Al_2O_3 coatings fabricated using plasma electrolysis. Surface and Coatings Technology (2002)., 149-245.

[7] Alsrayheen, E, Campbell, B, Mcleod, E, Rateick, R, & Birss, V. Exploring the effect of alkaline silicate solution composition on the ac/dc spark anodization of Al-Cu Alloys. Electrochimica Acta (2012)., 60-102.

[8] Venugopal, A, Panda, R, Manwatkar, S, & Sreekumar, K. Rama Krishna L., Sundararajan G. Effect of micro arc oxidation treatment on localized corrosion behaviour of AA7075 aluminum alloy in 3.5% NaCl solution. Trans. Nonferrous Met. Soc. China (2012).

[9] Guo, H, & An, M. Xu Sh., Huo H., Microarc oxidation of corrosion resistant ceramic coating on a magnesium alloy. Materials Letters (2006)., 60-1538.

[10] Chen, F, Zhou, H, & Yao, B. Qin Zh., Zhang Q. Corrosion resistance property of the ceramic coating obtained through microarc oxidation on the AZ31 magnesium alloy surfaces. Surface & Coatings Technology (2007)., 201-4905.

[11] Yerokhin, A. L, Nie, X, Leyland, A, Matthews, A, & Dowey, S. J. Plasma electrolysis for surface engineering. Surface and Coatings Technology (1999)., 122-73.

[12] Moon, S, & Jeong, Y. Generation mechanism of microdischarges during plasma electrolytic oxidation of Al in aqueous solutions, Corrosion Science (2009)., 51(2009), 1506-1512.

[13] Polat, A, Makaracib, M, & Ustac, M. Influence of sodium silicate concentration on structural and tribological properties of microarc oxidation coatings on 2017A aluminum alloy substrate. Journal of Alloys and Compounds (2010)., 504-519.

[14] Zabielski, C. V, & Levy, M. Study of Type II and Type III Anodized Al in Aqueous DS2 Solutions. U.S. Army Research Laboratory Environmental Effects. In: Proceedings of the TRI-Service conference on corrosion. Plymouth, Massachusetts 12-14 May (1992). Avavilable on http://namis.alionscience.com/conf/tscc/search/pdf/AM026095.pdf., 5052-0.

[15] Tseng ChCh., Lee J.-L., Kuo Tz.-H., Kuo Sh.-N., Tseng K.-H. The influence of sodium tungstate concentration and anodizing conditions on microarc oxidation (MAO) coatings for aluminum alloy. Surface & Coatings Technology (2012)., 206-3437.

[16] Raj, V. Mubarak Ali M. Formation of ceramic alumina nanocomposite coatings on aluminium for enhanced corrosion resistance. J. Mat. Proc. Technology (2009). , 209-5341.

[17] Wei, C. B, Tian, X. B, Yang, S. Q, Wang, X. B, Fu, R. K. Y, & Chu, P. K. Anode current effects in plasma electrolytic oxidation. Surface & Coatings Technology (2007). , 201-5021.

[18] Li, Y, Yao, B, Long, B. Y, Tian, H. W, & Wang, B. Preparation, characterization and mechanical properties of microarc oxidation coating formed on titanium in $Al(OH)_3$ colloidal solution. Applied Surface Science (2012)., 258-5238.

[19] Wang, Y. M, Jiang, B. L, Lei, T. Q, & Guo, L. X. Microarc oxidation coatings formed on Ti_6Al_4V in Na_2SiO_3 system solution: Microstructure, mechanical and tribological properties. Surface & Coatings Technology (2006)., 201-82.

[20] Pauporté, T, Finne, J, Kahn-harari, A, & Lincot, D. Growth by plasma electrolysis of zirconium oxide films in the micrometer range. Surface & Coatings Technology (2005)., 199-213.

[21] Cheng, Y, Matykina, E, Arrabal, R, Skeldon, P, & Thompson, G. E. Plasma electrolytic oxidation and corrosion protection of Zircaloy-4. Surface & Coatings Technology (2012)., 206-3230.

[22] Wang, Y. Jiang Zh., Yao Zh. Preparation and properties of ceramic coating on Q235 carbon steel by plasma electrolytic oxidation. Current Applied Physics (2009)., 9-1067.

[23] Xin ShG., Song L.-X., Zhao R.-G., Hu X.-F. Properties of aluminium oxide coating on aluminium alloy produced by micro-arc oxidation. Surface & Coatings Technology (2005)., 199-184.

[24] Shen, D, & Wang, J. Y.-L., Nash Ph., Xing G.-Zh. Microstructure, temperature estimation and thermal shock resistance of PEO ceramic coatings on aluminium. J. materials processing technology (2008)., 205-477.

[25] Snizhko, L. O, Yerokhin, A. L, Pilkington, A, Gurevina, N. L, Misnyankin, D. O, Leyland, A, & Matthews, A. Anodic processes in plasma electrolytic oxidation of aluminium in alkaline solutions. Electrochimica Acta (2004)., 49-2085.

[26] Timoshenko, A. V. Magurova Yu.V. Investigation of plasma electrolytic oxidation processes of magnesium alloy MAunder pulse polarisation modes. Surface & Coatings Technology (2005). , 2-1.

[27] Boinet, M, Verdier, S, Maximovitch, S, & Dalard, F. Plasma electrolytic oxidation of AM60 magnesium alloy: Monitoring by acoustic emission technique. Electrochemical properties of coatings. Surface & Coatings Technology (2005). , 199-141.

[28] Hsiao, H, Tsung, Y, Ch, H, Tsai, W, Anodization, T, & Of, A. Z. D magnesium alloy in silicate-containing electrolytes. Surface & Coatings Technology (2005). , 199-127.

[29] Duan, H. Yan Ch., Wang F. Effect of electrolyte additives on performance of plasma electrolytic oxidation films formed on magnesium alloy AZ91D. Electrochimica Acta (2007). , 52-3785.

[30] Liang, J, Guo, B, Tian, J, Liu, H, Zhou, J, & Xu, T. Effect of potassium fluoride in electrolytic solution on the structure and properties of microarc oxidation coatings on magnesium alloy. Applied Surface Science (2005). , 252-345.

[31] Yerokhin, A. L, Nie, X, Leyland, A, & Matthews, A. Characterisation of oxide films produced by plasma electrolytic oxidation of a Ti-6Al-4V alloy. Surface & Coatings Technology (2000). , 130-195.

[32] Vander Voort GF., editor. ASM Handbook, and Microstructures. (2004). ASM International., 09-Metallography

[33] MIT educational materialshttp://web.mit.edu/2.813/www/readings/Ellingham_diagrams.pdfaccessed on August 31, (2012).

[34] Wang ZhWu L., Cai W., Shan A, Jiang Zh. J. Effects of fluoride on the structure and properties of microarc oxidation coating on aluminium alloy. Alloys and Compounds (2010). , 505-188.

[35] Wang, K, Koo, B, Lee, H, Ch, G, Kim, Y, Lee, J, & Byon, S. -H. E. Effects of electrolytes variation on formation of oxide layers of 6061 Al alloys by plasma electrolytic oxidation. Trans. Nonferrous Met. Soc. China (2009). , 19-866.

[36] Gnedenkov, S. V, Khrisanfova, O. A, Zavidnaya, A. G, Sinebrukhov, S. L, Gordienko, P. S, Iwatsubo, S, & Matsui, A. Composition and adhesion of protective coatings on aluminium. Surface and Coatings Technology (2001). , 145-146.

[37] Lin, C. S, & Fu, Y. C. J. Characterization of anodic films of AZ31 magnesium alloys in alkaline solutions containing fluoride and phosphate anions. J. Electrochem. Soc. (2006). BB424., 417.

[38] Duncanson, A, & Stevenson, R. W. H. Some Properties of Magnesium Fluoride crystallized from the Melt. Proc.Phys.Soc. (1958). , 72, 1001.

[39] Table of Dielectric ConstantsASI Instruments Web Site. http://wwwasiinstr.com/technical/Dielectric%20Constants.htmaccessed on November 3, (2012).

Chapter 5

Physicochemical Analysis and Synthesis of Nonstoichiometric Solids

V. P. Zlomanov, A.M. Khoviv and A.Ju. Zavrazhnov

Additional information is available at the end of the chapter

http://dx.doi.org/10.5772/54815

1. Introduction

Technological progress is intimately associated with creation of new materials, such as composites, piezoelectrics, ferroelectrics, semiconductors, superconductors and nanomaterials with preset functional properties. For the synthesis of these materials, it is necessary to study the chemical processes that lead to the composition, structure, and accordingly desired properties. Investigation of the interrelation between the composition, structure and properties of matter and determination of synthetic conditions for obtaining substances with preset composition and structure are the major problems of physicochemical analysis. Although significant progress has been made in understanding several challenges remain for further advancements. These challenges and new approaches include some definitions – stoichiometry, nonstoichiometry, deviation from stoichiometry, substance, phase, component as well as the use of phase diagrams in selecting conditions for the synthesis of nonstoichiometric solids. Since nonstoichiometry is associated with defects, attention is also paid to their classification and formation. Synthesys of solid involves control over phase transformations. For this reason some features of the *P–T–x* phase diagrams are discussed. The following inportant features of *P–T–x* phase diagrams will be considered: highest maximal melting point $T_{m,AB}^{\max}$ of the solid compound S_{AB}, noncoincidence of the solid, liquid and vapor compositions ($x^L \neq x^S \neq x^V$) at this temperature, factors determining the value of the homogeneity range, some features of the terms congruent and incongruent phases and phase processes. Criteria for evaluating the homogeneity of nonstoichiometric solids are also considered.

The term *Physicochemical analysis* was introduced in [1] and defined as the field of chemistry dealing with the interrelation between the composition and properties of matter [1]. The foundations of physicochemical analysis were in [1-6].

It has been realized that it is necessary to study the thermodynamic properties of solids and the phase diagrams of the systems in which these solids occur. Because the properties of solids depend significantly on their composition, great attention has been focused on the physicochemical analysis and foundations of the directed synthesis.

2. The essence of physicochemical analysis — Some definitions

Investigation of the interrelation between the composition, structure, and properties as well as determination of synthetic conditions for obtaining solids with preset composition and structure are the the basic problems of physicochemical analysis.

The subject of chemistry is the conversions of substances. What is a substance, and what is its conversion? A substance is a multitude of interacting particles possessing certain characteristics: composition, particle size, structure, and the nature of chemical bonding. It is these characteristics that determine the properties of the substance (Figure 1).

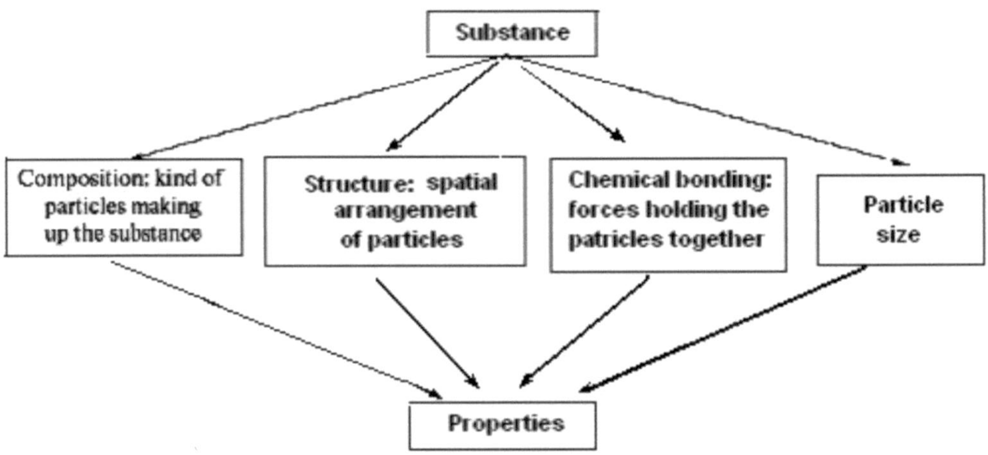

Figure 1. Interrelation between the basic characteristics and properties of a substance.

Composition is defined as the kinds of particles constituting the substance. For example, a sodium chloride crystal is built of sodium and chloride ions occupying cationic sites (Na^{+}_{Na}) and anionic sites (Cl^{+}_{Na}), respectively. The constituent particles can be not only atoms or ions, but also molecules (e.g., I_2 molecules in iodine crystals and water molecules in ice), coordination polyhedra (e.g., SO_4^{2-} tetrahedra in potassium sulfate), and other entities.

Structure is some ordered arrangement of the above particles in space.

The properties of a crystal, such as the lattice energy and electrical, optical, and chemical properties, are determined by the composition and structure of the crystal. Different spatial arrangements of the same particles, such as carbon atoms in diamond and graphite, are characterized by different lattice energies and, accordingly, different properties, including melting and boiling points and hardness.

The chemical bond is understood as the forces binding the particles together. These forces arise from the the Coulomb interaction of electrons and nuclei. Depending on the electron distribution among nuclei, there can be ionic, covalent, and metallic bonds.

The particle interaction energy depends on the particle size. At the nanometer level (1–100 nm), it changes markedly, making possible the formation of new physical and chemical properties of the substance. This is explicable in terms of surface physics and chemistry (dependence of the surface energy on the particle size).

A conversion of a substance is a change in one or several characteristics of the substance (Figure1). This process is accompanied by a change in energy (dU) in the form of heat transfer (δQ) or work execution (δA):

$$dU = \delta Q - \delta A. \tag{1}$$

The main purpose of directed synthesis is to obtain substances with the preset composition, structure, and, hence, properties.

The direct synthesis of solids includes control of phase transitions. Therefore, thermodynamic data characterizing phases and their transitions are necessary to estimate the optimum synthetic conditions.

Now let us consider some specific features of the concepts of a phase and a component.

2.1. Phase and component

A substance is made up of particles or their interacting sets with a certain structure and chemical bonding. Energy is an equivalent or measure of these interactions. In thermodynamics, the state of a system is defined using a set of variables, or coordinates, such as pressure P, temperature T, and volume V. For a thermodynamic state of a system characterized by a set of coordinates (intensive thermodynamic properties), Gibbs suggested the term *phase of matter* [7, 8]. This definition emphasizes that the phase of matter is size- and shape-independent. Later, the word combination *phase of matter* gave way to the term "*phase.*"

The equation of state of a phase in terms of pressure (P), temperature (T), and composition (x) is written as (2)

$$dG = -SdT + VdP + \sum_{i=1}^{k} \left(\frac{\partial G}{\partial x_i} \right)_{P,T,x_j} dx_i, \tag{2}$$

where G, S, and V are the molar Gibbs energy, molar entropy, and molar volume of the phase, espectively; x_i is the mole fraction of the i-th component k; x_j means the constancy of the mole fractions of all components but not the i-th component. The phases can be individual solid, liquid, gaseous compounds, or solid solutions, including nonstoichiometric compounds, whose thermodynamic properties are described by equations of state like Eq. (2) and are continuous functions of P, T, and x.

Note the following specific features of the concept of a phase. Firstly, existing phases should be distinguished from coexisting phases. The properties (Gibbs energy G, enthalpy, etc.) of an existing phase, such as a solution containing different concentrations of the same substance, are continuous functions of composition. By contrast, the properties of coexisting phases (i.e., phases that are in equilibrium) are equal in all parts of the system. The second specific feature is associated with the definition of a phase as the homogeneous part of a heterogeneous system. This definition is inexact because it does not specify whether homogeneous parts of a heterogeneous system are one phase or different phases. Thirdly, the concept of a phase is broader than the concept of a physical state (gas, solid or liquid). Finally, the fourth specific feature of the concept of a phase is associated with the issue of the minimum set of particles describable in terms of one equation of state like Eq. (2). From the standpoint of kinetic molecular theory, this set should be sufficiently large to allow regular particle energy distribution. Estimates demonstrate that the number of particles must be at least several tens. The properties of such a phase will be affected by surface tension [9].

The components of a system are the types of particles constituting this system. They are called constituents, and their number is designated n. If the concentrations of the n constituents are related by m independent equations, then $k = n - m$ is the number of independent constituents, or, simply, the number of components. For example, the system $NH_3(g) + HCl(g) = NH_4Cl(s)$, which consists of three substances ($n = 3$), whose concentrations are related by one equation ($m = 1$), is two-component since $k = n - m = 3 - 1 = 2$.

The components are subject to the following constraints:

1. their concentrations must be independent of one another;
2. they should completely describe the concentration dependence of the properties of the system;
3. their number should satisfy the electroneutrality principle.

2.2. Stoichiometry, nonstoichiometry, deviation from stoichiometry

The properties of a substance depend on its composition. The great focus of materials sience are the concepts of stoichiometry, nonstoichiometry, and deviation from stoichiometry.

The proportions in which substances react are governed by stoichiometric laws (stoichiometry). These laws, which characterize the composition of chemical compounds, were discovered by systematizing experimental data. The fundamental laws of stoichiometry include the law of constant composition and the law of multiple proportions.

The law of constant composition, established in the 19th century by the French chemist Joseph Louis Proust, states that the chemical composition of a compound is independent of the way in which this compound was obtained. The law of multiple proportions, formulated by the English chemist John Dalton in 1807, states that, when two elements combine with each other to form more than one compound, the mass fractions of the elements in these compounds are in a ratio of prime numbers. Both laws follow from atomistic theory and suggest that the saturation of the chemical bonds is necessary for the formation of a molecule from atoms. Any change in the number of atoms or their nature or arrangement indeed means the formation of a new molecule with new properties.

Are the law of constant composition and the law of multiple proportions always obeyed? They are valid only for the substances constituted by molecules. In fact, the composition of a substance can vary significantly, depending on the preparation conditions. It was long believed that only those chemical substances exist whose composition obeys the law of multiple proportions. They are stoichiometric and are called daltonides in honor of John Dalton. However, as methods of investigation were making progress, it turned out that the properties of solid inorganic substances, such as vapor pressure, electric conductivity, and diffusion coefficients, are composition dependence. In some composition range, the structure, i.e. the arrangement of the components in space remains invariable, while the component concentrations vary continuously. This range is called the homogeneity range or the nonstoichiometry range. These substances are referred to as nonstoichiometric or variable-composition compounds. Earlier, they were called berthollides in honor of Claude Louis Berthollet, Proust's compatriot. A nonstoichiometric compound can be treated as a solid solution of its components, such as cadmium and tellurium in the compound CdTe.

The homogeneity range is characterized by the corresponding deviation from stoichiometry. The stoichiometric composition of a solid compound, e.g., A*n*B*m*, where *n* and *m* are prime numbers, is the composition that obeys the law of multiple proportions. The deviation from stoichiometry or, briefly, nonstoichiometry (Δ) is defined as the difference between the ratio of the number of nonmetal atoms B to the number of metal atoms A in the real $A_nB_{m+\delta}$ crystal ($\delta \neq 0$) and the same ratio in the stoichiometric crystal A_nB_m [10]:

$$\Delta = \frac{m+\delta}{n} - \frac{m}{n} = \frac{\delta}{n}. \tag{3}$$

For the three-component system A–B–C, the composition of the solid phase $(A_{1-x}B_x)_{1-y}C_y$ is conveniently expressed in terms of the mole fraction of the binary compound (x) and nonstoichiometry (Δ).

In this case, the nonstoichiometry Δ can be viewed as the difference between the ratio of the equivalent numbers of nonmetal and metal atoms in the real crystal and the same ratio in the stoichiometric crystal. For example, for $(Pb_{1-x}Ge_x)_{1-y}Te_y$ crystals with a NaCl structure,

$$\Delta = y/(1-y) - 1/1 = (2y-1)/(1-y). \tag{4}$$

The mole fraction (molarity) of the binary compound determines the fundamental properties of nonstoichiometric crystals, including the band gap and heat capacity. The concentration of carriers—electrons and holes—and, accordingly, the galvanomagnetic and optical properties of nonstoichiometric crystals are also associated with nonstoichiometry.

3. Directed synthesis of nonstoichiometric solids

The strategy of directed synthesis of substances with the preset nonstoichiometry, structure, and properties is based on physicochemical analysis and includes the steps presented in Figure 2.

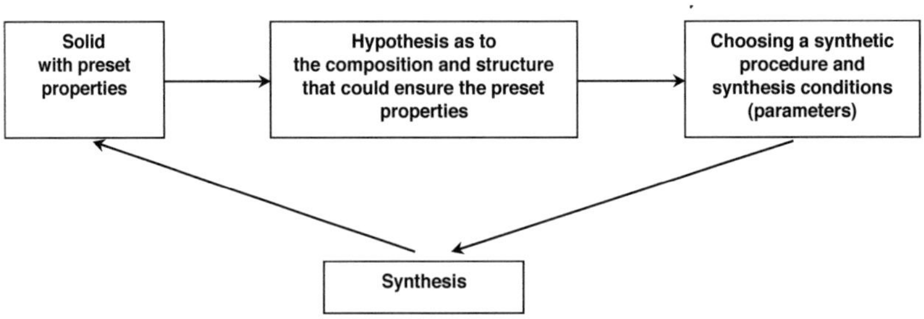

Figure 2. Steps of the directed synthesis of nonstoichiometric solids.

Synthesis is the consequence of processes involved in the conversion of the starting compounds into products. It includes selection and preparation of the starting chemicals (precursors), homogenization of the growth medium (melt, vapor, etc,), the nucleation of the desired phase, nucleus development (growth), and cooling (heating) of the product from the synthesis temperature to room temperature.

Each synthesis step depends on certain conditions or operating parameters. These include the chemical nature and composition of the growth medium, temperature, pressure, diffusion coefficients, heat and mass transfer, and the driving forces of chemical reactions (such as concentration, temperature, pressure, and chemical potential gradients). Figure 3 presents the most important operating parameters of the synthesis.

Provided that the analytical relationship between the rate of solid synthesis (growth) and the operating parameters is known, this process can be carried out under computer control. This problem has been solved for the Czochralski growth of silicon and germanium crystals and for the Bridgman growth of transition metal oxide crystals [10-14].

Solid synthesis involves phase transitions. For this reason, synthesis conditions are selected using phase diagrams (see section 4). The phase diagram of a system indicates the number of compounds forming in this system and the regions of their stability, specifically, the temperature, pressure, and composition intervals. Thus, using the phase diagram, it is possible to choose the medium (melt, vapor, etc.) for the synthesis of the desired solid, the synthesis conditions (temperature, pressure, and growth medium composition), and the way of carrying out the necessary conversions.

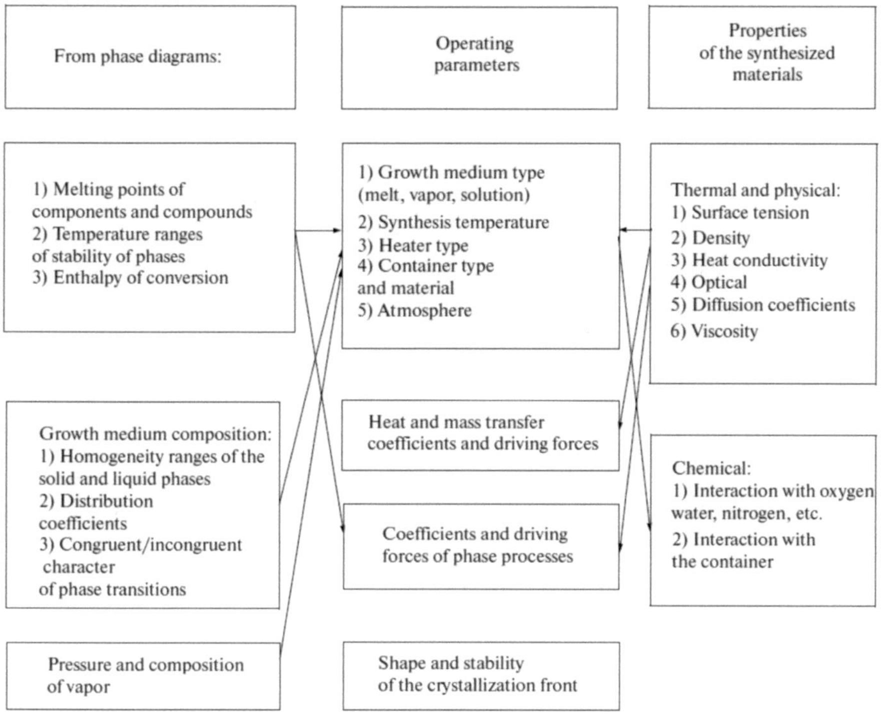

Figure 3. Operating parameters of solid synthesis.

The thermophysical and chemical properties of the starting substances, final synthesis products, and the growth medium to be used (melt, vapor, other solid phases) should be known along with thermodynamic data.

The conversion of a synthesis medium into a solid phase (crystallization) includes the nucleation and development of crystallization centers. Accordingly, it is related with heat and mass transfer and interfacial reaction kinetics. The general problem of analytically describing the crystallization process has not been solved. Furthermore, it is not always possible to evaluate the skill and equipment factors or to establish an unambiguous correlation between the properties of a nonstoichiometric phase and its molecular composition. In this sense, directed synthesis is regarded as being an art [13, 14].

4. Phase diagrams as a key to selecting conditions for synthesis of solid with well-defined nonstoichiometry

The synthetic conditions for nonstoichiometric solids (S_{AB}) coexisting with a vapor (V) and a melt (L) can be estimated using P–T–x diagrams or their T–x projections [15-17]. Let us consider the following features of the T–x diagrams: the highest (maximal) melting point $T_{m,AB}^{\max}$ of the solid S_{AB}; the compositions of the phases at this point; some features of the solidus, liquidus, and vapor lines; the nonvariant points of congruent melting (T_m^c), sublimation (T_S^c), and evaporation (T_e^c); and, finally, congruent and incongruent phases and phase processes.

4.1. Maximal melting point $T_{m,AB}^{max}$ of a nonstoichiomertic solid S_{AB}

The Gibbs energy of a phase in a two-component system A–B is given by Eq. (2), so the phase equilibria in this case are represented graphically in a four-dimensional space. This space is explored as four three-dimensional projections: *G–P–T*, *G–P–x*, *P–T–x*, and *G–T–x*.

To clarify the features of the *T–x* diagram let us consider the derivation of part of the *T–x* projection from the *G–T–x* diagram. For this purpose, it should be considered the relative positions of the *G*-surfaces of the solid phase S, liquid phase L, and vapor V. In order to determine the arrangement of these three surfaces, we will traverse them with isothermal planes. The projections of the sections of the *G*-surface on the *G–x* plane will appear as G^S, G^L, and G^V curves representing the dependence (5) of the Gibbs energy (per gram-atom) on composition at a constant temperature (Figure 4):

$$\left.\begin{aligned} G^S_{A_{1-x}B_x} &= (1-x_B^S)\mu_A^S + x_B^S\mu_B^S \\ G^L_{A_{1-x}B_x} &= (1-x_B^L)\mu_A^L + x_B^L\mu_B^L \\ G^V_{A_{1-x}B_x} &= (1-x_B^V)\mu_A^V + x_B^V\mu_B^V \end{aligned}\right\} \tag{5}$$

where μ_A^S and μ_B^S are the chemical potentials of the components A and B in the solid phase, μ_A^L and μ_B^L are those of A and B in the liquid phase, μ_A^V and μ_B^V are those of A and B in the vapor phase, and x_B is the mole fraction of the component B.

Let us consider the effect of temperature variation on the relative positions of the *G–x* curves. At $T_1 > T_{m,AB}^{\max}$ (Figure 4a), the solid phase S_{AB} is metastable and the two-phase system $L + V$ is stable because a common tangent line can be drawn for the corresponding *G* curves, indicating the equality of the chemical potentials of the components in the equilibrium phases.

As the temperature is decreased, the arrangement of the *G* curves changes. Since the temperature dependence of the Gibbs energy is determined by the entropy $\left(\frac{\partial G}{\partial T}\right)_{P,x} = -S$ and the entropies of the vapor and melt are higher than the entropy of the solid phase, $S^V > S^L > S^S$, the *G* curves shift upwards upon cooling and the G^V and G^L curves do so more rapidly than the G^S curve. As a consequence, a common tangent for the G^S, G^L, and G^V curves can appear at some temperature $T_2 = T_{m,\,AB}^{\max}$ (Figure 4b). This temperature $T_{m,\,AB}^{\max}$ is referred as the highest maximal melting point of the solid S_{AB}.

http://dx.doi.org/10.5772/54815

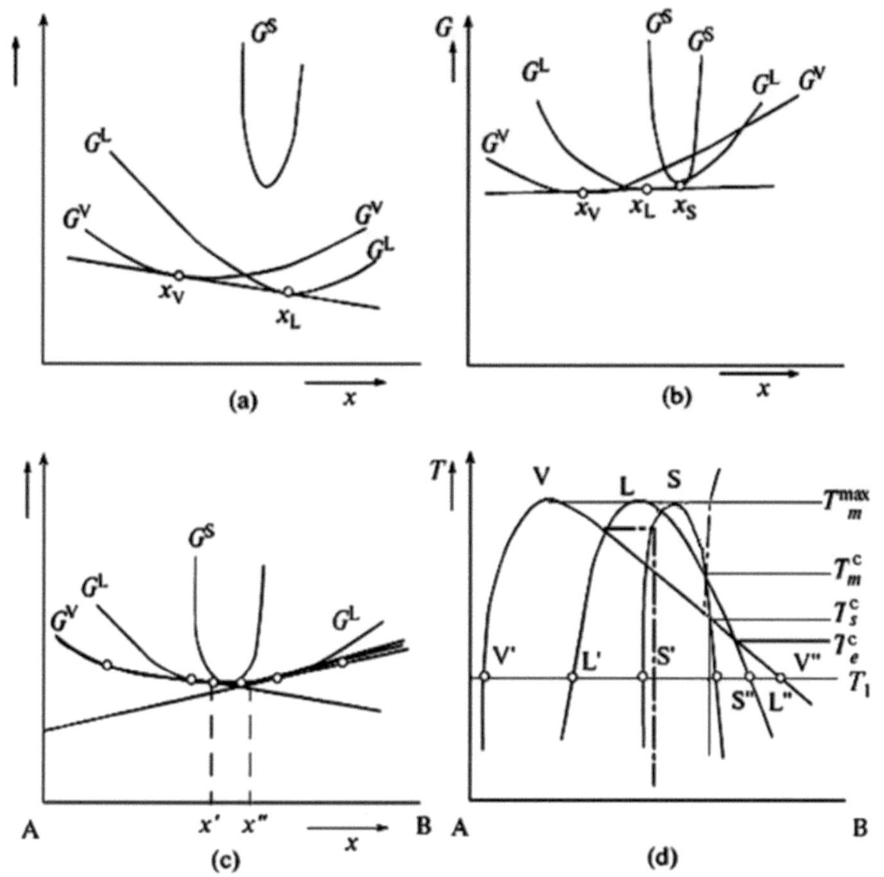

Figure 4. (*G*–*x*) sections of the *G*–*T*–*x* diagram: (a) $T_1 > T^{max}_{m,AB}$; (b) $T_2 = T^{max}_{m,AB}$; (c) $T_3 < T^{max}_{m,AB}$;(d) part of the *T*–*x* projection near $T^{max}_{m,AB}$.

From the definition of a phase as the totality of the parts of the system whose properties are described by the same equation of state, it follows that the properties of the system are homogeneous functions of composition, pressure, and temperature. Therefore, upon further cooling, for example, to $T_3 < T^{max}_{m,AB}$, it will be possible to draw two tangent lines for each of the G^S_{AB} , G^L, and G^V curves (Figure 4c). The coordinates of the equilibrium phases are designated as *V*, *L*, and *S* in Figure 4d. The subscripts (') and (") are given to the phase compositions to the left and right, respectively, of the composition corresponding to $T < T^{max}_{m,AB}$.

As the temperature changes, the G^V-, G^L- and G^S-curves, shifting upwards and to the sides, describe three surfaces: $G^V = G^V(T, x)$, $G^L = G^L(T, x)$ and $G^S = G^S(T, x)$ – and the projections of the tangency points on the *T*–*x* plane draw the solidus line S'–S", the liquidus line *L*'–*L*", and the vapor line *V*'–*V*" (Figure 4d). Thus, Figure 4d shows part of the *T*–*x* projection of the *P*–*T*–*x* diagram near the highest melting point of the S_{AB} compound. The points *V*', *L*', and S' and the points V'', L'', and S'', as examples, represent the compositions of the vapor V, solid phase S, and liquid phase L that are in equilibrium at the temperature $T = T_1$ and are called conjugate points. The lines VV'', LL'', and SS'' and the lines VV', LL', SS', as examples, formed by series of these points, are called conjugate lines. The lines connecting conjugate points, such as $V'L'S'$ and $V''L''S''$, are called tie lines. Note that, among the various intersection points between the vapor, liquidus, and solidus lines (Figure 4d), only the intersections points of the conjugate lines (VV'', LL'', SS'', VV', LL', SS') have a physical meaning, as distinct from those of the lines LL' и VV'', etc.

4.2. Specific features of the solids, liquidus and vapor lines

Here, let us consider the continuity of the solidus, liquidus, and vapor lines; the factors determining the homogeneity range; the issue of whether this range must include the stoichiometric composition; the causes of the retrograde character of the solidus line; and the concepts of a pseudocomponent and psevdobinary section in multicomponent systems.

The T–x and P–T projections of the P–T–x diagram of a two-component system having a compound AB [10, 15] are shown in Figure 5. The Gibbs energy of any phase is a homogeneous function of composition, so the solidus line $S'S''$, the liquidus line $L'L''$, and the vapor line $V'V''$, which represent the temperature dependence of the compositions of the equilibrium phases, are continuous lines having no inflection points.

The homogeneity range of a solid compound is bounded by the solidus line. It is determined by the coordinates of the tangency points of the common tangent line for the G curves of the equilibrium phases (Figure 4), i.e., by the equality of the chemical potentials of the components. Therefore, in the general case the homogeneity range (Δ in Eq. (3)) is determined by the relative positions of the G^S, G^L and G^V curves (Figure 4) (i.e., the properties of all coexisting phases) and by the shape of the $(G–x)_{P,T}$ curves. Thus the nonstoichiometry value depends not only the specific features of the nonstoichiometric solid, namely the radius value, electronic configuration and electronegativity of the constituents species, but on the properties of all coexisting phases.

In the case of the compound AB formed from solid components A and B,

$$A_S + B_S = AB_S;\ \Delta_f G^0 \tag{6}$$

the homogeneity range can be estimated using the following relationship [17]:

$$f(\delta' / k) - f(\delta''/ k) = -\Delta_f G^{\hat{i}}, \tag{7}$$

where $\Delta_f G°$ is the standard Gibbs energy of reaction (6), f is the monotonic function of composition δ, the subscripts ' and '' refer to the left and right sides of the homogeneity range, and k is a temperature dependent parameter. It follows from Eq. (7) that, the more negative the Gibbs energy $\Delta_f G°$, the larger the difference $f(\delta'/k) - f(\delta''/k)$ and, accordingly, the broader the homogeneity range $\delta'' - \delta'$. For crystals dominated by covalent bonding, such as those of the $A^{III}B^{V}$ compounds *GaAs, InP*, etc., $\delta'' - \delta'$ is on the order of thousandths of an atomic percent. For crystal with polar bonds (*CdTe, PbSe, SnTe*), the homogeneity range is between a few tenths of an atomic percent and several atomic percent.

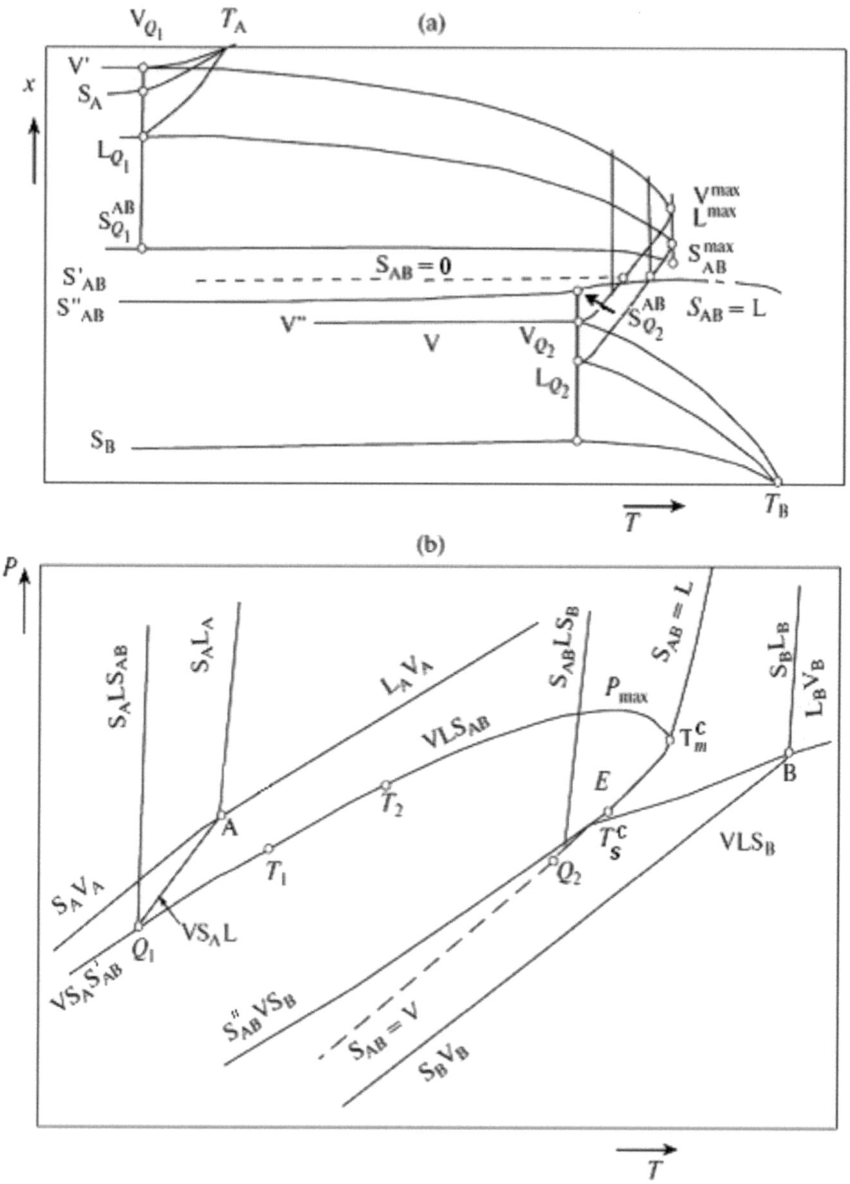

Figure 5. (a) *T–x* and (b) *P–T-* projections of the *P–T–x* diagram of a two-component system with a congruently melting compound (S_{AB}).

The homogeneity range may include the stoichiometric composition or not. Since the $\delta' + \delta''$ value defines the position of the center of the homogeneity range, the expression

$$f(\delta'/k) + f(\delta''/k) = [\mu_B(B) - \mu_A(A)]/RT - [\mu_B(\delta=0) - \mu_A(\delta=0)]/RT \tag{8}$$

describes the dependence of the position of the homogeneity range center on the difference between the Gibbs energies of the pure components (first term in (8)) relative to the isolated atoms in their ground states and on the difference between the chemical potentials of the components in the stoichiometric solid (second difference in (8). If the latter quantity is neglected in a series of crystallochemically similar compounds, the center of the homogeneity range, $(\delta' + \delta'')/2$ will shift in the direction in which the difference $\mu_B(B) - \mu_A(A)$ increases. This can easily be illustrated by examples of nontransition metal chalcogenides.

In the general case, the stoichiometric composition does not correspond to the minimum of the free energy of the solid phase and can fall outside the homogeneity range. When this is the case, the stoichiometric compound does not exist. For example, strictly stoichiometric ferrous oxide can be obtained only at high pressure.

The temperature dependence of the solidus $S'_{AB}S^{max}_{AB}S''_{AB}$ (Figure 5), which defines the boundaries of the homogeneity range of the nonstoichiometric compound S_{AB}, is described by the following equation [16, 18]:

$$\left(\frac{dx^S}{dT}\right)_P = \frac{(\partial S/\partial x)^S_{P,T} - (S^L - S^S)/(x^L - x^S)}{\left(\frac{\partial^2 G}{\partial x^2}\right)_{P,T}} \tag{9}$$

where S^L and S^S are the molar entropies and x^L and x^S are the compositions of the solid phase (*S*) and melt (*L*). The solubility of the components in the nonstoichiometric solid $A_{1-x}B_x = =A_{1/2-\delta}B_{1/2+\delta}$ can be retrograde: it can initially increase with increasing temperature ($(dx^S/dT)>0$) and, on passing though a maximum, decrease ($(dx^S/dT)_P < 0$). The maximum solubility $(x^S)_{max}$ is found from the extremum condition for function (9) ($(dx^S/dT)_P = 0$) since $\left(\frac{\partial^2 G}{\partial x^2}\right)_P > 0$, according to the phase stability criterion. The retrograde solubility is due to the difference between the rates at which the entropies S^L and S^S, including their configurational components, grow with increasing temperature.

The liquidus line ($L_{Q_1}L^{max}L_{Q_2}$) and vapor line ($V_{Q_1}V^{max}V_{Q_2}$) in Figs. 4 and 5 represent the temperature dependence of the compositions of the melt (*L*) and vapor (*V*) in equilibrium with the nonstoichiometric solid S_{AB}.

http://dx.doi.org/10.5772/54815

4.3. Noncoincidence of the phase compositions at the maximum melting point of a nonstoichiometric solid S_{AB}

Both pressure and temperature extrema at T = const and P = const, respectively, exists in a three-phase two-component system provided that the following condition is satisfied [16]:

$$(x^L - x^S)/(x^V - x^S) = (S^L - S^S)/(S^V - S^S) = (V^L - V^S)/(V^V - V^S) \qquad (10)$$

Thus, if the coexisting phases have different molar volumes V and entropies S, the composition of one of them will be a linear function of the compositions of the two others: $x^L = \beta x^S + (1 - \beta)x^V$ where β is a coefficient depending on the pressure and temperature and independent of the composition. Since $V^L \neq V^S \neq V^V$ and $S^S \neq S^L \neq S^V$ in the general case, at $T = T_{m,\,AB}^{\max}$ the composition of the solid phase S_{AB} does not coincide with the composition of the melt L and vapor V: $x^S \neq x^L \neq x^V$. This is in agreement with the phase rule:

$$c = k + 2 - r - \alpha, \qquad (11)$$

where k is the number of components, 2 is the number of external fields (baric and thermal), r is the number of phases, and α is the number of independent constraints imposed on the intensive variables (c). Indeed, if it is assumed that, at $T = T_{m,\,AB}^{\max}$ $x^S = x^L = x^V$, then $\alpha = 2$ in Eq. (11) and $c = 2 + 2 - 3 - 2 = -1 < 0$, which is impossible. Unfortunately, this mistake is frequently encountered in the literature [18-21].

In the case of lead, germanium, tin, and cadmium chalcogenides, the vapor phase consists mainly of a chalcogen, so it can be accepted that $x^S - x^V = -0.5$. At $P \cong 1$ atm, $V^S - V^V \cong 10^{-3} V^S$ and $V^S - V^L \cong 5 \bullet 10^{-2} V^S$. Therefore, $x^S - x^L = 2.5 \times 10^{-5}$; that is, the compositions of the phases at the highest melting point do not coincide, even though they are very similar [14, 22, 23].

4.4. Nonvariant congruent melting, sublimation, and evaporation points of the three-phase equilibrium $S_{AB} + L + V$

Since the solid, liquid, and vapor phases are characterized by different temperature and concentration dependences of the Gibbs energy, the following intersection points of the conjugate liquidus, solidus, and vapor lines can appear on the T–x projection. The temperature of the intersection point of the conjugate liquidus and solidus lines refers to solid and liquid phases of equal compositions ($x^S = x^L$) and is called the congruent melting point T_m^c of the S_{AB} phase. The temperature of the intersection point of the solidus and vapor lines refers to a solid phase and vapor of equal compositions ($x^S = x^V$) and is called the congruent sublimation point T_S^c. The temperature of the intersection point of the conjugate liquidus and vapor lines refers to a liquid phase and vapor of equal compositions ($x^S = x^V$) and is called the congruent evaporation point T_e^c. The temperatures T_m^c, T_S^c and T_e^c do not correspond to the stoichiometric composition ($\delta = 0$) of the solid phase $A_{1/2-\delta}B_{1/2+\delta}$, where $x = 1/2 + \delta$ and δ is the deviation from its stoichiometric composition *AB*. For example, congruently melting lead telluride contains $(2.8 \pm 0.3)\ 10^{-4}$ mol excess Te per mole of *PbTe* [22, 23]. Because the formation energies of the defects responsible for the nonstoichiometries $\delta > 0$ and $\delta < 0$ are different, the G^S curve (Figure 4a, 4b, 4c) is asymmetric relative to the $\delta = 0$ composition. Accordingly, the coordinates of the common tangency points of the *G* curves in the general case do not coincide with the $\delta = 0$ composition.

The equality of the compositions of two phases of the three ones involved in the equilibrium means the appearance of one more relationship $\alpha = 1$ between independent variables (degrees of freedom) in the phase rule expression (11). Therefore, the equilibrium at the points T_m^c, T_S^c and T_e^c is nonvariant: $c = 2+2-3-1 = 0$. The solid S_{AB}, which has a homogeneity range at these points, can be considered a pseudocomponent whose properties are composition-independent, and the sections through ternary, quaternary, and other multicomponent systems involving the S_{AB} solid can be considered as pseudobinary.

4.5. Congruent and inconguent phases and phase processes

The concept of congruence is of great significance in the synthesis of nonstoichiometric solids because, in the case of noncoincidence between the synthetics medium (vapor, melt) and solids compositions, there are fluxes of rejected material and the corresponding kinetic instability of the crystallization front.

A phase can be congruent and incongruent in different temperature intervals. A phase obtainable from the phases that are in equilibrium with it by mixing them in appropriate proportions is called a congruent phase. A phase that cannot be obtained from the coexisting phases is called incongruent [24, 25]. As an example, let us consider the phase S_{AB} (Figure 5a) in the three-phase equilibrium $S_{AB} + L + V$ at different temperatures. Above the congruent melting point ($T_{m,\circledast AB}^c < T < T_{m,\,AB}^{\max}$) the solid phase S_{AB} is incongruent. In the temperature range $T_S^c < T < T_{m,\circledast AB}^c$, it is congruent with respect to the vapor and melt. In designations of three phase equilibria, a congruent phase is written in the middle. Thus, in the former case, the three-phase equilibrium is designated VLS_{AB}; in the latter case, $VS_{AB}L$.

The concepts of congruent and incongruent phases should not be confused with the concepts of congruent and incongruent phase processes.

Phase processes are changes in the state of the system such that the masses of some phases increase owing to the decrease in the masses of others without changes in the intensive parameters (temperature, pressure, phase compositions) [26]. A phase process in which one phase forms or disappears is called congruent. A phase transition in which more than one phase forms or disappears is called incongruent.

The same phase, for example, S_{AB} (Figure 5a), can be involved both in congruent and in incongruent phase processes. At a temperature or pressure corresponding to the three-phase line Q_1Q_2 in Figure 5b, the solid phase S_{AB} melts congruently, yielding a melt

http://dx.doi.org/10.5772/54815

and a vapor: $S_{AB} = L + V$. The compositions of the phases involved in this congruent phase process are different and are represented by the solidus, liquidus, and vapor lines in the *T–x* projection (Figure 5a). The melt and solid phase compositions coincide only at the congruent melting point T_m^c . At the temperatures corresponding to the eutectic non-variant points T_{Q_1} and T_{Q_2} (Figure 5a), the S_{AB} phase is involved in the incongruent phase processes

$V + S_A = L + S_{AB}$ and $S_{AB} + L = V + S_B$.

Thus, S_{AB} phase can be said to melt congruently only in a certain temperature range.

Now let us consider the usage of the terms *congruently melting compound* and *incongruently melting compound* in the literature. Firstly, this differentiation is not strict, because the same compound (e.g., S_{AB}) can be involved in congruent and incongruent phase processes, depending on the temperature. Sometimes, a congruently melting compound is understood as a compound melting without dissociation or decomposition. However, a solid–melt phase transition is accompanied by the breaking and relaxation of chemical bonds in the crystal and by long-range disordering. Therefore, the term *melting without decomposition* is not quite correct, and it should be understood as the identity of the overall compositions of the coexisting phases. In the strict sense, the overall compositions of the phases coincide only at the congruent melting point $T^{c}_{m,\circledcirc AB}$, which is below the highest melting point $T^{\max}_{m,\,AB}$: $T^{c}_{m,\circledcirc AB} << T^{\max}_{m,\,AB}$.

An incongruently melting compound is sometimes understood as a solid compound that decomposes into a solid phase S_B and a liquid L upon melting (Figure 6). The compositions of the resulting phases differ from the composition of the parent phase. However, in some temperature range, such as $T_{Q_1} < T < T_P$, the "incongruently melting compound" is involved in the congruent melting process $S_{AB} = L + V$.

An essential feature differentiating congruently and incongruently melting compounds is that the highest melting point of a congruently melting compound is higher than the temperatures of the nearest nonvariant points: $T^{\max}_{m,\,AB} > T_{Q_1}$ and T_{Q_2} (Figure 5a).

The highest melting point of an incongruently melting compound is intermediate between the temperatures of the nearest nonvariant points of the system.

Supersaturation and synthesis of a nonstoichiometric solid at a fixed vapor pressure of the volatile component can be produced by cooling or, conversely, heating the three-phase system. The latter case corresponds to the temperature range $T_1 < T < T_2$ in Figure 5b. For example, cadmium telluride crystals were obtained by heating cadmium enriched melts [10]. Note that the composition of the crystals that were grown using the vapor–liquid–crystal technique always lies in the solidus line, i.e., at the boundary of the homogeneity range [15].

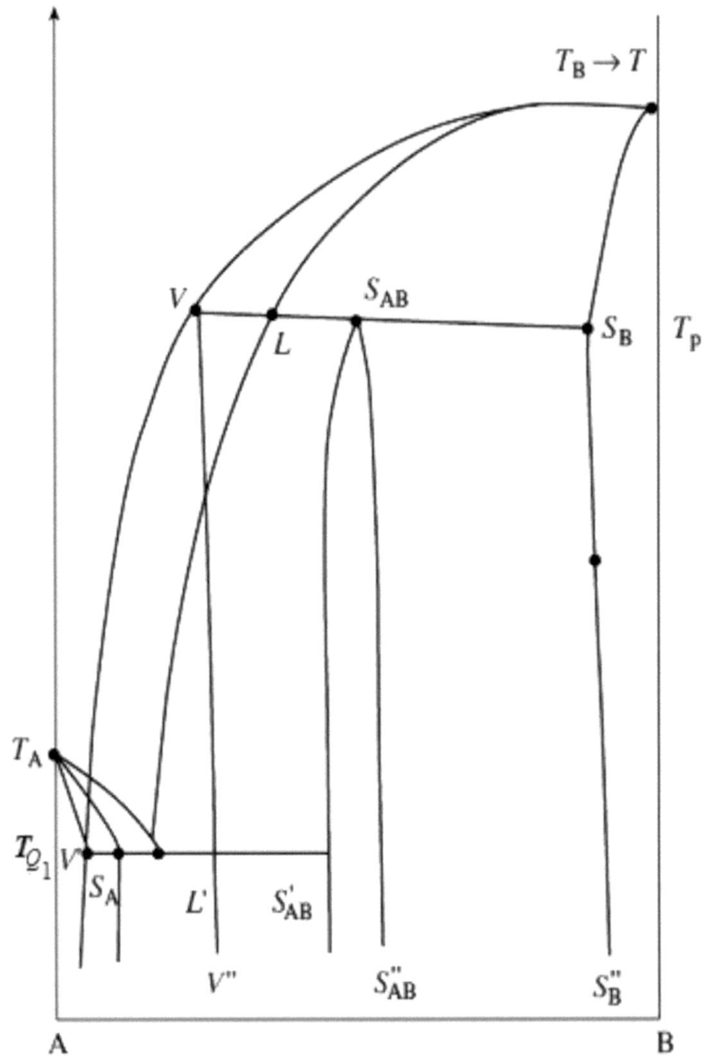

Figure 6. *T–x* projection of the *P–T–x* diagram of a two-component system with an incongruently melting compound (S_{AB}).

http://dx.doi.org/10.5772/54815

5. Nonstoichiometry and defects in solids

Solids like S_{AB} are grown from the vapor, melt (solution), or solid phases, which are called growth media. Crystallization can be viewed as the transfer of atoms of the components *A* and *B* from the growth medium (g.m.) to their regular sites A_A^x and B_B^x in the crystal lattice of the S_{AB} compound:

$$A^{g.m.} = A_A^x + V_B^x;\ \Delta G_1;\ \ K_1 = \frac{[A_A^x][V_B^x]}{a_A} = exp(-\Delta G_1 / kT), \tag{12}$$

$$\hat{A}^{g.m.} = B_B^x + V_A^x\ ;\ \Delta G_2;\ \ K_2 = \frac{[B_B^x][V_A^x]}{a_B} = exp(-\Delta G_2 / kT) \tag{13}$$

where [ ] designates concentrations and a_A and a_B are the activities of the components in a growth media. The generation of V_B^x and V_A^x vacancies via reactions (12) and (13) is explained by the principle of conservation of the ratio of the numbers of sites characteristic of a given crystal lattice.

Because of the size and energy differences, the Gibbs energies ΔG_1 and ΔG_2, the equilibrium constants of reactions (12) and (13), and, accordingly, the numbers of atoms A and B in the crystals turn out to be different: $K_1 \neq K_2$ and $[A_A^x] \neq [B_B^x]$. Thus, nonstoichiometry appears; that is, the difference between the *B*-to-*A* ratios in the real and stoichiometric crystals becomes nonzero.

Note that the properties of crystals are affected not by the nonstoichiometric atoms A_A^x and B_B^x that occupy their regular sites, but by the defects (resulting from a disordering of the ideal structure).

These defects may be the vacancies V_A^x and V_B^x or the interstitial atoms A_i^x and B_i^x

$$A^{g.m.} = A_i^x + \Delta G_3, \tag{14}$$

$$B^{g.m.} = B_i^x + \Delta G_4. \tag{15}$$

This circumstance is due to the fact that the A_A^x and B_B^x species do not change the energy structure of the crystal, but complete it in a way. Near the defects (V_A^x , V_B^x , A_i^x , B_i^x), the energy field and, accordingly, the electrical, mechanical and other properties of the crystal are altered (Figure 7). Thus, defects play an important role in the description of the real structure and properties of nonstoichiometric solids.

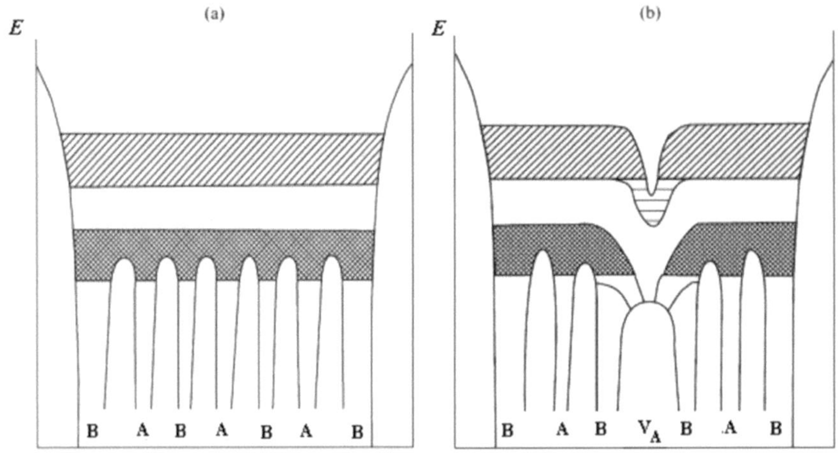

Figure 7. Model of the energy spectrum of (a) an ideal and (b) a nonstoichiometric crystal containing V_A vacancies.

6. Classification and formation of defects

Defect formation processes and defect classification are significant points in control of the defect composition of solids in their directed synthesis.

An ideal, or perfect, solid is one in which all particles making up the substance or structure elements (atoms, ions, molecules, etc.) occupy their regular sites in the lattice. Under heating, irradiation with a beam of high-energy particles, or mechanical treatment, the regular arrangement of particles over their sites undergoes disordering: some particles can leave their sites. The resulting disorder in the arrangement of particles over their normal sites is called defects [27].

In terms of geometry and size, all defects are divided into point and extended defects [27, 28] (Figure 8).

The size of point, or zero-dimensional, defects is comparable with the interatomic parameter. The zero-dimensional (OD) defects include electronic defects (holes, electrons, exitons), energy defects (phonons, polarons), and atomic point defects (APDs). The APDs in nonstoichiometric *AB* crystals include V_A^x and V_B^x vacancies (absence of atoms or ions in lattice sites); interstitial atoms A_i^x , B_i^x and F_i^x , where *F* - foreign atoms (the upper symbol *x* means the neutrality of the defect with respect to its environment), the lower symbol means the defect location.

The formation of an APD is an endothermic process requiring a small amount of energy: 0.5 $\leq E_f \leq$ 3 eV. Therefore, APDs are equilibrium defects and their concentration depends on synthesis conditions, namely, the temperature and the partial pressures of the components. The size of an APD is 0.1–0.5 nm; however, APDs polarize their environment in the crystal structure, causing slight displacements of neighborions, and largely determine the physical

http://dx.doi.org/10.5772/54815

and chemical properties (diffusion, electric conductivity, etc.) of the nonstoichiometric crystal.

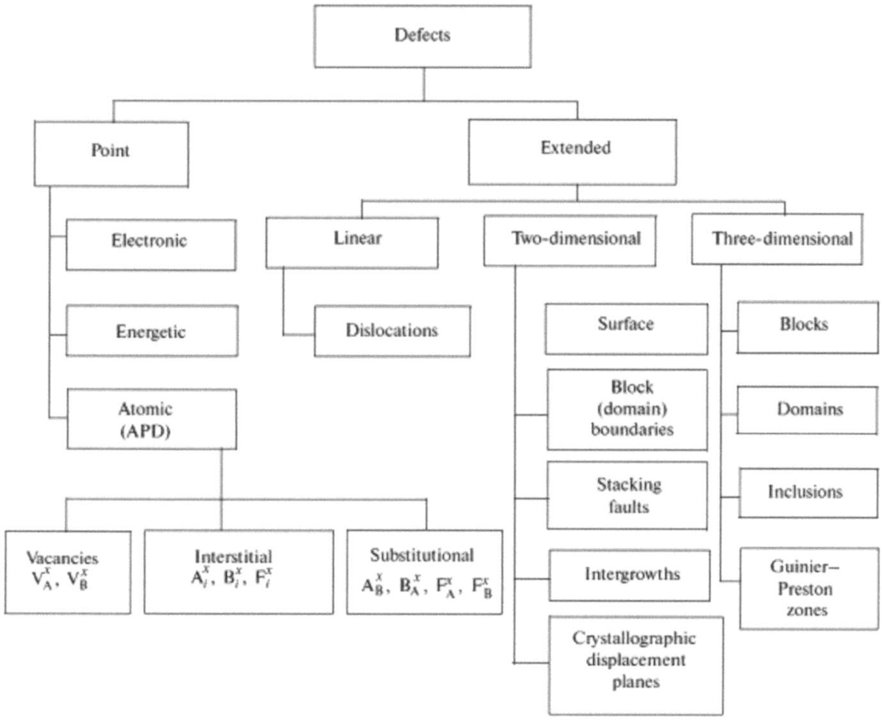

Figure 8. Classification of defects.

Vacancies, interstitial and antistructure (A_B^x , B_A^x) defects are classified as intrinsic defects of a crystal. The concentration (c_j) of these defects takes on a thermodynamically equilibrium value at any $T > 0$ K: $c_j \approx \exp(-W_j/kT)$, where Wj is the defect formation energy, $0 < W_j < 3$ eV. The higher the temperature, the higher the APD concentration c_j. As the crystal is cooled to room temperature, part of the APDs can annihilate via various mechanisms. However, many APDs persist even at exceptionally low reactions (12)–(15), but also from the irradiation of the crystal with high-energy (>1 MeV) electrons at very low temperatures.

Oppositely charged APDs can be attracted to one another to yield new APDs as electroneutral association species, such as ($V_A^- V_B^+$), ($V_A^- F_A^+$), (V_A^x)n. Dipole–dipole interaction leads to the formation of APD clusters, which can serve as nuclei of other phase in the nonstoichiometric crystal.

Extended defects include linear (1D), or one-dimensional, and two (2D)- and three-dimensional (3D) defects [29, 30] (Figure 8). Let us consider some specific features of these defects.

Linear defects, or dislocations, are similar to point and two-dimensional defects in the sense that their size is comparable with the unit cell parameter. In the third dimension, the dislocations are fairly long or even infinite. The simplest kind of dislocation is the edge dislocation.

The most important two-dimensional (2D) defects include solid surface, block (domain) boundaries, stadling faults and crystallographic displacement planes, which are surfaces in which coordination polyhedra of two contacting ideal are rearranged.

An example of periodic intergrowths is provided by the family of tungsten bronzes, A_xWO_3 (A = alkali metal). In these compounds, WO_3 layers are intergrown with hexagonal bronze bands. The formation of intergrown bronzes is likely favored by the synthesis conditions. The boundaries of intergrowths can be periodic (ordered) or nonperiodic (disordered).

Note the following specific features of two-dimensional defects. Firstly, their formation energy is fairly high (>3 eV) and they are kinetically stabilized nonequilibrium defects. The "frozen" state of these defects is responsible for their "memory" for preparation history. Secondly, these defects do not change the stoichiometry of the substance. Thirdly, planar defects result from APD interaction and exert a significant effect on on the reactivity and physical properties of nonstoichiometric solids.

The size of three-dimentional defects (Figure 8) exceeds the lattice constant in all three directions. These defects are, in essence, macroscopic imperfections of the crystal structure and are formed during crystal growth and subsequent processing. Three-dimentional defects include separate blocks; mosaics (totality of a large number of small- and large-angle boundaries); inclusions (microdeposits) resulting from phase transitions, such as the decomposition of a solid solution; magnetic domains (crystal zones with the same orientation of spins or electric dipoles); Guinier–Preston zones (parallel platelike formations as thick as a few unit cells, separated by different distances and having the same composition as the crystal); cavities; and cracks. Three-dimentional defects can be viewed as resulting from defect association and ordering. For example, pores can be considered to result from the association of a large number of vacancies. Bulk defects also include elastic tensile and compressive stresses.

Deviations from stoichiomrtry may be so large that defect interactions become significant leading to direct ordering, clustering, superstructure formation, long-range ordering, and the formation of new nonstoichiometric phases differing in symmetry, energy and other aspects from the parent phase. In such systems defects are intrinsic components of the crystal structure rather than being statistically distributed imperfections. The crystal-chemical and thermodynamic aspects of nonstoichiometric compounds with narrow and broad homogeneity range as well the approaches for controlling the nonstoichiometry and considered in [31].

7. Substance homogeneity criterion in physicochemical analysis

The Gibbs energy (G) of a crystal is a statistical quantity related to the distribution function. The mean G value determines the most likely distribution of zero-, one-, two-, and three-dimensional defects. Fluctuations around the mean value are possible, and, in a closed system (solid substance with a constant net composition at a fixed temperature), the configurational fluctuation can manifest itself as a change in composition within a small region as a consequence of a random motion of particles inside some volume element or to its surface. This brings up the question of what substance can be called homogeneous. The degree of heterogeneity of a solid phase is characterized by the statistical distribution of structure elements from which the crystal is built. These structure elements are the atoms of the components of the system in their regular sites, as well as various zero-, one-, two-, and three-dimensional defects in their regular positions (Figure 8).

A quantitative estimate of the degree of heterogeneity can be based on the following three types of distributions:

1. distribution of structure elements in some measurable volume,
2. distribution of these volumes in the crystal, and
3. distribution of measurement data and properties of the solid phase.

Let σ be the confidence interval and c_i be the concentration of structure elements in the ith microvolume. If

$$\left| c_i - 1/N\sum_{i=1}^{N} c_i \right| \le \sigma \tag{16}$$

the solid substance can be called homogeneous. If there is an i value at which this inequality is invalid, the solid phase should be considered to be heterogeneous. For practical use of a material, of significance are such deviations of the property in a given object from the weighted average value for the entire system that go beyond the confidence interval. In this sense, heterogeneity can be understood as the totality of property values, measured in all microvolumes, that fall outside the confidence interval. An analysis of generalized heterogeneity criteria in terms of an autocorrelation function was carried out by Nikitina et al. [32]. A thermodynamic analysis of defect ordering and an interpretation of the concentration dependences of physical properties taking into account short- and long- range order parameters were made using a model based on the above stricture elements or cluster components [33].

Some features of homogeneity criteria, physicochemical analysis, thermodynamics in materials science and inorganic crystal engineering are discussed in [34-37].

8. Conclusion

Technological progress is intimately associated with creation of new materials, such as composites, piezoelectrics, ferroelectrics, semiconductors, superconductors and nanomaterials with preset functional properties. For the synthesis of these materials, it is necessary to study the chemical processes that lead to the desired properties

Physicochemical analysis is the field of chemistry dealing with these processes, the interrelation between the composition, structure and properties of matter and determination of synthetic conditions for obtaining such substances. Several challenges and new approaches have been discussed. They include the concepts of a substance, phase, component, directed synthesis strategy as well as some definitions. Attention was paid to nonstoichiometry, classification and formation of defects.

Synthesis of nonstoichiometric solid involves control over phase transformations. The P–T–x phase diagram is a key to selecting conditions for synthesis of solid with well-defined nonstoichiometry. For this reason the following features of *P–T–x* phase diagrams were considered: highest melting point of a nonstoichiometric compound ($T^{max}_{m,AB}$), noncoincidence between the solid-, liquid-, and vapor-phase compositions ($x^L \neq x^S \neq x^V$) at this temperature, factors determining the nonstoichiometry range, the terms congruent and incongruent phases and phase processes. Criteria for evaluating the degree of heterogeneity of nonstoichiometric solids were also considered.

Glossary of Abbreviations

$T^{max}_{m,AB}$ maximal melting point of a nonstoichiometric compound AB

(Na^{+}_{Na}) cationic sites

(Cl^{-}_{Cl}) anionic sites

U internal energy

Q heat

A work execution

δ nonstoichiometry

T^{c}_{m} nonvariant point of congruent melting

T^{c}_{S} nonvariant point of congruent sublimation

T^{c}_{e} nonvariant point of congruent evaporation

S solid phase

L liquid phase

V vapor

μ^{S}_{A} chemical potential of the component A in solid phase

μ^{S}_{B} chemical potential of the component B in solid phase

μ^{L}_{A} chemical potential of the component A in liquid phase

μ^{L}_{B} chemical potential of the component B in liquid phase

μ_A^V chemical potential of the component A in vapor phase

μ_B^V chemical potential of the component B in vapor phase

x_B mole fraction of the component B

S^V entropy of vapor

S^L entropy of melt

S^S entropy of solid

G^V Gibbs energy of vapor

G^L Gibbs energy of liquid

G^S Gibbs energy of solid

$\Delta_f G°$ standard formation Gibbs energy

k number of components

r number of phases

α number of independent constraints imposed on the intensive variables (c)

A_A^x regular sites of atom A in crystal lattice of the S_{AB}

B_B^x regular sites of atom B in crystal lattice of the S_{AB}

K_1 equilibrium constant

K_2 equilibrium constant

V_B^x and V_A^x vacancies

A_i^x and B_i^x interstitial atoms

c_j concentration

W_j defect formation energy

σ confidence interval

$(V_A^- V_B^+)$ electroneutral association species

$(V_A^- F_A^+)$ electroneutral association species

Author details

V. P. Zlomanov[1], A.M. Khoviv[2] and A.Ju. Zavrazhnov[2]

*Address all correspondence to: zlomanov@inorg.chem.msu.ru

2 Department of Chemistry, Voronezh State University, Voronezh, Russia
1 Department of Chemistry, Moscow State University, Moscow, Russia

References

[1] Kurnakov N.S. Zadachi Inst. Physikochemical Analysis. Izv. Inst. Phys.Chem. Analysis. 1919; 1(1), 1-7.

[2] Kurnakov N.S. The Introduction to Physicochemical Analysis. 4th ed. Moscow: Akad. Nauk SSSR; 1940 [in Russian].

[3] Kurnakov N.S. Collection of Selected Works. Leningrad/Moscow: ORTI, Khimteoret; 1938; vol. II, p. 1-202 [in Russian].

[4] Kurnakov N.S. Selected Works. Moscow:Akad. Nauk SSSR; 1963; vol. II, 1-361 [in Russian].

[5] Anosov V.Ya., Pogodin S.A. The Principles of Physicochemical Analysis. Moscow: Akad.Nauk SSSR; 1947 [in Russian].

[6] Tamman G. The Manual on Heterogeneous Equilibria. Leningrad:ONTI, Khimteoret; 1935 [inRussian].

[7] Gibbs J.W. Thermodynamics: Statistical Mechanics. Moscow: Nauka; 1982.

[8] Voronin G.F. The Fundamentals of Thermodynamics. Moscow:Moskow State University; 1987 [in Russian].

[9] Fuks G.I. The Smallest Piece, the Smallest Drop. Khimya and Zhizn'. 1984; 2, 74-76 [in Russian].

[10] Kröger F,A. The Chemistry of Imperfect Crystals. Amsterdam: North-Holland Publ. Co.; 1973.

[11] Vasiliev Ya., Akhmetshin R., Borodlev Yu. BGO Crystal Growth by a Low Thermal Gradient Czochralski Technique. Nucl. Instr. Meth. Phys. Res. 1996; A379, 533.

[12] Golyshev V., Gonic M. et al. Heat Transfer in Growing $Be_4Ge_3O_{12}$ Crystals under Weak Convection. J. Cryst. Growth. 2004; 262, 202-214.

[13] Rosenberger R.F. Fundam. Cryst. Growth. New-York: Springer-Verlag; 1979.

[14] Gottshtain G. The Physical Foundation of Materials Science. Berlin-Heidelberg: Springer-Verlag; 2004.

[15] Zlomanov V.P., Novoselova A.V. P–T–x Diagrams of Metal–Chalcogen Systems. Moscow: Nauka; 1987 [in Russian].

[16] Storonkin A.V. The Thermodynamcis of Heterogeneous Systems. Leningrad: Leningrad State University; 1967; part 1 [in Russian].

[17] Brebrick R.F. Nonstoichiometry in Binary Semiconductor Compounds $M_{0.5-\delta}N_{0.5+\delta}$. Prog. Solid State Chem. 1966; 3, 213-263.

[18] Roozeboom W.B. Die heterogenen Gleichgewichte vom Standpunkte der Phasenlehre. Braunschweig: Vieweg; 1904.

[19] Ricci J.E. The Phase Rule and Heterogeneous Equilibrium. New York: Dover; 1966.

[20] Khaldoyanidi K.A. Phase Diagrams of Heterogeneous Systems. Novosibirsk: INKh RAN; 1991; part 1 [in Russian].

[21] Levinskii Yu.V. p–T–x Diagrams of Binary Metal Systems. Moscow: Metallurgiya; 1990; book 1 [in Russian].

[22] Peter K., Wenzel A., Rudolph P. The *p–T–x* Projection of the System *Cd–Te*. Cryst. Res. Technol. 1990; 25(10), 1107-1114.

[23] Avetisov I. Kh., Mel'kov A.Yu., Zinov'ev A.Yu., Zharikov E.V. Growth of Nonstoichiometric PbTe Crystals by the Vertical Bridgman Method Using the Axial-Vibration Control Technique. Crystallography Reports. 2005; 50, Suppl. 1, 124-129.

[24] Meuerhaffer W., Sanders A.Z. Z. für Phys.Chem. IV. Phys. Chem. 1899; 28, 453.

[25] W. Meuerhaffer W., Sanders A.Z. Z. für Phys.Chem. II. Phys. Chem. 1905; 53, 513-518.

[26] Münster A. Chemische Thermodynamik. Berlin: Academie – Verlag; 1969.

[27] Van Bueren H.G. Imperfections in Crystals. Amsterdam: North-Holland; 1960.

[28] Fistul' V.I. The Physics and Chemistry of the Solid State. Moscow: Metallurgiya; 1995 [in Russian].

[29] West A. Solid State Chemistry and Its Applications. Chichester: Wiley; 1984.

[30] Rabenau A., editor. Problems of Nonstoichiometry. Amsterdam: North Holland; 1970.

[31] Zlomanov V.P. Crystal Growth of Nonstoichiometric Compounds. Inorganic Materials. 2006, 42, Suppl. 1, 19-48.

[32] Nikitina V.G., Orlov A.G., Romanenko V.N. Problem of Nonhomogeneity Distribution of Atoms and Defects in Semiconductor Crystals. In: Growth of Semiconductor Crystals and Films. Novosibirsk: Nauka; 1981; 204 [in Russian].

[33] Men' A.N., Bogdanovich M.P., Vorob'ev Yu.P. et al. Composition–Imperfection–Properties of Solid Phases. Cluster Components Method. Moscow: Nauka; 1977 [in Russian].

[34] De Hoff Robert T. Thermodynamics in Materials Science. Taylor & Francis. Boca Raton. London. New-York. 2011. P. 531.

[35] Brammer L. Developments in Inorganic Crystal Engineering. Chem. Soc. Rev. 2004. V.33. P. 476-489.

[36] Shriver & Atkins. Inorganic Chemistry. Oxford University Press. Fourth Ed. 2006. 822 P.

[37] Ashby M., Sherclift H., Ceban D. Materials, Engineering, Science, Processing and Design. Elsevier. Amsterdam. Boston. 2011.

Chapter 6

Nanocrystalline Mn and Fe Doped ZnO Thin Films Prepared Using SILAR Method for Dilute Magnetic Semiconductor Application

Rathinam Chandramohan,
Jagannathan Thirumalai and
Thirukonda Anandhamoorthy Vijayan

Additional information is available at the end of the chapter

http://dx.doi.org/10.5772/54327

1. Introduction

Zinc oxide (ZnO) is a versatile material of compound semiconductors with excellent properties and extensive applications in electronics, optoelectronics, sensors, and catalyses (Das et al, 2007). ZnO thin films have attracted considerable attention because they can be tailored to possess high electrical conductivity, high infrared reflectance and high visible transmittance by different coating technique (Ryu et al, 2000). Some of the remarkable properties of ZnO are its wide direct-band gap of 3.37 eV, the binding energy of the exciton of ZnO (60 meV) which makes it an excellent material for excitonic devices (Wang et al, 2003). Other favourable aspects of ZnO include its broad chemistry leading to many opportunities for wet chemical etching, low power threshold for optical pumping, radiation hardness and bio-compatibility. Together, these properties of ZnO make it an ideal candidate for a variety of devices ranging from sensors through to ultra-violet laser devices and nanotechnology based devices such as displays. As fervent research into ZnO continues, difficulties such as the fabrication of p-type ZnO that have so far stated that the development of devices had overcome (Yang etal, 2008). Mitra et al (1998) has prepared Zinc Oxide thin films using chemical deposition technique. The structural, morphological properties of the prepared films are characterized using X-ray diffraction and scanning electron microscope. They have used Zn salts as precursor and successfully synthesized ZnO films. The growth of highly textured Zinc oxide (ZnO) thin films with a preferred (101) orientation has been prepared by employing chemical bath deposition using a sodium zincate bath on glass substrates has been reported by (Ramamoorthy et al, 2004).

(Natsume et al, 2000) have studied the d.c electrical conductivity and optical properties of zinc oxide film prepared by a sol-gel spin coating technique. The temperature dependence of the conductivity indicated that electron transport in the conduction band was due to thermal execution of donor electrons for temperatures from 250 to 300 K. (Chapparro et al, 2003) have proposed the spontaneous growth of ZnO thin films from aqueous solutions. An electroless – chemical process is proposed, consisting in the formation of the super oxide radical (O_2^-) followed by chemical reaction of two O_2^- with Zn $(NH_3)_4^{2+}$ cations. (Wellings et al, 2008) have deposited ZnO thin films from aqueous zinc nitrate solution at 80ºC onto fluorine doped tin oxide (FTO) coated glass substrates. Structural analysis, surface morphology, optical studies and electrical conductivity were studied and thickness of the ZnO films was found to be 0.40 μm. (Walter et al, 2007) have studied the characterization of strontium doped ZnO thin films on love wave filter applications. X-ray diffraction, scanning electron microscopy and atomic force microscopy studied the crystalline structure and surface morphology of films. The electrochemical coupling coefficient, dielectric constant, and temperature coefficient of frequency of filters were then determined using a network analyzer. (Vijayan et al, 2008, a, b; Chandramohan et al, 2010) have reported the preparation conditions for undoped ZnO using double dip technique and used them for gas sensor applications. They have also reported the synthesis of Sr doped ZnO using double dip technique and used them for gas sensor applications. Recently (Chandramohan et al, 2010) have synthesized Mg doped ZnO thin films using double dip chemical growth and reported the ferromagnetic properties of the films. Saeed et al. (1995) have deposited thin films of mono phase crystalline hexagonal ZnO from solutions of zinc acetate in the presence of ethylenediamine and sodium hydroxide on to glass microscope studies. Two distinct morphologies of ZnO were observed by scanning electron microscopy. The deposited films were specular and adherent. (Cheng et al 2006) have fabricated thin films transistors (TFTs) with active channel layers of zinc oxide using a low – temperature chemical bath deposition. Current voltage (I-V) properties measured through the gate reveal that the ZnO channel is n-type. (Sadrnezhaad et al 2006) have studied the effect of addition of Tiron as a surfactant on the microstructure of chemically deposited zinc oxide. Addition of tiron charges the surface morphology and causes to form the fine – grained structure. The obtained results indicate that increasing the number of dipping carves to progress the deposition process. (Piticescu, et al 2007) have studied the influence of the synthesis parameters on the chemical and microstructural characteristics of nanophases synthesized in the two methods. 'Al' doping tends to a lower material density and to a smaller gown size. Zhou et al (2007) have studied microstructure electrical and optical properties of aluminium doped zinc oxide films. The ZnO:Al thin films are transparent (~ 90%) in near ultraviolet and visible region A. with the annealing temperature increasing from 300ºC to 500ºC. The film was oriented more preferentially along the (002) direction, the grain size of the film increased, the transmittance also became higher and the electrical resistivity decreased. Bulk ZnO is quite expensive and unavailable in large wafers. So, for the time being, thin films of ZnO are relatively a good choice. Usually, the doped ZnO films with optimum properties (perfect crystalline structure, good conducting properties, high transparency, high intensity of luminescence) are obtained when they are grown on heated substrates and annealed after deposition at high temperature in oxygen atmosphere (Peiro et al, 2005 Lokhande et al, 2000; Srinivasan et al, 2006; Chou et al, 2005). However, for an extensive use in the commercial applications pure and doped ZnO films must be prepared at much lower substrate temperatures. Therefore, it is necessary to develop a low-temperature deposition technology for the growth of ZnO films. Many works are seen in the low temperature growth of this interesting ZnO system both undoped and metal doped (Tang et al, 1998, Cracium et al, 1994; Gorla et al, 1999; Kotlyarchuk et al, 2005) thin films and nano thin films.

Fe addition is expected to enhance the sensing ability of this system. When sensing of 300 ppm ethanol, the in situ extended X-ray absorption fine structure (EXAFS) spectra indicate that the bond distances of Zn-O and Fe-O are 1.90 and 1.98 Å, respectively and restored to 1.91 and 1.97 Å in the absence of ethanol (Hsu et al., 2007). The single-phase Fe-doped ZnO films ($x \leq 0.04$) exhibit ferromagnetism above room temperature (Wei et al., 2006). The Fe/ZnO composites display photoluminescence properties different from those of the ZnO entities generated similarly but in the absence of Fe powder: (i) a UV emission band is observed over the latter but not over the former, and (ii) the former shows an emission peak around 583 nm (wavelength) with intensity significantly stronger than that of the latter. Because of the encapsulation of the Fe nanoparticles, the Fe/ZnO composites are highly stable in air and high magnetization (Yang et al., 2009). Room temperature ferromagnetic Fe:ZnO film was prepared by chemical deposition of ZnO film and chemical introduction of Fe impurity in aqueous solutions. (Nielsen, et al., 2008) have reported on Magnetization measurements show clear ferromagnetic behavior of the magnetite layers with a saturation magnetization of $3.2\mu_B$/f.u. at 300 K. Their results demonstrate that the Fe3O4/ZnO system is an intriguing and promising candidate for the realization of multifunctional heterostructures. During recent years different techniques such as chemical vapor deposition (Hu, et al., 1991), magnetron sputtering (Jiang, et al., 2003), spray pyrolysis (Pawar, etal., 2005), sol–gel (Cheong et al., 2002), and molecular beam epitaxy (Kato et. al., 2002), have been used for deposition of zinc oxide films on various substrates. The ease with which the deposition can be made and the reproducibility of the method to produce doped oxide films in any laboratory make this method quite attractive (Kato et. al., 2002) (Vijayan et al, 2008, a, b) (Chandramohan et al, 2010). While the sol-gel method with double dipping has produced highly oriented thin films of Li, Mg doped ZnO thin films by (Liu et al., 2005), they have introduced a spin in the dipped films using spin coating set up to spread the films over substrate. In the method used in this work no spinning is required as the dehydrogenation of the films is done using a dip in hot water. The magnetic properties of the metal doped ZnO thin films grown by such a method has not been presented in detail to our knowledge. Nanostructured ZnO not only possesses high surface area, nontoxicity, good bio-compatibility and chemical stability, but also shows biomimeticand high electron communication features. A reagentless uric acid biosensor based on ZnO nanorodes can be synthesized from SILAR synthesized ZnO thin films. The enzyme is immobilized on ZnO nanocombs and nanorods to construct an amperometric biosensor for glucose biosensing. Besides such systems have prospective applications in Dye-sensitized solar cells (DSSCs), Sensor, optoelectronic devices, UV detector, Spintronics, Dilute Magnetic Semiconductor devices etc. SILAR is a modification of the chemical bath deposition technique. In CBD the solution is in contact with a substrate for a longer period and the reaction is slow and the formation is good (Hodes et. al., 2002). Accelerated by Bath temperature, agitation or bath conditions. The film formation results when ionic product exceeds solubility product. In SILAR originally called as multiple dip method by (Ristov, et al., 1987) and named SILAR (Nicolau, et al., 1988) the reaction Cationic and anionic precursor solutions separated for reaction at chosen conditions like temperature, pH, dipping rate, waiting time, number of dipping cycles. Time is important in ionic layer formations. Since deposition is in air too many microstructural variations are possible. The aim of this work is to investigate the influence of the preparation conditions on structural, optical and electrical properties of ZnO and incorporate magnetic properties by doping Mn/Fe leading to dilute magnetic semi conducting system using CBD method. Advantages are effectiveness and simplicity of the deposition equipment, high deposition rates, wide spectrum of deposition parameters for the control and the optimization of film properties, and film thickness. The sum of all these special features enables the growth of oxide thin films at low temperature substrates with perfect crystallinity. The present work is a preparation and characterization of undoped ZnO, Mn- doped ZnO (MZO) and Fe- doped ZnO (FZO) thin films by chemical deposition technique. In which the influence of solution concentration, solution pH value, film thickenss, annealing temperature and concentration of strontium and aluminium atoms of the grown films are investigated. In addition it demonstrates that any dopant can be used in principle along with the precursor to enable them to be included in the system. The technique can be tuned to get the desired morphology and nanocrystallites of desired sizes distributed over any type of substrate for various applications.

2. Synthesis, structure and morphology of the ZnO thin films

2.1. Synthesis

ZnO thin films were grown using a two-step chemical bath deposition technique using a solution comprising 0.1 M Zinc Sulphate (99% emerk), 0.2 M sodium hydroxide with a pH value of 9±0.2 deposited at bath temperature of 90 ºC under optimized condition. For Mg doped ZnO (MZO) thin films $Fe(SO4)_3$ was used at a concentration of 0.1mM. Before deposition, the glass substrates were cleaned by chromic acid followed by cleaning with acetone. The well-cleaned substrates were immersed in the chemical bath for a known standardized time followed by immersion in hot water for the same time for hydrogenation. The process of solution dip (step 1) followed by hot water dipping (step 2) is repeated for known number of times. The cleaned substrate was alternatively dipped for a predetermined period in sodium zincate bath and water bath kept at room temperature and near boiling point, respectively. According to the following equation, the complex layer deposited on the substrate during the dipping in sodium zincate bath will be decomposed to ZnO due to dipping in hot water. The proposed reaction mechanism is according to the following equations (Vijayan et al, 2008, a, b) (Chandramohan et al, 2010).

http://dx.doi.org/10.5772/54327

$$ZnSO_4 + 2\,NaOH \rightarrow Na_2ZnO_2 + H_2SO_4 \uparrow$$

$$Na_2ZnO_2 + H_2O \rightarrow ZnO + 2\,NaOH$$

Part of the ZnO so formed was deposited onto the substrate as a strongly adherent film and the remainder formed as a precipitate. The addition of MSO_4 in the ratio of Zn:M as 100:1 in the first dip solution leads to the formation of Mg/Fe doped zinc oxide nano thin films where M stands for Mg/Fe.

2.2. Structural analysis

The crystallographic structure of the films has been studied by X-ray diffraction (XRD). Fig. 1 shows the XRD spectrum of ZnO, MZO and FZO films deposited on the glass substrate under optimized condition. It can be seen from the XRD data, that all samples are polycrystalline and exhibit the single-phase ZnO hexagonal wurtzite structure [JCPDS (36–1451)]. All peaks in recorded range were identified.

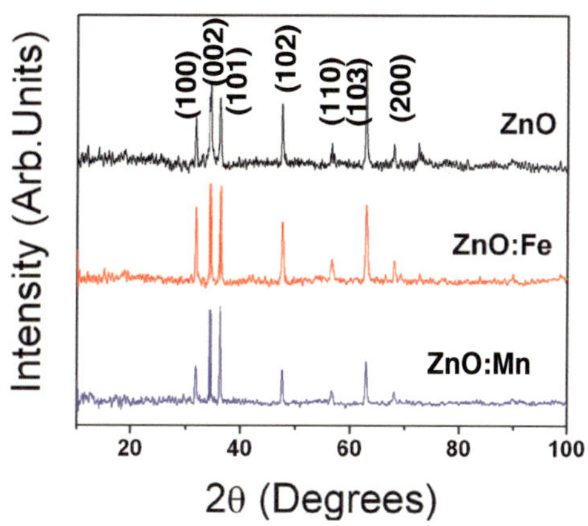

Figure 1. Typical XRD patterns of doped and undoped ZnO thin films

The XRD pattern clearly showed the polycrystalline nature of the ZnO, MZO and FZO films, whose *c*-axis was preferentially oriented normal to the glass substrate. In other words, those grains of undoped and doped films are mainly grown with c-axis vertical to the glass substrate. Hence, the multiple-coating or the piling up of each film was considered not to disturb the overall growth of the films with *c*-axis orientation. Therefore, the *c*-axis orientation may be a common phenomenon in the ZnO film deposition by the chemical process using organo-zinc compounds. Such preferred basal orientation is typically observed also in metal doped ZnO films (Vijayan et al, 2008, a, b) (Chandramohan et al, 2010). Moreover, from the recorded spectrums the minor diffraction peaks of (102) and (103) are approved of randomly oriented of the ZnO film (Roy, at al., 2004). The crystallite size was estimated to be 80 nm for undoped film, 26 nm for MZO and 20 nm for FZO films from the Debye Scherrer formula.

2.3. Morphological studies

Figures 2(a, b, c) shows the scanning electron micrograph (SEM) of ZnO, MZO and FZO films deposited at room temperature. The SEM micrograph of MZO thin film show the uniform polycrystalline surface of the film with a hexagonal morphology consistent with XRD result of P63mc crystal-structure with an average grain size of 300 nm. They are found to be single-crystalline in nature. It can be seen that, films grown at room temperature by varying Mn and Fe concentration consist of slightly agglomerated particles with less voids in the surface of the film with average grains 200 nm for undoped and 300 nm for doped ZnO films (Fig. 2(b)). This result confirmed that Mn and Fe doped into ZnO lattice and in good agreement line with XRD result.

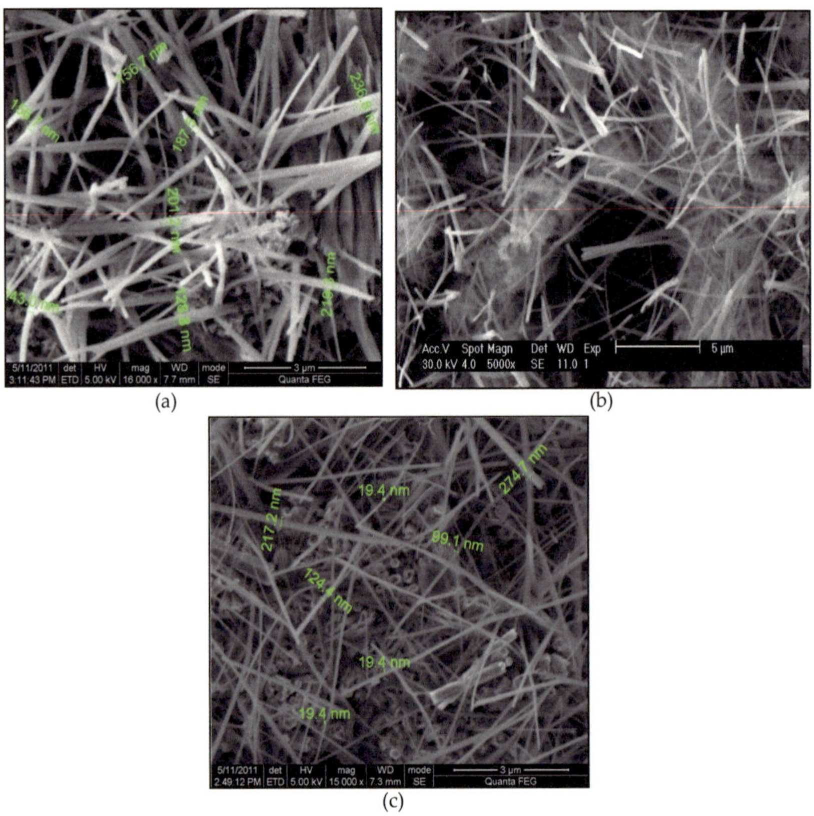

Figure 2. (a,b,c,) Typical SEM micrographs obtained for (a) undoped ZnO thin films and (b) Mn (2%), (c) Fe (2%) doped thin films prepared by SILAR

3. Optical properties of undoped and Fe doped ZnO thin films

Figure 3 shows the transmittance spectrum of the MZO, FZO thin films grown on glass substrate. ZnO is a non-stoichiometric oxide and is known to contain zinc-ion excess de-

http://dx.doi.org/10.5772/54327

fects based on the presence of either zinc interstitial or oxygen vacancies. The films have excellent transmittance and very low absorption and reflectance. The optical band gap of the FZO and MZO thin films has decreased on Fe and Mn doping, respectively. Assuming doping levels are well below Mott's critical density, the change in optical band gap can be explained in terms of Burstein–Moss band gap widening and band gap narrowing due to the electron–electron and electron–impurity scattering. At high doping concentrations, fermi level lifts into the conduction band. Due to filling of the conduction band, absorption transitions occurs between valance band and fermi level in the conduction band instead of valance band and bottom of the conduction band. This change in the absorption energy levels shifts the absorption edge to higher energies (blue shift) and leads to the energy band broadening. While on Mn and Fe doping into the ZnO matrix can explain the increase in shift in the band gap value indicating that either it may due to any charged defects or the charged defects formed had been neutralized by other defects. Hence, the blue shift in the band gap value by Mn and Fe doping suggest an increase in the n-type carrier concentration, most of the Fe ions must be incorporated as interstitial donors into the structure rather than substitution of acceptors.

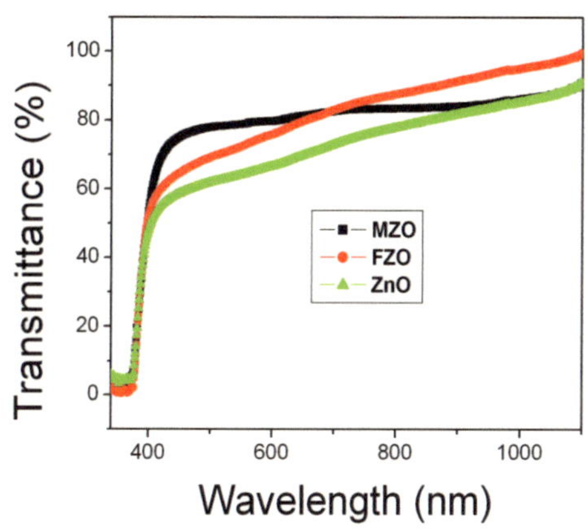

Figure 3. Optical transmittance of typical doped and undoped ZnO thin films

4. Magnetic properties of undoped and Mn, doped ZnO thin films

The ZnO thin films when studied using VSM show a Coercivity (Hci) 238.70G Under Field at Ms/20 G, its Hci, Negative -230.81 G and Hci, Positive 246.60 G, respectively. The undoped sample had Magnetization (Ms) 3.0830E-3 mu/cm^2 and Retentivity (Mr) 368.68E-6 emu/cm^2 respectively. The field at ms/2 0G specifies the absence of any soft or hard magnetic property. The magnetization M-H curve (Figure 4) also shows a behaviour anaalogus to a non magnetic material for undoped ZnO thin films. However the FZO and MZO sample showed the characteristics of a ferromagnetic behaviour magnetic field, and the saturated magnetization (Ms) is 697.32E-6 emu/cm^2. It's a well known fact that the magnetic properties of dilute magnetic systems can be explained by bound magnetic polaron (BMP) model (Hu, et al., 1992; Jiang, et al., 2003). According to this model an impurity site (donor or acceptor) plays the role of a trap and captures the carrier (electrons or holes) to form a bound polaron. These polarons are usually surrounded by the magnetic Mg ions. The polaron interaction with magnetic Mg ions causes the alignment fully or partially to generate the magnetic property of the system. Considering the morphology of the Metal doped ZnO films which had Zn defects and Fe^+/Mn^+ in the ZnO:Mn film. The ferromagnetism observed in the film can be explained by using the BMPs model. The magnetic exchange interaction between ZnO and Fe^+/Mn^+ occupying the same space is aligned with Fe^{1+}/Mn^{1+} spins, forming BMPs. Thus, the sample can exhibit ferromagnetism. Similar behavior of room temperature ferromagnetism is exhibited by Mn doped ZnO thin films grown by SILAR.

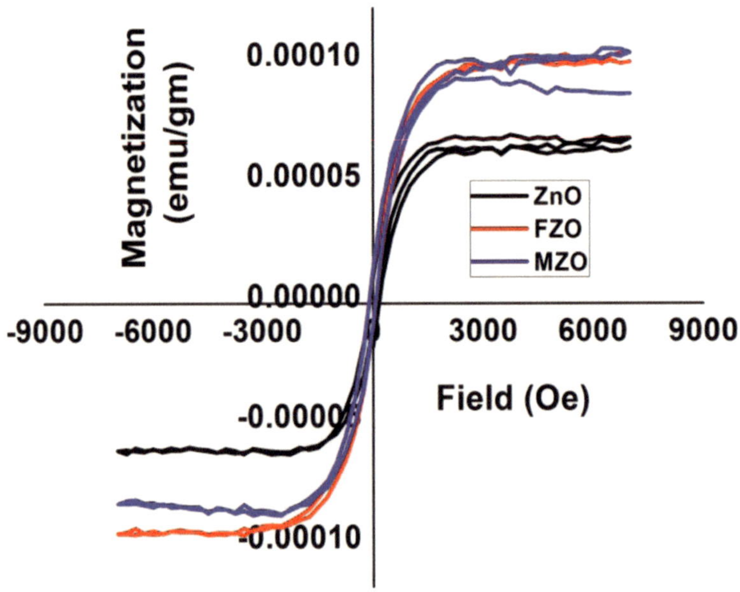

Figure 4. (a) The VSM spectra of undoped and Fe and Mn doped ZnO thin films grown by SILAR.

In summary, Polycrystalline, hexagonal ZnO, MZO and FZO thin films with (002) preferential orientation have been deposited from aqueous solutions using a modified two-step chemical bath deposition technique onto a glass substrate. The microstructure of the films are studied and reported. The studies revealed the potential of this SILAR method in creating and designing varieties of morphologies suitable for various applications. Optical absorption indicated the shift in band gap to 3.21 and 3.22 eV respectively for FZO and MZO films with respect to band gap of ZnO matrix and refractive index to be around 2.34 for FZO films and the bandgap decreased to 3.22 eV for MZO films with a refractive index around 2.3. The transmittance became higher for both MZO and FZO films with increase in doping concentration. In the doped ZnO films, the films were oriented more preferentially along the

(002) direction, the grain size of the films decreased, the transmittance also became higher and the electrical resistivities decreased. It is shown that doped ZnO thin films deposited with a CBD technique can have high temperature ferromagnetic property and this transition may be explained by the BMP model. Extensive characterizations on the structure, micro-structure optical and electrical properties have been made and the exotic choice available in this simple method has paved way for the synthesis of many similar systems by our group like Fe, Mg and Mn doped ZnO thin films and other TCO systems like CdO, etc. Also the properties of these thin film nanocrystallites can be tailored to suit variety of applications like, phosphors, display panels, thermal conduction and opto electronic devices. The technique is easy for automation and anticorrosive coatings can be coated employing doped ZnO systems on to various mechanical spares. The potential of this technique is yet to be exploited in full by the industrial community. The crystallite shape and size control is also feasible in this excellent method.

Acknowledgements

Dr.R.C. Thanks UGC, New Delhi for supporting part of this work through a major research project. Also Dr. R.C thanks Dr. P. Parameshwaran PMD, MSG, IGCAR, Kalpakkam India and C. Gopalakrishnan, Department of Nanotechnolgy, SRM University, Chennai extending SEM facilities.

Author details

Rathinam Chandramohan[1], Jagannathan Thirumalai[2] and
Thirukonda Anandhamoorthy Vijayan[2]

1 Department of Physics,Sree Sevugan Annamalai College,Devakottai, Tamil Nadu, India

2 Department of Physics, B.S. Abdur Rahman University, Vandalur, Chennai, Tamil Nadu, India

References

[1] Chandramohan, R.; Thirumalai, J.; Vijayan, T. A.; ElhilVizhian, S.; Srikanth, S.; Valanarasu, S. & Swaminathan, V. (2010). Nanocrystalline Mg Doped ZnO Dilute Magnetic Semiconductor Prepared by Chemical Route. Adv. Sci. Lett. 3., 3., (September & 2010) 319-322, ISSN: 1936-6612.

[2] Chapparro, M.; Maffiotte, C.; Gutierrez, M.T. & Herrero, J. (2003). Study of the spontaneous growth of ZnO thin films from aqueous solutions.Thin solid films., 431., 1., (May & 2003) 373-377, ISSN: 0040-6090.

[3] Cheong, K.Y.; Muti, N.; Ramanan, S.R. (2002). Electrical and optical studies of ZnO:Ga thin films fabricated via the sol–gel technique. Thin Solid Films. 410., 1-2., (May 2002), 142-146., ISSN: 0040-6090.

[4] Chou, T– L.; Ting, & J– M. (2005). Deposition and characterization of a novel integrated ZnO nanorods/thin film structure. Thin solid films. 494., 1-2., (January & 2006) 291-295, ISSN: 0040-6090.

[5] Cracium, V.; Elders, J.; Gardeniers, J.G.E. & Boyd, L. W. (1994). Characteristics of high quality ZnO thin films deposited by pulsed laser deposition. Appl. Phys. Lett., 65., 23., (October & 1994) 2963-2965., ISSN: 0003-6951.

[6] Das, S.; Chaudhuri, S. (2007). Mg^{2+} substitutions in ZnO–Al_2O_3 thin films and its effect on the optical absorption spectra of the nanocomposite. Appl. Surf. Sci., 253., 21., 8661-8668., ISSN: 0169-4332.

[7] Feng, Z–C.; Chen, C–F.; Kuo, C-T; Williams, K.; Shan, W. The 3rd Asian Conference on Chemical Vapor Deposition (3rd Asian-CVD), Taipei, Taiwan, November 12–14, 2004. Thin solid films., 498., 1-2., (March & 2006) 1, ISSN: 0040-6090.

[8] Gorla, C.R.; Emanetoglu, N.W.; Liang, S.; Mayo, W.E.; Cu, Y.; Wraback, M. & Shen, H. (1999). Structural, optical, and surface acoustic wave properties of epitaxial ZnO films grown on (011-2) sapphire by metalorganic chemical vapor deposition. J. Appl. Phys., 85., 5., 2595-2602., ISSN 0021-8979.

[9] Hodes, G.; Chemical Solution Deposition of Semiconductor Films, Marcel Dekker Inc., Oct. 2002.

[10] Hsu, H.H.; Paul Wang, H.; Chen, C.Y.; Jou C.J.G.; Wei, Y-L. (2007). Chemical structure of zinc in the Fe/ZnO thin films during sensing of ethanol, J. Elect. Spec. Rel Phen., 156-158., (December & 2007), 344-346., ISSN: 0368-2048.

[11] Hu, J.; Gordon, R. G. (1992). Textured aluminum-doped zinc oxide thin films from atmospheric pressure chemical-vapor deposition. J. Appl. Phys. 71., 2., (October & 1991), 880-891, ISSN 0021-8979.

[12] Hu, J.; Gordon, R.G.; (1992). Textured aluminum-doped zinc oxide thin films from atmospheric pressure chemical-vapor deposition. J. Appl. Phys. 71., 2., (October 1991), 880-890, ISSN 0021-8979.

[13] J. Cryst. Growth 237–239., Part 1., (December & 2001), 538-543., ISSN: 0022-0248.

[14] J. Cryst. Growth. 92., 1-2., (October & 1988) 128-142., ISSN: 0022-0248.

[15] J. S. Wellings, A. P. Samantilleke, P. Warren, S. N. Heavens and I. M. Dharmadasa. (2008). Comparison of electrodeposited and sputtered intrinsic and aluminium-doped zinc oxide thin films. Semicond. Sci. Technol. 23., 12., (August & 2008) 125003-125009, ISSN 0268-1242.

[16] Jiang, X.; Wong, F.L.; Fung, M.K.; Lee, S.T. (2003). Aluminum-doped zinc oxide films as transparent conductive electrode for organic light-emitting devices. Appl. Phys. Lett. 83., (July & 2003), 1875-1877, ISSN: 0003-6951.

[17] Jiang, X.; Wong, F.L.; Fung, M.K.; Lee, S.T.; (2003). Aluminum-doped zinc oxide films as transparent conductive electrode for organic light-emitting devices. Appl. Phys. Lett. 83., 9., (July & 2003), 1875-1877., ISSN 0003-6951.

[18] Kato, H.; Sano, M.; Miyamoto, K.; Yao, T. (2002). Growth and characterization of Ga-doped ZnO layers on a-plane sapphire substrates grown by molecular beam epitaxy.

[19] Kotlyarchuk, B.; Sarchuk, V. & Oszwaldowski, M. (2005). Preparation of undoped and indium doped ZnO thin films by pulsed laser deposition method. Cryst. Res. Technol., 40., 12., (December & 2005) 1118- 1123, ISSN: 0232-1300.

[20] Liu, C.; Yun, F.; & Morkoç, H, (2005). Ferromagnetism of ZnO and GaN: A Review J. Mater. Sci: Mat in Electronics. 16., 9., 555-597, ISSN: 0957-4522.

[21] Lokhande, B.J. & Uplane, M.D. (2000). Structural, optical and electrical studies on spray deposited highly oriented ZnO films. Appl. Surf. Sci., 167., 3-4., (October & 2000) 243-246, ISSN: 0169-4332.

[22] Mitra, P.; Chatterjee, A.P. & Maiti, H.S. (1998). Chemical deposition of ZnO films for gas sensors. J. Mater. Sci: Mat in Electronics., 9., 6., (December & 1998) 441-445, ISSN: 0957-4522.

[23] Natsume, Y. & Sakata, H. (2000). Zinc oxide films prepared by sol-gel spin-coating. Thin solid films., 372., 1-2., (September & 2000) 30-36, ISSN: 0040-6090.

[24] Nicolau, Y.F.; Menard, J.C.,; (1988). Solution growth of ZnS, CdS and $Zn_{1-}xCdxS$ thin films by the successive ionic-layer adsorption and reaction process growth mechanism.

[25] Nielsen, A.; Brandlmaier, A.; Althammer, M.; Kaiser, W.; Opel, M.; Simon, J.; Mader, W.; Goennenwein, S.T.B.; Gross, R. (2008). All Oxide Ferromagnet/Semiconductor Epitaxial Heterostructures. Appl. Phys. Lett. 93., 16., (October & 2008) 162510-3., ISSN: 0003-6951.

[26] Pawar, B.N.; Jadkar, S.R.; Takwale, M.G. (2005). Deposition and characterization of transparent and conductive sprayed ZnO:B thin films. J. Phys. Chem. Solids. 66., 10., (November & 2005) 1779-1782., ISSN: 0022-3697.

[27] Peiro, A. M.; Ayllon, J. A.; Pearl, J.; Domenech, X. & Domingo C. (2005). Microwave activated chemical bath deposition (MW-CBD) of zinc oxide: Influence of bath composition and substrate characteristics. J. Crystal. Growth., 285., 1-2., (November &2005) 6-16, ISSN: 0022-0248.

[28] Piticescu, R. R. Piticescu, R. M. & Monty, C. J. (2006). Synthesis of Al-doped ZnO nanomaterials with controlled luminescence. J. Europ. Cer. Soc. 26., 14., (March & 2006) 2979-2983, ISSN: 0955-2219.

[29] Ramamoorthy, K.; Arivanandhan, M.; Sankaranarayanan, K. & Sanjeeviraja, C. (2004). Mater. Chem. Phys., 85., 2-3., (June & 2004) 257-262, ISSN: 0254-0584.

[30] Ristov, M.; Sinadinovski, G.; Grozdanov, I.; Mitreski, M. (1987).Chemical deposition of ZnO films. Thin Solid Films. 149., 1., (May & 1987), 65-71., ISSN: 0040-6090.

[31] Roy, V. A. L.; Djurisic, A. B.; Liu, H.; Zhang, X. X.; Leung, Y. H.; Xie, M. H.; Gao, J.; Lui, H. F.; Surya, C. (2004). Magnetic properties of Mn doped ZnO tetrapod structures. Appl. Phys. Lett. 84, 5., (2004), 756-759., ISSN: 0003-6951.

[32] Ryu, Y. R.; Zhu, S.; Budai, J. D.; Chandrasekhar, H. R.; Miceli, P. F.; White, H.W. (2000). Optical and structural properties of ZnO films deposited on GaAs by pulsed laser deposition. J. Appl. Phys. 88., 1., 201-204, ISSN: 0021-4922.

[33] Sadrnezhaad, S.K. & Vaezi, M.R. (2006). The effect of addition of Tiron as a surfactant on the microstructure of chemically deposited zinc oxide. Mat. Sci. Engg: B., 128., 1-3., (March & 2006) 53-57, ISSN: 0921-5107.

[34] Saeed, T. & Brien, P. O'. (1995). Deposition and characterisation of ZnO thin films grown by chemical bath deposition. Thin solid films., 271., 1-2., (December & 1995) 35-38, ISSN: 0040-6090.

[35] Srinivasan, G. & Kumar, J. (2006). Optical and structural characterisation of zinc oxide thin films prepared by sol-gel process. Cryst. Res. Technol. 41., 9., (September & 2006) 893- 896, ISSN: 0232-1300.

[36] Tang, Z.K.; Wang, G.K. L.; Yu, P.; Kawaraki, M.; Ohtomo, A.; Koinuma, H. & Segawa, Y. (1998). Appl. Phys. Lett. 72., 3270, ISSN: 0003-6951.

[37] Vijayan, T. A.; Chandramohan, R.; Valanarasu, S.; Thirumalai, J.; Venkateswaran, S.; Mahalingam, T. & Srikumar, S.R. (2008). Optimization of growth conditions of ZnO nano thin films by chemical double dip technique. Sci. Tech. Adv. Mater., 9., (April & 2008) 035007, ISSN: 1468-6996.

[38] Vijayan, T. A.; Chandramohan, R.; Valanarasu, S.; Thirumalai, J.; Subramanian, S. P. (2008). Comparative investigation on nanocrystal structure, optical, and electrical properties of ZnO and Sr-doped ZnO thin films using chemical bath deposition method. J. Mater. Sci., 43., 6., (March & 2008) 1776–1782, ISSN: 0022-2461.

[39] Vijayan, T. A.; Chandramohan, R.; Valanarasu, S.; Thirumalai, J.; Subramanian, S. P. (2008). Comparative investigation on nanocrystal structure, optical, and electrical properties of ZnO and Sr-doped ZnO thin films using chemical bath deposition method. J. Mater. Sci., 43., 6., (March & 2008) 1776–1782, ISSN: 0022-2461.

[40] Wang, L.; Giles, N. C. (2003). Temperature dependence of the free-exciton transition energy in zinc oxide by photoluminescence excitation spectroscopy. J. Appl. Phys. 94., 2., 973-978., ISSN: 0021-4922.

[41] Wei, X. X.; C Song, K W Geng, F Zeng, B He and F Pan. (2006). Local Fe structure and ferromagnetism in Fe-doped ZnO films. J. Phys.: Condens. Matter. 18., 31., (July & 2006) 7471-7479., ISSN 0953-8984.

[42] Yang, Y.; Tay, B.K.; Sun, X.W.; Han, Z.J.; Shen, Z.X.; Lincoln, C.; & Smith, T, (2008). Nanoelectronics Conference, INEC 2008 2 nd IEEE International, Nanyang Technical University, Singapore 24-27 March 2008.

[43] Yang, Z. X.; Zhong, W.; Au, C. T.; Du, X.; Song, H. A.; Qi, X. S.; Ye, X. J.; Xu M. H.; Du, Y. W. (2009). Novel Photoluminescence Properties of Magnetic Fe/ZnO Composites: Self-Assembled ZnO Nanospikes on Fe Nanoparticles Fabricated by Hydrothermal Method. J. Phys. Chem. C., 113., 51., (November & 2009) 21269 – 21273., ISSN: 1932-7447.

[44] Zhou, H–M.; Yi, D–Q.; Yu, Z–M.; Rang, L.; Xiao, Li, J. (2007). Preparation of aluminum doped zinc oxide films and the study of their microstructure, electrical and optical properties. Thin solid films., 515., 17., (June 2007) 6909-6914, ISSN: 0040-6090.

Chapter 7

Layers of Inhibitor Anion – Doped Polypyrrole for Corrosion Protection of Mild Steel

Le Minh Duc and Vu Quoc Trung

Additional information is available at the end of the chapter

http://dx.doi.org/10.5772/54573

1. Introduction

1.1. Theoretical background

Almost metals are in contact with wet atmosphere or another aggressive medium such as seawater. Therefore, the corrosion process always occurs on the metal surface. This is also a challenge for scientists to control and reduce the enormous damages due to corrosion. The term 'corrosion' refers to deterioration of materials due to the chemical reactions with the environment. Corrosion is involved in the conversion of the surface of metals in contact with corrosive medium into another insoluble compound. Corrosion is also defined as 'the undesirable deterioration' of a metal or an alloy i.e. an interaction of the metal with its environment affecting the main properties of the metal. Corrosion protection is required for a long life and economical use of equipment in technical processes [1, 2].

Corrosion preventing technology has many options, for instance cathodic protection, anodic protection, use of corrosion inhibitors, forming the precipitates on the metal surface and acting as passive layers, organic coatings etc. Among the methods to prevent corrosion of metals, protection by conducting polymers has been investigated extensively in the recent years [3-18]. This is considered as a possible alternative for friendly-environment coating because an electrochemical process could eliminate the use of toxic chemicals. There are many publications related to conducting polymer in corrosion protection. Conducting polymer can decrease the corrosion rate of many metals such as iron, mild steel, aluminium, magnesium and others [19-25].

Conducting polymer has been investigated extensively for corrosion protection of metal. It is observed that a conducting polymer film alone cannot protect an unnoble metal completely. With a galvanic coupling experiment it could be shown that the polymethylthiophene film did not act as a redox mediator, passivating the steel substrate within the defect and reoxidising itself by dissolved oxygen [26]. Polypyrrole could not provide anodic protection for iron [9, 27]. Conducting polymers like polyaniline, polypyrrole (PPy) etc. can improve the corrosion protection of unnoble metals but it is impossible that the porous films protect the metal surface completely. It is expected that protective properties of polypyrrole can be improved by dopant anion [28].

Counter anions, the so-called dopant anions play an important role in the development of physical properties and morphology. These actions of the anions could be:

- Electroneutralising: dopant anions neutralise the positive charges on the polymer backbone during synthesis of conducting polymers.
- Changing the morphology: the size of the dopant anion can control the microstructure and the porosity of the polymer film.
- Improving the conductivity: the interaction between the positive charges of polymer and anions can influence the conductivity of the polymer films.
- Stabilising the polymer films.
- Compatibility with other polymeric matrices.
- Corrosion inhibition: small dopant anions can be released from the polymer coating when the coating is reduced. If these anions have some inhibiting properties they can provide for some additional protection.

1.2. Mechanism for corrosion protection by polypyrrole

Corrosion at metal surfaces is a severe industrial problem. A large amount of metal is wasted by corrosion. It can cause tremendous economic damages. Minimising this corrosion can save substantial money and prevent accidents due to equipment failure. Corrosion has and continues to be the research object of scientists [1, 2].

Corrosion is an electrochemical process in nature. An anode (negative electrode), a cathode (positive electrode), an electrolyte (environment), and a circuit connecting the anode and the cathode are required for corrosion to occur. For simplicity, the corrosion process of iron in aqueous environment is discussed as a typical example.

The general reaction that occurs at the anode is the dissolution of metal atoms as ions:

$$Fe = Fe^{2+} + 2e$$

Electrons from the anode flow to the cathode area through the metallic circuit and force a cathodic reaction (or reactions) to occur. Depending on the pH of the electrolyte, different cathodic reactions can occur. In alkaline and neutral aerated solutions, the predominant cathodic reaction is

$$O_2 + 2H_2O + 4e \rightarrow 4OH^-$$

In aerated acids, the cathodic reaction could be

$O_2 + 4H^+ + 4e \rightarrow 2H_2O$

In deaerated acids, the cathodic reaction usually occurs is

$2H^+ + 2e \rightarrow H_2$

The corrosion product formed on iron surface in the presence of oxygen is:

$Fe^{2+} + 2OH^- \rightarrow Fe(OH)_2$

This hydrous ferrous oxide ($FeO.nH_2O$) or ferrous oxide $Fe(OH)_2$ composes a diffusion barrier layer on the surface. This layer is green to greenish black in colour. In the presence of oxygen Fe^{2+} is oxidised to Fe^{3+}. Ordinary rust is the product of this step. The formal reaction equation is

$4Fe(OH)_2 + O_2 + 2H_2O = 4Fe(OH)_3$

Hydrous ferric oxide is orange to red-brown in colour. It exists as nonmagnetic Fe_2O_3 (hematite) or as magnetic Fe_2O_3. $Fe_3O_4.nH_2O$ often forms a black intermediate layer between hydrous Fe_2O_3 and FeO. Hence, rust films can consist of up to three layers of iron oxides in different states of oxidation [29].

Beck et al. suggested a model of corrosion protection by PPy [25]. The initial fast corrosion was a superposition of cathodic film reduction and anodic oxidation. Cathodic process was the driving process. The second step was caused of nucleophilic molecules dissolved in the solid. Both processes were of pseudo-first-order.

Jude O. Iroh et al. suggested a corrosion protection mechanism of iron by PPy on the basis of EIS results [10]. The double bonds and the polar –NH group in the ring caused the strong adsorption of PPy and improved corrosion protection. PPy coating was acting as diffusion barrier and was inhibiting charge transfer.

Su and Iroh reported a large shift of the corrosion potential (E_{corr}) nearly 600 mV for PPy(oxalate) on steel compared to bare steel [10]. Reut et al. also recorded this shift of corrosion potential. The different shifts of E_{corr} could be explained by the different pretreatment of the substrate [30]. It was concluded from these results that PPy produced the significant ennobling of steel.

But controversial results were published by Krstajic et al. [9]. It was found that PPy(oxalate) did not provide anodic protection of mild steel in 0.1 M H_2SO_4. PPy was undoped in a short time. The dissolution of the steel continued in the pores of the coating. The mechanism of corrosion protection of steel by PPy is not yet fully understood and it is likely to change with the corrosion conditions [30].

PPy film doped with inhibitor anions such as molybdate were synthesised on mild steel in a one-step process. Corrosion tests indicated a significant improvement of the protective performance of PPy film. PPy coatings even prevented corrosion in defect of the coating. Now the corrosion protection mechanism of PPy with small defect on mild steel will be discussed.

A defect on PPy coating deposited on mild steel substrate may be produced either because of manufacturing or due to damage. When this occurs, bare mild steel is exposed to the corrosive atmosphere. The oxidation of substrate occurs. The following corrosion or anodic reaction takes place at the bottom of the defect:

$$Fe = Fe^{2+} + 2e \quad (1)$$

The open circuit potential of the coated sample drops down to the corrosion potential of iron if the sample is immersed in NaCl (see Figure 11).

The electrons produced in (1) are consumed in cathodic reactions as followed:

$$O_2 + 2H_2O + 4e = 4OH^- \quad (2)$$

$$PPy_{OX}\left(DOP\right) + ne = PPy_{RED} + DOP^{n-} \quad (3)$$

Where PPy_{OX}, PPy_{RED} are the oxidised and reduced states of polypyrrole, respectively. DOP^{n-} is a dopant anion.

The PPy film is reduced (reaction 3) as a result of the galvanically coupling to metal substrate. Reaction (2) takes place within the defect as well as on PPy film if Ppy is conductive and can mediate the electron transfer [31, 32].

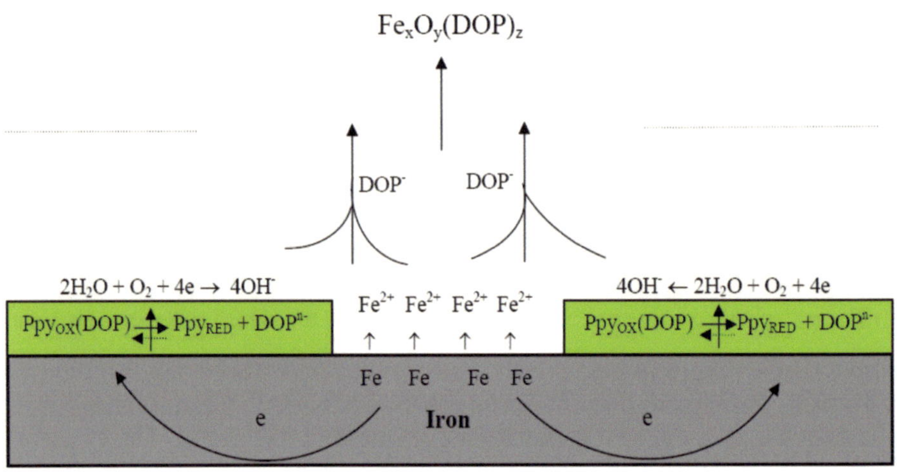

Figure 1. Model for the mechanism of corrosion protection of PPy proposed by Plieth et al. [34-36].

The dopant anions DOP^{n-} form insoluble salts or complexes [$Fe_xO_y(DOP)$] with iron ions and can prevent further corrosion. In other words, the defect is repaired by the inhibitor anions produced during the reduction of PPy film.

Based on the model of Kinlen [33], the proposed mechanism of corrosion protection of PPy is developed as in figure 1 [34-36]. However, from the results of the OCP and EIS measurement re-oxidation of conducting polymer by of oxygen could not be found.

In most studies on the corrosion protection of mild steel by PPy, the role of the dopant anion as corrosion inhibitor has been investigated somewhere. The use of large anions such as polystyrensulfonate, dodecyl sulphate could improve the corrosion protection of the PPy film by preventing the penetration of chloride [20, 21]. Corrosion is not inhibited if the coating has small defects. The defect is protected only if inhibitor anions can diffuse to the defect. Thus, the mobility of dopant anions is one of the important parameters. The release of dopant anions from the polypyrrole film is the first step of corrosion inhibition. Further studies presented in this chapter are necessary to understand the role of dopants in PPy film for corrosion protection. In addition, the organic coatings (such as epoxy) containing nanocomposites based on suitable anion doped PPy seems to be an useful solution for application to replace the toxic cromate paitings. Therrefore, in this chapter the corrosion protection of mild steel of epoxy coatings using doped PPy nanocomposites is also presented.

2. Experimental and analytical methods

2.1. Chemicals

Pyrrole monomers (Aldrich, 98-99%), $LiClO_4$ (Fluka, P.A), $(C_4H_9)_4NBr$ (Merck, P.A), Na_2MoO_4 (Aldrich), Na_2MoO_4 (Acros) and NaCl (J.T.Baker, P.A). Clay obtained from Di Linh mine, Vietnam, was refined by suspension method and then was sodized. Epoxy resin was received from Dow (D.E.R 324); hardener DETA (Diethylentriamin) was also perchased from Dow (USA) with amount of 10%.

2.2. Equipment

The following equipment was employed:

* EG&G-263A Model potentiostat/galvanostat.

* IM6 and Zenium impedance measurement system of ZAHNER-Elektrik. The ZAHNER simulation software was integrated in this system. The frequency range used was 100 kHz – 0.1 Hz.

* The network analyser Advantest R3753BH was connected to the EG&G-263A Model potentiostat/galvanostat to measure the impedance of the quartz electrode in EQCM.

* SEM pictures were obtained with Zeiss DSM 982 Gemini microscope (Carl Zeiss, Germany).

* Raman Spectrometer Series 1000 – Renishaw.

2.3. Cell of measurements

2.3.1. Cell for electropolymerisation

The cell consisted of two parts which can be screwed together by Teflon screws. The working electrode (WE) was placed between of them. A silicon ring was used for sealing. The surface of the WE was 0.64 cm^2. The Pt sheet counter electrode (CE) was placed in a narrow slot. The distance between anode and cathode was 2.5 cm. The reference electrode was connected with the working electrode through a salt bridge.

2.3.2. Cell for EIS measurement

The EIS cell consisted of two parts which could be fixed to each other with four metal screws at the corners. The working electrode was placed between these parts. A silicon gasket defined the immersion surface of working electrode (0.125 cm^2). A glass container was placed on the upper part to hold the electrolyte. Pt net was the counter electrode.

2.4. Pretreatments of substrates

The substrates were pretreated as below:

- Mild steel (20 x 20 mm^2): Polishing with emery paper No 600; rinsing in ethanol in an ultrasonic bath for about 15 minutes; then drying in an N_2 stream.
- Passive film on mild steel (20 x 20 mm^2): Treating as described above; immersion in 0.1 M Na_2MoO_4 for 60 minutes, 30^{0}C; potentiostatic passivation at 0.5 V_{SCE} for 1 hour; rinsing in distilled water and drying in an N_2 stream.

2.5. Condition for electrochemical polymerisation of polypyrrole

The PPy films were generated galvanostatically on the pretreated mild steel surface at a current density of 1.5 mA.cm^{-2} in an aqueous solution of 0.1 M pyrrole (from Aldrich 98%, keep at 4°C and distil in argon atmosphere before using) and 0.01 M sodium molybdate (pH=4.8). After forming, the sample was rinsed in distilled water and dried in nitrogen atmosphere. The film thickness was about 1–1.2 μm. For investigation of the release behaviour of molybdate anion during the reduction of PPy, the films were electrodeposited on Pt.

2.6. Condition for chemical preparation of molybdate anions doped polypyrrole/ montmorillonite nanocomposites PPy(MoO_4)/MMT

Nanomposites were prepared following a procedure described in Ref. [37, 38]. Firstly, a dispersion was prepared by mixing (for 30 min.) of 3.0 g Na^+-MMT, 0.87 g dopant and 1.0 mL pyrrole monomers in distilled water. Then 3.94 g $(NH_4)_2S_2O_8$ was added to the dispersion during stirring. The colour of the mixture was changed from grey to black. After 2 hours of stirring, the solid product was cleaned by distilled water, filtered and dried at 40-50^{0}C for 6 hours under low pressure.

2.7. Preparation of epoxy coatings

Epoxy coatings containing molybdate anions doped PPy/montmorillonite nanocomposites PPy(MoO_4)/MMT were prepared by spray. The thickness of the coatings are 50-60μm. The amounts of PPy(MoO_4)/MMT in the coatings are 2; 3; 4% (by weight).

2.8. Electrochemical impedance spectroscopy and its interpretation

Electrochemical Impedance Spectroscopy (EIS) primarily characterises a conducting polymer system in terms of electrical properties. The electrochemical behaviour of the polymer film is substituted by an equivalent circuit. The typical Bode plot and the equivalent circuit of a Ppy film are shown in Figure 2 and in Figure 3, respectively.

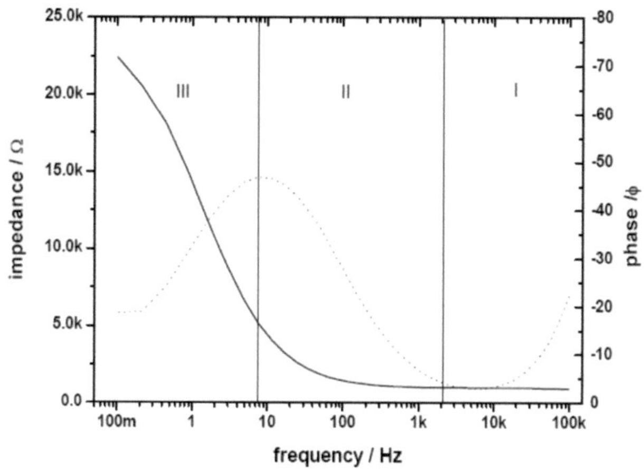

Figure 2. Bode of PPy film: Impedance (solid line) and phase angle (dot line)

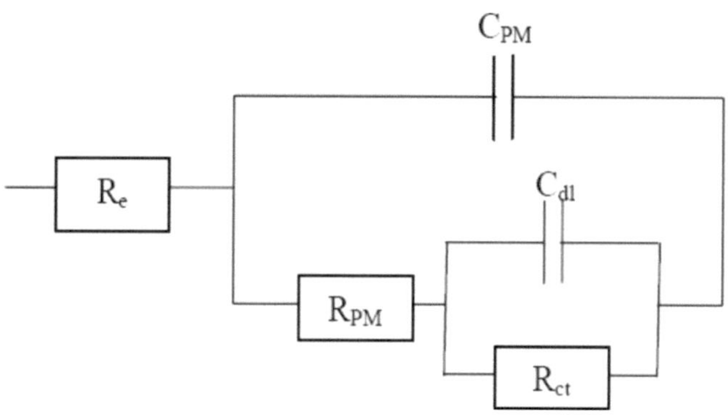

Figure 3. Equivalent circuit for fitting the impedance spectrum of PPy films [34-36].

The obtained impedance spectra of the polymer film can be divided in three regions:

- Region I (*high frequency > 1 kHz*): characterises the behaviour of the electrolyte. Phase angle is nearly zero (dotted line in Figure 2). The resistance of the electrolyte is described by Re in the equivalent circuit (Figure 3).

- Region II (*middle frequency*): shows the properties of the polymer film. The PPy film behaves as dielectricum with a capacitive impedance. The behaviour is simulated by a capacitance C_{PM} parallel to the resistance of the polymer film R_{PM}.
- Region III (*low frequency*): represents the interface polymer/substrate. C_{dl} and R_{ct} are the capacitance of double layer and the charge transfer resistance of the interface, respectively.

The experimental EIS data could be modelled by the equivalent circuit in Figure 3 using the fitting procedure of Zahner software.

3. Results and discussions

3.1. Molybdate anions doped polypyrrole films for corrosion protection

3.1.1. Electropolymerisation of pyrrole on mild steel

Electrochemical polymerisation of pyrrole on metals such as Fe, Zn, Al, and Mg is prevented by the oxidation of these metals because the oxidation potentials are lower than that of pyrrole. The dissolution of metals is so large that the PPy film has no adhesion to the substrate. This problem can be overcome by many methods [39], one of them is the proper selection of the electrolyte.

The role of molybdate as an inhibitor stored in the PPy film as well as the use of the PPy to protect mild steel from corrosion is discussed in the part of theoretical background.

3.1.2. Behaviour of mild steel in molybdate solution

Figure 4 shows the open circuit potential (OCP) vs. time curves of the mild steel electrode immersed in aerated and deaerated aqueous solution of 0.01 M Na_2MoO_4.

As seen in Figure 4, in both cases, the OCP shifts rapidly to positive potentials towards the passive region of mild steel. This potential rise can be ascribed to the reaction between molybdate anions and mild steel as soon as the electrode is immersed into the solution. The insoluble product can block the surface and make the surface potential more positive. This effect can be seen in deaerated medium. In the presence of oxygen, however, the OCP shift is faster and achieves larger positive potentials.

In other words, MoO_4^{2-} compound acts as an oxidant and passivates the surface of mild steel surface shortly even if without oxygen. Surface analysis of mild steel exposed to molybdate by XPS, AES and an electron microprobe confirms the presence of FeO.OH in combination with MoO_3 [40]. In contrast, the passivation of mild steel needs oxygen in solutions containing inhibitor such as oxalate, succinate, and phthalate [41].

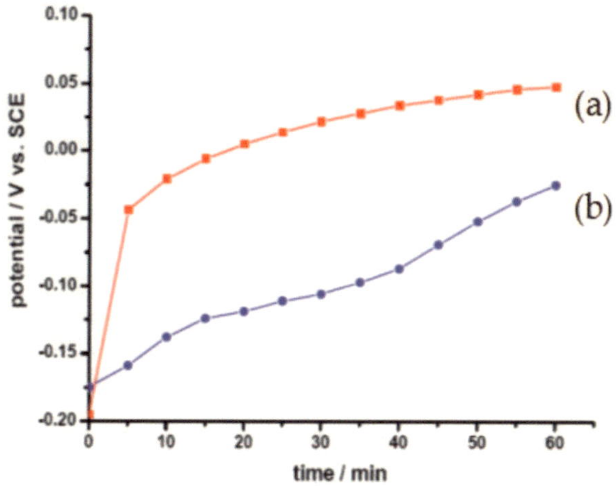

Figure 4. OCP – time curve of mild steel in aerated (a) and in deaerated (b) 0.01 M Na_2MoO_4

3.1.3. Electropolymerisation of pyrrole on mild steel

The PPy films were generated galvanostatically at a current density of 1.5 mA cm^{-2} in an aqueous solution of 0.1 M pyrrole and 0.01M sodium molybdate (pH=4.8). The total charge passed was 0.9 C cm^{-2}.

After formation, the sample was rinsed in distilled water and dried in nitrogen atmosphere. The potential-time curve for galvanic deposition of PPy(MoO_4) on mild steel is shown in Figure 5.

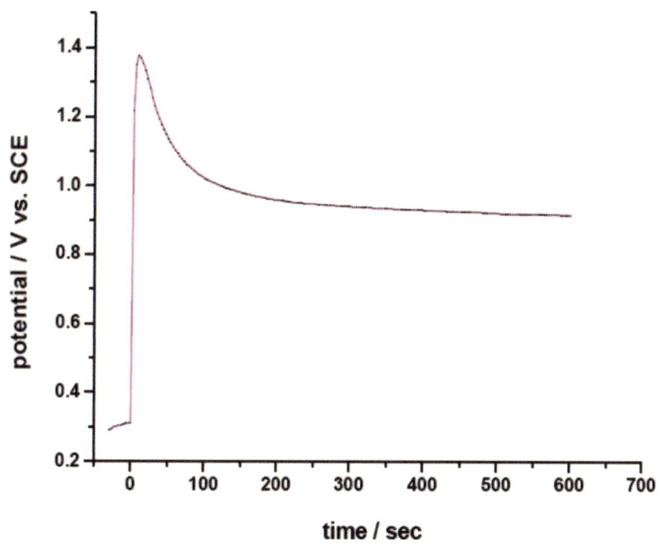

Figure 5. Electrochemical polymerization of pyrrole on mild steel (i = 1.5 mA.cm^{-2}; 0.01 M MoO_4^{2-}; pH = 4.8; 0.1 M pyrrole)

The polymerisation starts at a constant current of 1.5 mA.cm^{-2} for which the potential increases rapidly and then decreases slowly. After 100 seconds, the potential is stabilised at about 0.9 V_{SCE} which corresponds to the oxidation potential of pyrrole.

This behaviour indicates that the dissolution of mild steel is prevented. Pyrrole can be oxidised on mild steel. However, the oxidation potential of pyrrole is higher than normal. The reasons may be: i) the barrier effect of the passivating layer of molybdate on the surface ii) the lower conductivity of the electrolyte.

It should be noted that the polymerisation in presence of molybdate on mild steel occurs without an induction period. This is in contrast to other procedures which have been reported [20, 42, 43]. The electropolymerisation of pyrrole in oxalate aqueous solution is a typical example to elucidate the role of molybdate in the polymerisation. An induction time is present during the polymerisation as shown in Figure 6.

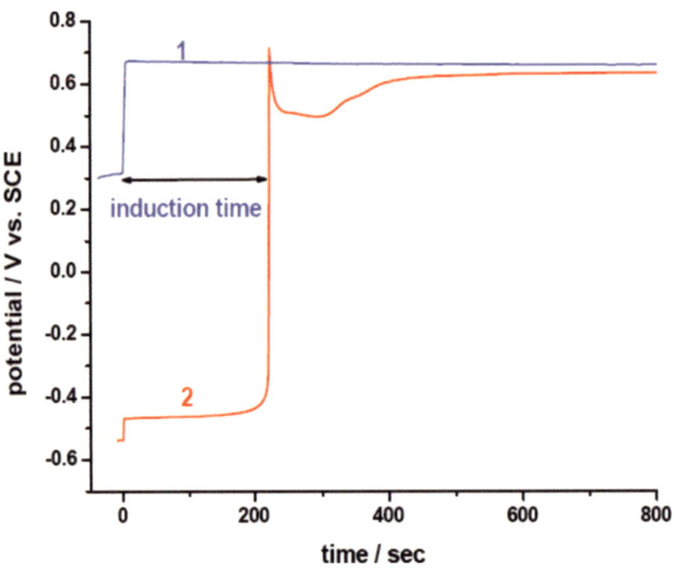

Figure 6. Potential – time curve for the electrodeposition of PPy on Pt(1) and on mild steel (2) from 0.1 M $H_2C_2O_4$; 0.1 M pyrrole, at 1 mA.cm^{-2}

Obviously, the surface of mild steel needs nearly 200 seconds to be passivated in oxalic acid ($H_2C_2O_4$). The induction time is attributed to the active dissolution of mild steel. This active range is assigned at the negative potential. Next, the formation of Fe-oxalate results in the potential shift towards positive potentials high enough for oxidation of pyrrole. Finally, the positive potential levels off until the end of polymerisation. The behaviour is different from that of the electropolymerisation on Pt. Pyrrole can be oxidised immediately on Pt after applying the current through the cell.

3.1.4. Characterisation of PPy(MoO_4)/mild steel film

3.1.4.1. Film morphology

Figure 7 shows SEM micrographs of PPy films doped with MoO_4^{2-} on mild steel (left) and on Pt (right). The total consumed charge for deposition was about 0.9 C cm^{-2}. The films formed on mild steel are homogenous, compact, but thinner than on Pt. The typical cauliflower structure of PPy films on mild steel is observed.

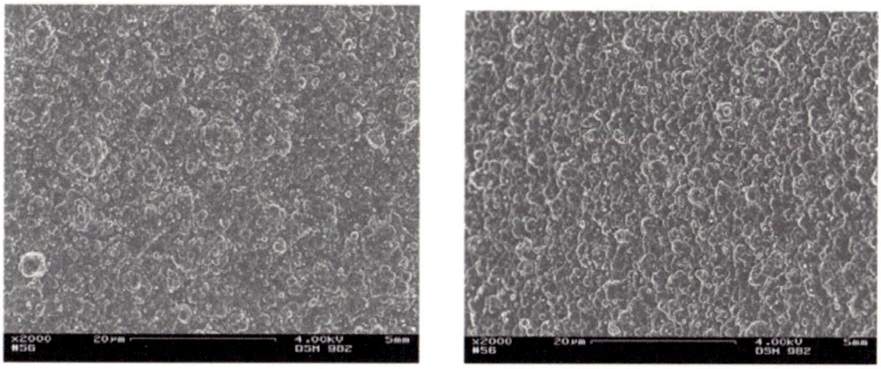

Figure 7. SEM pictures of PPy(MoO_4) on mild steel (left) and on Pt (right)

3.1.4.2. Thickness of PPy films on mild steel

A PPy film was formed on mild steel at the condition mentioned above, and then covered by conducting resin. A cross-section of the sample was made by cutting and polishing with 1200 emery paper. SEM and Energy Dispersive X-ray (EDX) measurements were carried out to determine the thickness of the PPy film.

As seen in Figure 8, three parts can be distinguished clearly. The Ppy film is determined in the middle part. The thickness of the Ppy film was estimated to about 1.2 μm, the charge consumed was 0.9 C cm^{-2}. The result is reproducible.

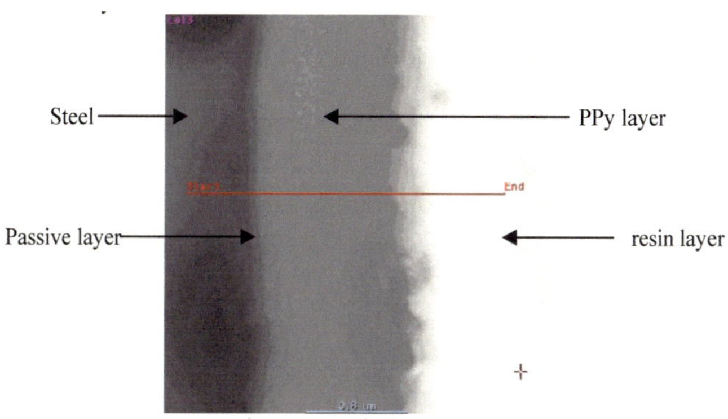

Figure 8. Figure 8. Cross-section SEM of PPy(MoO_4)/mild steel total charge passed is 0.9 C.cm^{-2}

It has been known that total charge for electropolymerisation of pyrrole is about 0.4 C $cm^{-2}\mu^{-1}$ on inert electrodes such as platinum. The value obtained in the current experiment is 0.75 C $\mu^{-1}cm^{-2}$. The reason of this difference may be: a part of charge is used for the passivation of mild steel with molybdate.

3.1.4.3. XPS analysis

XPS spectra for molybdate in PPy(MoO_4) on mild steel presented in [44] showed the XPS surface analysis of the PPy(MoO_4) on mild steel. PPy was electrodeposited under similar conditions mentioned above.

The complex spectra of Mo 3d peaks correspond to the chemical states. The peaks are assigned: $Mo1_{3d5}$ (231 eV), $Mo1_{3d3}$ (233 eV), $Mo2_{3d5}$ (234.5 eV) and $Mo2_{3d3}$ (236 eV) [45]. The XPS spectra are supposed that molybdate should be in two compounds: $[MoO_4]^{2-}$ (62%) and $[Mo_7O_{24}]^{6-}$ (38%). Both types of molybdate are doped in PPy.

One should take into account the fact that molybdate exist in different forms depending on the pH of solution [46]. This relation can be presented:

$$\underset{pH=6.5-4.5}{[MoO_4]^{2-}} \Leftrightarrow \underset{pH=2.9}{[Mo_7O_{24}]^{6-}} \Leftrightarrow \underset{pH<2}{[Mo_8O_{26}]^{4-}} \Leftrightarrow [Mo_{36}O_{112}]^{8-} \Leftrightarrow \underset{pH=0.9}{MoO_3 \cdot 2H_2O}$$

In this condition of electropolymerisation (pH about 4.8), the presence of $[MoO_4]^{2-}$ and $[Mo_7O_{24}]^{6-}$ in polymer film is obvious.

3.1.5. Anion release from Ppy(MoO_4)/mild steel

Electrochemical behaviour of PPy on mild steel was characterised by EIS combined with cyclic voltammetry. Figure 9 shows the change of the resistance R_{PM} and the capacitance C_{PM} of PPy(MoO_4)/mild steel with the potential.

In the potential range from 0.6 V - 0.2 V_{SCE}, the PPy film is conductive; R_{PM} is small (about 20 Ω). Following the negative scan, R_{PM} increases gradually. The reduction of the PPy film begins at 0.2 V_{SCE}. The behaviour is reversible (see reverse scan) i.e. R_{PM} decreases nearly to the original value. The PPy(MoO_4) film is now in the oxidised state.

The change of the film capacitance C_{PM} is inversely proportional to that of R_{PM}. It decreases in the negative scan and increases in the positive scan gradually. It is observed, however, that the film capacitance does not return to the original value. The explanation for this phenomenon is that the transport of solvent (water) accompanies by the anion exchange. It leads to the conformation changes of the conducting polymer.

Hence, it can be concluded that:

- PPy(MoO_4) can be synthesised electrochemically on active metals like mild steel, in a one-step process. The dissolution of mild steel can be prevented with molybdate. The film is homogenous and adhesive on mild steel.

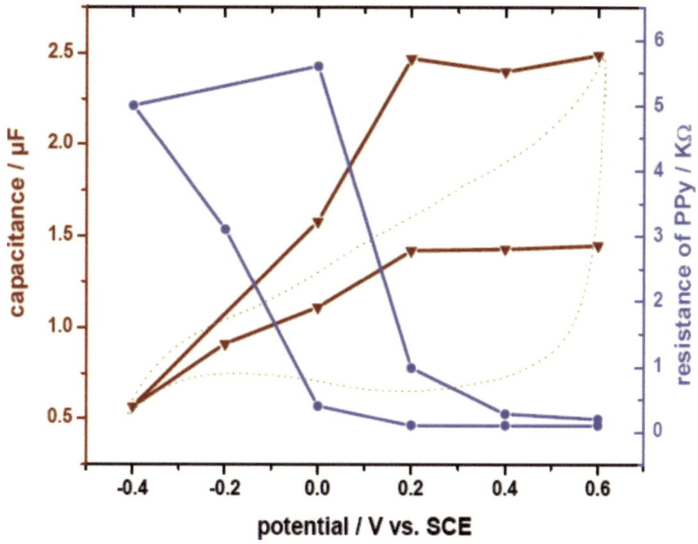

Figure 9. Change of R_{PM} and C_{PM} during reduction of $PPy(MoO_4)$/mild steel in 0.1 M $(Bu)_4NBr$, N_2, Dotted line is the CV

- $PPy(MoO_4)$ film formed on mild steel has the same redox property as on Pt. During reduction in $(Bu)_4NBr$ solution the decrease of film capacitance is observed. It means that molybdate releases from the film.
- There is no evidence of film decomposition. Corrosion test of $PPy(MoO_4)$ on mild steel

3.1.6. Polarisation curves

The protective effect of $PPy(MoO_4)$ film on mild steel was examined in aerated 0.1 M NaCl. The corrosion potential E_{corr} and corrosion current density i_{corr} was determined by extrapolation of anodic and cathodic Tafel lines.

Figure 10 indicates that E_{corr} shifts towards positive potentials (about 500 mV) and i_{corr} decreases about one order in magnitude as compared to the bare mild steel electrode. The polymer film can prevent the metal surface from corrosion. The inhibiting efficiency E of $PPy(MoO_4)$ is obtained by equation:

$$E = \frac{i^o - i}{i^o} \%$$

Where i^o, i denote the corrosion current of bare mild steel and $PPy(MoO_4)$/mild steel, respectively. The inhibiting efficiency E is about 98% for $PPy(MoO_4)$ on mild steel.

3.1.7. OCP measurement

The corrosion behaviour of mild steel covered by $PPy(MoO_4)$ films was investigated by OCP-time measurements. The samples were immersed in a 0.1 M NaCl solution as corrosive medium and the OCP was recorded versus time. The protection time is characterised by the

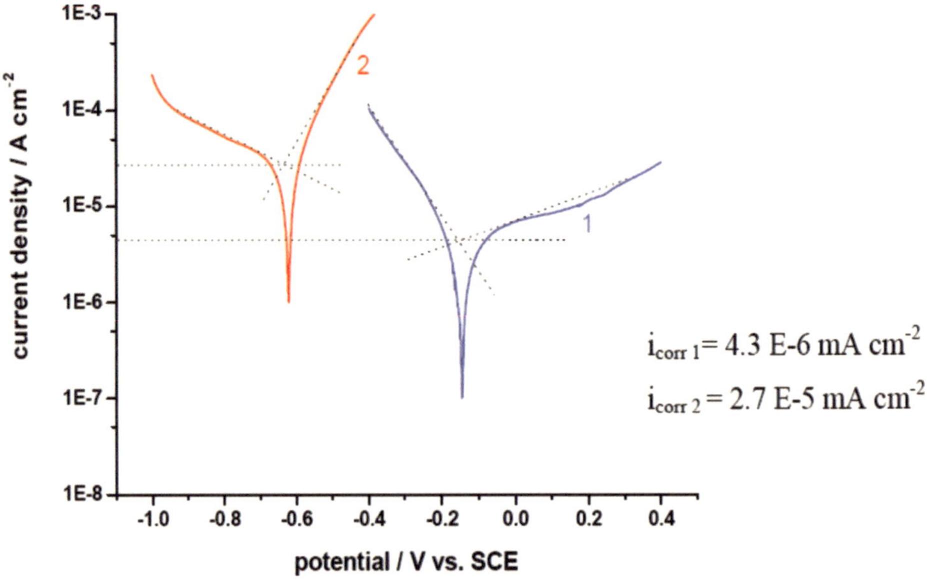

Figure 10. Potentiodynamic curve of PPy(MoO_4)/mild steel film (1) and bare mild steel (2) in 0.1 M NaCl, 1 mV.s^{-1}

time during which the OCP of the covered electrode remains in the passive state of mild steel before it drops down to the corrosion potential of unprotected mild steel.

The OCP-time curve of PPy(MoO_4) coated on mild steel in NaCl is presented in Figure 11. The OCP is initially positive at about 0.35 V_{SCE} which corresponds to redox potential of PPy. The mild steel electrode maintains in its passive state for about 7 hours. Then, the potential sharply decreases to a second plateau at about -0.2 V_{SCE} and is stable there for about 5 hours. In this plateau, chloride anions have reached the metal surface through the pores of PPy film. The anodic reaction can take place and the polymer is reduced partially. This reduction causes the molybdate anion release which is needed to slow down the corrosion rate. This may be a reason why OCP is stable at plateau 2. The release of molybdate from the PPy film in this second plateau was confirmed by EIS measurement shown in Figure 12. The reduction of PPy film causes the increase of the film resistance (marked by an arrow) and the release of molybdate causes the decrease of the film capacitance.

Finally, OCP decreases towards the corrosion potential of the mild steel because of the large concentration of chloride in the polymer/mild steel interface, PPy cannot protect mild steel any longer.

This second plateau is only observed if the dopant anions of the conducting polymer have some ability to inhibit the corrosion reaction of mild steel [9, 20]. If the anions cannot give this protection, the second plateau is missing and the potential falls down to the corrosion potential of mild steel at the end of the first plateau [47].

PPy(MoO_4) has shown the protective ability for mild steel. The corrosion potential is kept at the second plateau where mild steel is in a passive state.

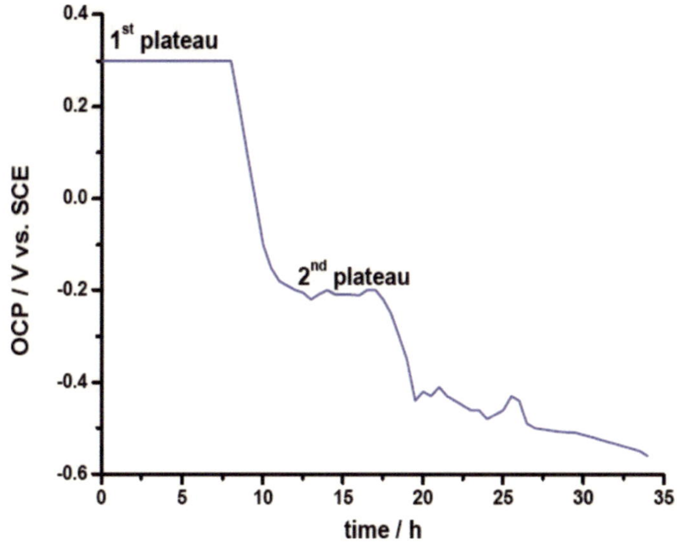

Figure 11. OCP-time curve of PPy(MoO_4)/mild steel (thickness of 1.5 μm) in 0.1 M NaCl

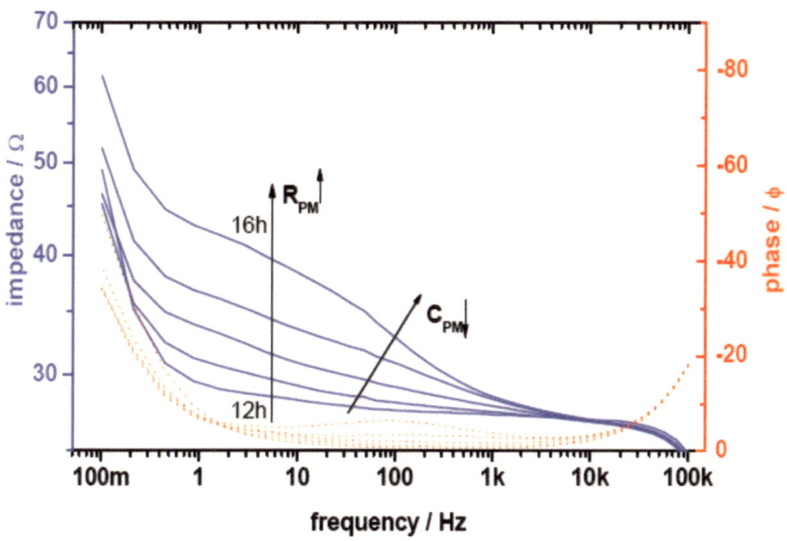

Figure 12. EIS spectra of the PPy(MoO4) film in the second plateau of the OCP in 0.1 M NaCl solution. Impedance (solid line) and phase angle (dotted line)

3.1.8. The role of molybdate passive layers in corrosion protection

It is clear that a passive layer of molybdate is formed on mild steel during the electropolymerisation process. This layer can reduce the oxidation of mild steel and facilitates the polymerisation of pyrrole. In order to clarify the role of molybdate in this passive layer in corrosion protection, a PPy film with the non-inhibitive anions ClO_4^- was electrodeposited on mild steel passivated with molybdate. The corrosion test was carried out in 0.1 M NaCl solution.

A mild steel electrode was passivated in molybdate solution with the following procedure: immersion in 0.1 M Na_2MoO_4 for 60 minutes, 30°C; potentiostatic passivation at 0.5 V_{SCE} for 1 hour; rinsing in distilled water and drying under N_2 stream [48]. The PPy(ClO_4) film was electrodeposited under these conditions: 0.1 M $LiClO_4$, 0.1 M pyrrole monomer, i = 1 mA cm^{-2}. The OCP of PPy(ClO_4) was recorded in 0.1 M NaCl.

Figure 13 shows the OCP-time curve of PPy(ClO_4)/passive layer/mild steel in 0.1 M NaCl. At the beginning, the OCP is still in the passive range of mild steel. PPy(ClO_4) can protect the substrate from corrosion. Nevertheless, this protection only remains for a short time (about 100 seconds). The penetration of chloride through the film is very fast and breaks down the passive film of molybdate formed in the pretreatment procedure. The OCP decreases sharply to the active potential range.

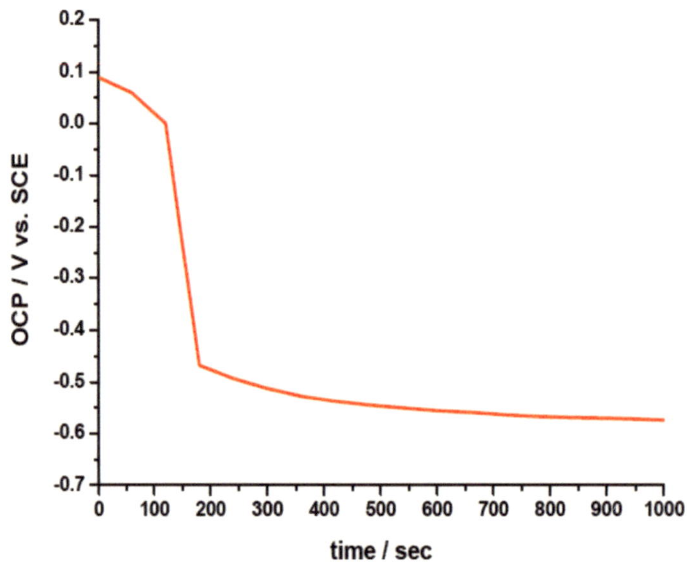

Figure 13. OCP-time vurve of PPy(ClO_4)/passive layer/mild steel in 0.1 M NaCl

This OCP measurement indicates that the passive film of molybdate on the mild steel cannot prevent the penetration of chloride and cannot reduce the corrosion rate. Molybdate in a passive layer under the polymer film does not play any role for corrosion protection.

3.1.9. The possibility of self-healing with PPy(MoO_4) film on mild steel

The self-healing action of PPy(MoO_4) film was investigated. On a fresh PPy(MoO_4)/mild steel film, a small defect (about 0.04 mm^2) was made with a needle. OCP–time curve was recorded in aerated 0.1 M NaCl as seen in Figure 14.

http://dx.doi.org/10.5772/54573

After immersion, the potential decreases immediately and then levels off at a potential of about -0.1 V for 4 hours. This observation could be explained as follows: the dissolution of mild steel at defect occurs immediately after the PPy film contacts with the corrosive environment. Because of the galvanical connection with mild steel, PPy will be reduced to compensate the redox process. This reduction is a driving-force for molybdate anions to be released from the PPy near the defect. A passive compound $Fe_xMo_yO_z$ is produced and it acts as inhibitor in the defect. It results in the maintenance of the potential for a certain time. The fluctuation of the OCP, numerous small spikes of potential are observed, is explained as the breakdown of the passive film by chloride and the re-passivation by MoO_4^{2-} in the defects. A small defect of the PPy film is protected from corrosion by the PPy(MoO_4) film.

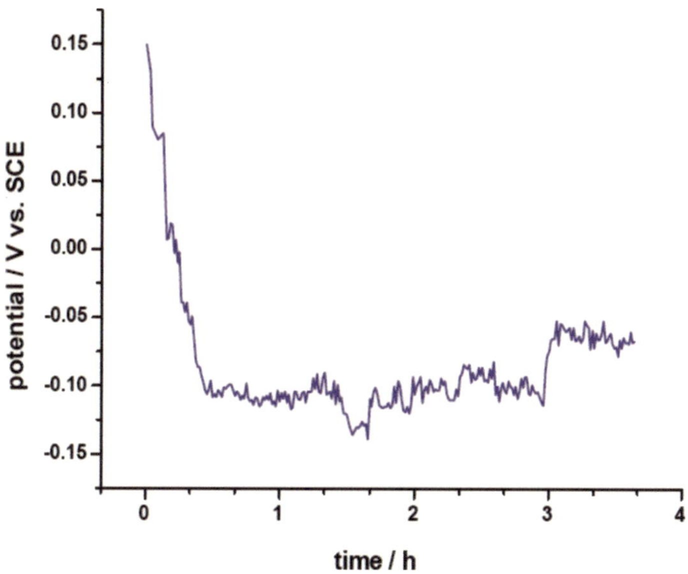

Figure 14. OCP of PPy(MoO_4)/mild steel with a defect (about 0.04 mm^2) *in aerated 0.1 M NaCl*

The same experiment was carried out with a PPy film doped with PF_6^- as non-inhibitive anion. After passivating with molybdate, a mild steel electrode was covered with PPy in 0.1 M $(Bu)_4NPF_6$, 0.1 M pyrrole in dichloromethane at 1.5 mA cm^{-2}. A small defect was made with a needle (0.04 mm^2) on the fresh film. As corrosive medium 0.1 M NaCl was also used.

The polarisation curves shown in Figure 15 are obtained on two PPy films with different dopant anions, namely PF_6^- (curve 1) and molybdate (curve 2). Although there is a molybdate passive layer, the corrosion potential of PPy(PF_6) is still in the active range at the beginning of the experiment. No shift of the corrosion potential is observed here. This behaviour shows that the defect is attacked continuously by chloride. PPy(PF_6) film cannot protect and repair this defect.

On the contrary, a positive shift of the corrosion potential and the decrease of the corrosion current are observed on PPy(MoO_4)/mild steel (curve 2). The defect is passivated for 4 hours.

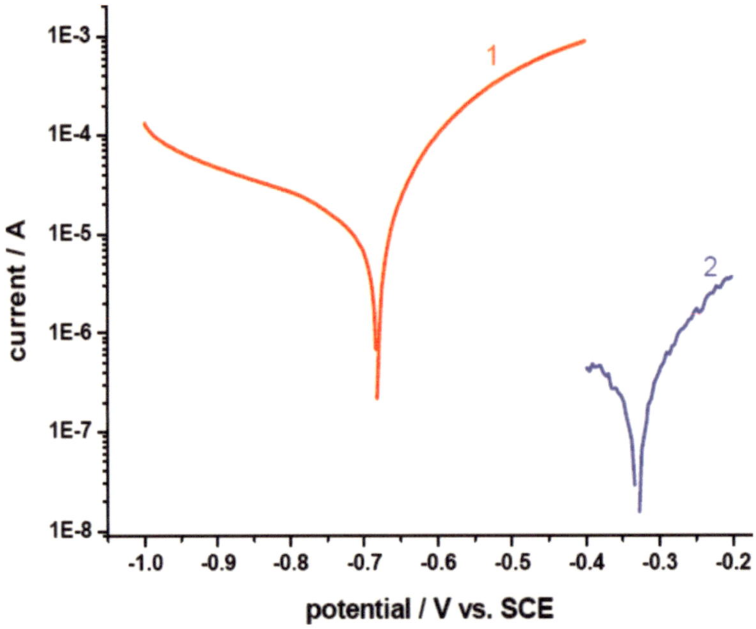

Figure 15. Fe/molybdate passive layer/PPy (PF_6), i=1.5 mA.cm^{-2}, in dichloromethane, about 1.5 µm; (2): Fe/PPy(MoO_4), ca. 1.5 µm, after dipping 4 hours

The corrosion potential remains in the passive range. The corrosion current for PPy(MoO_4) is nearly two orders of magnitude smaller than that of PPy(PF_6).

These results show that molybdate within a polymer plays an important role in self-healing of a defect. Passive layers containing molybdate on mild steel can reduce the dissolution of the active metal during the polymerisation but cannot act as corrosion inhibitor. The self-healing action for mild steel only takes place on PPy films doped with corrosion inhibitors such as molybdate.

3.1.10. Delamination

The corrosion process of PPy(MoO_4) on mild steel was investigated with the Scanning Kelvinprobe (SKP). The experimental set-up and the SKP measurements were made in MPI (Max-Planck Institute for Iron, Dusseldorf - Germany). The artificial defect was prepared on a part of PPy film shown in Figure 16.

The used SKP tip had a diameter of 100µm. Top-coat was polyacryl resin (BASF) applied on the film in order to avoid the difficulties resulting from the folding of the film during the delamination experiment. The humid atmosphere (93 – 95%) was controlled during the experiment.

PPy was formed on mild steel under the following conditions: 0.1 M pyrrole + 0.01 M molybdate aqueous electrolyte; current density 1.5 mA cm^{-2}. The SKP measurements were obtained in 0.1 M KCl solution. SKP experiment of PPy(MoO4)/mild steel in 0.1 M KCl presented in [34]

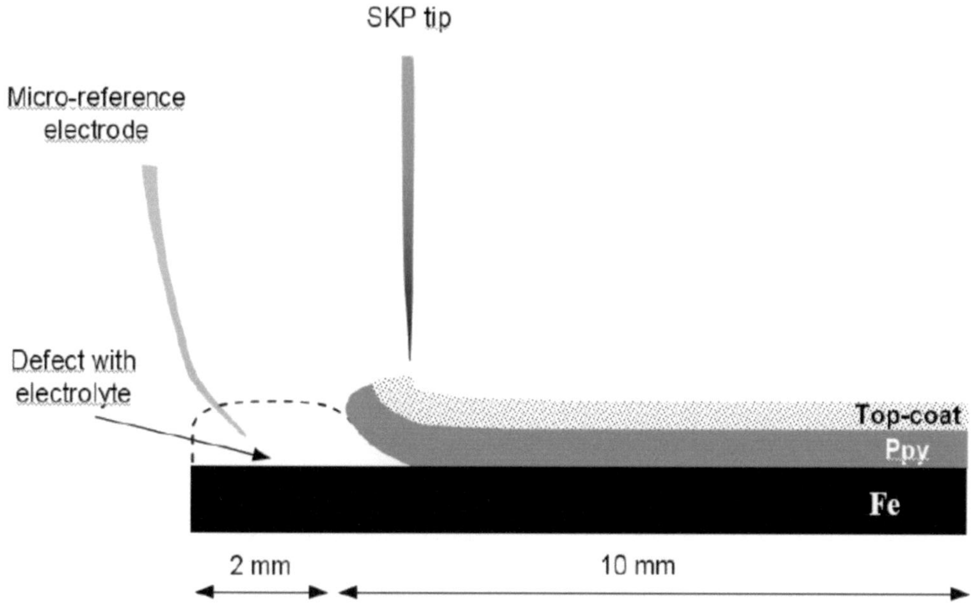

Figure 16. Preparation of iron electrode coated by PPy(MoO_4) folm for SKP experiment [34]

showed the profile of corrosion potential E_{corr} as function of the distance defect border for different times after contact of the defect with electrolyte. The features must be mentioned: (i) the region next to the defect where E_{corr} is similar corrosion potential of bare mild steel (-0.45 V_{SHE}). (ii) a region of abrupt increase low to high values of E_{corr}. This region shifts from left to right i.e. away from border with increasing time; (iii) a region where the adhesion is not yet lost [49].

It can be seen that the delamination of PPy film is very fast in KCl (about 1600 μm for 2 hours). The corrosion potential of the delaminated area remains in the active range potential of corroding iron (about 0.45 V_{SHE}). The passivation of the defect does not take place.

Because of their size, cations K^+ can move into the PPy film easier than the release of molybdate from the film during delamination process. The transport of anions in electrochemical experiment is only few micron while the delamination over hundreds micron, that leads to the predominately cation incorporation into the film for charge compensation. The amount of molybdate is not enough to passivate the large defect.

The delamination of PPy(MoO_4) film on mild steel investigated further in 0.1 M $(Bu)_4NCl$ solution is shown in Figure 17. The delamination is much slower than that in small cation solution. Now the delamination front reaches 1600μm after 33 hours instead of 2 hours in KCl solution. The reduction of PPy at the defect is slowed down. Obviously, the size of cation in electrolyte had effects on the delamination process. The incorporation of cation $N(Bu)_4^+$ is hindered because of their size. The release of molybdate anions is predominant.

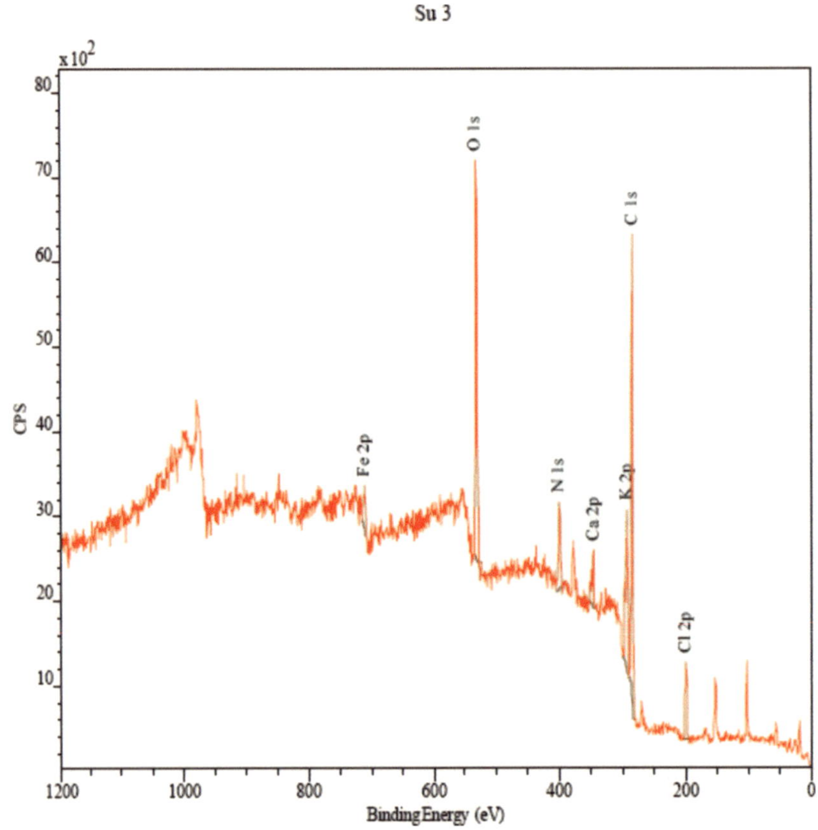

Figure 17. XPS measurement of PPy(MoO_4) on mild steel after delamination in 0.1 M $(Bu)_4NCl$ [44]

The release of molybdate in delamination process is confirmed by the XPS experiment of PPy(MoO_4)/mild steel in 0.1 M $(Bu)_4NCl$ [34]. After delamination, the PPy film was peeled off and molybdate was detected by XPS in the film. No signals of molybdate were found, only iron was seen in the spectra. The presence of iron may come from the corrosion process. The molybdate has moved to the defect during the corrosion process. The same SKP experiment was repeated and the amount of molybdate within the PPy film was measured by XPS at the defect and at the interface polymer/metal. The result is presented in Figure 18.

It can be seen that there is a difference of the amount of molybdate at the defect and at the interface polymer/substrate. At the defect, molybdate anions are expelled from the polymer due to the reduction of PPy when the defect is connected with the substrate galvanically. The amount of molybdate is consumed to form the passive layer and to suppress the cathodic delamination. Nevertheless, the PPy film is still in the oxidised state in the undelaminated region of PPy (point 2, Figure 18). Molybdate anions still remain in the PPy film. This is the reason why the quantity of molybdate in the PPy film at the defect is smaller than at the interface where the PPy protects the metal.

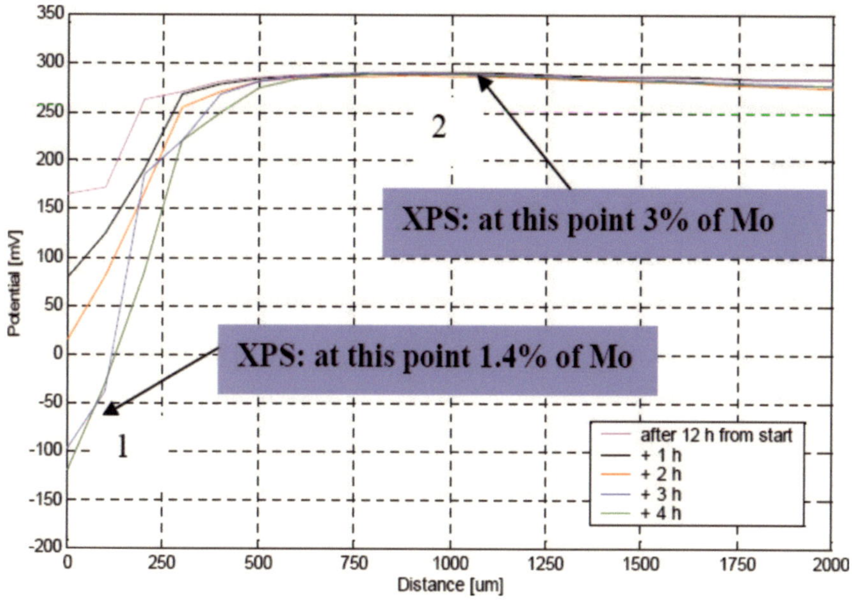

Figure 18. Quality of molybdate in the PPy film at the defect (1) and at the interface polymer/substrate (2) after delamination in 0.1 M $(Bu)_4NCl$ [44]

3.1.11. Raman spectroscopy of PPy(MoO_4)

During the corrosion process, PPy will be reduced because a galvanic cell is established between PPy and mild steel. Raman spectroscopy was used to characterise the state of PPy, oxidised or reduced.

PPy(MoO_4)/mild steel was prepared in aqueous solution of 0.1 M pyrrole + 0.01 M molybdate at 1.5 mA cm^{-2}. Three states of PPy (fully oxidised, partially reduced and totally reduced) were determined through the OCP obtained by dipping samples in 0.1 M NaCl. Raman spectra of these samples are shown in Figure 19.

Several bands are representative for the oxidised state. The band 1600 cm^{-1} belongs to the inter-ring (C=C) of oxidized PPy (0.1 V). It shifts towards low wavenumbers in more negative potential (1597 cm^{-1} at –0.4 V and 1594 cm^{-1} at –0.6 V). The bands 1052 cm^{-1}, 1083 cm^{-1} are assigned to the C-H in plane deformation [50, 51]. They are also shifted to lower wavenumbers during the corrosion process. The Raman peaks of the dopant anion shift from 931 cm^{-1} to 920 cm^{-1} in the reduced state [52]. The PPy(MoO_4)/mild steel film is in the oxidised state while it protects mild steel substrate and is progressively changed to the reduced state during the protection progress.

The properties of PPy(MoO_4) can be summarised:

- PPy(MoO_4) films covered on mild steel have the effect of corrosion inhibition. The E_{corr} is in the passive range of potential in chloride solution. The potential shift is nearly 500 mV_{SCE} compared to bare mild steel. At the same time, i_{corr} decreases one order of magnitude when

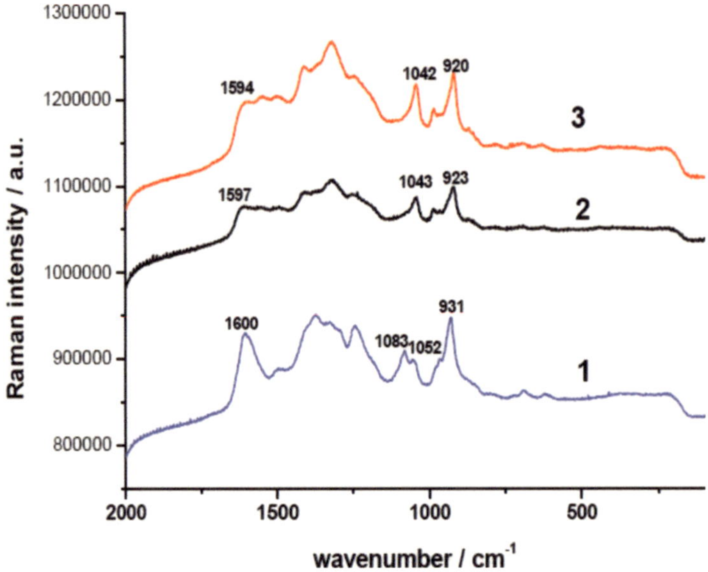

Figure 19. Raman spectra of PPy in different states: (1): complete oxidation (after formation), (2): partial reduction and (3): complete reduction

PPy(MoO_4) covers on mild steel. The protective efficiency is fairly high (about 98%), indicating a good property of PPy(MoO_4) in corrosion protection for mild steel.

- The self-healing effect of PPy(MoO_4) film can be observed by OCP measurement. In contrast, PPy(PF_6) cannot prevent corrosion of mild steel even with no defect. Molybdate in corrosion protection of mild steel is acting as anodic inhibitor.

- The release of the molybdate is necessary for self-healing of defect by PPy. The reduction of the PPy film during the corrosion process is the driving-force to release inhibitor anions. Raman experiments show that PPy changes from the oxidised state to the reduced state during the corrosion process. This observation is reported elsewhere [50, 53]. XPS results confirm this observation.

- The EQCM and EIS results in aqueous solution indicate that the reduction of PPy(MoO_4) is accompanied by a mixed anion/cation transport and by water uptake. However, it is possible to show that the medium size molybdate anion can be released from the film to improve the protective properties of PPy film deposited on mild steel.

- The Scanning Kelvinprobe experiments of PPy(MoO_4) film show that the release of molybdate can be found when the film is reduced in a large size cation solution. The delamination can be stopped. On the contrary, the delamination of PPy(MoO_4) film is fast in a small cation solution. The incorporation of cations is predominant. Therefore, the delamination cannot be hindered [34].

- The release behaviour of molybdate from the PPy film depends much on the size and mobility of cations in the electrolyte.

3.2. Epoxy coatings containing molybdate anions doped polypyrrole/montmorillonite nanocomposites

3.2.1. SEM images

Figure 20 shows SEM micrographs of PPy (a), MMT (b) and PPy(MoO_4)/MMT nanocomposite (c). As shown in Figure 20, PPy is black powder (Figure 20a) and MMT consists of platelet particles accumulating each other and as crystals (Figure 20b). Compared with MMT (Figure 20b), the hackly surface of PPy(MoO_4)/MMT nanocomposite can be seen due to the deposition of the PPy onto the layer surface of MMT (Figure 20c). At high magnification, it is easier to see the flaky structure of MMT (Figure 20c). One can observe (as shown in Figure 20c at the tip of the arrow) the antenna-like PPy "stretching out" from the layer surface of MMT.

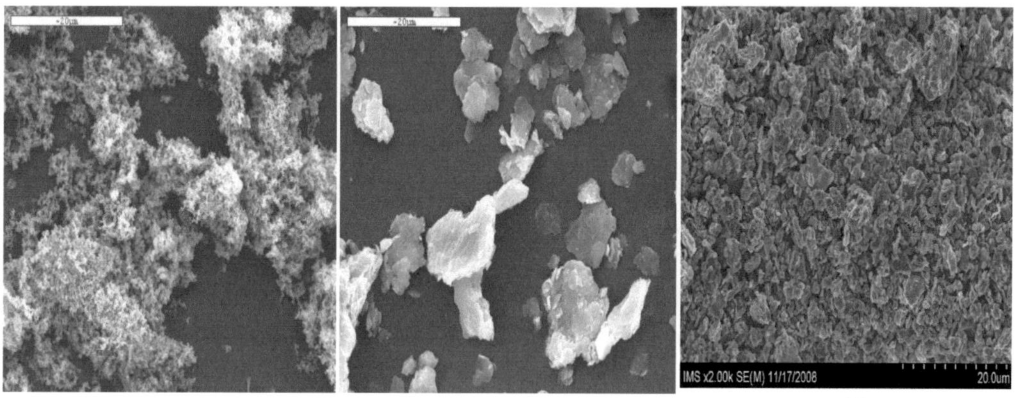

(a) (b) (c)

Figure 20. SEM images of PPy (a), MMT (b) and PPy/MMT nanocomposite (c)

3.2.2. XRD patterns

The XRD patterns of the materials before and after polymerization are shown in Figure 21. At first, Na^+-MMT was mechanically stirred for 30 min as reference. However, there are no changes in the XRD patterns of MMT before and after stirring. Therefore, the stirring does not affect the crystallinity of the MMT itself. The diffraction peak of Na^+-MMT was observed at $2\theta = 7.0°$, therefore, the basal spacing of Na^+-MMT was 1.22 nm (Fig. 21a). The intercalation of pyrrole monomers and dopant into MMT is shown in Figure 21b. The basal spacing increased from 1.22 nm to 1.58 nm ($2\theta = 5.8°$), indicating the expansion of the interlayer space (d-expansion) by 0.36 nm; and the successful intercalation by the mechanical intercalation method. The diffraction peaks of the products after polymerization were shifted to a higher angle than those before polymerization as shown in Figure 21c, indicating the synthesis of PPy in the clay layers. As a result, the the basal spacing of monomer-absorbed MMT changed from 1.58 nm to 1.42 nm ($2\theta = 6.0°$). They are in the agreement with other publications [54].

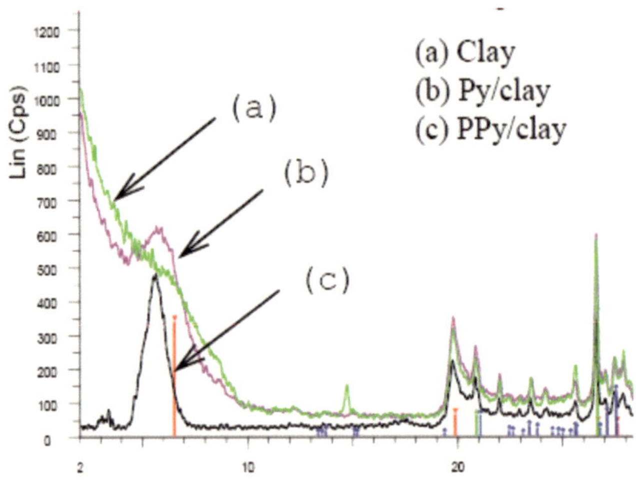

Figure 21. XRD patterns of MMT, monomer-absorbed MMT and PPy/MMT nanocomposites

3.2.3. *Raman spectra*

Figure 22 presents the Raman spectrum of PPy/MMT nanocomposite measured at 514 nm with 1 mW laser power. Table 1 gives the assignments of some Raman bands and compares the frequencies of the various vibration modes collected on the PPy/Ag spectra and with those theoretically calculated by Faulques *et al.* [55]. According to the theoretical values reported by Faulques et al. [55], the vibration frequency of the C=C double-bond of the PPy in the oxidized form (at 1593 cm^{-1}) is greater than that in the reduced form (Table 1). This evedence shows that molipdate anion was doped into PPy.

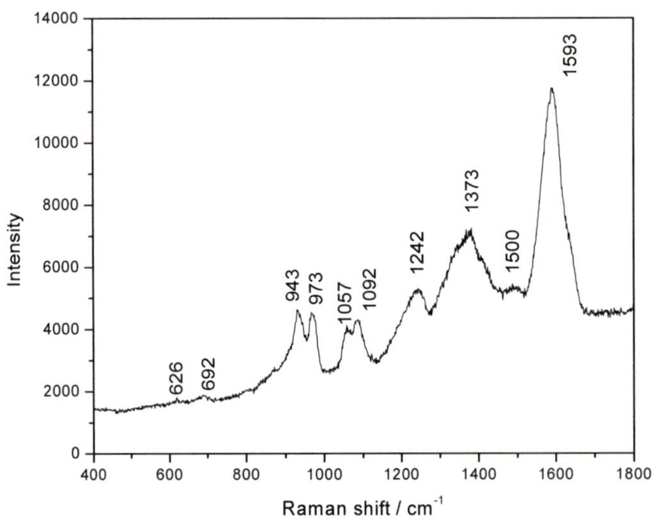

Figure 22. Raman spectrum of PPy(MoO_4)/MMT nanocomposite

Wavenumber (cm^{-1})			Assignment
Theoretical calculation [55] (oxidized PPy)	PPy on Ag [56] (oxidized PPy)	PPy(MoO_4)/MMT (oxidized PPy)	
1676.6	1584	1593	C=C ring stretching
1524.4	1414	1500	C-N stretching
1307.2	1327	1373	
-	1258	1242	
1049.4	1046	1057; 1092	C-H in plane deformation
955.8	989	973	C-H stretching
-	938	943	

Table 1. PPy/MMT nanocomposites and their components

3.2.4. EDX spectra

In order to determine the presence of molipdate anion in the synthetic nanocomposites EDX spectra were used. The EDX result of C6 is presented in Figure 23. The amount of element Mo in the nanocomposite PPy/MMT nanocomposite is 5,55%.

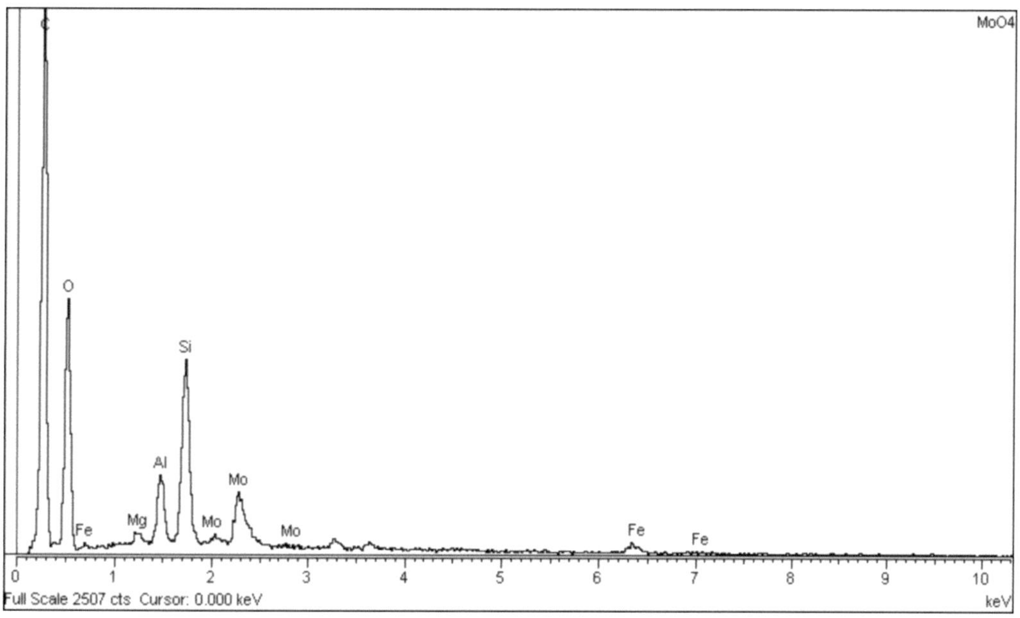

Figure 23. EDX spectra of PPy(MoO_4)/MMT nanocomposites

As seen in Figure 23 and Table 2, the amont of element Mo is 2.93% weight. It means that Na_2MoO_4 occupied 6.29% weight. In addition, the presence of other elements Si, Al, Mg, Fe, O in MMT and C in PPy component were presented in Fugure 23. Other peaks corresponding to hydrogen and nitrogen did not disappeared in EDX spectra.

Element	% Weight	% Element
C	54,53	64,15
O	37,07	32,74
Mg	0,27	0,16
Al	1,39	0,73
Si	3,33	1,68
Fe	0,46	0,12
Mo	2,93	0,43

Table 2. Amount of elements in PPy(MoO_4)/MMT nanocomposites

3.2.5. Thermal analyses

Thermal analyses of PPy(MoO_4)/MMT nanocomposites have been carried out. Figure 24 shows the thermal analyses curves of PPy(MoO_4)/MMT nanocomposites. Under 120^{0}C, the weight reduce origined from water inside samples. The strong reduce in this temperature range can be explained by the hydrophilic property of MMT and the oxidized state of PPy. It is also the source of the wide band between 4000 and 2500 cm^{-1} in the IR spectra. In the range of 120-330^{0}C, the weight reduces are very small, corresponding to the decomposition of redundant monomers, oligomers. At higher temparature (300-700^{0}C), the change of weight is attributed to the decomposition of the oxidized PPy. From Figure 24, the amounts of PPy in the PPy(MoO_4)/MMT nanocomposites are approximately calculated to be 16%.

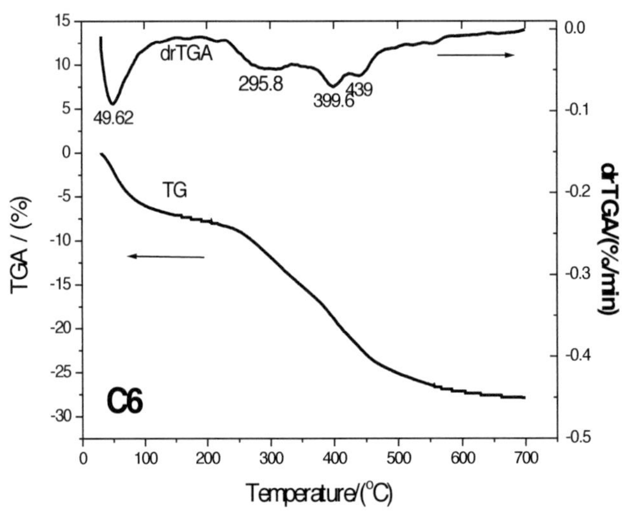

Figure 24. TGA curves and PPy(MoO_4)/MMT nanocomposite

3.2.6. Tests of corrosion protection

Figure 25 shows the OCP-time curves of steels covered by epoxy coatings contaning PPy(MoO_4)/MMT nanocomposite. It shows that the OPC of steel covered by epoxy was corroded after 2 weeks of immersion in NaCl solution. The OCP of the steel covered by epoxy coatings PPy(MoO_4)/MMT nanocomposite shows more possitive. Firstly the OCP of these samples gradually decreased and then increased (Figure 25). The reduction of OCP could be explained by the penetration of the corrosion medium. These results are in good agreement with that preasented in Figure 12. However, the OCP of these samples then incresed. These results shows the same phenomena of the sample PPy(MoO_4)/mild steel presented in Figure 11. When epoxy coating containing 2% PPy(MoO_4)/MMT, the OCP moved to anodic region from 7th day to 21st day and then keft plateau untill 42nd day. After that the OCP of A3 reduced to the value of -0.389VSCE on 56th day. In the case of epoxy containing 3% PPy(MoO_4)/MMT (sample A3), the OCP moved to anodic region from 7th to 28th day and keft plateu untill 42th day. On the 56th day, the OCP approached the value of -0.296V_{SCE}. When epoxy coating containing 4% PPy(MoO_4)/MMT, the OCP of the sample keft the constant value in the first week showing that the barier role of coating containing MMT. Then the OCP moved ossitively from 21st day to 28th day. This region could be explained by the by the penetration of the corrosion medium as weel as the situation of samples A2 and A3. After 42nd day, the OCP of sample A1 increased to -0.0087VSCE and keft plateau. This reason could be explained by the role of molybdate inhibitor anions. These rerults are in good agreement with that of PPy(MoO_4)/mild steel presented in section 3.1.

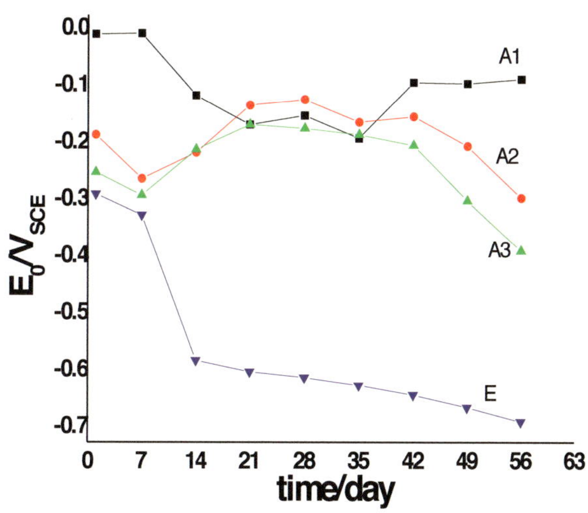

(A1: Epoxy containing 4% PPy(MoO_4)/MMT ; A2: Epoxy containing 3% PPy(MoO_4)/MMT ; A3: Epoxy containing 2% PPy(MoO_4)/MMT ; E: epoxy)

Figure 25. OCP-time curves of steel samples coverd epoxy coatings

In comparision with epoxy coating (sample E), all three epoxy coatings containing PPy(MoO_4)/MMT could prevent mild steel from corrosion (Figure 25).

Figure 26 shows the impedance – time curves at 1Hz of steels covered by epoxy coatings containing. At low frequency (1 Hz) the impedance of the system equals to the value of resistance of epoxy coating (Figure 3). Generally, the imdepance of epoxy coatings gradually reduced during time of immersion. These results could be explained by the penetration of the corrosion medium. Epoxy coating had lowest impedance because of no nanocomposites inside. The samples of A2 and A3 had the impedances of around 10^6 Ω and it was stable throughout experiments. The sample A4 epoxy coating containing 4% PPy(MoO_4)/MMT had highest impedance value at 1 Hz. It even increased after the 56th day (Figure 26). This reason could be explained by the role of molydate dopants.

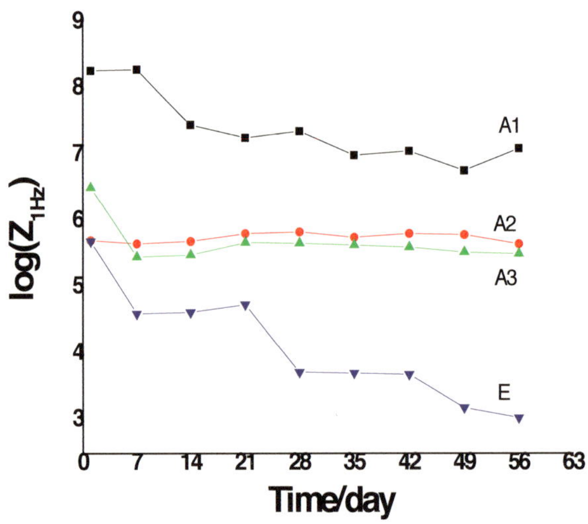

(A1: Epoxy containing 4% PPy(MoO_4)/MMT ; A2: Epoxy containing 3% PPy(MoO_4)/MMT ; A3: Epoxy containing 2% PPy(MoO_4)/MMT ; E: epoxy)

Figure 26. Impedance – time curves of steel samples coverd epoxy coatings

4. Conclusion

Inhibitors used as dopant anions in polymer films are responsible for the anticorrosion behaviour of PPy. The PPy film can work as self-repairing for small defects of the film. Molybdate was built-in into the film as dopant anion. The results of XPS revealed that molybdate exist in two types: $[MoO_4]^{2-}$ (62%) and $[Mo_7O_{24}]^{6-}$ (28%). The film was conductive, homogenous, and compact. Cyclic voltammograms have shown that the film was active. By scanning the potential, the Ppy film changed from the oxidised to the reduced state and at the same time the anions were released from the polymer. To support this observation, Electrochemical Impedance Spectroscopy (EIS), Electrochemical Quartz Crystal Microbalance (EQCM) and Raman Spectroscopy were used. EIS has indicated the change of the resistance R_{PM} and the capacitance C_{PM} of the PPy film during reduction. EQCM has shown: the mass of the Ppy film decreased in the cathodic region and increased in the anodic region. The anion flux was also observed in Scanning Kelvinprobe (SKP) and X-ray Photoelectron Spectroscopy (XPS) experiments. The release of anion is one of the important factors of the corrosion protection property of the PPy film. However, the release behaviour of molybdate anions depends much on the size of cations in the electrolyte. At negative potentials, the incorporation of cations is predominant. Probably, molybdate is a medium size anion and not mobile enough to compete with the cations in the electrolyte for the moving in or out of the PPy film.

It should be also noted that PPy(MoO_4) can be electrodeposited on mild steel in a one-step process. No induction period was observed during polymerisation. PPy(MoO_4) shifted the corrosion potential of mild steel into the passive range. The corrosion current decreased simultaneously. The role of molybdate in corrosion protection was also investigated for Ppy films with small defects. The Open Circuit Potential (OCP) showed the fluctuations around -0.2 V_{SCE}. It means that the defect was passivated / repassivated continuously for 4 hours. It was assumed that molybdate was released from Ppy, move to the defect and act as corrosion inhibitor by forming a complex with iron ion.

The application of PPy(MoO_4)/MMT nanocomposites in corrosion protection for mild steel was also investigated in this work. Mixture of core-shell particles with a polymer was used as primer coatings. The positive effect on the corrosion protection of the coating was illustrated by the increase of the coating resistance and the stabilisation of the film capacitance during immersion in corrosive medium. These rerults show the promissing application potential to exchange the paints containing toxic cromate.

Author details

Le Minh Duc[1] and Vu Quoc Trung[2*]

*Address all correspondence to: vuquoctrungvn@netnam.vn

1 Faculty of Chemical Engineering, Danang University of Technology, Danang, Vietnam

2 Faculty of Chemistry, Hanoi National University of Education, Hanoi, Vietnam

References

[1] C. M. A. Brett, A. M. O. Brett, Electrochemistry: Principles, Methods and Application, 1993.

[2] R. W. Revie, Uhlig's Corrosion Handbook, John Wiley & Sons Inc, 2000.

[3] G.G. Wallace, G.M. Spinks, L.A.P. Kane-Maguire, P.R. Teasdale, Conductive Electroactive Polymers: Intelligent Materials Systems, CRC press, New York, 2003.

[4] W.K. Lu, R.L. Elsenbaumer, in: L. Rupprecht (Ed.), Conducting Polymers and Plastics in Industrial Applications, SPE, Norwich, NY, 1999, 195.

[5] N. Ahmad, A.G. MacDiarmid, Synth. Met., 78 (1996), 103.

[6] T. Schauer, A. Joos, L. Dulog, C.D. Eisenbach, Prog. Org. Coat., 33 (1998), 20.

[7] R.N. Rothon, Adv. Polym. Sci., 139 (1999), 67.

[8] A.M. Thayer, Chem. Eng. News, 81 (35) (2003), 15.

[9] N. V. Krstajic, B. N. Grgur, S. M. Jovanovic, V. Vojnovic, Electrochimica Acta, 42 (1997), 1685.

[10] Jude O. Iroh, W. Su, Electrochimica Acta, 46 (2000), 15.

[11] Michael Rohwerder, Adam Michalik, Electrochimica Acta, 53 (2007) 1300–1313

[12] G. Mengoli, M.T. Munari, P. Bianco, M.M. Musiani, J. Appl. Polym. Sci., 26 (12) (1981), 4247.

[13] D.W. Deberry, J. Electrochem. Soc., 132 (5) (1985), 1022.

[14] B. Wessling, Mater. Corros., 47 (1996), 439.

[15] J. Reut, A. O° pik, K. Idla, Synth. Met., 102 (1999), 1392.

[16] T.D. Nguyen, M. Keddam, H. Takenouti, Electrochem. Solid-State Lett., 6 (2003), B25.

[17] M. Rohwerdera, Le Minh Duc, A. Michalika, Electrochimica Acta, 54 (2009), 6075.

[18] G. Williams, H.N. McMurray, Electrochem. Solid State Lett., 8 (2005), B42.

[19] Victoria Johnston Gelling, Michelle M. Wiest, Dennis E. Tallman, Gordon P. Bierwagen, Gordon G. Wallace, Prog. in Org. Coatings, Volume, 43 (1-3) (2001), 149.

[20] H. N. T. Le, B. Garcia, C. Deslouis, Q. L. Xuan, J. of App. Electrochem., 32 (2002) 105.

[21] N. T. H. Le, B. Garcia, A.Pailleret, C. Deslouis, Electrochimica Acta 50, (2005), 1747.

[22] A. Michalik, M. Rohwerder, Z. Phys. Chem., 219 (2005), 1547.

[23] G. Paliwoda-Porebska, M. Stratmann, M. Rohwerder, U. Rammelt, L. Minh Duc, W. Plieth, J. Solid State Electrochem., 10 (2006), 730.

[24] G. Paliwoda, M. Stratmann, M. Rohwerder, K. Potje-Kamloth, Y. Lu, A.Z. Pich, H.-J. Adler, Corros. Sci., 47 (2005), 3216.

[25] F. Beck, U. Barsch, R. Michaelis, J. of Electroanalytical Chemistry, 351 (1993), 169.

[26] U. Rammelt, P. T. Nguyen, W. Plieth, Electrochimica Acta, 48 (2003), 1257.

[27] U. Rammelt, P. T. Nguyen, W. Plieth, Electrochimica Acta, 46 (2001), 4251.

[28] M. Kendig, M. Hon, L. Warren, Progress in Organic Coating, 47 (2003), 183.

[29] Pierre R. Roberge, Handbook of corrosion engineering, McGraw-Hill, 2000.

[30] G. M. Spinks, A. J. Dominis, G. G. Wallace, D. E. Tallman, J. of Solid State Electrochem., 6 (2002), 85.

[31] P. J. Kinlen, D. C. Silverman, C. R. Jeffreys, Synthetic Metals, 85 (1997), 1327.

[32] J. He, V. J. Gelling, D. E. Tallman, G. P. Bierwagen, G. G. Wallace, J. of Electrochem. Soc., 147 (2000), 3667.

[33] P. J. Kinlen, V. Menon, Y. Ding, J. of Electrochem. Soc., 146 (1999), 3690.

[34] G. Paliwoda-Porebska, M. Rohwerder, M. Stratmann, U. Rammelt, L. M. Duc, W. Plieth. J. of Solid State of Electrochem., 10 (2006), 730.

[35] U. Rammelt, L. M. Duc, W. Plieth, J. of Appl. Electrochem., 35 (2005),1225.

[36] W. Plieth, A. Bund, U. Rammelt, S. Neudeck, L.M.Duc, Electrochimica Acta, 2005.

[37] Yongqin Han, Polymer composite, 30 (1) (2009), 66.

[38] S.H. Hong, B.H. Kim, J. Joo, J.W. Kim, Hyung J. Choi, Current Applied Physics, 1 (6), (2001), 447.

[39] S. U. Rahman, M. S. Ba-Shamakh, Synthtic Metals, 140 (2004), 207.

[40] V. S. Sastri, Corrosion inhibitor- Principles and Applications, Wiley, 1998.

[41] G. Reinhard, M. Radtke, U. Rammelt, Corrosion Science, 33 (1992), 307.

[42] F. Beck, R. Michaelis, F. Schloten, B. Zinger, Electrochimica Acta, 39 (1994), 229.

[43] W. Su, Jude O. Iroh, Electrochimica Acta, 44 (1999), 2173.

[44] G.Paliwoda-Porebska, PhDThesis, Bochum Universtity, Germany, 2005.

[45] E. Almeida, T. C. Diamantino, M. O. Figueiredo, C. Sa, Surface and Coatings Tech., 106 (1998), 8.

[46] T. V. Vernitskaya, O. N. Efimov, A. B. Gavrilov, Russ. J. of Electrochem., 30 (1994), 1022.

[47] M.C. Bernard, S. Joiret, A. Hugo-Le Goff, P.V. Phong, J. of Electrochem. Soc., 148 (2001), B12.

[48] K. Aramaki, Corrosion science, 42 (2000), 1975.

[49] A. Leng, H. Streckel, M. Stratmann, Corrosion Science, 41 (1999), 547.

[50] H. N. T. Le, M. C. Bernard, B. G. Renaud, C. Deslouis, Synthetic Metals, 140 (2004), 287.

[51] Y. Chuan Liu, B. J. Hwang, Synthetic Metals, 113 (2000), 203.

[52] F. Chen, G. Shi, M. Fu, L. Qu, X. Hong, Synthetic Metals, 132 (2003), 125.

[53] T. D. Nguyen, T. Anh Nguyen, M. C. Pham, Electroanalytical Chemistry, 572 (2004), 225.

[54] S. Yoshimoto, F. Ohashi, Y. Ohnishi and T. Nonami, Chem. Commun., 17, (2004), 1924.

[55] A. Faulques, W. Wallnoefer, H. Kuzmany, J. Chem. Phys., 90 (12), (1989), 7585.

[56] M. Bazzaoui, E. A. Bazzaoui, L. Martins and J. I. Martins, Synthetic Metals, 128 (1), (2002), 103.

Chapter 8

Mechanochemical Synthesis of Magnetite/Hydroxyapatite Nanocomposites for Hyperthermia

Tomohiro Iwasaki

Additional information is available at the end of the chapter

http://dx.doi.org/10.5772/54344

1. Introduction

Hydroxyapatite ($Ca_{10}(PO_4)_6(OH)_2$, HA), which is a calcium phosphate ceramic, has been widely used as a biomaterial in various applications (e.g., artificial bone and dental root, cosmetic foundation, etc.) because of its high biocompatibility and chemical stability. Moreover, many attempts are being made to give new functions to HA by incorporating effective components into a HA matrix. In particular, magnetite (Fe_3O_4)-incorporated HA (Fe_3O_4/HA) nanocomposites have attracted much attention as a promising material for hyperthermia therapy of malignant bone tumor [1–4]. Recently, Fe_3O_4/HA composites have also been used as adsorbents [5–7] and catalysts [8,9].

Fe_3O_4/HA composites can be synthesized conventionally by mixing HA powder with Fe_3O_4 nanoparticles which are prepared individually [1–3,5–10]. The conventional synthesis methods have disadvantages: reaction time required for completing the formation of HA and Fe_3O_4 is relatively long, subsequent heat treatments for long periods of time are required for aging and crystallization. Thus, the synthesis of Fe_3O_4/HA composites generally consist of multi-step processes. Therefore, a simple method which can provide Fe_3O_4/HA composites rapidly is needed to be developed.

In this chapter, a mechanochemical method for the simple synthesis of Fe_3O_4/HA nanocomposites is presented. In this method, superparamagnetic Fe_3O_4 nanoparticles are first prepared mechanochemically from ferric hydroxide [11], and then the mechanochemical synthesis of HA from dicalcium phosphate dihydrate ($CaHPO_4$ $2H_2O$) and calcium carbonate ($CaCO_3$, calcite) is performed [12–14], followed by the aging for a short period of time. These mechanochemical treatments are sequentially performed in a single horizontal tumbling ball mill at room temperature under wet conditions. The wet mechanochemical process can also contribute to the distribution of Fe_3O_4 nanoparticles in the HA matrix, which can result in a good hyperthermia property. In addition, the use of horizontal tumbling ball mills is reasonable for the synthesis of Fe_3O_4/HA nanocomposites because the device structure is simple, the handling is easy, the energy consumption is relatively low, and the scale-up is easy [15]. The influence of conditions on the formation of Fe_3O_4/HA nanocomposites was investigated and the hyperthermia property was examined. The details are described below.

2. Mechanochemical synthesis of hydroxyapatite nanoparticles

First of all, the synthesis of HA nanoparticles containing no Fe_3O_4 nanoparticles was investigated to optimize the synthesis process of Fe_3O_4/HA nanocomposites. In all the experiments presented in this chapter, the chemicals of analytical grade were used as received without further purification. Typically, 30 mmol of $CaHPO_4$ $2H_2O$ and 20 mmol of $CaCO_3$, corresponding to the stoichiometric molar ratio in the formation reaction of HA expressed by Equation (1) [14], were added to 60 ml of deionized and deoxygenated water.

$$6CaHPO_4{\cdot}2H_2O + 4CaCO_3 \rightarrow Ca_{10}\left(PO_4\right)_6\left(OH\right)_2 + 14H_2O + 4CO_2 \qquad (1)$$

The resulting suspension was subjected to a mechanochemical treatment using a horizontal tumbling ball mill, as illustrated in Figure 1. The suspension was placed in a Teflon-lined milling pot with an inner diameter of 90 mm and a capacity of 500 ml. Zirconia balls with a diameter of 3 mm were used as the milling media; the charged volume of the balls (including voids among the balls) was 40% of the pot capacity. The wet milling was performed at room temperature in air atmosphere under atmospheric pressure for a designated period of time. The rotational speed was 140 rpm, corresponding to the ideal critical rotational speed. After milling, the precipitate was isolated from the suspension by centrifugation, washed with acetone, and dried at room temperature in air. As a control experiment without milling, the starting suspension was vigorously stirred at room temperature for 24 h.

The samples obtained under various conditions were characterized according to standard methods. The powder X-ray diffraction (XRD) pattern of samples was obtained by an X-ray diffractometer (RINT-1500, Rigaku; CuKα radiation, 40 kV, 80 mA, 2θ=5°–50°, scanning rate: 1.0°/min). Figure 2 shows the XRD pattern of samples obtained in different milling times. As the milling time increased, the diffractions indicating the presence of $CaHPO_4$ $2H_2O$ and $CaCO_3$ decreased. Simultaneously, the diffractions indicating HA appeared. In particular, a drastic change was observed between 1 h and 3 h. On the contrary, when stirred for 24 h without milling, the XRD pattern (not shown) hardly changed from the beginning, which was almost the same as that before milling as shown in Figure 2a. These results indicate that the milling promoted the solid phase reaction expressed by Equation (1). However, after milling for 12 h, the XRD pattern was almost the same and the diffraction at 2θ=29.4°, indicating the presence of $CaCO_3$, still remained even in 24 h.

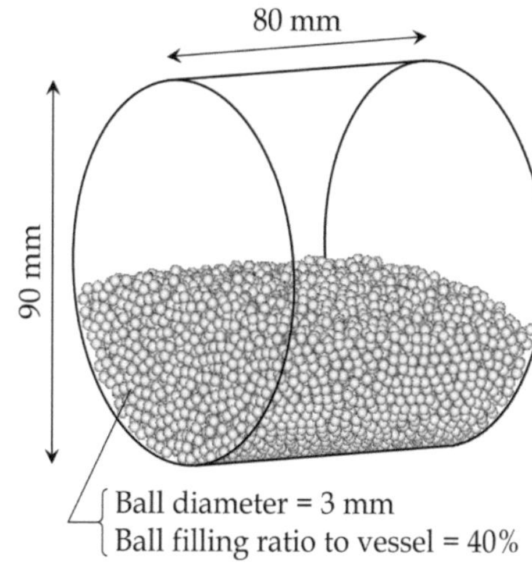

Figure 1. Schematic illustration of horizontal tumbling ball mill used in this work.

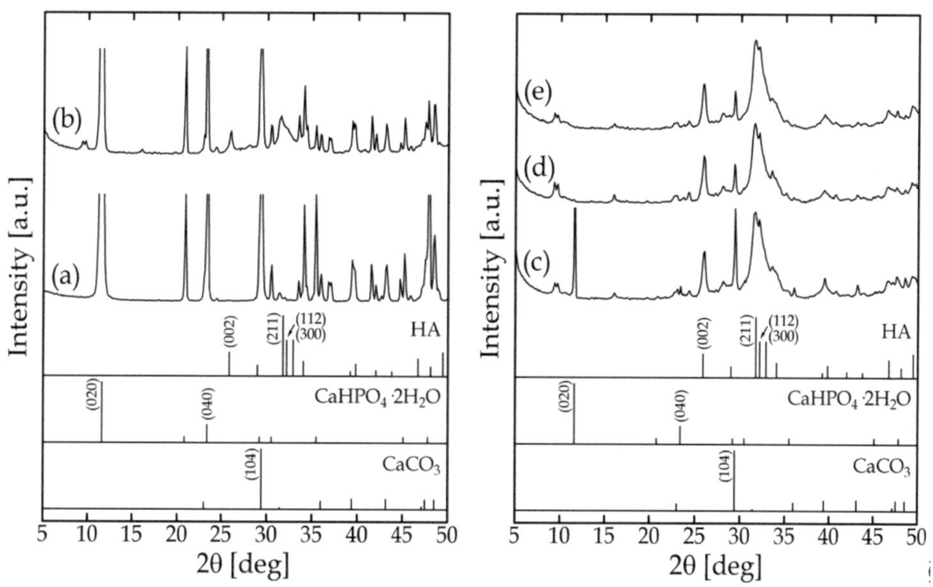

Figure 2. XRD pattern of samples (a) before milling and after milling for (b) 1 h, (c) 3 h, (d) 12 h, and (e) 24 h.

The differential scanning calorimetry (DSC) was performed using a thermal analyzer (SDT2960, TA Instrument) with an argon flow rate of 100 ml/min. The temperature was raised from ambient temperature to 900 °C at a rate of 20°C/min. Figure 3 shows the results of DSC analysis for the raw materials and the samples. In the sample obtained in 1 h (Figure 3d), the endothermic peaks were clearly observed at around 200°C and 750°C, which resulted from the elimination of water of crystallization in $CaHPO_4\ 2H_2O$ and the thermal decomposition of $CaHPO_4\ 2H_2O$ and $CaCO_3$. Although the peaks relating to $CaHPO_4\ 2H_2O$ disappeared as the milling time, the peak resulted from the thermal decomposition of $CaCO_3$ remained even in 12 h. Accordingly, it was found that the milling was not sufficient to complete the formation reaction of HA.

The morphology of samples was observed by field emission scanning electron microscopy (FE-SEM; JSM-6700F, JEOL). Figure 4 shows typical SEM images of samples. In a milling time of 1 h, coarse particles coated with fine particles of about 100 nm were observed. From the particle size analysis of $CaHPO_4\ 2H_2O$ and $CaCO_3$ by the laser diffraction/scattering method (SALD-7100, Shimadzu), the median sizes were determined to be 16.2 μm for $CaHPO_4\ 2H_2O$ and 2.0 μm for $CaCO_3$. In general, horizontal tumbling ball mills are difficult to produce nanoparticles for short milling times. Therefore, coarse and fine particles could be the raw materials and HA, respectively. As the milling time increased, coarse particles disappeared and the number of HA nanoparticles increased. However, even after 12 h, a little number of coarse particles was found.

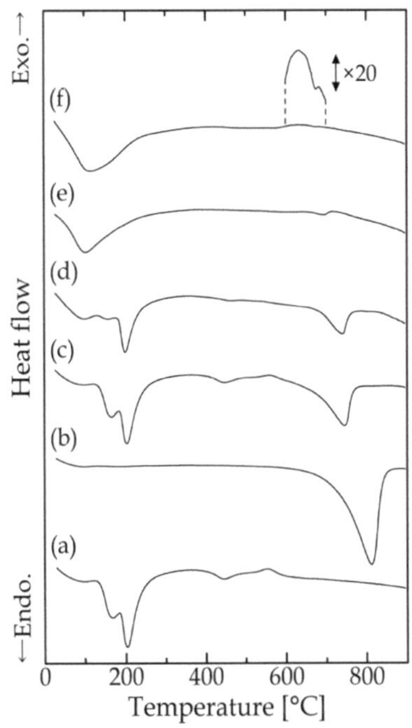

Figure 3. DSC curve of (a) $CaHPO_4\ 2H_2O$, (b) $CaCO_3$, and samples (c) before milling and after milling for (d) 1 h, (e) 3 h, and (f) 12 h.

In order to complete the formation reaction of HA, the heat treatment (aging) was performed after milling. For investigating the effect of heating on the formation of HA, the un-milled suspension of $CaHPO_4$ $2H_2O$ and $CaCO_3$ was heated under various conditions of temperature and time. Figures 5, 6, and 7 show the XRD patterns of samples obtained without milling after heating at 40, 60, and 80°C, respectively. When the suspension was heated at 40°C, the formation reaction of HA hardly took place. As increasing in the temperature, the reaction was promoted and could complete at 80°C in 8 h. Thus, when without milling, higher heating temperatures and longer heating times are needed for the formation of HA.

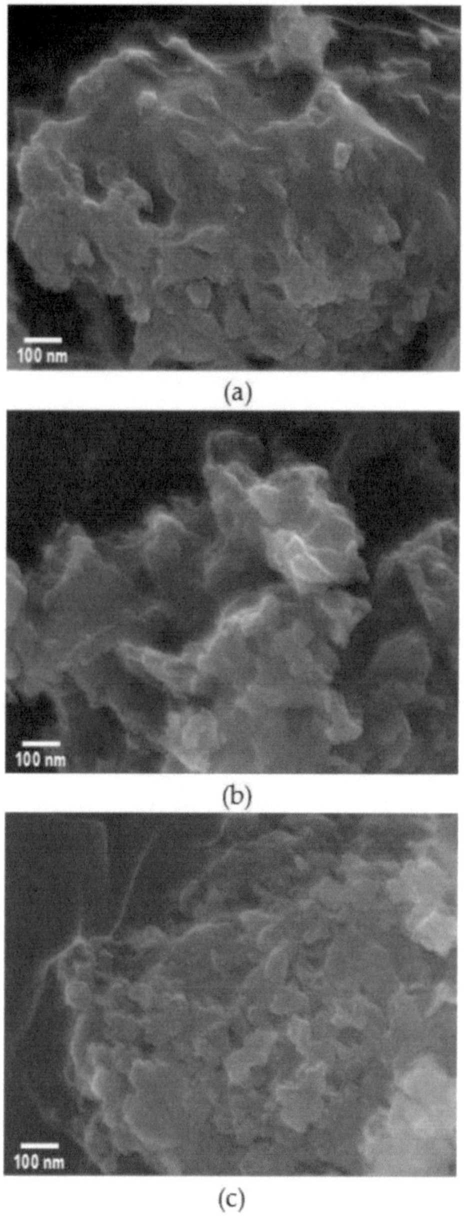

Figure 4. SEM image of samples obtained after milling for (a) 1 h, (b) 3 h, and (c) 12 h.

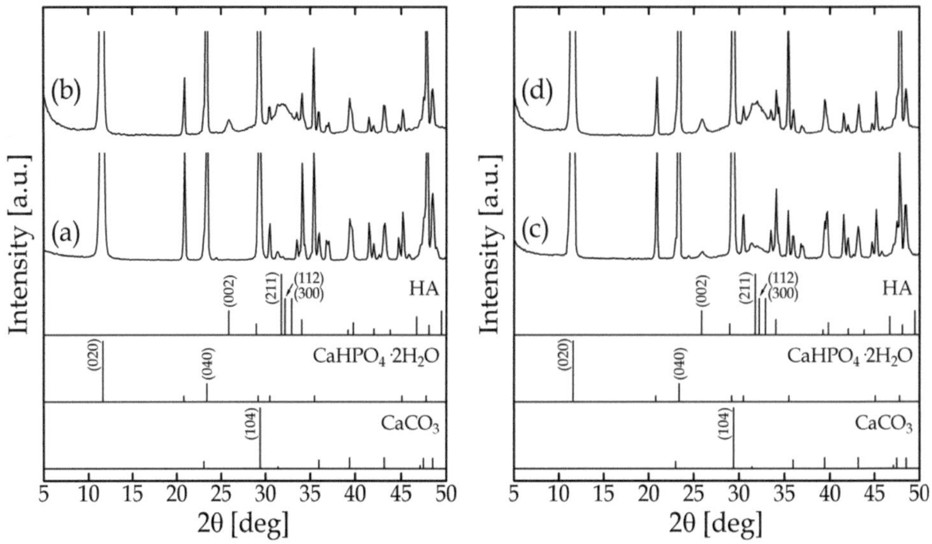

Figure 5. XRD pattern of un-milled samples (a) before heating and after heating at 40°C for (b) 3 h, (c) 5 h, and (d) 8 h.

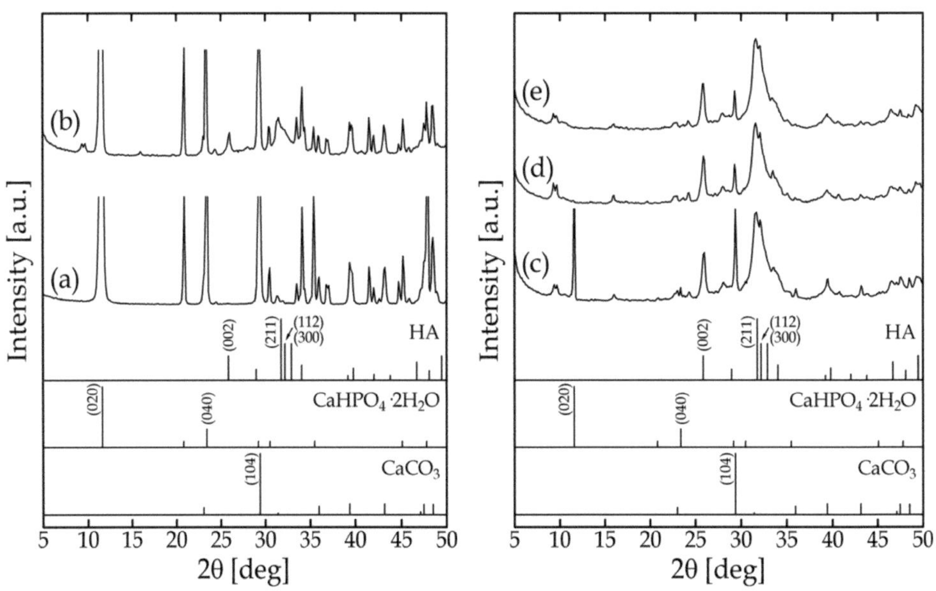

Figure 6. XRD pattern of un-milled samples (a) before heating and after heating at 60°C for (b) 1 h, (c) 3 h, (d) 5 h, and (e) 8 h.

http://dx.doi.org/10.5772/54344

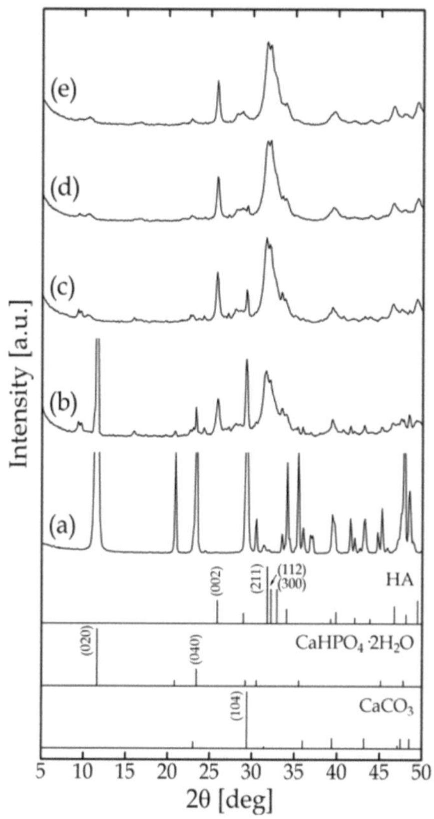

Figure 7. XRD pattern of un-milled samples (a) before heating and after heating at 80°C for (b) 1 h, (c) 3 h, (d) 5 h, and (e) 8 h.

Next, the effect of milling of the suspension before heating on the formation of HA was investigated. Figures 8–14 show the XRD patterns of samples obtained under various conditions of milling time, heating temperature, and heating time. It was found that longer milling times, higher heating temperatures, and longer heating times promoted the formation reaction of HA. In particular, as shown in Figure 10c, when the heating was performed at 80°C, only the milling for 1 h and the following heating for 1 h provided a single phase of HA. The SEM images of samples obtained by milling for different times under constant heating conditions of 80°C and 1 h are shown in Figure 15. When heating at 80°C for 1 h, a typical morphology of HA was observed regardless of the milling time. However, the particle size intended to decrease as the milling time increased. Consequently, the combination of milling and heating of the suspension of $CaHPO_4$ $2H_2O$ and $CaCO_3$ can produce efficiently HA for short periods of time.

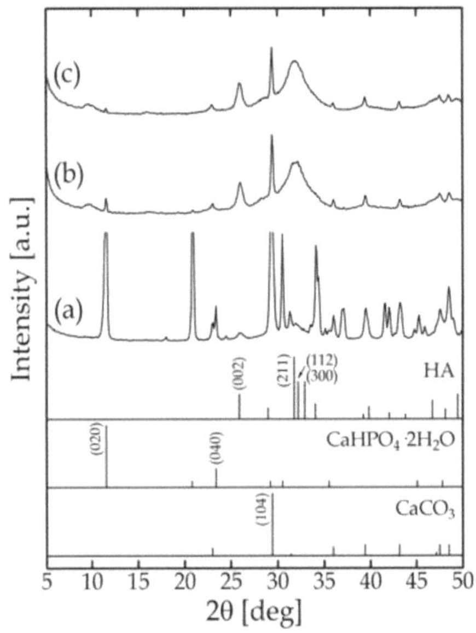

Figure 8. XRD pattern of 1 h-milled samples (a) before heating and after heating at 40°C for (b) 1 h, and (c) 5 h.

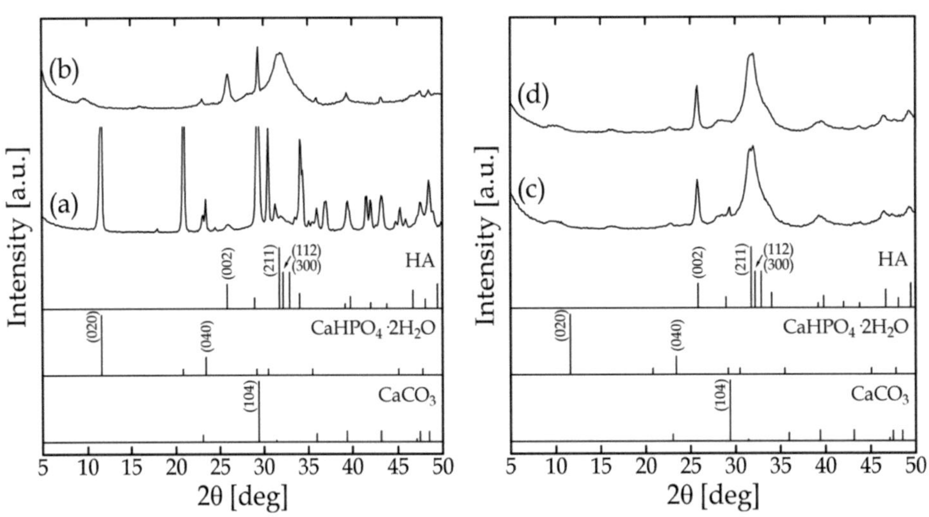

Figure 9. XRD pattern of 1 h-milled samples (a) before heating and after heating at 60°C for (b) 1 h, (c) 3 h, and (d) 5 h.

http://dx.doi.org/10.5772/54344

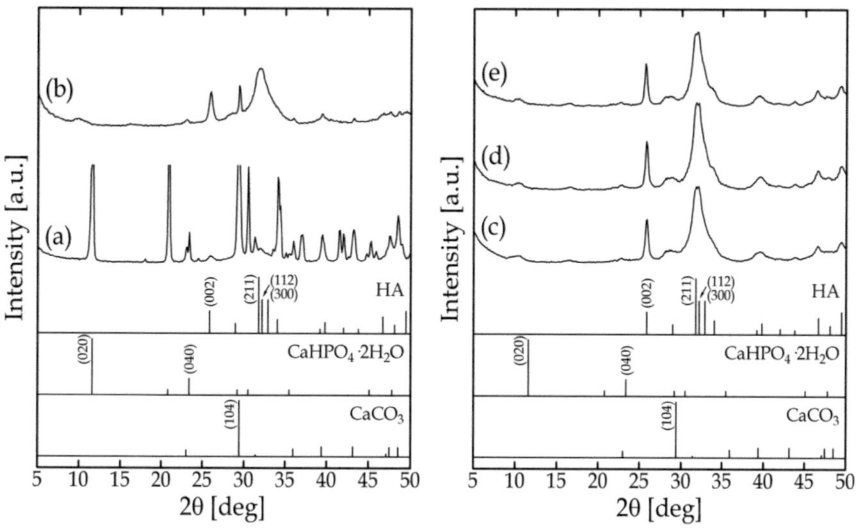

Figure 10. XRD pattern of 1 h-milled samples (a) before heating and after heating at 80°C for (b) 30 min, (c) 1 h, (d) 3 h, and (e) 5 h.

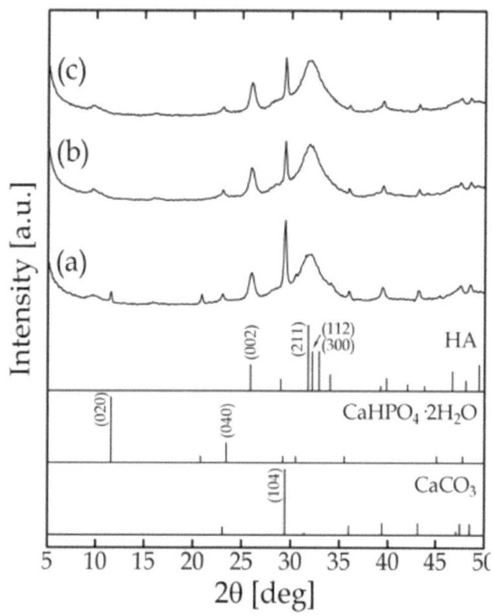

Figure 11. XRD pattern of 3 h-milled samples (a) before heating and after heating at 40°C for (b) 1 h, and (c) 5 h.

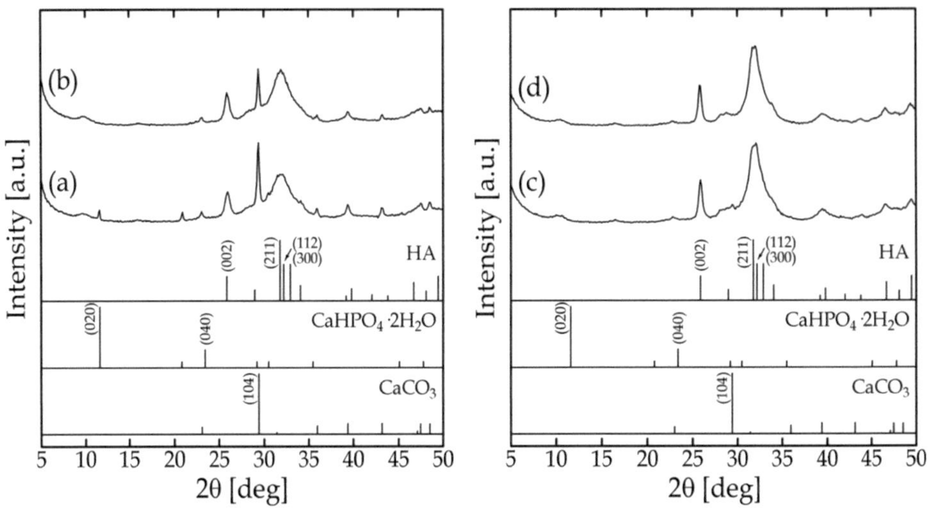

Figure 12. XRD pattern of 3 h-milled samples (a) before heating and after heating at 60°C for (b) 1 h, (c) 3 h, and (d) 5 h.

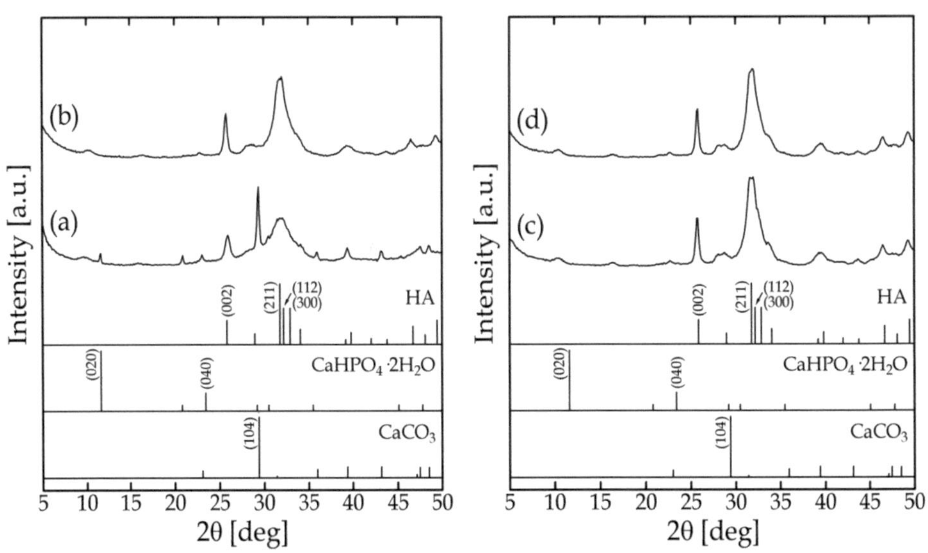

Figure 13. XRD pattern of 3 h-milled samples (a) before heating and after heating at 80°C for (b) 1 h, (c) 3 h, and (d) 5 h.

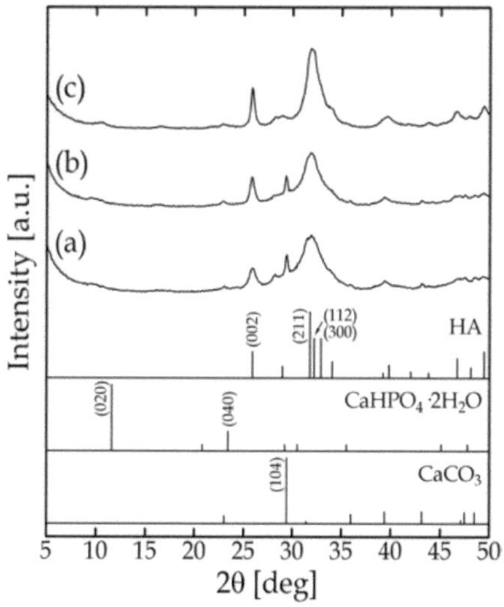

Figure 14. XRD pattern of 12 h-milled samples (a) before heating and after heating at 80°C for (b) 30 min, and (c) 1 h.

3. Synthesis and hyperthermia property of magnetite/hydroxyapatite nanocomposites

In the synthesis of Fe_3O_4/HA nanocomposites, first a suspension of superparamagnetic Fe_3O_4 nanoparticles was prepared according to a mechanochemical method reported in elsewhere [11]. This method provides Fe_3O_4 from ferric hydroxide (goethite) in the absence of a reducing agent; goethite is reduced to ferrous hydroxide by mechanochemical effects and the solid phase reaction between ferrous hydroxide and goethite generates Fe_3O_4 [16]. Subsequently, HA nanoparticles were synthesized in the suspension of Fe_3O_4 nanoparticles in the same container by the mechanochemical method mentioned above.

4.5 mmol of ferric chloride hexahydrate ($FeCl_3$ $6H_2O$) was dissolved in 60 ml of deionized and deoxygenated water. To precipitate amorphous ferric hydroxides (mostly, goethite), a proper amount of 1.0 M sodium hydroxide (NaOH) solution was dropped into the solution which was magnetically stirred under a continuous flow of argon at room temperature. The pH was adjusted to higher than 13. A brown suspension thus prepared was placed in a gas-tight milling pot (inner diameter 90 mm, capacity 500 ml) made of 18%Cr–8%Ni stainless steel. Stainless steel balls (diameter 3.2 mm) were used as the milling media. The charged volume including the voids among the balls was about 40% of the pot capacity. The pot was purged of air, filled with argon, and sealed. The milling was performed at room temperature for 11 h. The rotational speed was 140 rpm, corresponding to the ideal critical rotational speed.

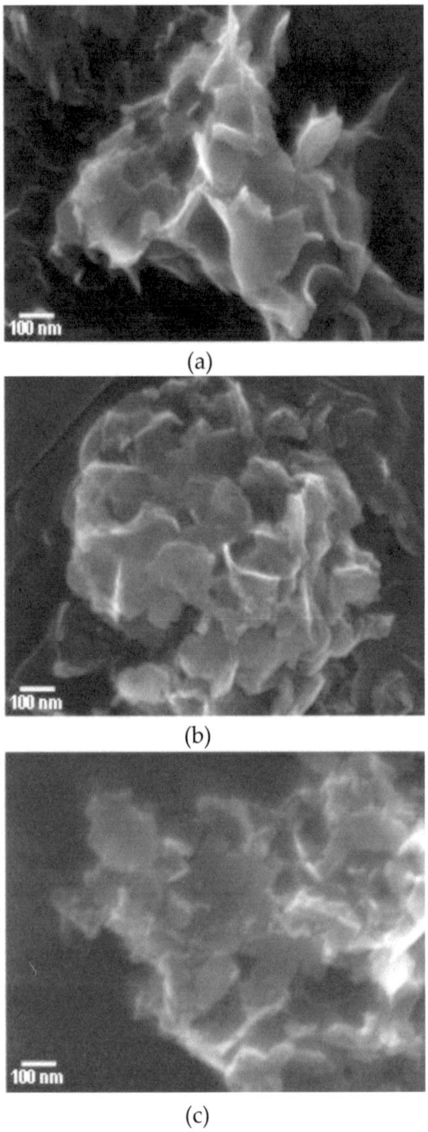

Figure 15. SEM image of samples obtained by milling for (a) 1 h, (b) 3 h, and (c) 12 h, followed by heating at 80°C for 1 h.

The XRD pattern of Fe_3O_4 nanoparticles thus prepared is shown in Figure 16. The Fe_3O_4 nanoparticles had a high crystallinity and an average crystallite size of 11.7 nm which was calculated from the full width at half-maximum (FWHM) of the Fe_3O_4 (311) diffraction peak at $2\theta=35.5°$ using Scherrer's formula. The lattice constant was determined to be 8.387 Å from several diffraction angles showing high intensity peaks, which was close to the standard value of Fe_3O_4 (8.396 Å) as compared to that of maghemite (8.345 Å). Figure 16 also shows that no reflections indicating formation of other compounds were observed. This indicates the Fe_3O_4 nanoparticles were high purity.

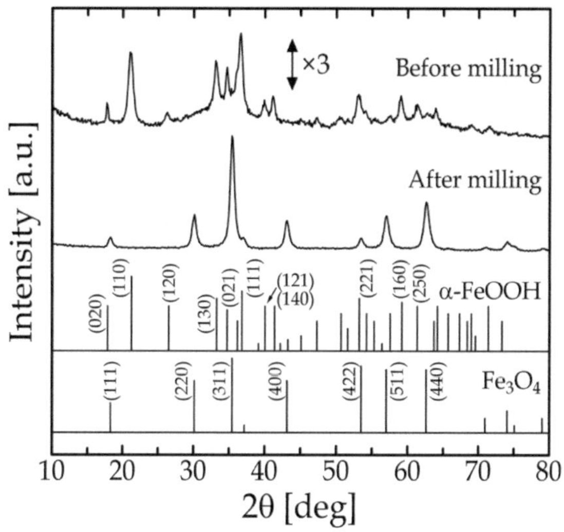

Figure 16. XRD pattern of Fe_3O_4 nanoparticles prepared by mechanochemical method.

As shown in Figure 17, the SEM image indicated that the Fe_3O_4 nanoparticles had a diameter of approximately 10–20 nm, which almost agreed with the average crystallite size (11.7 nm). The hydrodynamic size (number basis) was measured by dynamic light scattering (DLS; Zetasizer Nano ZS, Malvern Instruments) for a dispersion, as shown in Figure 18. The median diameter was determined to be 16.4 nm from the size distribution, which was also near the average crystallite size. These results reveal that the Fe_3O_4 nanoparticles have a single-crystal structure.

The magnetic property (magnetization-magnetic field hysteretic cycle) was analyzed using a superconducting quantum interference device (SQUID) magnetometer (Quantum Design model MPMS) at room temperature in the rage of magnetic field between –10 kOe and 10 kOe. Figure 19 shows the magnetization-magnetic field curve. The Fe_3O_4 nanoparticles had a low coercivity (4 Oe), showing superparamagnetism. The saturation magnetization (78 emu/g) was a little lower than that of the corresponding bulk (=92 emu/g) because of the smaller size [17].

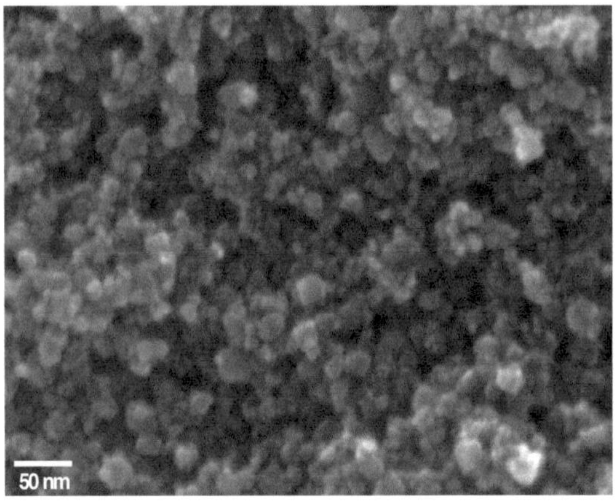

Figure 17. SEM image of Fe_3O_4 nanoparticles prepared by mechanochemical method.

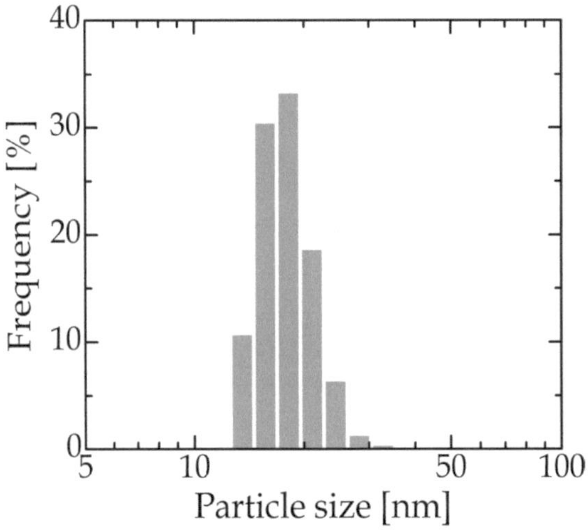

Figure 18. DLS particle size distribution of Fe_3O_4 nanoparticles prepared by mechanochemical method.

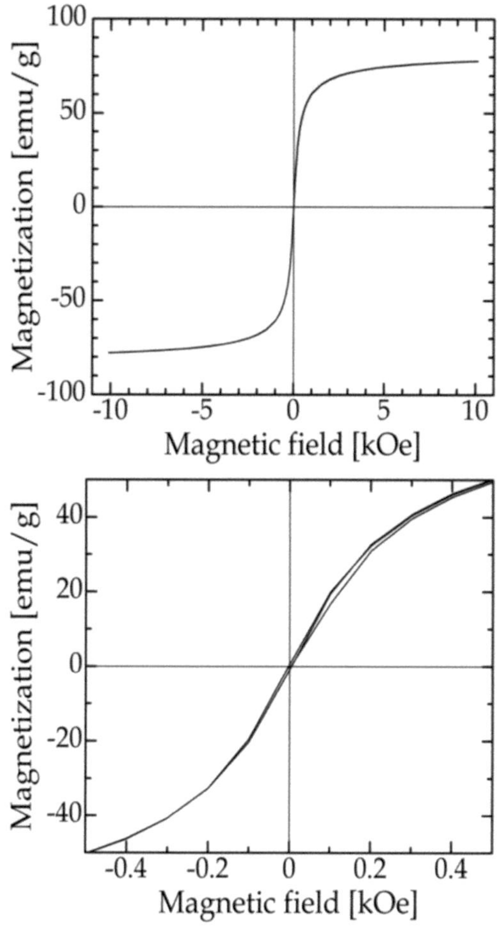

Figure 19. Magnetization-magnetic field curve of Fe_3O_4 nanoparticles prepared by mechanochemical method.

After the suspension of Fe_3O_4 nanoparticles was prepared, the milling pot was opened, and then predetermined amounts of $CaHPO_4$ $2H_2O$ and $CaCO_3$ were added to the suspension. Their amounts were adjusted so that the mass concentration of Fe_3O_4 nanoparticles in the Fe_3O_4/HA nanocomposite was 10, 20, and 30 mass%. In order to prevent the oxidation of Fe_3O_4 during milling, the pot was purged of air, filled with argon, and sealed prior to milling. The suspension was milled at a rotational speed of 140 rpm for 1 h at room temperature, followed by the heating at 80°C for 1 h.

Figure 20 shows the XRD pattern of Fe_3O_4/HA nanocomposites with different Fe_3O_4 concentrations. It was confirmed that the nanocomposites consisted of Fe_3O_4 and HA having no by-products regardless of the Fe_3O_4 concentration. The average crystallite sizes of Fe_3O_4 and HA were calculated from the FWHM of the Fe_3O_4 (311) plane at 2θ=35.5° and the HA (002) plane at 2θ=25.9°, respectively, using Scherrer's formula, and listed in Table 1. The average crystallite sizes of Fe_3O_4 and HA were almost constant regardless of the concentration of Fe_3O_4 in the Fe_3O_4/HA nanocomposites.

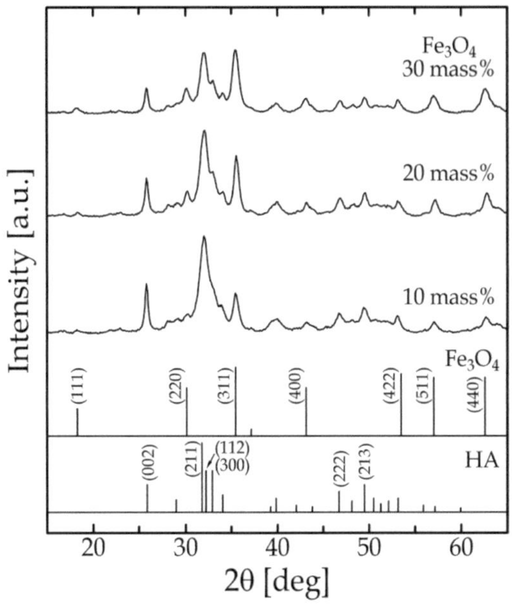

Figure 20. XRD pattern of Fe_3O_4/HA nanocomposites with different Fe_3O_4 concentrations.

Fe_3O_4 concentration	Crystallite size of Fe_3O_4	Crystallite size of HA
10 mass%	11.3 nm	20.3 nm
20 mass%	12.8 nm	18.8 nm
30 mass%	9.8 nm	17.8 nm

Table 1. Crystallite sizes of Fe_3O_4 and HA in Fe_3O_4/HA nanocomposites.

Figure 21 shows the SEM image of nanocomposite containing 30 mass% Fe_3O_4 as an example. The Fe_3O_4 nanoparticles with a diameter of about 20 nm were distributed homogeneously in the HA matrix without forming large aggregates. It was confirmed that nanometer-sized Fe_3O_4/HA composite particles were successfully synthesized.

The magnetic hyperthermia property was evaluated using an apparatus reported in elsewhere [18]. A proper amount of Fe_3O_4/HA nanocomposite powder sample was placed in a polystyrene tube with a diameter of 16 mm, and packed by tapping the tube. The packing volume was constant at 0.8 cm^3 regardless of the Fe_3O_4 concentration. The temperature increase was measured in an AC-magnetic field using an optical fiber thermometer. The frequency and amplitude of the AC-magnetic field were 600 kHz and 2.9 kA/m, respectively. Figure 22 shows the temperature increase for the nanocomposites in the AC-magnetic field. As the Fe_3O_4 concentration increased, the temperature increased more rapidly. When the Fe_3O_4 concentration was 30 mass%, the temperature increase of 40°C was achieved only after

http://dx.doi.org/10.5772/54344

about 20 sec. This result supports that the Fe_3O_4/HA nanocomposites synthesized by this mechanochemical process exhibit a good hyperthermia property [1–4].

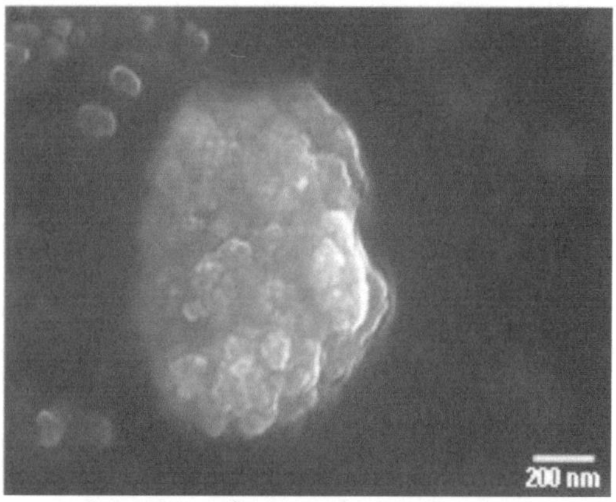

Low-magnification observation

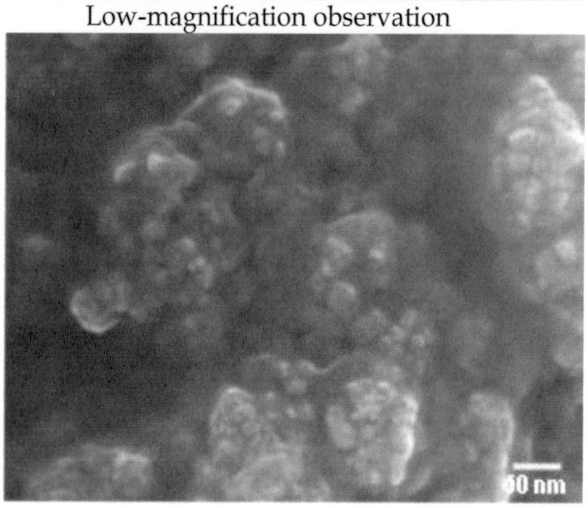

High-magnification observation

Figure 21. SEM images of 30 mass% Fe_3O_4/HA nanocomposites.

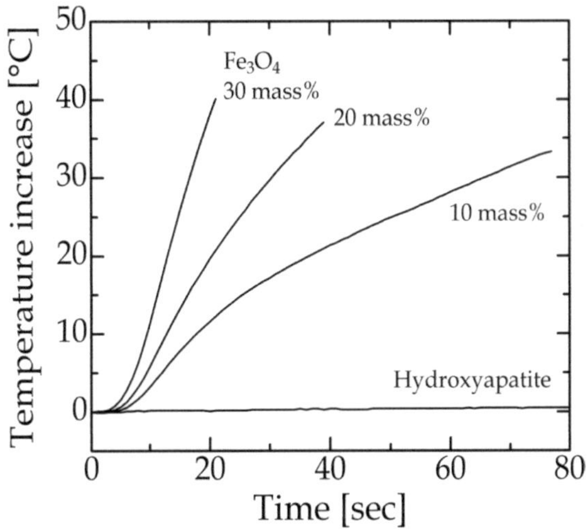

Figure 22. Temperature profiles of HA nanoparticles and Fe_3O_4/HA nanocomposites with different Fe_3O_4 concentrations in AC-magnetic field.

4. Conclusion

A mechanochemical method for the simple synthesis of Fe_3O_4/HA nanocomposites has been developed, in which superparamagnetic Fe_3O_4 nanoparticles and HA nanoparticles are sequentially prepared in a single horizontal tumbling ball mill at room temperature under wet conditions. First, the synthesis process of HA (containing no Fe_3O_4) was optimized. The obtained HA samples were characterized by XRD, DSC, and SEM. The influence of conditions on the formation of HA nanoparticles was investigated. Mechanochemical effects induced during wet milling promoted the reactions between $CaHPO_4$ $2H_2O$ and $CaCO_3$ forming HA even at room temperature. The combination of milling and heating (aging) of the suspension of $CaHPO_4$ $2H_2O$ and $CaCO_3$ can produce efficiently HA for short periods of time. The optimum operating conditions in the synthesis of HA were determined as follows: a rotational speed of 140 rpm, a milling time of 1 h, an aging temperature of 80°C, and an aging time of 1 h. Next, the synthesis of Fe_3O_4/HA nanocomposites was investigated. The mechanochemically synthesized Fe_3O_4 nanoparticles, of which the median diameter was 16 nm, had a high crystallinity and a high saturation magnetization of 78 emu/g, and showed superparamagnetism. The wet mechanochemical process also contributed to the distribution of Fe_3O_4 nanoparticles in the HA matrix. The Fe_3O_4/HA nanocomposites were confirmed to have a good hyperthermia property through the measurement of temperature increase in an AC-magnetic field. For example, the 30 mass% Fe_3O_4/HA nanocomposites showed the temperature increase of 40°C after about 20 sec under a frequency of 600 kHz and an amplitude of 2.9 kA/m. Consequently, the Fe_3O_4/HA nanocomposites thus synthesized were found to be a promising material for hyperthermia therapy.

http://dx.doi.org/10.5772/54344

Acknowledgements

The author would like to thank Professor Kenya Murase of Osaka University for his support in measuring temperature profile of nanocomposites.

Author details

Tomohiro Iwasaki*

Department of Chemical Engineering, Osaka Prefecture University, Japan

References

[1] Murakami S, Hosono T, Jeyadevan B, Kamitakahara M, Ioku K. Hydrothermal synthesis of magnetite/hydroxyapatite composite material for hyperthermia therapy for bone cancer. Journal of the Ceramic Society of Japan 2008;116(9) 950–954.

[2] Andronescu E, Ficai M, Voicu G, Ficai D, Maganu M, Ficai A. Synthesis and characterization of collagen/hydroxyapatite: magnetite composite material for bone cancer treatment. Journal of Materials Science: Materials in Medicine 2010;21(7) 2237–2242.

[3] Inukai A, Sakamoto N, Aono H, Sakurai O, Shinozaki K, Suzuki H, Wakiya N. Synthesis and hyperthermia property of hydroxyapatite-ferrite hybrid particles by ultrasonic spray pyrolysis. Journal of Magnetism and Magnetic Materials 2011;323(7) 965–969.

[4] Tampieri A, D'Alessandro T, Sandri M, Sprio S, Landi E, Bertinetti L, Panseri S, Pepponi G, Goettlicher J, Bañobre-López M, Rivas J. Intrinsic magnetism and hyperthermia in bioactive Fe-doped hydroxyapatite. Acta Biomaterialia 2012;8(2) 843–851.

[5] Dong L, Zhu Z, Qiu Y, Zhao J. Removal of lead from aqueous solution by hydroxyapatite/magnetite composite adsorbent. Chemical Engineering Journal 2010;165(3) 827–834.

[6] Wang X. Preparation of magnetic hydroxyapatite and their use as recyclable adsorbent for phenol in wastewater. Clean—Soil, Air, Water 2011;39(1) 13–20.

[7] Xie H, Li X, Cheng C, Wu D, Zhang S, Jiao Z, Lan Y. Kinetic and thermodynamic sorption study of radiocobalt by magnetic hydroxyapatite nanoparticles. Journal of Radioanalytical and Nuclear Chemistry 2012;291(2) 637–647.

[8] Yang Z, Gong X, Zhang C. Recyclable Fe_3O_4/hydroxyapatite composite nanoparticles for photocatalytic applications. Chemical Engineering Journal 2010;165(1) 117–121.

[9] Liu Y, Zhong H, Li L, Zhang C. Temperature dependence of magnetic property and photocatalytic activity of Fe_3O_4/hydroxyapatite nanoparticles. Materials Research Bulletin 2010;45(12) 2036–2039.

[10] Covaliu CI, Georgescu G, Jitaru I, Neamtu J, Malaeru T, Oprea O, Patroi E. Synthesis and characterization of a hydroxyapatite coated magnetite for potential cancer treatment. Revista de Chimie 2009;60(12) 1254–1257.

[11] Iwasaki T, Sato N, Kosaka K, Watano S, Yanagida T, Kawai T. Direct transformation from goethite to magnetite nanoparticles by mechanochemical reduction. Journal of Alloys and Compounds 2011;509(4) L34–L37.

[12] Silva CC, Pinheiro AG, Miranda MAR, Góes JC, Sombra ASB. Structural properties of hydroxyapatite obtained by mechanosynthesis. Solid State Sciences 2003;5(4) 553–558.

[13] Shu C, Yanwei W, Hong L, Zhengzheng P, Kangde Y. Synthesis of carbonated hydroxyapatite nanofibers by mechanochemical methods. Ceramics International 2005;31(1) 135–138.

[14] Wu SC, Hsu HC, Wu YN, Ho WF. Hydroxyapatite synthesized from oyster shell powders by ball milling and heat treatment. Materials Characterization 2011;62(12) 1180–1187.

[15] Iwasaki T, Yabuuchi T, Nakagawa H, Watano S. Scale-up methodology for tumbling ball mill based on impact energy of grinding balls using discrete element analysis. Advanced Powder Technology 2010;21(6) 623–629.

[16] Iwasaki T, Sato N, Nakamura H, Watano S. Aqueous-phase synthesis of crystalline magnetite nanoparticles by a new mechanochemical method. In: proceedings of the 5th Asian Particle Technology Symposium, APT2012, 2–5 July 2012, Singapore.

[17] Lee J, Isobe T, Senna M. Magnetic properties of ultrafine magnetite particles and their slurries prepared via in-situ precipitation. Colloids and Surfaces A: Physicochemical and Engineering Aspects 1996;109, 121–127.

[18] Murase K, Oonoki J, Takata H, Song R, Angraini A, Ausanai P, Matsushita T. Simulation and experimental studies on magnetic hyperthermia with use of superparamagnetic iron oxide nanoparticles. Radiological Physics and Technology 2011;4(2) 194–202.

Section 2

Characterization Techniques

Chapter 9

Nonlinear Image Filtering for Materials Classification

Giovanni F. Crosta

Additional information is available at the end of the chapter

http://dx.doi.org/10.5772/55633

1. Introduction

The Ruderman - Bialek *paradigm*

An image, or a general pattern of (Euclidean) dimension two or three, can be often analysed and classified according to its autocorrelation properties. By means of a well-known theorem of Fourier analysis (the Wiener - Khinchin - Einstein theorem), the autocorrelation function relates to the power spectral density (of the image, of the pattern).

With reference to image analysis, D. L. Ruderman and W. Bialek, in the introductory paragraph of a paper of theirs [1], wrote

> *efficient signal processing systems*
> *take advantage of statistical structure in their input signals, ...*
> *to generate compact representations of seemingly complex data.*

Their declared focus was on *early visual processing*, occurring in the central nervous system of mammals. Nonetheless, their statement was to affect subsequent research on statistics-based image analysis and classification.

More recently, progress in image analysis and in automated image understanding systems has benefitted from the formalisation of statistical learning theory [2], [3], [4] and from developments in functional analysis [5], [6].

Image classification and pattern recognition

The need for analysing and understanding (e.g., classifying) images with possibly high throughput is felt everywhere in the natural sciences, in engineering and in medicine: optical and electron microscopes and diagnostic instruments acquire and produce large amounts of images.

Two tasks can be assigned: *a*) classification, which is usually stated in broad terms, and *b*) pattern recognition, where specific features, shapes, or objects are located and matched to a library. The former task can rely on the statistical properties of the image encoded by the power spectral density function. This function replaces the original image for statistics-based classification. Pattern recognition, instead, requires processing at a higher, often cognitive, level.

The satisfactory performance of classification and, respectively, recognition is still a challenge in terms of reliability and efficiency. Namely, there are two main obstacles: at the system input, faults in the overall design of experiments and, at the system output, biased judgment.

The "compact representation of seemingly complex data" requires a method to filter information out. Unlike a procedure which may have been specifically designed for some application and need not work in another case, a method reflects, and is affected by, prior knowledge about what is being processed.

In other words, a **method** always relates to a **model**. The R. C. CONANT - W. R. ASBHY [7] paradigm (*sensū lato*) on the relation between system modelling and system control, comes to mind here.

Feature extraction

In the realm of this article, the RUDERMAN - BIALEK paradigm has served as a guideline for obtaining quantitative image features (or descriptors) of statistical type. The procedure relies on the comparison between the power spectral density of the image and some model function of the spatial wavenumber or wave-vector.

Feature extraction usually complies with some requirements, dictated in turn by the materials or body organs which images represent, and by the imaging technique. The requirements which have been taken into account herewith are summarised as follows.

- Independence of the coarse (very low spatial frequency) details in the image and possibly of the absolute intensity scale. This requirement is typical in optical or epifluorescence microscopy, where background need not be uniform.
- Capability to separate image structure from image texture e.g., in terms of relative power spectral densities. Structure may correspond to spatial organisation arising from morphogenesis e.g., a cell colony, or from other processes e.g., turbulence. (The so-called "coherent structures" [8].) Separation of structure from texture is one of the fundamental tasks of image understanding, as formalised by the OSHER-RUDIN paradigm [6], Ch. 1.

A simple morphological model is obtained by regarding a gray scale image as the graph of a (scalar, real valued) function which depends on two variables. If the function is FOURIER-transformed and the corresponding power spectral density, say $|G[.]|^2$, is plotted against spatial wavenumber, u, along a given ray on the $\{u_1, u_2\}$ plane, then the roll-off of $\log[|G[u]|^2]$ *vs.* $\log[u]$ leads to an estimate of the surface fractal dimension [9], Ch. 4]. The **m**odel here is represented by the roll-off law.

A more satisfactory model starts from a remark: the function $\log[|G[u]|^2]$ generally exhibits **deviations** from the simple-minded affine law with respect to $\log[u]$; in fact, deviations may

http://dx.doi.org/10.5772/55633

characterize the image or the image class. Rings in diffraction patterns (e.g., X-ray selected area diffraction of polycrystalline materials) are the most intuitive example. In 1977 the first author [10], in analyzing deposits of micron-sized particles on glass substrates by coherent light, devised a texture analysis method based on the occurrence of local maxima in the power spectral density profiles (i.e., rings) as a function of wave-number. In this frame of mind it makes sense to focus on the relative departure e.g., a difference on a log scale, of power spectral density from a given model, and classify images accordingly: this is the basic idea behind the **non-linear** filtering algorithm named **spectrum enhancement**.

Scope

The purpose of this article is twofold

- to outline the principles of image feature extraction by means of the spectrum enhancement algorithm;
- to describe applications of the algorithm to morphological analysis of nano-composite materials.

The first application deals with the surface morphology of a compounded elastomer, tread rubber, worn by abrasion and corrosion, the second application with the morphology of nano-dispersions in a thermoplastic polymer.

2. The *spectrum enhancement* algorithm

Given an image $g[.]$, its "enhanced spectrum" is a function $h[.]$ of spatial wave-number u, obtained by a sequence of operations on the power spectral density of $g[.]$.

Broadly speaking, by "spectrum enhancement", $\sigma\eta$ for short, one understands a sequence of linear and non-linear operations carried out in the spatial frequency domain, whereby either the complex amplitude or the power spectral density of the given image are compared to a reference function. Comparison emphasizes ("enhances") differences between the **observed** properties, carried by the image, and the **expected** ones, represented by the reference function, which plays the role of a model. The $\sigma\eta$-algorithm of interest herewith [11] relies on the following definitions.

Direct domain, image flop-flip

Let $\mathcal{Q}\Omega$ denote a square of sidelength L and $\mathcal{T}$ the surface of the torus obtained by glueing the opposite sides of $\mathcal{Q}\Omega$ together. Let $\vec{x} \equiv \{x_1, x_2\} \in \mathcal{Q}\Omega$ be the position vector in the direct domain and $\vec{u} \equiv \{u_1, u_2\}$ be the spatial frequency vector (a.k.a. wave vector) in the reciprocal domain. Assume the grayscale image is modeled by a scalar function $\mathcal{Q}g[x]$, which is continuous on $\mathcal{T}$. One way of obtaining such a $\mathcal{Q}g[\vec{x}]$ from an image $g[.]$ defined in a square tile Ω of sidelength $\frac{L}{2}$ is the application of the twofold $\mathcal{Q}[.]$ = flop[flip[.]] reflection. Next let $\mathcal{Q}\Omega$ be discretized by a square grid of step-length ℓ. In practice, each square mesh corresponds to a pixel.

Reciprocal domain

The FOURIER transform $G[\vec{u}]$ of $\mathcal{Q}g[\vec{x}]$ is supported at grid nodes in the square

$$0 \leq |u_1|, |u_2| \leq u_{max} = \frac{L}{2\ell} - 1 \text{ cycles/image.} \tag{1}$$

and is distribution-valued. In particular $G[\mathbf{O}] = a_{0,0}\delta[\mathbf{O}]$, where obviously $|a_{0,0}| > 0$ for any non-degenerate image. As a consequence of the continuity of $\mathcal{Q}g[\vec{x}]$ on $\mathcal{T}$, the graph of $G[\vec{u}]$ exhibits no "cross artefact" [9], Ch. 4].

Let $\vec{u}$ be represented in polar coordinates $\vec{u} \equiv \{u, \theta\}$, where $u = |\vec{u}|$ is wavenumber and θ the polar angle, with $0 \leq \theta \leq 2\pi$.

Denote by $G|[\vec{u}]|^2$ the power spectral density. Let

$$\Theta := \{\theta_b \leq \theta \leq \theta_e\} \tag{2}$$

denote an arc of half-width θ_H, where $0 < \theta_H \leq \frac{\pi}{2}$. Typically $\Theta = [-\theta_H, \theta_H]$ or $\Theta = [\frac{\pi}{2} - \theta_H, \frac{\pi}{2} + \theta_H]$. Its length, $|\Theta|$, is given by $|\Theta| = 2u\theta_H \ (\leq \pi u)$.

Obtaining functions of wave number

DEF. 1 (*arc-averaged spectral density*). The *scaled, arc-averaged spectral density profile* is the function $s[.]$ of u alone defined in $0 \leq u \leq u_{max}$ (cycles/image) according to

$$s[u] = \frac{1}{|\Theta|} \int_{\Theta} 10\text{Log}_{10} \left[\frac{|G[\vec{u}]|^2}{|a_{0,0}|^2} \right] u d\theta, \tag{3}$$

where the integral reduces to a finite sum over the grid nodes belonging to Θ.

Let $m[u]$ be a *model spectral density*. For example choose the continuous function parameterized by p and defined by

$$m^{(p)}[u] := 0 \ , \ 0 \leq u \leq 1 \ , \ m^{(p)}[u] := -10\text{Log}_{10}[u^p] \ , \ u \geq 1 \text{ cycles/image,} \tag{4}$$

where p (> 0) is the *model exponent*.

DEF. 2 (*log-enhanced spectrum*). The $m^{(p)}$*-log enhanced spectrum* $h^{(p)}[u]$ is defined by

$$h^{(p)}[u] := s[u] - m^{(p)}[u] \ , \ 0 \leq u \leq u_{\text{max}}. \tag{5}$$

The above operations are algebraic sums of logarithms. Since $s[0] = 0$ and $m^{(p)}[0] = 0$, $\forall p$, then $h^{(p)}[.]$ complies with $h^{(p)}[0] = 0$. Intuitively, the function $h^{(p)}[.]$ represents **deviations** of $s[.]$ from the model $m^{(p)}$. The values of L, $u_{\max}$, Θ and p are determined by the intended application.

From now on the word "tile" will designate the function $g[.]$ supported in Ω i.e., before the flop[flip[.]] operation.

As it will be shown next, the enhanced spectrum is related to spatial differentiation of integer order of the image, $\mathcal{Q}g[\vec{x}]$, followed by non linear operations.

Spectrum enhancement vs. spatial differentiation

Def. 3 (*enhanced spectrum*). The $m^{(p)}$-(plain)*enhanced power spectral density* is defined by

$$H^{(p)}[\vec{u}] := |\vec{u}|^p \frac{|G[\vec{u}]|^2}{|a_{0,0}|^2} + \delta[\vec{u}], \tag{6}$$

where $|\vec{u}|^p := (u_1{}^2 + u_2{}^2)^{\frac{p}{2}}$.

Thm. 1. (*Spectrum enhancement and spatial differentiation of integer order*). Assume the image is not degenerate and that all partial derivatives of $g[.]$ up to a suitable order exist as tempered distributions.

a) If the model exponent satisfies $p/2 = N\ (> 0)$, integer, then $H^{(p)}[\vec{u}]$ has the representation

$$H^{(p)}[\vec{u}] = \frac{1}{|a_{0,0}|^2} \sum_{n=0}^{N} \binom{N}{n} \left| \mathcal{F}\left[\frac{\partial^N \mathcal{Q}g}{\partial^{(N-n)} x_1 \partial^n x_2} \right] \right|^2 + \delta[\vec{u}]. \tag{7}$$

b) Let $p/2 = N + \lambda$ such that $N + 1 \in \mathbf{N}$, $0 < \lambda < 1$. Then

$$H^{(p)}[\vec{u}] = \frac{(u_1{}^2 + u_2{}^2)^{\lambda}}{|a_{0,0}|^2} \sum_{n=0}^{N} \binom{N}{n} \left| \mathcal{F}\left[\frac{\partial^N \mathcal{Q}g}{\partial^{(N-n)} x_1 \partial^n x_2} \right] \right|^2 + \delta[\vec{u}]. \tag{8}$$

c) In either case, if all Fourier coefficients satisfy

$$|a_{l,m}| \geq \epsilon > 0 \tag{9}$$

then the relation between $H^{(p)}[.]$ and $h^{(p)}[.]$ is

$$h^{(p)}[u] = \frac{10}{|\Theta|} \int_{\Theta} \mathrm{Log}_{10}[H^{(p)}[\vec{u}]] u d\theta. \tag{10}$$

A relation between spectrum enhancement and the spatial differentiation of fractional order when $p/2$ is not an integer has also been obtained ([11], Thm. 2).

The role of spectrum enhancement as a non linear image filter has thus been specified by Thm. 1.

Fractal analysis vs. spectrum enhancement

Unlike linear regression of the log[G] *vs.* log[u] plot used in fractal analysis [9], $\sigma\eta$ serves to separate structure from texture as suggested by the already mentioned OSHER-RUDIN paradigm [6]. Namely, $\sigma\eta$ can be tuned to emphasise the low frequency behaviour of $h^{(p)}[.]$, which corresponds to image **structure**.

Polynomial interpolation, image feature vector

The values of the raw enhanced spectrum generally form an oscillatory sequence. Whereas differences between nearby values of $h^{(p)}[.]$ need not be informative (or, the other way 'round, may be too difficult to explain), "trend" exhibited by $h^{(p)}[.]$ may be more significant and can be obtained by polynomial interpolation.

A polynomial $q[.]$ of suitable degree, d, can be obtained from $h^{(p)}[.]$ via singular value decomposition [12]. Each tile is eventually represented by the graph of $q[.]$, supported in $0 \leq u \leq u_{\max}$, where discrete wavenumbers have non-negative integer values (EQ. 1). One may want to restrict the support to the wave-number interval according to

$$(0 \leq)u_L \leq u \leq u_H(\leq u_{\max}). \tag{11}$$

The variables and parameters on which the graph of $q[.]$ depends are arranged as an n-tuple i.e., a vector $\vec{\psi}$ of n real-valued components

$$\vec{\psi} := \{\cdots, \theta_b, \theta_e, p, d, u_L, u_H\}. \tag{12}$$

Only some entries of the n-tuple are shown: the endpoints of Θ (EQ. 2), the model exponent p of EQ. 4, the polynomial degree d, as well as u_L and u_H of EQ. 11.

As a consequence one states

DEF. 4 (*The $\sigma\eta$ -derived feature vector*). Let

$$M := u_H - u_L + 1, \tag{13}$$

and

$$u_m := u_L + \frac{u_H - u_L}{M-1} m, \text{ with } m \text{ integer}, 0 \leq m \leq M-1, \tag{14}$$

then the *$\sigma\eta$ -derived feature vector* of an image is the M-dimensional vector $\vec{w} = [w_m]$, the m-th entry of which reads

$$w_m := q[u_m; \vec{\psi}], \;\; 0 \leq m \leq M-1. \tag{15}$$

The dependence of $\vec{w}[.]$ on $\vec{\psi}$ may be occasionally emphasised by writing

$$\vec{w} = \vec{w}[\vec{\psi}]. \tag{16}$$

http://dx.doi.org/10.5772/55633

Applications of $\vec{w}[\vec{\psi}]$

As the Reader may have already deduced, the n-tuple $\vec{\psi}$ can **control** the $\sigma\eta$ algorithm, hence image feature extraction.

The most straightforward application of $\sigma\eta$ consists of choosing only a few n-tuples and focussing on some properties of the corresponding polynomials, $q[\cdot;\vec{\psi}]$. An example is provided by the morphological analysis of worn tread rubber (SECTION 3).

In a more elaborate application, $\vec{\psi}$ plays the role of a (vector-valued) control variable in the training and validation of an image classifier. An example is provided by the morphological analysis of nano-dispersions (SECTION 4).

Still more articulate applications include classifier training followed by image recognition in complex experiment designs. This issue is not covered herewith.

3. Application to the morphology of worn tread rubber

3.1. Rubber and tread wear

Rubber wear

The term wear referred to tire rubber, in particular tread rubber, includes a variety of processes which, according to the classification by ALAN G. VEITH [13] and, years earlier, by B. BRISCOE [14], can be divided into wear of cohesive and interfacial type. These processes are of interest to industrial designers, product engineers and environmental scientists. Treadwear occurs between tire and road and is essential to the performance of a tire. In addition to determining the roadworthiness of a vehicle, treadwear affects fuel efficiency. It also has an impact on the environment, because it contributes to the so-called non exhaust emissions from vehicle traffic, in both gaseous and particulate form. Compound rubber, of which a tire is made, is the result of scientific research and technological advancement which have characterized the past 100 years. In essence, compounding consists of adding, under suitable conditions, sub-micrometer-sized particles of a filler, such as carbon black or amorphous silica, to an elastomer matrix, such as a styrene-butadiene block copolymer (*SBR*), natural rubber, . . . , or mixtures thereof.

In designing laboratory or field experiments involving tread wear, regardless of the final goal, scientists of any discipline, from chemistry to mechanical engineering to toxicology, shall bear in mind four basic facts:

- rubber, as a polymer, has a complex thermodynamical behaviour [15];
- tire rubber is a composite material originating from filler dispersion and distribution within the elastomer;
- different additives in the compound play different roles at different times in different parts of the tire during its life cycle;
- many properties and processes, including fracture, are rate-dependent.

Abrasive, corrosive and fatigue wear

The processes of interest herewith are abrasive wear and a kind of corrosive wear.

Quoting VEITH [13], **abrasive wear** is "caused by cutting-rupture action of sharply cornered asperities on the sliding counterface" among which road pavement or "third bodies (particles)". Hence it belongs to the cohesive type, being "controlled by the rupture strength or energy (toughness) of the wearing material".

Instead, **corrosive wear** results "from direct chemical surface attack", under albeit very mild conditions, and as such is a type of interfacial wear, because it involves reactions originating "in very thin surface layers" [13].

During normal use, tire tread undergoes both abrasive and **fatigue wear**. The latter is "caused by rapid or gradual material property changes, that give rise to cracks and, with their growth, to a loss of material" [13]. As by-products, gases, aerosols and material particles in a broad range of sizes are released to the environment.

The material particles, generally called tire wear particles (*TWP*), carry information about the thermo-mechanical degradation of compound rubber and the interaction between tread and road. The *TWP* release rate of a vehicle is a function of the tread wear rate, ρ_{tw}, the ratio of lost tread mass over traveled distance. For a four-tire passenger car, $50 \leq \rho_{tw} \leq 200$ mg/km [16].

The contribution of *TWP* to total suspended particulate matter is estimated to range from 2% to 10%, with minor contribution to PM_{10} and to $PM_{2.5}$ [17]. Once released, *TWP* undergo degradation (corrosive wear) by environmental agents: this can be regarded as a secondary process.

A thorough investigation of all processes involving treadwear is a broad scope, resource intensive project, where interdisciplinary expertise is mandatory, if the risk of gross errors in experiment design is to be kept under control.

Whenever the fracture or corrosion of materials are involved, quantitative morphological information is needed. In this SECTION, the morphology of particles originated from laboratory tread wear is analysed and an attempt is made at relating morphological features to elemental microanalytical data.

3.2. Experimental

Tread abrasion material

In laboratory simulation experiments, tire tread abrasion particles can be obtained from different equipment, which include steel brush abraders (material labeled *TrBP*) and high severity abraders (material labeled *hsWP*). This article focuses on *TrBP*. However, some properties of *hsWP* are described for comparison.

The steel brush abrader operates in air and is used for coarse balancing newly manufactured tires. The passenger car tire rotates at $v_T \simeq 9$ m/s (tangential speed) against the brush, which exerts local pressure p in the range $10^5 < p < 10^6 \text{N/m}^2$. (Average pressure is orders of magnitude lower.) Steel brush abrasion does not compare to tread wear on the road, because

a) it causes degradation of the elastomer mass by mechanical fracture only,
b) it does not cause any significant thermal degradation, because the average tread temperature never exceeds 50 C,
c) it does does not involve contact between tread and pavement, hence no foreign materials contaminate rubber debris.

High severity abraders, instead, are intended to simulate the tire - road interaction. A much higher pressure is exerted on the rubber and the tread surface temperature is remarkably higher. The *hsWP* material was obtained by a Gent - Nah-type abrader [18], where a drum clad by tread rubber rotates at $v_T = 8 \cdot 10^{-2}$m/s against a steel blade. The latter exerts a force $F \simeq 35$ N, resulting in p in the range $10^6 < p < 10^7 \mathrm{N/m^2}$ and surface temperature $T \simeq 120$ C. For this reason *hsWP* is believed to represent real world *TWP*.

Sample images of *TrBP* and *hsWP* are shown by Figure 1.

Differences in the surface chemistry of *TrBP* and *hsWP* are best understood by means of X-ray photoelectron (*XP*) spectroscopy [19]. Some *XP* spectra of *TrBP* and *hsWP* are shown by Figure 2. The caption provides the details.

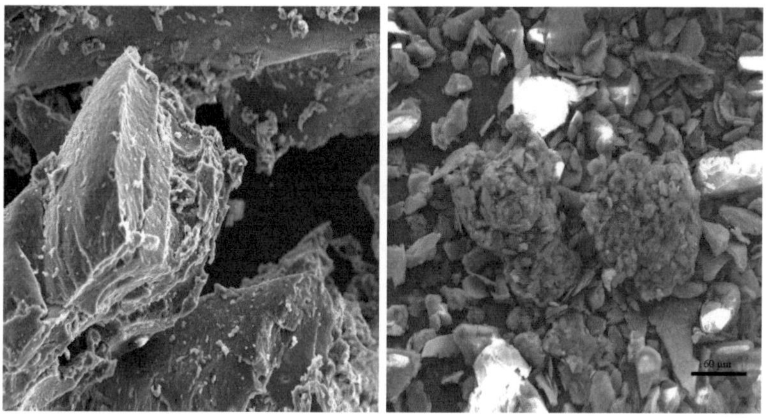

Figure 1. Materials *TrBP* and *hsWP*.
Scale bar: 60μm. Square side (both panels) $\simeq$ 400μm. Coating: Au. Images from scanning electron microscopy as described below and set at 500x magnification.
(Left panel) Material *TrBP*, made of sub-millimeter sized rubber particles.
(Right panel) Material *hsWP*: the two clusters (centre - left, centre - right) are $\simeq$ 100μm-sized rubber particles clad by micron-sized talc particles. Talc serves as anti-smear agent during abrader operation.

Leaching tests

From the arguments presented in the previous Subsection, *TWP* originated from vehicle traffic are most likely going to remain on the ground and be leached out by rain and surface waters at pH values ranging from 3 to 7.5. In principle, leaching can be simulated by laboratory tests as part of a more comprehensive experiment design.

In spite of the differences between *TWP* (*hsWP* in particular) and *TrBP* outlined above, a laboratory leaching test can be informative. Experiments were set up at the then *Department of Environmental Sciences*, University of Milan - Bicocca, Milan, IT. In order to simulate leaching by water in a pond, *TrBP* from the abrader were poured into half-filled, 5-litre glass flasks containing water buffered at pH = 7.5. The flasks were fastened to the steel table

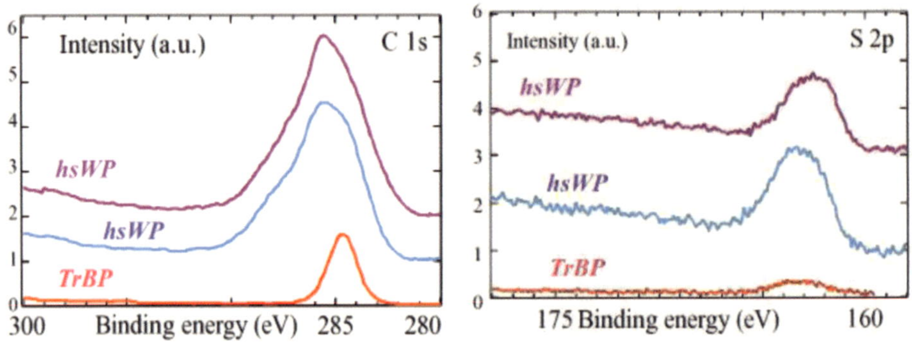

Figure 2. C $1s$ and S $2p$ X-ray photoelectron spectra of *TrBP* and *hsWP*.
Spectra obtained at the *Lawrence Livermore National Laboratory*, Livermore, CA, by means of a *Physical Electronics 5400 ESCA* using Mg $K\alpha$ radiation (1253.6 eV). The left panel displays C $1s$ energy spectra for one *TrBP* specimen (bottom graph) and two *hsWP* specimen (middle and top graphs). The right panel pertains to S $2p$ spectra.
The following bonds are found in both materials: on the left panel, C-H peaking @ 284.6 eV, on the right panel, the polysulfide group (-S-S-)n @ 162.9 $\pm$0.3 eV, which is typical of vulcanized rubber. In addition, the C $1s$ spectra of *hsWP* include contributions by C-O @ 285.6 eV, O-C-O @ 287.0 eV and O-C=O @ 288.5 eV, which broaden peaks remarkably. From the S $2p$ spectra, multiple R-S-O-R bonding states @ 165 eV can be assumed to exist on the surface of *hsWP*. Oxygen bonds are evidence of thermomechanical degradation, not observed in *TrBP*.

of a *LaboShake*™apparatus, which rotated at $\simeq$ 100 rpm on the horizontal plane, at room temperature. Samples of the reference material (fresh *TrBP*, labelled *e0*) and the 24h- and 48h- leached *TrBPs*, respectively labelled *e1* and *e2*, were then examined by the experimental methods described below and images analysed by the algorithm outlined in SUBSECTION 3.3.3.

Electron microscopy and microanalysis

Materials were analysed at the Electron Microscopy Laboratory, *Center for Advanced Materials*, University of Massachusetts-Lowell, Lowell, MA by means of an *Amray 1400* scanning electron microscope (*SEM*) equipped with a *Tracor Northern TN 3205* energy dispersive X-ray (*EDX*) spectrometer. Specimen preparation complied with the laboratory standards. *TrBP* were directly applied to *Shinto Paint*™carbon tape on top of Al studs, without depositing any conductive coating. The electron optics magnification was set at 5,000 and the resulting 640^2 pixel *SEM* images were saved in TIF. *EDX* spectra were acquired and saved by *DTSA* software.

Sample images of the *TrBP* material are shown by FIGURE 3.

3.3. Surface roughness and *EDX* data *vs.* leaching time

Spectrum enhancement

The $\sigma\eta$ algorithm is applied to the images of *TrBP*. One 640^2 pixel image per material is available, from which four 512^2 pixel tiles are obtained. Tiles represent different portions of the original with overlap.

For reasons which will be explained below, the following n-tuple (as defined by EQ. 12) is chosen

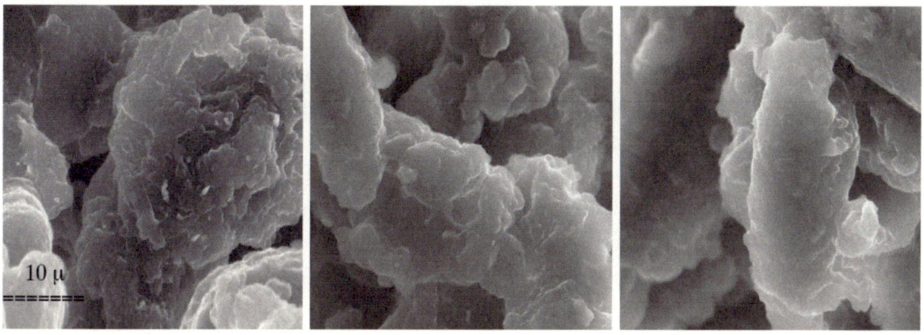

Figure 3. Tire tread particles from a steel brush abrader (*TrBP*).
Length of square side (all panels) $\frac{L}{2} \simeq 40\mu\mathrm{m}$. Tile size = 512^2 pixels. Coating: none.
Particles are sub-millimeter sized. Before the leaching experiment (left panel) surfaces show coarse and fine features i.e., carry structure and texture. After 24h leaching in water at room temperature and pH = 7.5 (centre panel), and more so after 48h (right panel), particle surfaces lose texture and become smoother.

$$\vec{\psi} := \{\theta_b = -\frac{\pi}{4}, \theta_e = \frac{\pi}{4}, p = 2.6, d = 9, u_L = 0, u_H = 255\}. \tag{17}$$

The graphs of the corresponding polynomials $q[.;\vec{\psi}] - q[0;\vec{\psi}]$ are assumed to describe surface morphology. Three such graphs, one per material, are displayed by Figure 4. The figure caption carries more details.

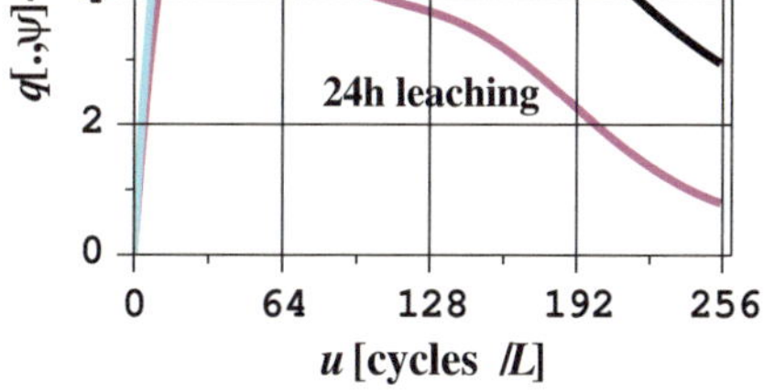

Figure 4. Application of the $\sigma\eta$ algorithm to tiles of leached *TrBP*.
Wavenumber, u, is given in cycles per sidelength, L, where $L = 80\mu\mathrm{m}$, is the side-length of the flop[flip[.]]-ed tile, in compliance with Eq. 1.
Black curve: tile #1 from material *e0* (reference); magenta curve: tile #3 from material *e1* (leached for 24h); cyan curve: tile #2 from material *e2* (leached for 48h).
In all three examples, the graphs of $q[u;\vec{\psi}] - q[0;\vec{\psi}]$ strictly increase at low wavenumbers, exhibit a relative maximum, then decrease with different slopes.
A positive value of $q[u;\vec{\psi}] - q[0;\vec{\psi}]$ at a given u means a power spectral density higher than that of the model $m^{(p)}[u]$ (Eq. 4), hence a more "structured" or "textured" image.

Qualitative interpretation of enhanced spectra

Qualitatively, the graphs of FIGURE 4 are interpreted as follows. The reference material, tile #1 of *e0* (black curve), shows a broad peak in $16 \leq u \leq 200$ cycles/L, which is due to structure (features ranging from 5 down to 2 μm) and texture (features down to 0.4 μm). In the 24- and 48h-leached materials, tile # 3 of *e1* (magenta, lowest curve) and tile #2 of *e2* (cyan, middle curve), surface texture disappears. The narrower peaks centered at 32 cycles/L come from coarse features only (5 down to 1.5 μm). The relation between wavenumber values and feature sizes is further explained by TABLE I below. The absolute ordinate values of $q[.;\vec{\psi}]$ do not carry information.

A surface roughness index

A quantitative indicator of surface morphology, of structure in particular, can be thus obtained from integrating $q[u;\vec{\psi}] - q[0;\vec{\psi}]$ over a suitable interval.

DEF. 5 (*The $\sigma\eta$ -derived surface roughness index*). Let
$u_4 := 384$ cycles/image;
$u_5 :=$ argmax $q[u;\vec{\psi}]$, in $0 < u \leq u_4$, only if $q[u_5]$ is a relative maximum, else $u_5 = 0$;
if $u_5 > 1$ then

$$\{u_3 := \text{argmax } q'[u;\vec{\psi}], \text{ in } 0 \leq u \leq u_5\}$$
$$\text{and}$$
$$u_6 := \text{argmin } q[u;\vec{\psi}], \text{ in } u_5 < u \leq u_4\}$$

else

$$\{u_3 = 0 \text{ and } u_6 = 0\}.$$

As a consequence, the roughness index ρ is defined by

$$\text{if } u_3 > 0 \text{ then } \{\rho := \sum_{u=u_3}^{u_6} (q[u;\vec{\psi}] - q[u_6;\vec{\psi}])\Delta u \ \}, \text{else } \rho := 0. \tag{18}$$

REM. 1 (*Justification*). The value of u_4 is chosen high enough to account for both surface structure and texture, but not too high in order to leave out high frequency image noise. For the index ρ to be strictly positive, a proper maximum of $q[u;\vec{\psi}]$ has to exist in $1 < u \leq u_4$. The values of u_5 and u_6 define the left and, respectively, right endpoints of the "bell" about the maximum at u_4. The index ρ, whenever > 0, is the area under such a bell.

Wavenumber values and feature sizes

Some representative values of the above defined u_5, u_3 and u_6 in cycles/image (of side length L) and of ρ are shown TABLE I, as well as the spatial periods ℓ_5 and ℓ_6, in μm, which correspond to u_5 and u_6, respectively, by recalling $L = 80\mu$m.

With the exception of u_6 of material *e0*, wave number values vary by one or two units from tile to tile of the same material. The variation of ρ is more significant: standard deviations are

material	tile	u_5	ℓ_5	u_3	u_6	ℓ_6	ρ
e0	1	44	$\simeq 1.8$	0	382	$\simeq 0.2$	1515
e1	1	32	2.5	0	383	$\simeq 0.2$	1099
e2	2	31	$\simeq 2.6$	0	245	$\simeq 0.33$	358

Table 1. Typical values of the quantities of Def. 5: u_5, u_3 and u_6 in cycles/image, spatial periods ℓ_5 and ℓ_6, in μm, and ρ in arbitrary units.

displayed in FIGURE 5 below. The ℓ_5 column of TABLE I suggests coarsening of the "dominant" feature, from 1.8 to 2.6 μm, caused by elastomer leach out. The computed values of ρ confirm the trend as well as the qualitative remarks about surface structure and texture given above (before DEF. 5).

How to choose $\vec{\psi}$ and why

REM. 2 (*On the selection of $\vec{\psi}$ and on the dependence of ρ on $\vec{\psi}$*).
The values u_H and u_L of EQ. 17 are selected in order to include the coarsest image features ($u_L = 0$) and to filter out image noise ($u_H = 384$). Since the computation of ρ is controlled by $\vec{\psi}$, the entries θ_b, θ_e, p and d are chosen among a few 4-tuples (e.g., p= 2.2, 2.4, 2.6, 2.8; d= 7, 9, 11) in order to maximise the morphological discrimination of the materials through ρ.

Elemental microanalysis

Since the same materials had been analysed by an *EDX* probe, elemental microanalytical data for S, Si, and Zn were available. The microanalytical data of FIGURE 5 (right axis) were collected at three sampling points for each specimen. The operating conditions were: incident electron beam energy = 200 keV, detector resolution = 143 eV, take-off angle = 68 deg, analyser channel width = 10 eV. Detector live times varied from one sampling point to another, ranging between 4.9 to 9.0 s. The background count rates were estimated in the neighbourhood of the spectral lines of interest, Si $K\alpha$, S $K\alpha$, and Zn $K\alpha$, then subtracted from the corresponding peak count rates.

Surface morphology and microanalytical data

The sample-averaged values of elemental concentrations and of ρ are displayed by FIGURE 5 *vs.* leaching time.

The graphs of FIGURE 5 describe the corrosive wear of *TrBP*. As the surface becomes smoother (decreasing ρ), the relative S content decreases as well i.e., the vulcanized elastomer loses its cross-links; at the same time SiO_2, the mineral filler, expressed as Si, becomes more and more exposed, whereas Zn increases slightly.

Main result

In other words, FIGURE 5 correlates surface morphology to microanalysis during the corrosive wear of the compounded elastomer.

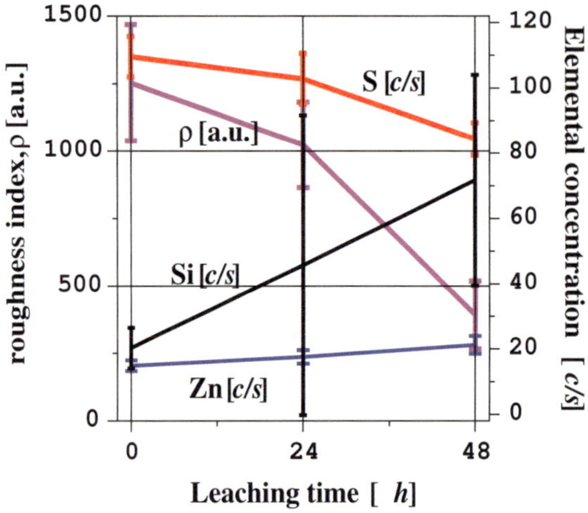

Figure 5. Correlation between surface morphology and microanalysis.
Left axis: surface roughness index, ρ (arbitrary units). Right axis: elemental concentrations (detector counts per second). All quantities derive from the leaching experiment of *TrBP* and have been averaged over each material.

4. Application to the morphology of nano-dispersions

4.1. Filler dispersion and distribution

Polymer nano-composites

A polymer nano-composite is a heterogeneous medium which is characterised, from a geometrical point of view, by the same methods developed for random media [21]. The host material is a polymer and the filler consists of nano-particles having suitable shape, which shall comply with three requirements, all related to geometry.
a) At least one dimension shall be smaller than 100nm.
b) Filler nano-particles shall not form agglomerates i.e., shall be **dispersed**.
c) This property shall be exhibited everywhere in the material volume i.e., filler **distribution** shall be uniform.

Interest in the quantitative assessment of dispersion and distribution arises because they alter, in fact improve, many properties of the materials as compared to the raw polymer.

Let the nano-composite occupy a domain Ω and $\mathcal{U} \subset \Omega$ be a test subdomain of suitable size. Let U (= meas[$\mathcal{U}$]) be the volume of $\mathcal{U}$ and V_F be the total volume of the filler particles in $\mathcal{U}$.

Dispersion indicator

As suggested by experimental evidence, theoretical arguments and computer simulations [22] - [23], polymer chains interact with the filler inside a hull of thickness t which surrounds the filler particle. Alterations of the geometry and the physical properties of the chains occurring inside the hull affect the material as a whole. Let V_H be the total volume of hulls in $\mathcal{U}$. An indicator of dispersion in $\mathcal{U}$ is defined by

$$Q := \frac{V_H}{V_F}. \tag{19}$$

Let r be the smallest dimension (e.g., radius of gyration) of the particle and $\delta := \frac{t}{r}$. For example, if the three assumptions hold:
*A*1) t depends on the polymer properties, not on the size of the filler particle (typically $3 \leq t \leq 10$ nm);
*A*2) filler particles are spheres of radius r;
*A*3) individual filler spheres, not aggregates, are found in $\mathcal{U}$,
then

$$Q = \delta^3 + 3\delta^2 + 3\delta, \tag{20}$$

and the value of $Q[\delta]$ can be assumed to represent the degree of dispersion in $\mathcal{U}$.

With particles of different shape the corresponding relation between δ and Q can be found. Filler-host interaction is stronger if dispersion is higher: hence the interest in determining, or estimating Q by some method.

Representative elementary domain

A high Q is a necessary requirement for a good nano-composite, but not the only one. Namely, the quantities V_F and r make sense in statistical terms. If the same degree of uniformity as inside $\mathcal{U}$ is observed everywhere else in the nano-composite, then the values of V_F and r determined in $\mathcal{U}$ characterize the material and one says distribution to be good. In this case, the function $Q[\delta]$ calculated by e.g., Eq. 20 is valid everywhere, hence $\mathcal{U}$ is a representative elementary domain [21] of the heterogeneous medium and U its representative elementary volume (*REV*). Conversely, if either r or V_F fluctuate from one location to another because of non-uniform distribution, then a larger domain is needed to represent the composite. The Debye algorithm (Ch. 1 of [21]]) is applied to estimate the *REV*. Very roughly speaking, for a fixed confidence level supplied to the Debye algorithm, a smaller *REV* means better distribution. The practical implication of inhomogeneity at the micrometer (or sub-millimeter) scale is a material which performs poorly or unreliably.

Since dispersion and distribution can be derived from the geometrical properties of the nano-composite, and the latter can be seen by microscopy, the role of image analysis in materials characterization should be evident.

4.2. Materials, visual scoring

The nano-composite materials [24] of interest herewith were prepared from polyethylene terephthalate and alumina nanoparticles (Al_2O_3, approximately spherical of average diameter = 48 nm). A Haake *Rheomix 600* torque rheometer was used, hence the material is labelled H for short. Four different sets of processing parameters (temperature, torque, feed rate) gave rise to as many classes (1 to 4), hereinafter regarded as the classes of belonging.

Unfortunately, the parameter values were not disclosed. Material samples were cut by a diamond knife microtome and examined using a transmission electron microscope in bright field mode at 100 keV. Recording took place on photographic film. Hardcopies were scanned, giving rise to "master images". A number of tiles (8 to 9) can be read out of each master image. Sample tiles of the material H classes are shown by FIGURE 6.

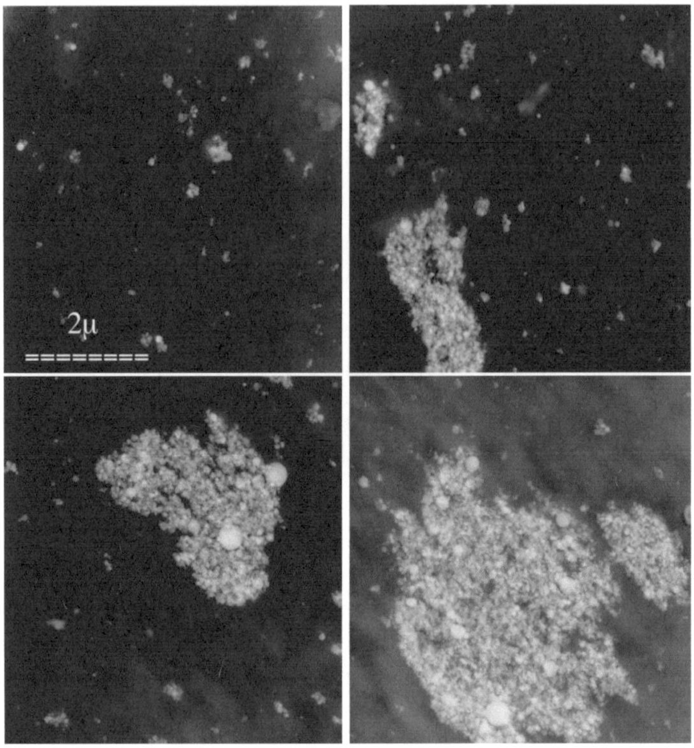

Figure 6. Al_2O_3 nano-dispersions in polyethylene terephthalate.
Square side (all panels): 5.3μm. From top left to bottom right, tiles represent classes 1 to 4.

In the past [24], images of each class had been visually rated by dispersion and distribution and ranked by an expert according to a subjective criterion. Another expert looked for a particle scoring method which could justify visual rating. A method was found and the corresponding skewness index, β, was determined, which met expectation. As a consequence, each class of belonging comes with its value of β.

4.3. Classification by $\sigma\eta$

Scope

The purpose of this SUBSECTION is to address the following question:
can $\sigma\eta$ lead to an automated classifier capable of replicating the expert's, β-based, ranking ?

Classifier flow chart

The application of the $\sigma\eta$ algorithm to material H images is more complex than the previous one (§ 3).

Since master images were *a priori* known to belong to four classes, the algorithm is requested to assign each tile to its own class with the highest possible score.

In the first place, the request can indeed be addressed, if one recalls how $\sigma\eta$ can be controlled by $\vec{\psi}$ of EQ. 12. Next, one has to design a scheme by which
• prior knowledge about the class of belonging is passed on to the algorithm,
• the class assigned by the algorithm is compared to that of belonging and class assignment is displayed,
• the classifier success rate is measured and the best performing $\vec{\psi}$ determined.

The first requirement includes the formation of training sets. The second relies on multivariate statistics (principal components analysis). The third is based on the classifier training matrix, $\mathbf{M}_T[\vec{\psi}]$, where the row index is the class of belonging, the column index the assigned class and each entry is a count; exact classification corresponds to a diagonal matrix; non-zero off diagonal entries mean misclassification.

The $\sigma\eta$ algorithm thus becomes part of a feedback loop [25], which includes multivariate statistical analysis and supervised training. An outline of the procedure is provided by FIGURE 7 and its caption.

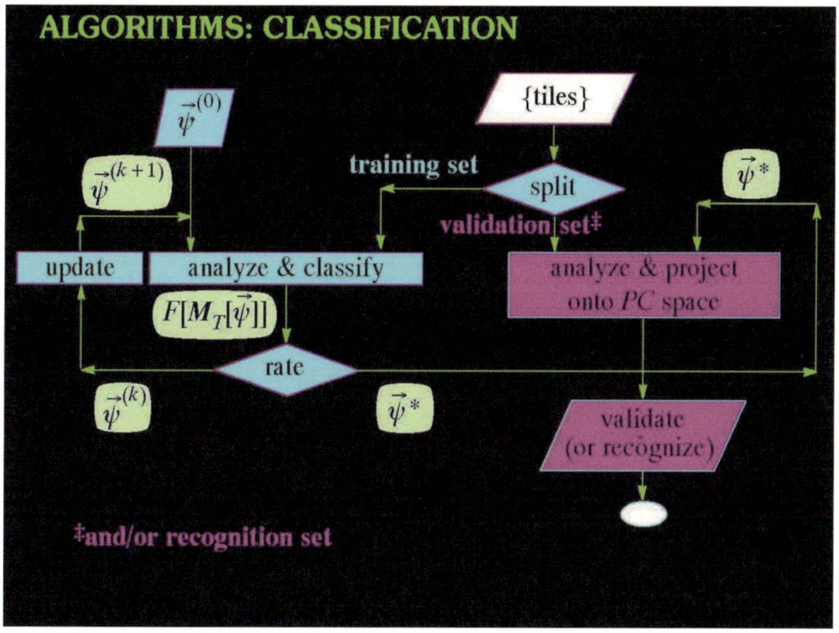

Figure 7. Flow chart of the $\sigma\eta$-based classification algorithm.
(Top right) The data set consists of tiles (§2) from all materials to be classified. (Left arrow) Some tiles, of known class of belonging, are chosen to form the training set and undergo morphological analysis (center left). This process is iterative (feedback loop on left) and is based on the $\sigma\eta$ algorithm. The latter depends on a number of control parameters (the vector $\vec{\psi}$). Supervised classification relies on principal components (*PC*) analysis and consists of
• entering the initial vector $\vec{\psi}^{(0)}$ (top left);
• analysing tiles, by determining the feature vectors $\vec{w}[\vec{\psi}]$ of Eq. 15, which represent tiles hereinafter;
• applying principal components analysis to the $\vec{w}[\vec{\psi}]$'s and assigning each tile to a class;
• forming the classifier matrix $\mathbf{M}_T[\vec{\psi}]$, and rating classification accordingly, by a suitable figure of merit, $F[\mathbf{M}_T[\vec{\psi}]]$.
If rating is poor, $\vec{\psi}$ is updated (leftmost block); if rating is acceptable, the resulting $\vec{\psi}*$ is saved (right arrow) for validation and recognition (rightmost blocks).

Classifier output display

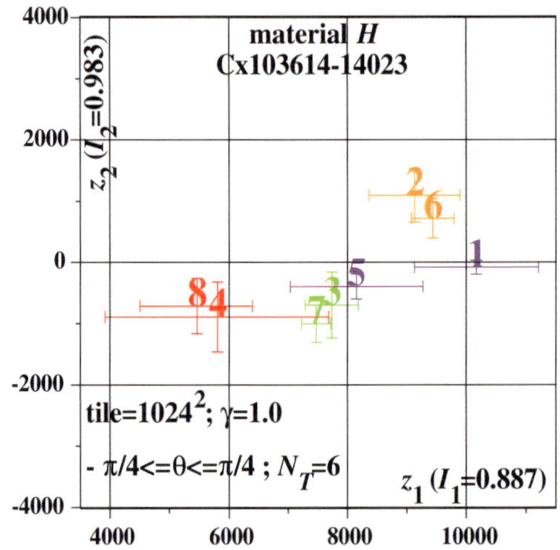

Figure 8. Classifier output.
The output is displayed on the plane of the first two principal components, z_1 and z_2. I_1 is the sample variance explained by z_1, I_2 the variance explained by z_1 and z_2 together. The training class centroids are numbered 1 to 4, their counterparts in validation are 5 to 8. According to error bars, classes 3 and 4 are not discriminated in training. Validation disclaims the discrimination of class 1 (centroid 5) from classes 3 and 4.

Interpretation of enhanced spectra

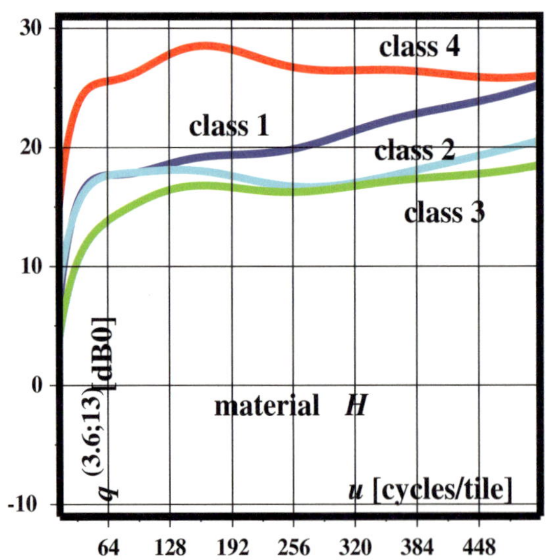

Figure 9. Sample spectra of material H tiles.
Wavenumber interval: $13 \leq u \leq 511$; degree of $q[.]$: $d = 13$; ordinate scale: dB such that $q[0] = 0$ dB. Further comments in the main text.

With reference to FIGURE 9, the $q[.]$ from class 1 (top left panel of FIGURE 6) rises steeply up to $u = 64$ cycles/tile, then exhibits a slowly increasing trend: this corresponds to the lack of relevant filler structure and texture. The $q[.]$ from class 3 (bottom left panel of FIGURE 6) reflects agglomerates. The graphs representing the other two materials can also be easily interpreted: a graph of $q[.]$ with a uniform trend (class 1) comes from a feature-less image; instead, local maxima in $60 \leq u \leq 150$ as in class 3 and 4 (bottom right panel of FIGURE 6) are due to particles of size ranging between 15 and 45 nm, regardless of aggregation or agglomeration.

A class with feature-poor tiles, such as class 1, is more likely to yield inconsistent training *vs.* validation results (FIGURE 8).

From the principal components plane display and from enhanced spectra, the dispersion and distribution properties of the nano composite can be at least **qualitatively** inferred. Namely, **dispersion** can be rated by the occurrence of agglomerates (e.g., class 1 vs. classes 3 and 4). Class 1 turns out to have the poorest **distribution** (longest class error bars in Figure 3).

Correlation between visual scoring and automated classification

The first principal component, z_1, of the class centroids, as determined by training and validation, (FIGURE 8) when regressed against the visual scoring index, β, yields the result of FIGURE 10.

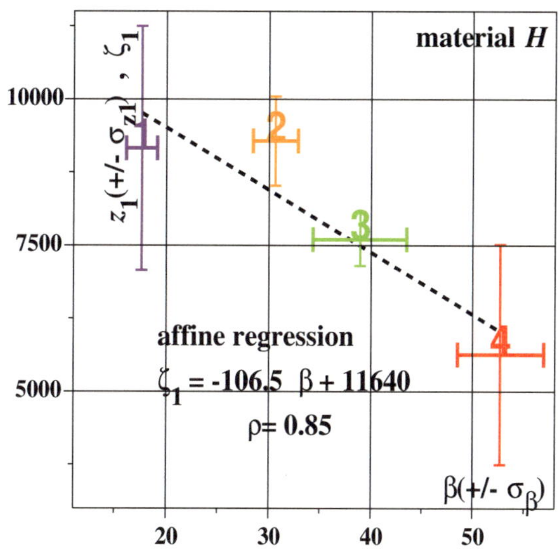

Figure 10. Correlation between morphological indices β and z_1.
The index β derives from visual scoring, z_1 from spectrum enhancement classification. The centroid z_1 coordinates, as determined by both training **and** validation, have been included into affine regression to make the result more realistic.

5. General discussion and conclusion

The core of image spectrum enhancement ($\sigma\eta$) is spatial differentiation of a suitable order, including a fractional one, followed by non linear transformations. When the control parameters are optimized, enhanced spectra seem to adequately describe the morphology of the image by separating structure from texture, which in turn are (statistical) properties of the image, or of the image set, as a whole. Image classification based on spectrum enhancement follows accordingly. If one is interested in structure and texture, then classification most likely succeeds. Indeed, the algorithm has been shown to perform in a satisfactory way on a variety of image sets, originated from as many different processes (nanodispersion, growth of tubulin microfilaments, formation of cell colonies, light scattering by material particles). Instead, spectrum enhancement, as any frequency domain method, is inappropriate to exactly locate isolated features or details.

The applications to materials science described in this article differ by complexity of the algorithm and by the degree of "assimilation" to other experimental data.

The morphology of *TrBP* has been described in a very simple way by means of the surface roughness index, ρ (DEF. 5). Graphs of enhanced spectra of the investigated materials (FIGURE 4) have been related to surface structure and texture (TABLE I). Finally, ρ has been related to elemental microanalytical data from *EDX* spectroscopy (FIGURE 5).

Possible developments include: the analysis of other types of *TWP* and studies of fracture dynamics.

Images of other *TWP* materials by the $\sigma\eta$ algorithm is possible, provided the particle surfaces are visible. Namely, coarse particles from treadwear tests are clad by minerals (from anti-smear agents or from road pavement), as shown by the right panel of FIGURE 1. Micrometer sized particles are more easily imaged and analysed.

Quantitative morphology of wear debris is relevant to the characterisation of rate dependent fracture mechanisms and therefore to assess the reliability of a given product.

The application to nanodispersions has included the development of an automated classifier, capable of discriminating materials to some extent (FIGURE 8). Enhanced spectra of single tiles have been interpreted in relation to particle dispersion and the formation of aggregates (FIGURE 9). Correlation between the automated classifier and visual scoring has been obtained (FIGURE 10). Knowledge of the mixer parameters might have deepened the understanding of automated morphological analysis.
Developments are needed at least in two directions. In the first place a relation shall be found between the Q of EQ. 19 and enhanced spectra. Namely, Q describes dispersion if evaluated locally (i.e., estimated from a single image), whereas its statistical properties describe distribution, and shall be estimated from an image set. Another issue of interest is the estimation of the representative elementary domain, $\mathcal{U}$, from $\sigma\eta$ and other experimental techniques.

Acknowledgments

The author thanks the Editor and the Publisher for the invitation and for the acceptance of this chapter.

The experimental data analysed herewith are the result of a collaboration network. Credits are given to the following people, with gratitude.

For SUBSECTION 3.2:
to Dr. LUCA MARTIGNONI (now with Galgano & Associati, Milan, IT) for the leaching

http://dx.doi.org/10.5772/55633

experiments performed on *TrBP* at the then *Department of Environmental Sciences*, Università degli Studi Milano - Bicocca, Milan, IT (*UniMiB*);
to Dr. Claudia Regazzoni (then at *Pirelli Pneumatici S.p.A.*, Milan, IT) for providing the treadwear materials;
to Dr. Art J. Nelson of the Scientific Directorate of *Lawrence Livermore National Laboratory*, for X-ray photoelectron spectroscopy;
to Dr. Tigran Dolukhanyan (now with MaxTor, Bedford, MA), Dr. Bongwoo Kang, and Prof. Changmo Sung (former Laboratory Director, now with Hyosung Co., Seoul, KR), of the Electron Microscopy Laboratory, *Center for Advanced Materials*, University of Massachusetts-Lowell, Lowell, MA.

For Subsection 4.2:
to Prof. Joey L. Mead, *Department of Plastics Engineering*, University of Massachusetts-Lowell, Lowell, MA, and to Dr. Jun S. Lee, formerly with the same Department.

The financial support of *UniMiB*'s Fondo di Ateneo per la Ricerca throughout the years is gratefully acknowledged.

Research on the $\sigma\eta$ algorithm would have not been possible without the support by the service contract no. 21861-2004-03 F1SC ISP IT - *Automatic recognition of light and electron microscope images by means of artificial intelligence techniques and functional analysis methods* granted by the *IHCP*, Joint Research Centre of the European Commission under the supervision of Laura Gribaldo, M. D.

Author details

Giovanni F. Crosta

Università degli Studi di Milano-Bicocca, Department of Earth- and Environmental Sciences, Inverse Problems & Mathematical Morphology Unit, Milan, Italy

References

[1] Ruderman D L, Bialek W. Statistics of Natural Images: Scaling in the Woods. *Phys. Rev. Lett.* 1994; **73**(6) 814-817.

[2] Vapnik V N. *The Nature of Statistical Learning Theory*. Berlin: Springer; 1995.

[3] Cucker F, Smale S. On the mathematical foundations of learning. *Bull. of the American Mathematical Society (N.S.)* 2001; 39(1) 1-49.

[4] Poggio T, Smale S. The Mathematics of Learning: Dealing with Data. *Notices of the American Mathematical Society* 2003; 50(5) 537-544.

[5] Chan T F, Shen J H, Vese L. Variational PDE Models in Image Processing. *Notices of the American Mathematical Society* 2003; 50(1) 14-26.

[6] Meyer Y. *Oscillating Patterns in Image Processing and Nonlinear Evolution Equations - The Fifteenth Dean Jacqueline B. Lewis Memorial Lectures*, Providence, RI: American Mathematical Society; 2001.

[7] Conant R C, Ashby W R. Every good regulator of a system must be a model of that system. *Int. J. Systems Sci.* 1970; 1(2) 89-97.

[8] Lesieur M. *Turbulence in Fluids*. Berlin: Springer; 2006.

[9] Russ J C. *Fractal Surfaces*. New York: Plenum; 1994.

[10] Crosta G F. Texture analysis of phosphor screens. *J Phys E: Scientific Instruments* 1977; 10 187-190.

[11] Crosta G F. Image analysis and classification by spectrum enhancement: new developments. In: Astola J T, Egiazarian K O. (eds.) *Image Processing: Algorithms and Systems VIII - Proceedings of SPIE* 7532. Bellingham, WA: SPIE; 2010. p. 75320L-01–75320L-12,

[12] Press W H, Teukolsky S A, Vetterling W T, Flannery B P. *Numerical recipes in C - The art of scientific computing. Second edition*. Cambridge, UK: Cambridge University Press; 1992.

[13] Veith A G. A review of important factors affecting treadwear. *Rubb. Chem. Technol.* 1992; 65 601-658.

[14] Briscoe B J. Wear of polymers: an essay on fundamental aspects. *Tribol. Int.* 1981; 14(4) 231-242.

[15] ver Strate G. Structure characterization in the science & technology of elastomers. In: Eirich F R. (ed.) *Science and technology of rubber*. New York, NY: Academic; 1978. p. 112-ff.

[16] http://www.kfz-betrieb.vogel.de/fahrzeugtechnik/raederundreifen/articles/ 138429/

[17] ChemRisk, Inc., DIK, Inc. State of knowledge report for tire materials and tire wear particles. Tire Industry Project Group; 2008.

[18] Gent A N, Nah C. Abrasion of rubber by a blade abrader: effect of blade sharpness and test temperature for selected compounds. *Rubber Scie. Tech.* 1996; 69 819-850.

[19] Crosta G F, Nelson A J, Camatini M C. Characterizing the thermomechanical degradation of a filled elastomer by morphology and X ray photoelectron spectroscopy. *Materials Research Society Symposium Proceedings* 2002; 702 U6.10.1-U6.10.6. DOI: http://dx.doi.org/10.1557/PROC-702-U6.10.1

[20] Crosta G F. Morphological classification and microanalysis of tire tread particles worn by abrasion or corrosion. In: Postek M T, Newbury D E, Platek S F, Joy D C, Maugel T K. (eds.) *Scanning Microscopies 2011: Advanced Microscopy Technologies for Defense, Homeland Security, Forensic, Life, Environmental, and Industrial Sciences - Proceedings of SPIE* 8036. Bellingham, WA: SPIE; 2011. p 80360H.01-80360H.10.

[21] Bear J, Bachmat Y. *Introduction to modeling of transport phenomena in porous media*. Dordrecht, NL: Kluwer Academic Publishers; 1990.

[22] Schadler L S, Kumar S K, Benicewicz B C, Lewis S L, Harton S E. Designed interfaces in polymer nanocomposites: A fundamental viewpoint. *MRS Bulletin* 2007; 32(4) 335-340.

[23] Klos J, Pakula T. Computer simulations of chains end-grafted onto a spherical surface. Effect of matrix polymer. *Macromolecules* 2004; 37(21) 8145-8151.

[24] Kim D, Lee J S, Barry C M F, Mead J L. Microscopic measurement of the degree of mixing for nano-particles in polymer nano-composites by TEM images. *Microscopy research and techniques* 2007; 70(6) 539-546.

[25] Crosta G F, Lee J S. Nanodispersion, nonlinear image filtering, and materials classification. In: Postek M T, Newbury D E, Platek S F, Joy D C, Maugel T K. (eds.) *Scanning Microscopies 2011: Advanced Microscopy Technologies for Defense, Homeland Security, Forensic, Life, Environmental, and Industrial Sciences - Proceedings of SPIE* 8036. Bellingham, WA: SPIE; 2011. p. 80360I.01–80360I.10.

Chapter 10

Multilayered Wire Media: Generalized Additional Boundary Conditions and Applications

Alexander B. Yakovlev, Yashwanth R. Padooru,
George W. Hanson and Chandra S.R. Kaipa

Additional information is available at the end of the chapter

http://dx.doi.org/10.5772/36937

1. Introduction

The history of homogenization methods describing the interaction of electromagnetic waves with materials/matter (formed by a large number of periodic metal-lattices/atoms) goes a long way back (see, for a detailed historical review, the books by [1] and [2]). Typically, these methods are applied when the size of the material inclusions is small compared to the wavelength of the incident wave. In such cases, the microscopic fluctuations are averaged out to obtain smooth and slowly varying macroscopic quantities that can be used to characterize the long range variations of the electromagnetic waves [3].

In recent years, there has been an increased interest in homogenization methods characterizing artificial materials (such as metamaterial structures). In particular, materials that constitute wire media have attracted special attention, due to their ability in enabling interesting phenomena such as negative refraction [4, 12] and sub-wavelength imaging [6, 7], among others. It has been recently shown in [8] that wire media exhibits strong spatial dispersion at microwaves, and that the constitutive relations between the macroscopic fields and the electric dipole moment are non-local [5]. Due to the non-local character of the material, solving the reflection and transmission problems at interfaces associated with wire media becomes difficult. This is because the non-local character of the material enables it to support extra (or extraordinary) waves, which in general are not supported by materials with local responses. To overcome this, uniquely solved ABCs [9–11, 13] which are pertinent to the specific problem (composition of the structure) can be introduced.

In [10], the scattering problem of a wire-medium slab (consisting of long parallel array of thin-metallic wires normal to the interface) was solved by deriving an ABC at the interface of the wire medium and a dielectric material. Here the ABC was derived based on the fact that the microscopic current must vanish at the tip of the wires and that the macroscopic fields must satisfy the ABC at the interface. The above result was extended to study the reflection characteristics of a textured surface [9], wherein the metallic wires are connected to a ground plane. In this scenario, the ABC was derived by taking into account that the microscopic electric charge density vanishes at the connection of wire end to the ground plane. Later, the ABC developed in [9] was applied to characterize the reflection and transmission properties of the single-layer [14, 15] and multilayer mushroom-type structures [17] (composed of metallic patches). It should be noted that these structures (single-layer and multilayer) were shown to suppress spatial dispersion in wire media. This is because the presence of metallic patches at the wire ends diminishes charge buildup in such a way that, upon homogenization, the mushroom structure can be treated as a uniaxial continuous Epsilon-Negative (ENG) material loaded with a capacitive grid of patches. Further, in [13] generalized additional boundary conditions (GABCs) have been derived for wire media terminated with distributed loads (metallic patch arrays acting as parallel loads to the wires) and lumped loads (arbitrary impedance insertions acting as series loads to the wires) or a combination of both at the junction, with the latter case presented in [18, 19]. Although, the GABCs derived in [13] are applied at the wire-to-patch connection with the finite size of the patch (with certain restrictions imposed on the size of the gap between the patches with respect to the separation of adjacent patch arrays), these boundary conditions are valid only for perfect electric conductor terminations. However, when the metallic terminations (patches) are thin (resistive) (or for no patch case [10]), the charge accumulation and diffusion at the wire-to-patch interface (or charge accumulation at the open wire end interface) becomes important (spatial dispersion effects have to be considered), necessitating a new additional boundary condition at this interface [16], which takes into account the finite conductivity of the material at the connection points. Upon homogenization, these charge effects are reflected in the nonlocal slab permittivity. The ABC developed in the later case (thin patch) is a generalized form from which one can easily obtain the ABCs derived in [9] and [10].

The ABC derived in [16] can only be applied to a single-layer wire-medium slab terminated with either PEC or thin resistive (metal/graphene) patches or a combination of both. It is derived under the hypothesis that the material adjacent to it is either free space or a dielectric filled material. However, for a more general configuration shown in Fig. 1 (where the metallic wires of one wire-medium slab are connected to another wire medium with a thin resistive patch at the junction), the ABC proposed in [16] cannot be applied.

In our recent paper [26] we further extend the theory of [16] and study the reflection properties of a more general case of a multilayer mushroom-type structure composed of thin resistive patches with a typical configuration shown in Fig. 1. Based on charge conservation, new ABCs are derived at the interface of two uniaxial wire mediums with thin imperfect conductors at the junction. The scattering problem is solved by imposing the classical boundary conditions (at the thin resistive patch interface and at the ground plane), and the new additional boundary conditions obtained at the wire-to-patch junctions. To illustrate the application of the homogenization model with the new ABCs, we characterize the reflection properties of the multilayer structure, demonstrating that such a configuration with proper choice of the geometrical parameters acts as an absorber. Interestingly, it is noticed that the presence of vias results in the enhancement of the absorption bandwidth and an improvement in the absorptivity performance for increasing angles of the obliquely incident TM-polarized plane wave. The results obtained using the homogenization model for the proposed structure are validated against full-wave numerical simulations.

http://dx.doi.org/10.5772/36937

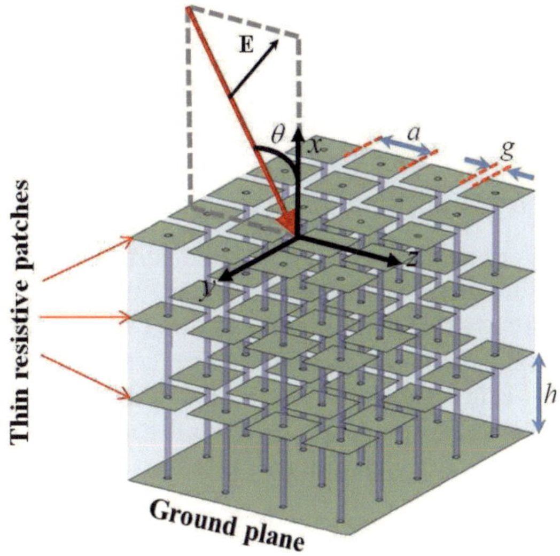

Figure 1. Geometry of a multilayer mushroom structure formed by periodically loading a grounded wire medium with thin resistive patches.

The chapter is organized as follows. In Section 2, at first, a review of the ABCs for the wire media is provided. Then, generalized ABCs at the wire-to-thin-resistive-patch junction are derived and the formalism of the nonlocal homogenization model is presented for the analysis of the reflection characteristics of the multilayer mushroom-type structure. The results of the single-layer, two-layer, and three-layer mushroom structures composed of thin resistive patches are discussed in Section 3. Finally, concluding remarks are given in Section 4.

2. Homogenization of multilayered mushroom-type HIS structures

In this section, at first a brief review of the recently derived ABCs [9, 10, 16] for wire media with applications to mushroom structures [14–17] is given (see Section 2.1). It should be noted that the configurations studied in [14–16] are single-layer mushroom structures (wire media loaded with patch arrays) with/without ground planes, and the one studied in [17] was a multilayer mushroom structure without a ground plane. In all these cases, except the one studied in [16], the vias, the patches, and the ground plane have all been assumed to be perfect electric conductors (PECs). In [16], the vias, and the ground plane are assumed to be PECs, however, the patch is a thin 2-D material. In [26] we study a multilayer structure (shown in Fig. 1) that is backed by a perfect electrically conducting ground plane, and assume that the vertical wires are PEC conductors and that the patches are arbitrary thin resistive materials. Secondly, we show the derivation of ABCs for the interface of two uniaxial wire mediums with thin imperfect conductors at the junction (see Section 2.2). Finally, the reflection problem of the multilayer structure (shown in Fig. 1) will be solved for the obliquely incident TM-polarized plane waves (see Section 2.3).

Referring to [8], it is known that the wire-medium slab is strongly spatially dispersive, and supports three different modes: transverse electric (TE^x), transverse magnetic (TM^x), and transverse electromagnetic (TEM) modes. Since TE incident waves do not interact with the wires, the study is restricted to only TM incident waves. In what follows, the term *microscopic* refers to currents and fields in the microstructure of the medium, i.e., on the wires and patches of the actual physical structure. The term *macroscopic* refers to fields averaged over the lattice period, i.e., the fields in the equivalent homogenized (continuous) medium. In the following, a time variation of the form $e^{j\omega t}$ is assumed and suppressed.

2.1. Additional boundary conditions for wire media

The wire medium consists of an array of long metallic parallel wires arranged in a periodic lattice as shown in Fig. 2. The wires are oriented along the x-direction and are embedded in a host medium with permittivity ε_r. Solving the scattering problem for the wire-medium slab (which supports three modes) with the regular classical boundary conditions is impossible. To overcome this, an ABC is necessary. The ABC is derived by identifying some property of the wire-medium slab which gives a relation between the macroscopic fields, which in turn provides an extra degree of freedom to solve the scattering problem.

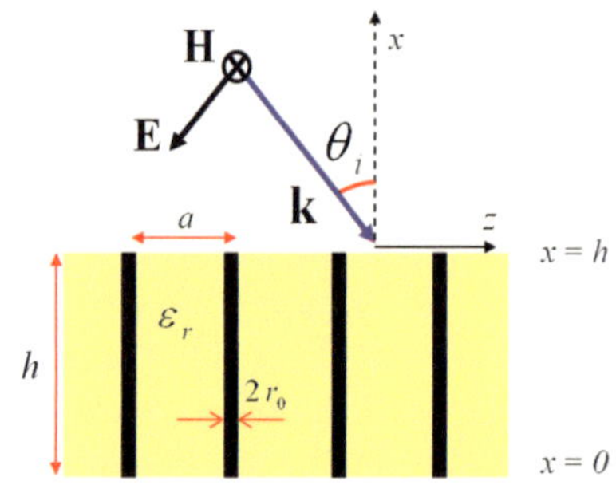

Figure 2. Geometry of a wire-medium slab with thickness h illuminated by a TM- polarized plane wave.

In [10], the authors showed that when the wire medium is adjacent to a nonconductive medium (such as air) and that the wires are thin, the microscopic current must vanish at the wire end $x = x_0$,

$$J_c(x_0) = 0 \tag{1}$$

and that the macroscopic field must satisfy the ABC at the interface with air [11],

$$\varepsilon_r \mathbf{E} \cdot \hat{x}|_{\text{wire medium side}} = \mathbf{E} \cdot \hat{x}|_{\text{air side}} \tag{2}$$

or, equivalently, the macroscopic field condition

$$k_0 \varepsilon_r E_x(x_0) - k_z \eta_0 H_y(x_0) = 0\,. \tag{3}$$

However, the ABC (3) is not valid when the wire medium is adjacent to a conducting material, such as a perfect electrically conducting ground plane (shown in Fig. 3). In such a scenario, the microscopic current (J_c) at the wire-to-patch interface does not vanish and thus the ABC (3) is no longer applicable. The authors in [9, 10] have shown that it is relatively simple to derive the boundary condition for the wire-to-PEC interface. More specifically, they proved that the electric density of surface charge, σ_c, in a wire must vanish at the connection with the PEC ground plane,

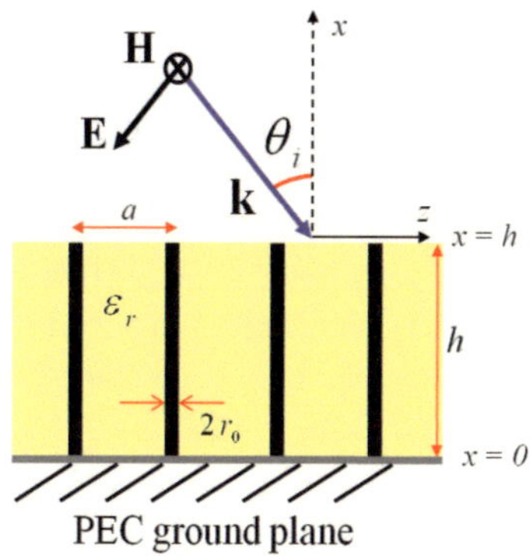

Figure 3. Geometry of a wire-medium slab connected to a ground plane.

$$\sigma_c = 0 \,(\text{at the interface}) \tag{4}$$

which necessitates that

$$\frac{\mathrm{d}J_c(x)}{\mathrm{d}x}|_{x=0^+} = 0 \tag{5}$$

or, in terms of macroscopic fields,

$$\left(k_0\varepsilon_r \frac{\mathrm{d}E_x(x)}{\mathrm{d}x} - k_z\eta_0 \frac{\mathrm{d}H_y(x)}{\mathrm{d}x} \right)|_{x=0^+} = 0\,. \tag{6}$$

The above derived ABC (6) has been successfully applied by many researchers to various configurations of interest in which wire media were connected to one or many conducting elements (2-D array of patches). The most important of these configurations is the so-called mushroom structure [21] (wire media loaded with PEC pacthes and backed by a ground plane) which can be used in the design of high-impedance surfaces [14, 15] (shown in Fig. 4(a)) and metamaterials with negative refraction [17] (shown in Fig. 4(b)), among others.

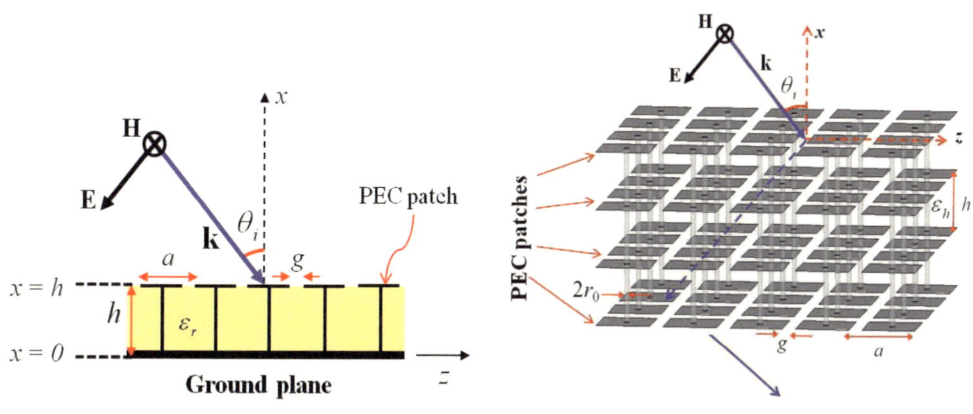

Figure 4. Mushroom-type wire medium structure: (a) Single-layer structure. (b) Multilayered structure.

However, this ABC applies only to a wire-to-PEC interface. When the skin depth is such that the field penetrates throughout the material, the PEC model is a very poor approximation of the actual physics. In this case, we need an ABC for a wire connected to an imperfect conductor, or, more generally, to an arbitrary material characterized by its complex conductivity. The same problem occurs when trying to model the ground plane or patch as a 2-D material, such as graphene or a 2-D electron gas. To answer this, the authors in [16] derived a generalized ABC for the wire-thin-metal-patch interface. The ABC is derived based on the principle of conservation of charge and that the microscopic wire current must satisfy the boundary condition,

$$\left(J_c(x) + \frac{\sigma_{2\mathrm{D}}}{j\omega\varepsilon_0\varepsilon_r} \frac{\mathrm{d}J_c(x)}{\mathrm{d}x} \right) |_{x=h^-} = 0 \tag{7}$$

or, in terms of macroscopic field condition,

$$\left(1 + \frac{\sigma_{2\mathrm{D}}}{j\omega\varepsilon_0} \frac{\mathrm{d}}{\mathrm{d}x} \right) \left[k_0 E_x(x) - \frac{k_z \eta_0}{\varepsilon_r} H_y(x) \right] |_{x=h^-} = 0. \tag{8}$$

In the limiting case of $\sigma_{2\mathrm{D}} \to 0$, we have the wire medium (bed-of-nails) result (3), and for $\sigma_{2\mathrm{D}} \to \infty$, we have the PEC patch result (6). Hence, (7) is the generalized form of ABC for the wire-thin-metal-patch interface (or for any conductivity of the thin-resistive patch).

The ABC (7) is valid as long as the material adjacent to it is nonconductive (such as air). If there is another spatially dispersive material (such as wire media) adjacent to it (see Fig. 1), then the ABC given by (7) is no longer valid. In the next subsection we derive a generalized ABC for the two uniaxial wire media with a thin resistive patch at the junction [26]. However, when two wire mediums are connected with a PEC patch at the junction (see Fig. 4(b), for the multilayer structure with PEC patches), GABCs similar to those proposed in [13] (for the PEC patches) have to be considered at the connection $x = x_0$:

$$\left.\frac{\mathrm{d}J_c(x)}{\mathrm{d}x}\right|_{x_0^+} + \left.\frac{\mathrm{d}J_c(x)}{\mathrm{d}x}\right|_{x_0^-} = \frac{2C}{C_0}\left[J_c(x_0^+) - J_c(x_0^-)\right] \tag{9}$$

$$\left.\frac{\mathrm{d}J_c(x)}{\mathrm{d}x}\right|_{x_0^+} - \left.\frac{\mathrm{d}J_c(x)}{\mathrm{d}x}\right|_{x_0^-} = 0 \tag{10}$$

where C is the capacitance of the wires and C_0 is the capacitance of the metallic patch which depends on the period a and gap g (the values of C and C_0 are defined in [13]). These ABCs are accurate for moderate and large gaps between the patches, provided the distance h between the metallic patches in adjacent layers is much greater than g. When the gap between the patches reduces and $C_0 \to \infty$, we have a perfect electric conducting ground plane with the ABCs (9), (10) reduced to $\mathrm{d}J_c(x)/\mathrm{d}x|_{x_0^+} = \mathrm{d}J_c(x)/\mathrm{d}x|_{x_0^-} = 0$, which is the same expression obtained above from (5). The PEC-ABC (5) is applicable at the junction of wire-PEC-wire interface, because the electric surface charge density on the wires will vanish independently at the junction making the fields in one media independent of the other. However, for the

http://dx.doi.org/10.5772/36937

case of thin-resistive patch at the junction, this is not the case, where charge diffusion and accumulation takes place with the fields in one wire media interacting with the ones in the other. Hence an ABC is necessary for this kind of interface (wire-thin-resistive-patch-wire).

2.2. Additional boundary condition for the junction of uniaxial wire media with thin resistive patch at the interface

Consider a plane wave incident on the configuration shown in Fig. 5, which consists of two uniaxial wire mediums with a thin resistive sheet (or in general, an arbitrary 2-D material such as graphene or a 2-D plasma characterized by a complex surface conductivity) placed at the interface $x = x_0$. Let a be the lattice period, σ_{2D} be the complex *surface* conductivity of the thin resistive patch (such as graphene patches with the surface conductivity given in [22]), r_1 and r_2 be the wire radii with $r_{1,2} \ll a$, and $\varepsilon_{r,1}$ and $\varepsilon_{r,2}$ be the corresponding dielectric host material for the two uniaxial wire media. For thin materials with *bulk* conductivity σ_{3D}, the surface conductivity can be written as $\sigma_{2D} = \sigma_{3D}t = 1/R_s$, where $t \ll \delta$ is the material thickness, $\delta = \sqrt{2/\omega\mu_0\sigma_{3D}}$ is the skin depth, and R_s is the sheet resistance. To understand how a thin resistive material (with bulk conductivity) is made see [29]. The resistive sheets can also be realized using commercially available resistive materials.

Due to the presence of discontinuities at the junction (sheet and different properties of the wire media, i.e., different wire radii and host permittivities) one can expect irregularities in the charge and the current distributions close to the junction. Let $J_{w,1}$ and $J_{w,2}$ be the microscopic current densities on the surface of the wires in mediums 1 and 2, and ρ_{s1} and ρ_{s2} be the surface charge densities on the PEC wires with radii r_1 and r_2, respectively, which are given by $\rho_{s1}(x) = \varepsilon_0\varepsilon_{r,1}E_{n1}(x)$ and $\rho_{s2}(x) = \varepsilon_0\varepsilon_{r,2}E_{n2}(x)$, where E_{n1} and E_{n2} are the normal components of the microscopic electric fields at the wire surfaces. On the thin conductive sheet, assumed local and isotropic, the microscopic current and the field are related as $J_s(y,z) = \sigma_{2D}E_t$, where J_s is the surface current density and E_t is the tangential electric field on the sheet. It should be noted that the tangential fields on the sheet in mediums 1 and 2 are assumed to be continuous at $x = x_0$, i.e.,

$$E_{t1}(x_0^-) = E_{t2}(x_0^+) = E_t \ . \tag{11}$$

Considering that, at the wire-to-sheet and sheet-to-wire connection points (x_0^+ and x_0^-) the electric fields normal to the wires are the same as the tangential fields on the thin resistive sheet, we can write

$$E_{n1}(x_0^-) = E_{t1}(x_0^-) = \rho_{s1}(x_0^-)/(\varepsilon_0\varepsilon_{r,1}) \tag{12}$$

$$E_{n2}(x_0^+) = E_{t2}(x_0^+) = \rho_{s2}(x_0^+)/(\varepsilon_0\varepsilon_{r,2}) \ . \tag{13}$$

From the continuity equation for the wires $\rho_{si} = -(1/j\omega)\mathrm{d}J_{w,i}(x)/\mathrm{d}x$ ($i = 1,2$) one can write the surface charge densities at the connection points (x_0^+ and x_0^-) as

$$\rho_{s1}(x_0^-) = -(1/j\omega)(\mathrm{d}J_{w,1}(x)/\mathrm{d}x)|_{x_0^-} \tag{14}$$

$$\rho_{s2}(x_0^+) = -(1/j\omega)(\mathrm{d}J_{w,2}(x)/\mathrm{d}x)|_{x_0^+} \ . \tag{15}$$

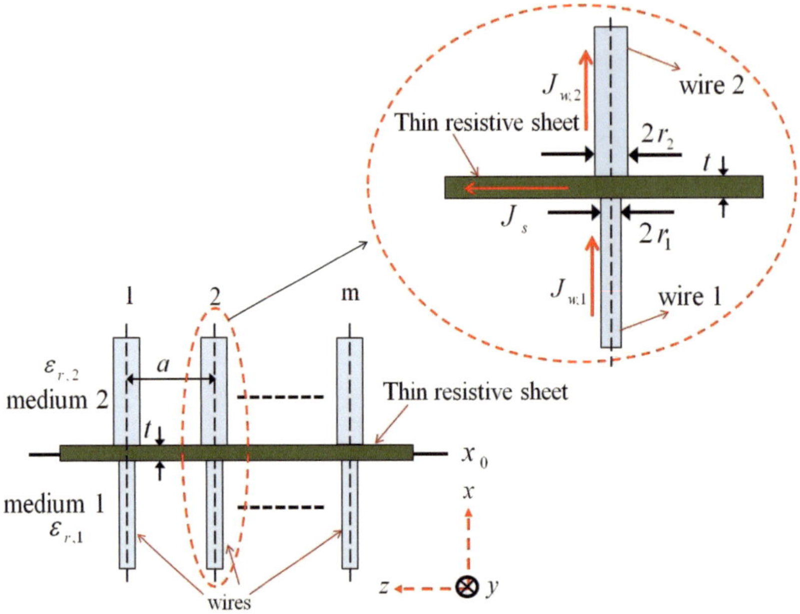

Figure 5. Geometry of a junction of two wire mediums with a thin resistive sheet at the interface.

Applying Kirchoff's current law (conservation of charge) at the junction of two wire mediums with thin resistive sheet at the interface (from Fig. 5), we have

$$J_s = J_{w,1} - J_{w,2}\,. \tag{16}$$

Using (11), the surface current density can be expressed as

$$\begin{aligned} J_s &= \sigma_{2D} E_t = \sigma_{2D} E_{t1}(x_0^-) = \sigma_{2D} E_{t2}(x_0^+) \\ &= \sigma_{2D}[E_{t1}(x_0^-) + E_{t2}(x_0^+)]/2\,. \end{aligned} \tag{17}$$

Equating now (16) and (17), we have at the connection points

$$\sigma_{2D}[E_{t1}(x_0^-) + E_{t2}(x_0^+)]/2 = [J_{w,1}(x_0^-) - J_{w,2}(x_0^+)]\,. \tag{18}$$

Substituting the tangential fields (12)-(13) in (18), we can write

$$\sigma_{2D}[\rho_{s1}(x_0^-)/(\varepsilon_0\varepsilon_{r,1}) + \rho_{s2}(x_0^+)/(\varepsilon_0\varepsilon_{r,2})]/2 = [J_{w,1}(x_0^-) - J_{w,2}(x_0^+)]\,. \tag{19}$$

Now, using the surface charge densities of the two wires (14)-(15), in (19) we obtain the ABC

$$\frac{\sigma_{2D}}{2j\omega\varepsilon_0}\left[\frac{1}{\varepsilon_{r,1}}\frac{dJ_{w,1}(x)}{dx}\Big|_{x_0^-} + \frac{1}{\varepsilon_{r,2}}\frac{dJ_{w,2}(x)}{dx}\Big|_{x_0^+}\right] + [J_{w,1}(x_0^-) - J_{w,2}(x_0^+)] = 0. \tag{20}$$

Using the fact that the tangential fields are continuous at the thin conductive sheet interface (11), we obtain the second ABC for the microscopic wire current,

$$\frac{1}{\varepsilon_{r,1}}\frac{\mathrm{d}J_{w,1}(x)}{\mathrm{d}x}\Big|_{x_0^-} - \frac{1}{\varepsilon_{r,2}}\frac{\mathrm{d}J_{w,2}(x)}{\mathrm{d}x}\Big|_{x_0^+} = 0\,. \tag{21}$$

In the next section, it will be shown that the ABCs (20) and (21) along with the classical boundary conditions will completely characterize the reflection properties of the multilayer mushroom-type HIS structure (shown in Fig. 1). Also, it is worth noting that the conditions (20) and (21) derived in this section are rather general and applicable to the cases of different conductivities of the thin conductive sheet at the wire-medium junction. In the limiting case of $\sigma_{2D} \to 0$ (transparent sheet), we have a continuous wire-medium slab with simple continuity conditions for the current: $J_{w,1}(x_0^-) = J_{w,2}(x_0^+)$ and $\varepsilon_{r,1}^{-1}\mathrm{d}J_{w,1}(x)/\mathrm{d}x|_{x_0^-} = \varepsilon_{r,2}^{-1}\mathrm{d}J_{w,2}(x)/\mathrm{d}x|_{x_0^+}$. For $\sigma_{2D} \to \infty$, we have a PEC conductor with the ABC for the wire microscopic currents given by $\mathrm{d}J_{w,1}(x)/\mathrm{d}x|_{x_0^-} = \mathrm{d}J_{w,2}(x)/\mathrm{d}x|_{x_0^+} = 0$, i.e., the derivative of each of the wire currents is independently zero at the connection points. This is consistent with the result of the single-sided wire-medium junction with a PEC conductor [10].

For the limiting case of the same host material on either side of the thin resistive sheet interface at x_0 (i.e., $\varepsilon_{r,1} = \varepsilon_{r,2} = \varepsilon_r$), the ABCs (20) and (21) can also be obtained by enforcing the continuity of surface charge densities and using the Kirchoff's current law at the connection points x_0^+ and x_0^- (the junction of wire media with thin resistive sheet at the interface), i.e., $\rho_{s1}(x_0^-) = \rho_{s2}(x_0^+) = \rho_s = \varepsilon_0\varepsilon_r J_s/\sigma_{2D} = (\rho_{s1}(x_0^-) + \rho_{s2}(x_0^+))/2$ and $J_{w,1} = J_s + J_{w,2}$.

It is interesting to note that (20) and (21) yield two independent ABCs. At first sight, this may seem inconsistent with the result of [16], which considered a single ABC to model the interfaces of wire media with an imperfect conductor. However, the reason why we obtain an extra ABC is quite simple. Here we consider the junction of two different wire mediums; that is, we have a spatially dispersive material on both sides of the interface (a double-sided wire-medium junction). Quite differently, the configuration considered in [16] consists of a single-sided wire-medium junction because one of the semi-spaces separated by the interface is free space. To show this, consider the two ABCs (20) and (21). Rewritting them, we have

$$\left[\frac{1}{\varepsilon_{r,1}}\frac{\mathrm{d}J_{w,1}(x)}{\mathrm{d}x}\Big|_{x_0^-} + \frac{1}{\varepsilon_{r,2}}\frac{\mathrm{d}J_{w,2}(x)}{\mathrm{d}x}\Big|_{x_0^+}\right] = -\frac{2j\omega\varepsilon_0}{\sigma_{2D}}\left[J_{w,1}(x_0^-) - J_{w,2}(x_0^+)\right] = 0 \tag{22}$$

$$\frac{1}{\varepsilon_{r,1}}\frac{\mathrm{d}J_{w,1}(x)}{\mathrm{d}x}\Big|_{x_0^-} - \frac{1}{\varepsilon_{r,2}}\frac{\mathrm{d}J_{w,2}(x)}{\mathrm{d}x}\Big|_{x_0^+} = 0\,. \tag{23}$$

By solving (22) and (23) and rearranging the terms, we can write

$$\frac{1}{\varepsilon_{r,1}}\frac{\mathrm{d}J_{w,1}(x)}{\mathrm{d}x}\Big|_{x_0^-} = -\frac{j\omega\varepsilon_0}{\sigma_{2D}}\left[J_{w,1}(x_0^-) - J_{w,2}(x_0^+)\right]\,. \tag{24}$$

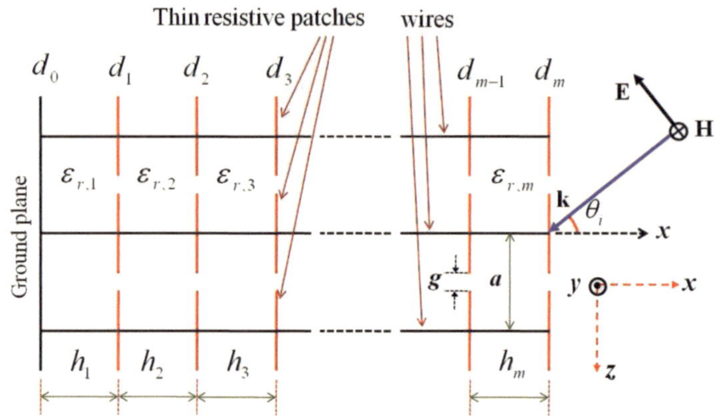

Figure 6. Schematic of a generic multilayer mushroom structure formed by periodically loading grounded wire medium with thin resistive square patches (side view).

Assuming that there is no second wire medium, we have $J_{w,2} = 0$, then (24) can be expressed as

$$\frac{1}{\varepsilon_{r,1}} \frac{\mathrm{d}J_{w,1}(x)}{\mathrm{d}x}\Big|_{x_0^-} = -\frac{j\omega\varepsilon_0}{\sigma_{2\mathrm{D}}} J_{w,1}\left(x_0^-\right). \tag{25}$$

or,

$$J_{w,1}(x_0^-) + \frac{\sigma_{2\mathrm{D}}}{j\omega\varepsilon_0\varepsilon_{r,1}} \frac{\mathrm{d}J_{w,1}(x)}{\mathrm{d}x}\Big|_{x_0^-} = 0 \tag{26}$$

which corresponds exactly to the ABC derived in ([16], Eq. (5)). Therefore, while a single ABC is sufficient to describe the electrodynamics of a single-sided wire-medium junction, the general case of a double-sided junction requires two ABCs due to the increased number of degrees of freedom (i.e., extra waves can be generated on both sides of the junction). Hence, (20) and (21) are generalizations of the simpler case studied in [16].

2.3. Nonlocal homogenization of the multilayered mushroom-type HIS structure

In this section, we show how the ABCs (20) and (21) derived in the previous section are necessary to calculate the reflection properties of a multilayered mushroom-type HIS structure for a obliquely incident TM-polarized plane wave (with the geometry as that shown in Fig. 6). Each of the wires with via radii $r_l \ll a$ (where a is the period of the patches and vias) are embedded in a dielectric host media (which is homogeneous and isotropic) of thickness h_l, characterized by relative permittivity $\varepsilon_{r,l}$ and permeability of free space, and are loaded with 2-D periodic thin resistive patches of conductivity $\sigma_{2\mathrm{D},l}$ at the interfaces $d_l, l = 1, 2, \ldots, m$. Here, we assume that the wires are lossless (PEC).

The wire-medium slab is characterized by the nonlocal dielectric function [8, 9]

$$\varepsilon_{\mathrm{eff},l} = \varepsilon_0\varepsilon_{r,l}[\varepsilon_{xx,l}(\omega, k_x)\hat{x}\hat{x} + \hat{y}\hat{y} + \hat{z}\hat{z}] \tag{27}$$

where $\varepsilon_{xx,l}(\omega, k_x) = 1 - k_{p,l}^2/(k_{h,l}^2 - k_x^2)$, $k_{h,l} = k_0\sqrt{\varepsilon_{r,l}}$ is the wavenumber in the host material, k_0 is the wavenumber in free space, $k_{p,l}$ is the plasma wavenumber which depends on the period and radius of the vias: $k_{p,l}^2 = (2\pi/a^2)/[\ln(a/2\pi r_l) + 0.5275]$, and k_x is the x-component of the wave vector $\mathbf{k} = (k_x, 0, k_z)$. Let $J_{w,l}$ be the currents induced on the metallic wires. It is known that for a TM plane-wave incidence, the wire medium excites both TEM and TM^x modes, and thus, following [9], the magnetic field in all space is given by:

$$\eta_0 H_y = \mathrm{e}^{\gamma_0(x-d_m)} + R\mathrm{e}^{-\gamma_0(x-d_m)}, \text{ air side: } x > d_m \tag{28}$$

$$\eta_0 H_y^{(l)} = A_{\mathrm{TM},l}^{+}\mathrm{e}^{\gamma_{\mathrm{TM},l}(x-d_{l-1})} + A_{\mathrm{TM},l}^{-}\mathrm{e}^{-\gamma_{\mathrm{TM},l}(x-d_{l-1})} + B_{\mathrm{TEM},l}^{+}\mathrm{e}^{\gamma_{\mathrm{TEM},l}(x-d_{l-1})} + B_{\mathrm{TEM},l}^{-}\mathrm{e}^{-\gamma_{\mathrm{TEM},l}(x-d_{l-1})}, \text{ wire medium slab: } d_{l-1} < x < d_l \tag{29}$$

where $d_l = h_1 + h_2 + \ldots + h_l$, $l = 1, 2, \ldots, m$, $d_0 = 0$, $\eta_0 = \sqrt{\mu_0/\epsilon_0}$ is the free-space impedance, R is the reflection coefficient, $A_{\mathrm{TM},l}^{\pm}$, $B_{\mathrm{TEM},l}^{\pm}$ are the amplitude coefficients of the TM and TEM fields, $\gamma_0 = \sqrt{k_z^2 - k_0^2}$, $\varepsilon_{xx,l}^{\mathrm{TM}} = 1 - k_{p,l}^2/(k_z^2 + k_{p,l}^2)$, $\gamma_{\mathrm{TEM},l} = jk_0\sqrt{\epsilon_{r,l}}$, $\gamma_{\mathrm{TM},l} = \sqrt{k_{p,l}^2 + k_z^2 - k_{h,l}^2}$, and $k_z = k_0 \sin\theta_i$. The corresponding electric fields can be expressed as follows:

$$E_z = \frac{-j\gamma_0}{k_0}\left[\mathrm{e}^{\gamma_0(x-d_m)} - R\mathrm{e}^{-\gamma_0(x-d_m)}\right], \text{ air side: } x > d_m \tag{30}$$

$$E_z^{(l)} = \frac{-j\gamma_{\mathrm{TM},l}}{\varepsilon_{r,l}k_0}\left[A_{\mathrm{TM},l}^{+}\mathrm{e}^{\gamma_{\mathrm{TM},l}(x-d_{l-1})} - A_{\mathrm{TM},l}^{-}\mathrm{e}^{-\gamma_{\mathrm{TM},l}(x-d_{l-1})}\right] - \frac{j\gamma_{\mathrm{TEM},l}}{\varepsilon_{r,l}k_0}\left[B_{\mathrm{TEM},l}^{+}\mathrm{e}^{\gamma_{\mathrm{TEM},l}(x-d_{l-1})} - B_{\mathrm{TEM},l}^{-}\mathrm{e}^{-\gamma_{\mathrm{TEM},l}(x-d_{l-1})}\right], \text{ wire medium slab: } d_{l-1} < x < d_l \tag{31}$$

$$E_x^{(l)} = \frac{k_z}{\varepsilon_{xx,l}^{\mathrm{TM}}k_0\varepsilon_{r,l}}\left[A_{\mathrm{TM},l}^{+}\mathrm{e}^{\gamma_{\mathrm{TM},l}(x-d_{l-1})} + A_{\mathrm{TM},l}^{-}\mathrm{e}^{-\gamma_{\mathrm{TM},l}(x-d_{l-1})}\right], \text{ wire medium slab: } d_{l-1} < x < d_l\,. \tag{32}$$

To calculate the unknown coefficients, R, $A_{\mathrm{TM},l}^{\pm}$, $B_{\mathrm{TEM},l}^{\pm}$, we impose boundary conditions at $x = 0, d_1, d_2, \ldots, d_m$. Since, there are m dielectric layers and m interfaces, we have the total number of unknowns as $4m+1$ (i.e., four unknowns in each layer corresponds to $4m$, and the remaining unknown is R). Hence, $4m+1$ boundary conditions are necessary to calculate the $4m+1$ unknown coefficients. At the thin resistive patch interfaces ($x = d_l^{\pm}$, $l = 1, \ldots, m$), the macroscopic two-sided impedance boundary conditions establish that the tangential electric ($E_z^{(l)}$) and magnetic fields ($H_y^{(l)}$), can be related via a sheet impedance, i.e.,

$$E_z^{(l)}|_{x=d_l^+} = E_z^{(l)}|_{x=d_l^-} = Z_{g,l}\left(H_y^{(l+1)}|_{x=d_l^+} - H_y^{(l)}|_{x=d_l^-}\right) \tag{33}$$

where $Z_{g,l}$ is the grid impedance of the thin conductive patches [16, 23, 24] given by

$$Z_{g,l} = \frac{a}{(a-g)\sigma_{2D,l}} - j\frac{\pi}{2\omega\varepsilon_0(\varepsilon_{r,l}^{qs})a\ln(\csc\frac{\pi g}{2a})} \tag{34}$$

where $\varepsilon_{r,l}^{qs} = (\varepsilon_{r,l} + \varepsilon_{r,l+1})/2$ for interior patches $(l = 1,2,\ldots,m-1)$ and $\varepsilon_{r,m}^{qs} = (\varepsilon_{r,m} + 1)/2$ for the patch located at the upper interface $(l = m)$. This gives $2m$ boundary conditions. At the ground plane interface $(x = 0^+)$, we have two more boundary conditions [9]: i) tangential macroscopic total electric field vanishes $(E_z^{(1)}|_{x=0^+} = 0)$ and ii) derivative of current is zero $(\mathrm{d}J_{w,1}(x)/\mathrm{d}x|_{x=0^+} = 0)$ or in terms of macroscopic fields [10]

$$\left[k_0\varepsilon_{r,1}\frac{\mathrm{d}E_x^{(1)}(x)}{\mathrm{d}x} - k_z\eta_0\frac{\mathrm{d}H_y^{(1)}(x)}{\mathrm{d}x}\right]|_{x=0^+} = 0\,. \tag{35}$$

Following [16], the boundary condition at the top patch interface, $x = d_m^-$, can be written as

$$J_{w,m}(d_m^-) + \frac{\sigma_{2D,m}}{j\omega\varepsilon_0\varepsilon_{r,m}}\frac{\mathrm{d}J_{w,m}(x)}{\mathrm{d}x}|_{d_m^-} = 0 \tag{36}$$

or, equivalently, the macroscopic field condition

$$\left(1 + \frac{\sigma_{2D,m}}{j\omega\varepsilon_0}\frac{\mathrm{d}}{\mathrm{d}x}\right)\left[k_0E_x^m(x) - \frac{k_z\eta_0}{\varepsilon_{r,m}}H_y^m(x)\right]|_{d_m^-} = 0. \tag{37}$$

This gives the total number of $2m+3$ conditions, clearly insufficient to calculate the $4m+1$ unknown coefficients, which makes apparent the need of the ABCs derived in Section 2.2.

At the junction of two wire mediums with thin conductive patches at the interfaces $(x = d_l^{\mp}, l = 1,\ldots,m-1)$ it is necessary to impose the ABCs (20) and (21) (with the assumption that the gap between the patches is small), in addition to the boundary condition (33)

$$\frac{\sigma_{2D,l}}{2j\omega\varepsilon_0}\left[\frac{1}{\varepsilon_{r,l}}\frac{\mathrm{d}J_{w,l}(x)}{\mathrm{d}x}|_{d_l^-} + \frac{1}{\varepsilon_{r,l+1}}\frac{\mathrm{d}J_{w,l+1}(x)}{\mathrm{d}x}|_{d_l^+}\right] + [J_{w,l}(d_l^-) - J_{w,l+1}(d_l^+)] = 0 \tag{38}$$

$$\frac{1}{\varepsilon_{r,l}}\frac{\mathrm{d}J_{w,l}(x)}{\mathrm{d}x}|_{d_l^-} - \frac{1}{\varepsilon_{r,l+1}}\frac{\mathrm{d}J_{w,l+1}(x)}{\mathrm{d}x}|_{d_l^+} = 0\,. \tag{39}$$

In terms of macroscopic fields, (38) and (39) can be rewritten as

$$\left(1+\frac{\sigma_{2D,l}}{2j\omega\varepsilon_0}\frac{\mathrm{d}}{\mathrm{d}x}\right)\left[k_0E_x^{(l)}(x) - \frac{k_z\eta_0}{\varepsilon_{r,l}}H_y^{(l)}(x)\right]|_{d_l^-} = \left(1-\frac{\sigma_{2D,l}}{2j\omega\varepsilon_0}\frac{\mathrm{d}}{\mathrm{d}x}\right)\left[k_0E_x^{(l+1)}(x) - \frac{k_z\eta_0}{\varepsilon_{r,l+1}}H_y^{(l+1)}(x)\right]|_{d_l^+} \tag{40}$$

http://dx.doi.org/10.5772/36937

$$\frac{\mathrm{d}}{\mathrm{d}x}\left[k_0 E_x^{(l)}(x) - \frac{k_z \eta_0}{\varepsilon_{r,l}} H_y^{(l)}(x)\right]|_{d_l^-} = \frac{\mathrm{d}}{\mathrm{d}x}\left[k_0 E_x^{(l+1)}(x) - \frac{k_z \eta_0}{\varepsilon_{r,l+1}} H_y^{(l+1)}(x)\right]|_{d_l^+}. \quad (41)$$

Since there are $m-1$ layers of two-sided wire-medium junctions, we have $2(m-1)$ boundary conditions and hence, the total number of boundary conditions are equal to $4m+1$. Using the boundary conditions (33), (35), (8), (40), and (41), we can easily obtain a linear system for the $4m+1$ unknowns of the problem. This system can be solved either numerically or analytically for the unknown field coefficients, $A^{\pm}_{\mathrm{TM},l}$ and $B^{\pm}_{\mathrm{TEM},l}$, and the reflection coefficient R.

3. Numerical results

To illustrate the application of the proposed homogenization model, in this section we study the reflection properties of different multilayered mushroom-type HIS structures. To test the model, at first we analyze a single-layer wire medium loaded with an arbitrary material (a thin copper patch with finite bulk conductivity and a graphene patch characterized by its complex surface conductivity) at one end and a ground plane at the other. Next, we study the prospects of the multilayered mushroom-type HIS structure being used as an absorber. All the results obtained using the homogenization model are tested against the full-wave numerical simulations.

3.1. Single-layer mushroom structure with thin metal/graphene patches

As a first example, a single-layer mushroom-type HIS structure with geometry shown in Fig. 7 is chosen. In this configuration, the patches are copper and have a thickness of 60 nm. The parameters of the structure are: $a = 2$ mm, $g = 0.2$ mm, $h = 1$ mm, $r_0 = 0.05$ mm, and $\varepsilon_r = 10.2$. The analysis is performed for an obliquely incident TM-polarized plane wave. Fig. 8(a) shows the comparison of reflection magnitude behaviors calculated using HFSS and the proposed homogenization model for a TM-polarized plane wave incident at 30° to the normal. Also, in Fig. 8(a) we have included the result obtained using the wire-PEC ABC ([10], or (6)). Clearly, one can notice the difference between the results obtained using the ABC of the wire-PEC interface and the new ABC (GABC, wire-thin-metal interface - see (7) or (26)). In fact, the result obtained using the GABC is in good agreement with the HFSS [31] result.

Fig. 8(b) shows the behavior of reflection magnitude for a metal patch with $\sigma_{3\mathrm{D}} t = 0.058$ S (e.g. $t = 20$ nm and $\sigma_{3\mathrm{D}} = 2.9 \times 10^6$ S/m). The remaining parameters are the same as those considered in the previous example. Again, the results obtained using the new ABC (wire-thin-metal interface) are in good agreement with the HFSS results, whereas the results obtained using the old ABC (wire-PEC interface, (6)) deviate from the HFSS results.

Fig. 9 shows the reflection magnitude behavior for the mushroom structure loaded with graphene patches with a chemical potential of $\mu_c = 0.5185$ eV [see [22] for the surface conductivity of graphene] for a plane wave incident at $\theta = 45°$. For example, at $f = 11.96$ GHz the complex surface conductivity, $\sigma_{2\mathrm{D}} = 0.0304 - j0.0011$ S. Clearly, excellent agreement is seen between the GABC and the HFSS results. As is obvious, the ABC-PEC (6)

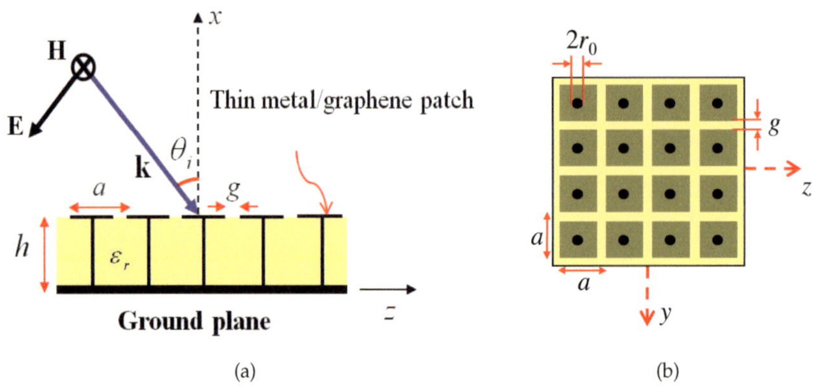

Figure 7. (a) Mushroom-type wire-medium structure with thin metal/graphene patches: (a) Side-view showing incident TM plane wave and (b) Top view of the structure.

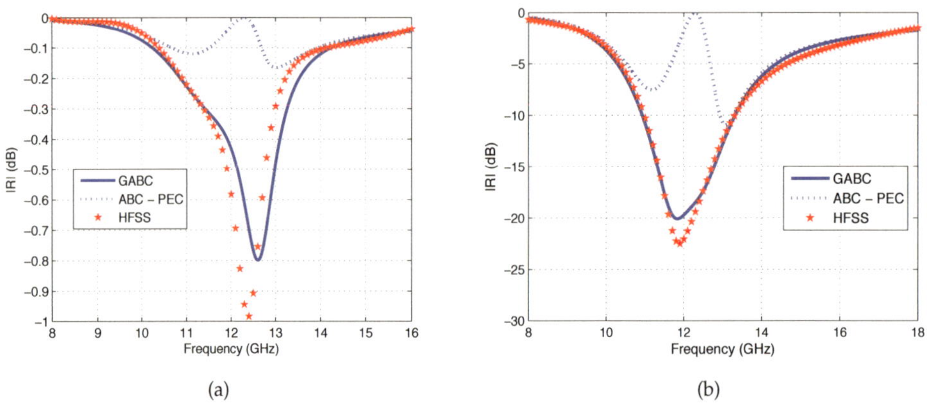

Figure 8. Reflection coefficient for a TM-polarized plane wave incident at $\theta = 30°$. (a) Copper patches with thickness 60 nm. (b) A material patch with $\sigma_{3D}t = 0.058$ S.

and local model [14] results give significant errors since this ABC assumes that the surface charge at the tip of the wires or at the wire-patch interface vanishes (4). This may hold true for the wire-PEC interface, but does not apply for the wire-thin-metal/graphene interface. This is because at the wire-thin-metal interface, charge accumulation and diffusion takes place and the fields completely penetrate the metal. Hence, a new ABC (GABC) is required to obtain the correct result.

Referring to Fig. 9, one can notice that the nonlocal model result (ABC-PEC) and the local model result agree well with each other. This is because, for the large chemical potential 0.5185 eV considered in this example, it seems that the spatial dispersion effects in the wire media are reduced, and the mushroom structure can be treated as a uniaxial continuous Epsilon-Negative material loaded with PEC patches (although this is not the case with the GABC, which sees the interface in a correct manner).

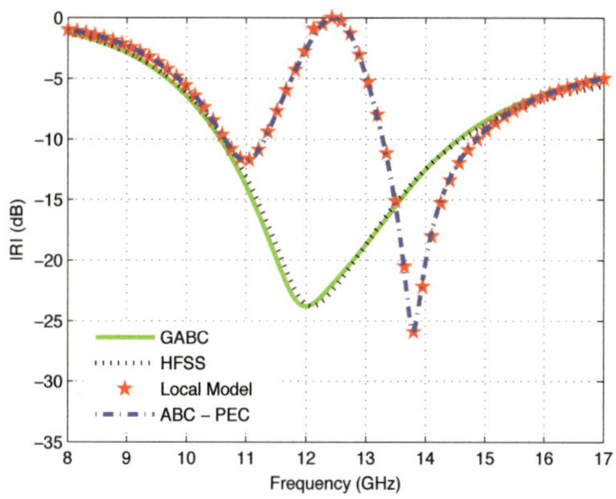

Figure 9. Reflection coefficient of the mushroom structure loaded with graphene patches with $\theta = 45°$.

Overall Figs. 8 and 9 show the effectiveness of the new ABC. Hence, when a PEC wire is connected to a thin-metal patch, as is the case considered here, the new ABC is the good choice to obtain the correct result.

To understand the behavior (effects of spatial dispersion) of the grounded wire-medium slab loaded with graphene patches, in Fig. 10 we plotted the normalized wire current along the vias for different values of bias (μ_c). It can be noticed that for zero bias ($\mu_c = 0$) the current is quite nonuniform and as the bias increases, the current becomes more and more uniform. This is because, for the case of zero bias the patch is almost transparent or in other words its conductivity (σ_{2D}) is very small (can be seen in Fig. 10) and behaves as a dielectric material rather than a metal. As the bias increases, the conductivity increases and the properties of the patch will be close to that of a metal (since, it is known that for a vias truncated with a PEC patch the current is uniform [14]). Hence, the current starts to become more and more uniform, indicating that the spatial dispersion effects are negligible and the wire-medium slab can be treated as a uniaxial continuous Epsilon Negative material [14, 15] loaded with patches. However, it should be noted that, despite the uniformity of the current for the case of $\mu_c = 0.5185$ eV, the homogenization model still needs an ABC to model the graphene patch mushroom structure, indicating that spatial dispersion effects are important for this structure. This is because, at the patch-to-wire interface, diffusion and accumulation of charge occurs (unlike at a wire-to-PEC interface, where surface charge vanishes) and an ABC is required to capture the physics.

By observing Figs. 8 and 9, one can notice a dip in the reflection magnitude curves for some values of the conductivity or the thickness of the metal patch. This behavior of showing reflection nulls allows the mushroom structures with thin-metal/graphene patches to be used in absorber applications. By properly selecting the parameters of the mushroom structure, one can easily obtain either a narrow band or even a wideband absorber. More on the design and analysis of realizing the absorbers is explored in the next section.

3.2. Design and analysis of multilayered mushroom-type HIS structures with applications to absorbers

In this section, we concentrate on the design and analysis of single-layer, two-layer, and three-layer mushroom structures for obliquely incident TM-polarized plane waves using the

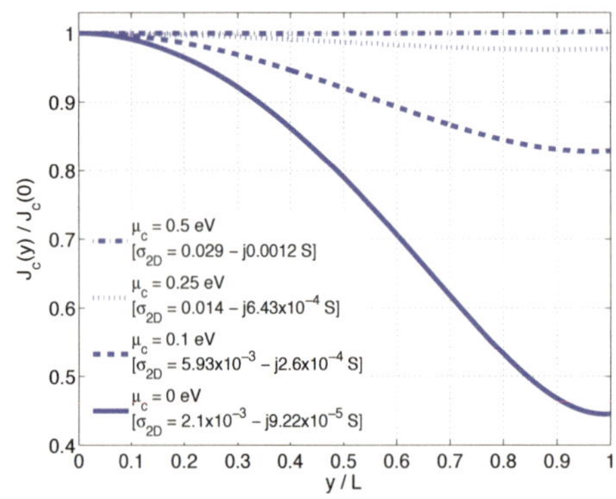

Figure 10. Normalized wire current for a graphene patch mushroom structure for different values of bias at $f = 14$ GHz.

proposed homogenization model. All the configurations are loaded with thin resistive sheets. The results obtained using the homogenization model are confirmed with the numerical HFSS simulations.

3.2.1. Single-layer mushroom HIS structure with thin resistive patches

Understanding the mechanism of the multilayered mushroom-type HIS structure as an absorber is quite a complicated task. Hence, to make it is easier to comprehend, we begin with the design and analysis of a simpler case, i.e., a single-layered mushroom HIS structure (with the geometry shown in Fig. 11). Although this structure is analyzed in the previous section, the aim here is to show that by proper selection of the mushroom structure parameters (such as periodicity, gap between the patches, height and permittivity of the dielectric substrate, radius of the vias, and resistivity of the patches) one can obtain remarkably attractive results (such as increase in the absorption bandwidth, enhancement in the absorption level, among others).

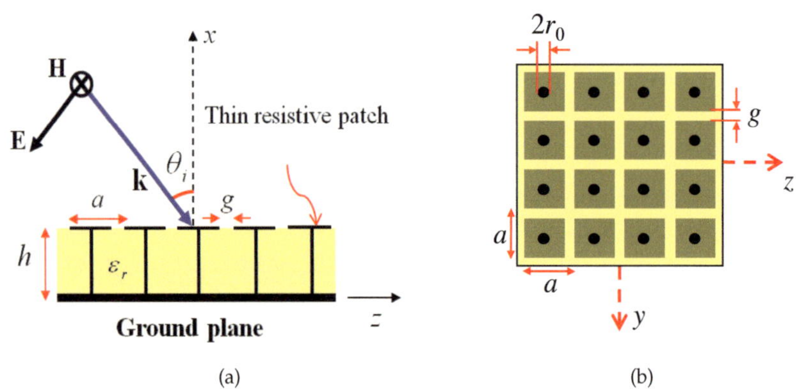

Figure 11. (a) Single-layer mushroom-type HIS absorber with thin resistive patches. (b) Top view of the structure.

To show this effect, we considered the following parameters in the design of the absorber: $a = 6.8$ mm, $g = 0.5$ mm, $h = 3.5$ mm, $r_0 = 0.08$ mm, $\varepsilon_r = 2.5$, and $R_s = 106.54\,\Omega$. The sheet resistance R_s, can be related to the grid impedance of the patch as follows: from (34) the grid impedance of the lossy patch array can be represented as a series RC circuit ($R_g - j/(\omega C_g)$), where the real value corresponds to R_g (resistance per unit cell) given by $a/((a-g)\sigma_{2D})$ or $R_s a/(a-g)$ and C_g is the capacitance of the patch grid whose value can be obtained from (34). Although the selection of a particular value of R_s is a tedious process, the procedure will be discussed later in this section. In order to show the advantage of the mushroom-type absorber, we compare its reflection coefficient behavior against a similar structure but without vias. Figs. 12(a) and 12(b) show the reflection magnitude curves for

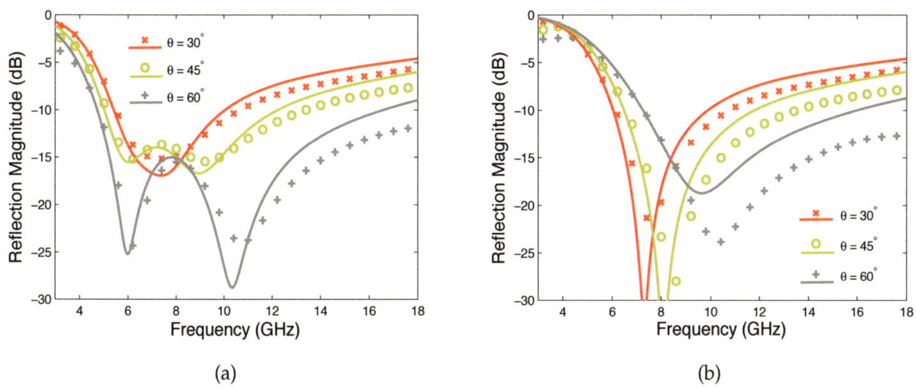

Figure 12. Comparison of analytical (solid lines) and full-wave HFSS results (crosses, circles, and plus signs) of the reflection coefficient for the single-layer HIS absorber excited by a TM-polarized plane wave at oblique angles of incidence θ: (a) with vias. (b) without vias.

30°, 45°, and 60° for a TM-polarized plane wave. Referring to the results shown in Fig. 12(a), one can see good agreements between the analytical results and the full-wave HFSS results. The analytical results here are obtained using either the ABC given by (7) or (26) obtained in Section 2.2 as a limiting case for a single-sided wire-to-patch junction. However, the analytical results shown in Fig. 12(b) are obtained using the circuit theory model given in [20, 24, 25]. By comparing Figs. 12(a) and 12(b) one can see that for the structure with vias, the absorption bandwidth increases for increasing angles of incidence (i.e., the structure with vias gives a better performance than the structure without vias), although, one can notice a decrease in the absorption level for 30° and 45°. Also, in Fig. 12(a) it is observed that the lower frequency bound of the absorption band (around 6 GHz) is stable, which is in complete contrast to the behavior of the structure without vias (Fig. 12(b)). This frequency stability can be attributed to the increased interaction of the incident wave with the vias [30].

To comprehend the nature of the mushroom structure and its reasons to act as a wideband absorber when compared to the structure without vias, we studied the reflection properties (phase and magnitude) of the mushroom structure for various sheet resistivities for a TM-polarized plane wave incident at 45° to the normal. Fig. 13 shows the reflection phase and magnitude for various resistive values of the patch ranging from 0 ($\sigma_{3D}t = \infty$, PEC case) to ∞ ($\sigma_{3D}t = 0$, transparent case) calculated using the nonlocal homogenization model discussed in Section 2.3. Starting with the PEC case, i.e., $R_s = 0$, the reflection phase behavior (shown in Fig. 13(a)) shows two resonances. One corresponding to 0° and the other corresponding to 360° (shown in red circles) acting as a HIS at the two frequencies. These resonances are such that one lies above the plasma frequency ($f_p/\sqrt{\varepsilon_r} = 6.28$ GHz) and the other lies below $f_p/\sqrt{\varepsilon_r}$, consistent with the result of the single-layer mushroom-type

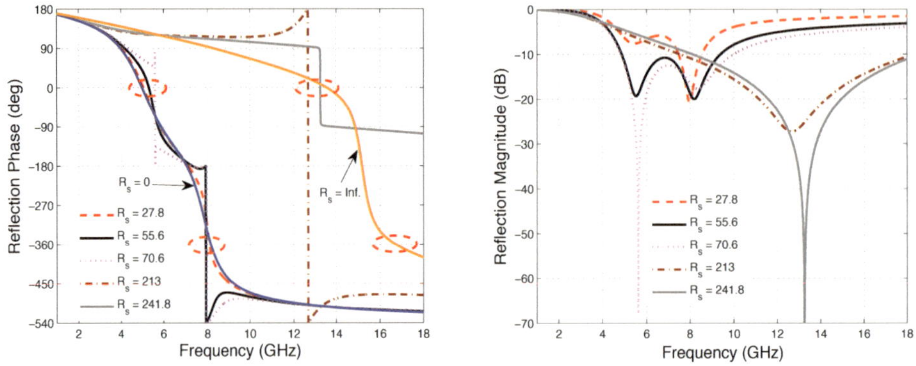

(a) (b)

Figure 13. (a) Phase and (b) Magnitude of the reflection coefficient of the single-layer mushroom structure for different values of sheet resistivity R_s (in Ω), $\theta = 45^{\circ}$.

HIS discussed in [14, 15] (although the dimensions and $f_p/\sqrt{\varepsilon_r}$ are different). With the increase in R_s, one can notice a deviation in the phase behavior from the actual HIS behavior. This deviation can be attributed to the change in the impedance of the mushroom structure due to change in the value of R_s. This deviation is referred to as perturbed HIS behavior. Corresponding to this in Fig. 13(b), the reflection magnitude curves show some deviations as R_s changes. The results in Fig. 13(b) can also be treated as the perturbed behavior of the HIS (the result of HIS when $R_s = 0$ is not included here, since its magnitude is unity).

At first glance, by noticing the results in Fig. 13(b) one might only comprehend that for different values of R_s there are different lossy patches, and hence, different reflection magnitudes. However, the results add meaning when they are seen with the increasing values of resistivities. Starting from $R_s = 27.8\ \Omega$, the reflection magnitude curve (dashed red curve) shows two partial nulls, which correspond closely to the two perturbed HIS resonances (shown in Fig.13(a)). With the further increase in the sheet resistivity, the two partial reflection nulls start to deviate in such a way that one of the null deepens and the other null either subdues or remains the same. As mentioned before, this is because as R_s changes, the grid impedance changes, which in turn changes the surface impedance of the mushroom structure. For example, for $R_s = 70.6\ \Omega$ one of the reflection nulls has reached its minimum at 5.612 GHz (perfect match to free-space impedance) and the other reflection null is at the around 20 dB level, which is same as that of the second reflection null of $R_s = 27.8$ Ω case. Similarly, one can only see one deep reflection null for $R_s = 241.8\ \Omega$ at 13.25 GHz. Based on these facts (how reflection nulls deviate), one can correlate the partial reflection magnitude curves (for the case of $R_s = 106.54\ \Omega$) in Fig. 12(a) with the perturbed HIS resonances. With the further increase in the sheet resistivity (for $R_s = \infty$, the patch array is fully transparent and we have a grounded wire-medium slab), the reflection phase behavior shown in Fig. 13(a) deviates further to the right and shows two resonances corresponding to 0° and 360° (acting as a HIS). Hence, for any value of R_s between 0 and ∞, the behavior of the structure should either resemble the properties of a perturbed HIS or a perturbed wire-medium slab. Thus, by utilizing the two resonances of the mushroom structure, and with the proper choice of the resistivity of the patch arrays one can increase the absorption bandwidth. This behavior of the mushroom structure which shows two resonances makes it more attractive to consider compared to the structure without vias (which has only one HIS resonance, even when perturbed).

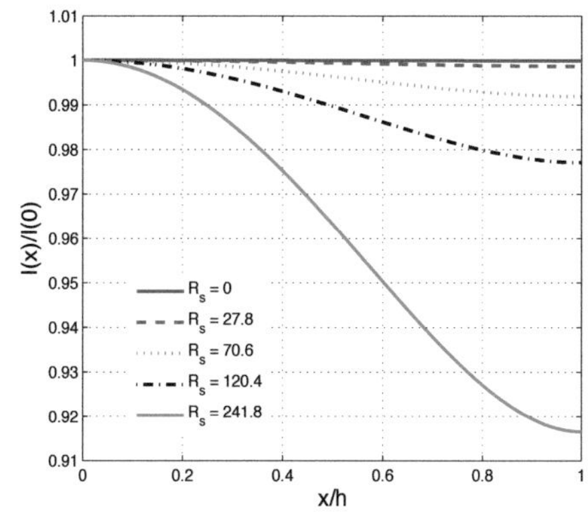

Figure 14. Normalized wire current in the single-layer mushroom structure for different values of sheet resistivity R_s (in Ω) at $f = 7\,\text{GHz}$, $\theta = 45°$.

To further validate that the behavior observed in Fig. 12(a) is in fact due to the perturbed HIS behavior of the mushroom structure with PEC patches, we plotted the current behavior for various R_s varying from 0 to ∞. Fig. 14 shows the normalized wire current in the single-layer mushroom structure at $f = 7\,\text{GHz}$. It can be observed that for small values of resistivity ($R_s = 0$ to 120 Ω), the current is uniform, which indicates that spatial dispersion effects are negligible (although requires the ABC (26)). This result is consistent with the result shown in Fig. 10, where the current distribution was uniform for large values of σ_{2D} indicating that spatial dispersion effects were almost negligible (although it required the ABC (26) to characterize the properties of the mushroom structure with graphene patches). Also, it should be noted that the value of R_s chosen in the design of absorber shown in Fig. 11 is in the range of 0 to 120.4 Ω, hence the current behavior will be uniform similar to the case of mushroom structure with PEC patches, [14, 15]. Since, the current distribution is uniform and the phase behavior is close to that of the mushroom structure with PEC patches, the magnitude behavior or resonances of the absorber shown in Fig. 12(a) are indeed the perturbed HIS resonances of the mushroom structure with PEC patches.

Although the increase in absorption bandwidth and the enhancement in the absorption level is due to the presence of vias and proper choice of R_s, the absorption mechanism occurs mainly due to the lossy patch array. Since, the amount of energy lost or energy absorbed by the patch array is evaluated based on the value of R_s, it is important to determine the right value of R_s in designing wideband absorbers. The most common way is to use numerical optimization techniques such as [27, 28], among others. However, discussion of these techniques is beyond the scope of this work. Since, the analytical model gives the results instantaneously, it provides a reliable, fast, and efficient solution in selecting an optimum sheet resistance of the patch arrays. The idea is to obtain a certain range for R_s (in the range between the PEC case and wire-medium slab case) for each of the angles of incidence, where perfect reflection nulls are noticed for one of the two perturbed HIS resonances of the structure.

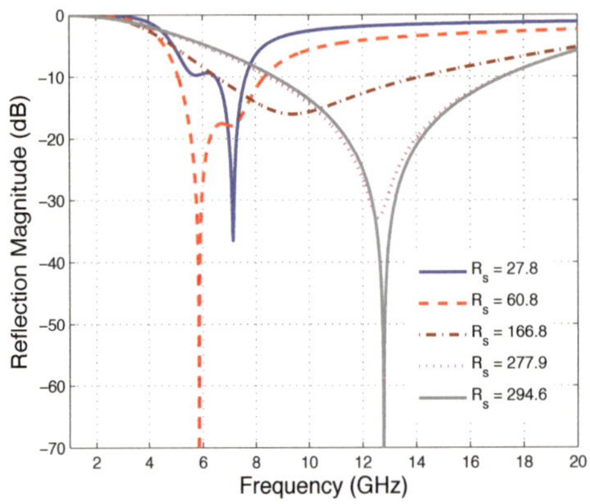

Figure 15. Reflection magnitude of the single-layer mushroom structure for different values of sheet resistivity R_s (in Ω), $\theta = 30^{\circ}$.

To obtain a range of values for R_s for different incident angles, we plotted the reflection magnitude curves (shown in Fig. 15) for various R_s varying from 0 to ∞ (results for $R_S = 0$ and ∞ are not included here) for a TM-polarized plane wave incident at 30° to the normal. Clearly, one can notice the deviation of the reflection nulls as R_s is increased. Then, we fix the range of R_s where perfect reflection nulls are obtained. Based on this fact, from Fig. 15, the range of R_s is 60.8 - 294.6 Ω. A similar procedure is repeated for $\theta = 60^{\circ}$ and the range of R_s is found to be 94.03-146.38 Ω. From Fig. 13(b), the range of R_s for 45° can be given as 70.6-241.8 Ω. Then, based on these values of R_S, a unique range that fits for all angles of incidence (up to 60°) can be found: $(94.03 - 146.38)\ \Omega$ for the case under study. A further optimization procedure limited to the above range gives us an optimum value of $R_s = 106.54\,\Omega$, which is the one used in the absorber design in Fig. 11.

3.2.2. *Two-layer mushroom HIS structure with thin resistive patches*

The goal of this section is two-fold. The first being the applicability of the derived ABCs (Section 2.2) for the interface of two uniaxial wire media with thin resistive patches at the junction. The second aim is to improve the performance of the single-layer mushroom HIS absorber. To achieve these goals, we first design an absorber over a wide range of frequencies (wider than the single-layer structure) using the homogenization model (discussed in Section 2.3) and compare its results with the full-wave HFSS simulations.

Figure 16 shows the two-layer mushroom HIS absorber, which consists of two wire media embedded in a dielectric host media, loaded with thin resistive patch arrays, and backed by a ground plane. The parameters of the dielectric slabs used in the design together with the dimensions and sheet resistivity values of the square patches are given in the caption of Fig. 16. Figures 17(a) and 17(b) show the reflection magnitude behavior of the two-layer HIS absorber with and without vias for oblique angles of incidence. At first instance, it is clear that the structure with vias offers a significant improvement in the bandwidth, and at the same time shows an enhancement in the absorption level when compared to the reference structure without vias. For example, for $\theta = 60^{o}$, it can be noticed that the 20 dB absorption bandwidth of the structure with vias covering the frequency band from 9.03 GHz to 25.29 GHz has a 55.5% bandwidth increase in comparison to the structure without vias with the

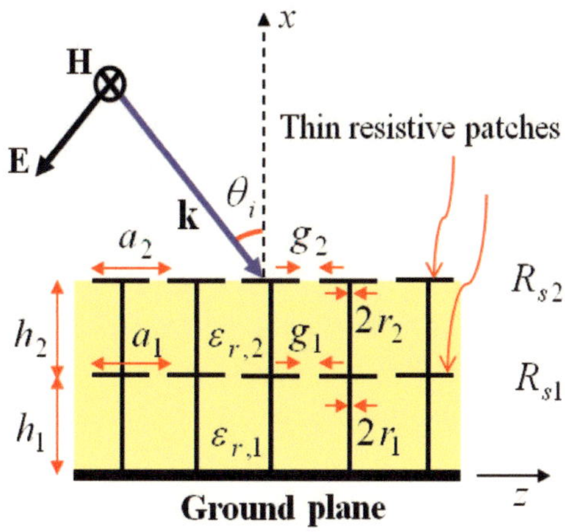

Figure 16. Two-layer mushroom-type HIS absorber with thin resistive patches. Structural parameters used in this work: $h_1 = h_2 = 3.2\,\text{mm}$, $\varepsilon_{r,1} = 2.2$, $\varepsilon_{r,2} = 1.33$, $r_1 = r_2 = 0.05\,\text{mm}$, $a_1 = a_2 = 5\,\text{mm}$, $g_1 = g_2 = 0.1\,\text{mm}$, $R_{s1} = 196\,\Omega$, $R_{s2} = 1078\,\Omega$.

frequency band from 12.67 GHz to 18.86 GHz. Referring to Fig. 17(a), it can be noticed that the results obtained using the nonlocal homogenization model agree very well with the HFSS results in the entire frequency band. This also verifies the new ABCs derived in Section 2.2 for the interface of two uniaxial wire media with thin resistive patch at the junction. This would not be possible using the PEC-ABC (discussed in Section 3.1 for the single-layer mushroom structure with thin metal/graphene patch), which indicates that the new ABCs are in fact important to capture the physics at the thin resistive patch junction of two uniaxial wire media.

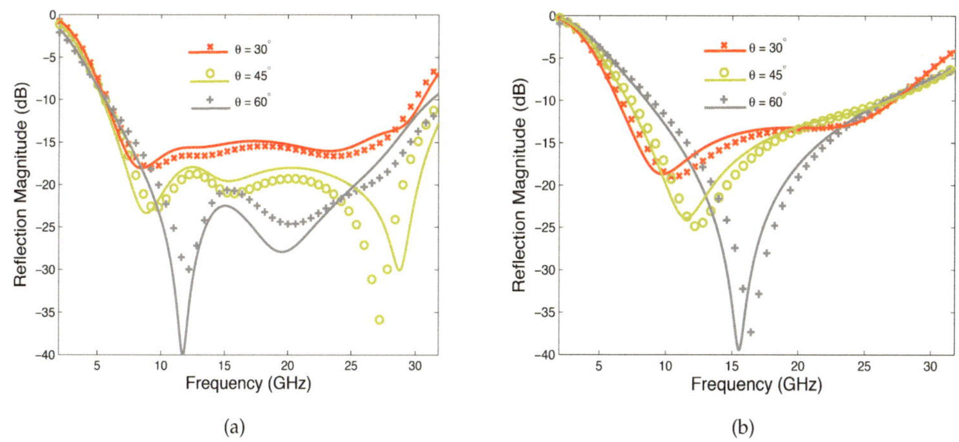

Figure 17. Comparison of analytical (solid lines) and full-wave HFSS results (crosses, circles, and plus signs) of the reflection coefficient for the two-layer HIS absorber excited by a TM-polarized plane wave at oblique angles of incidence θ: (a) with vias. (b) without vias.

Due to increased degrees of freedom in the two-layer structure when compared to the single-layer structure, one can easily see the differences in the relative bandwidth of absorption of Figs. 17(a) and 17(b), when compared to Figs. 12(a) and 12(b). Even though, the results of the two-layer mushroom structure show significant improvements in the bandwidth and the absorption level, there can be many design solutions (with different R_s) which can yield better results than that shown in Fig. 17(a). Hence, designing an absorber with optimum performance, particularly with many degrees of freedom, is a challenging task. The simpler way is to use the analytical procedure described in the previous section for designing a single-layer mushroom absorber. Although, this procedure may seem simple for the single-layer structure, implementing it for the two-layer structure is a tedious task and, hence, that design procedure is not explained here for the sake of brevity. The other way is to use the numerical optimization techniques [27, 28].

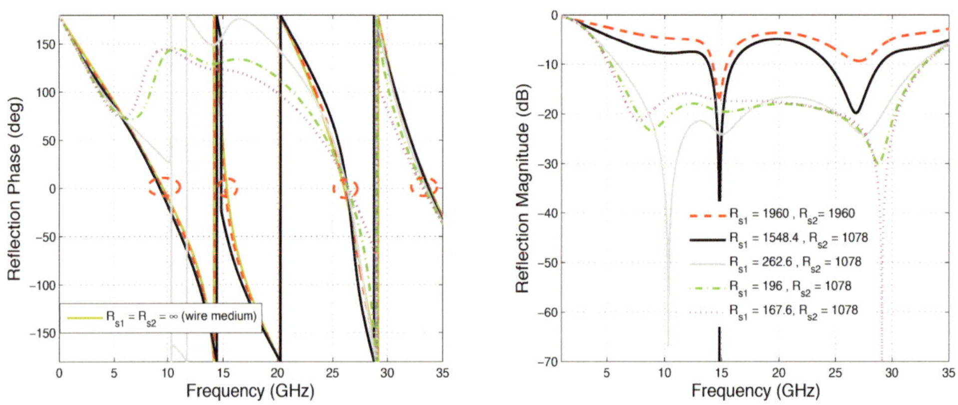

(a) (b)

Figure 18. a) Phase and (b) magnitude of the reflection coefficient of the two-layer mushroom structure for different values of sheet resistivity R_s (in Ω), $\theta = 45^\circ$.

In order to understand the nature of the resonances or the wideband behavior of the two-layer mushroom absorber, here we will use the strategy that was implemented to study the wideband behavior of single-layer mushroom absorber, i.e., studying the reflection magnitude behavior for various R_s ranging from 0 to ∞. Since, the value of sheet resistivity for the patch arrays in the top layer is high ($R_s = 1078\ \Omega$) and low in the bottom layer ($R_s = 196\ \Omega$), one can start the analysis either from the PEC case ($R_s = 0$ in both the layers) or the wire-medium case ($R_s = \infty$ in both the layers). Then, either decrease the sheet resistivities in both the layers at the same time (if the wire-medium case is considered) or increase the sheet resistivities in both the layers at the same time (if the PEC case is considered). The other way is start from either the PEC case or the wire-medium case, then fix R_s in one layer while increasing the R_s in the other (if PEC case is chosen) or fix R_s in one layer while decreasing the R_s in the other (if wire-medium case is chosen). Here, we employ the later strategy, because the goal is to explain the wideband behavior of the absorber with $R_{s1} = 196\,\Omega$ and $R_{s2} = 1078\,\Omega$. Figures 18(a) and 18(b) show the reflection phase and magnitude behavior of the two-layer mushroom HIS structure for different values of R_s. With the resistivity of the top patch array being fixed and by gradually decreasing the resistivity of the patch array in the bottom layer it is observed that the phase behavior starts to deviate from the wire-medium slab case (shown in Fig. 18(a)). It is also observed from Fig. 18(b) that the reflection nulls deviate for varying R_s, and for some cases there are perfect reflection nulls (similar to the behavior observed in a single-layer mushroom structure) corresponding to the zero phases of the perturbed HIS resonances of the wire-medium slab.

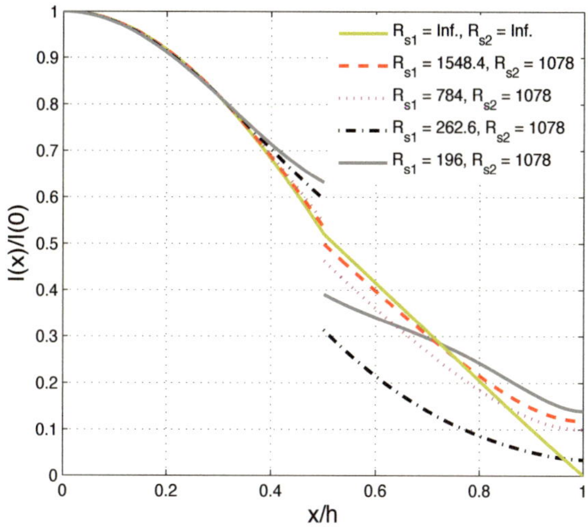

Figure 19. Normalized wire current in the two-layer mushroom structure for different values of sheet resistivity R_s (in Ω) at $f = 15$ GHz, $\theta = 45^o$. Here h is the total thickness of the two-layer structure given by $h_1 + h_2$.

This proves the fact that the partial reflection nulls/resonances shown in Fig. 17(a) , for the case of $R_{s1} = 196\,\Omega$ and $R_{s2} = 1078\,\Omega$ are associated with the perturbed HIS resonances of the actual wire-medium slab. It should be noted that the bandwidth enhancement is not only observed for higher angles of incidence, but also observed for smaller angles of incidence. The difference is that for small angles of incidence the electric field interaction with the vias is negligible, and the bandwidth enhancement is due to interactions between the patches in the adjacent layers, and for higher angles of incidence vias play a dominant role in widening the absorption band. Hence, by using the resonances of the mushroom structure, along with the proper choice of dimensions and resistivities of the patch arrays, and with good selection of the permittivities of the dielectric slabs (perforated with metallic vias) the absorption bandwidth can be enlarged, as compared to the case with no vias.

The behavior of current along the vias has also been studied to validate that the resonances are the perturbed HIS resonances of the wire-medium slab. Figure 19 shows the normalized wire current in the two-layer mushroom structure at $f = 15$ GHz. For the values of R_s used in Fig. 18, it can be observed from Fig. 19 that the current along the metallic vias varies significantly, which indicates that the spatial dispersion effects are not suppressed. The reason is that when the R_s is large (conductivity is small), the thin resistive patch resembles a dielectric rather than a metal, and charges accumulate at the tip of the double-sided wire-to-patch junction, which necessitates the ABCs derived in Section 2.2. This behavior is consistent with the results shown in Fig. 10 (Section 3.1) for small values of σ_{2D}, however, it should be noted that the structure analyzed in that section has a single-sided wire-to-patch junction. Since, the current distribution is nonuniform and the phase behavior is close to that of the wire-medium slab case, the resonances of the absorber are indeed the perturbed HIS resonances of the wire-medium slab.

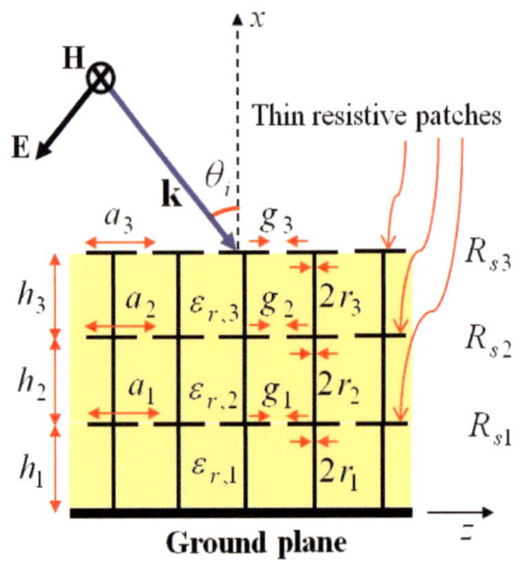

Figure 20. Three-layer mushroom-type HIS absorber with thin resistive patches. Structural parameters used in this work: $h_1 = 2.7$ mm, $h_2 = h_3 = 3.2$ mm, $\varepsilon_{r,1} = 3.2$, $\varepsilon_{r,2} = 1.8$, $\varepsilon_{r,3} = 1.33$, $r_1 = r_2 = r_3 = 0.05$ mm, $a_1 = a_2 = a_3 = 5$ mm, $g_1 = g_2 = g_3 = 0.1$ mm, $R_{s1} = 196\,\Omega$, $R_{s2} = 588\,\Omega$, $R_{s3} = 1176\,\Omega$.

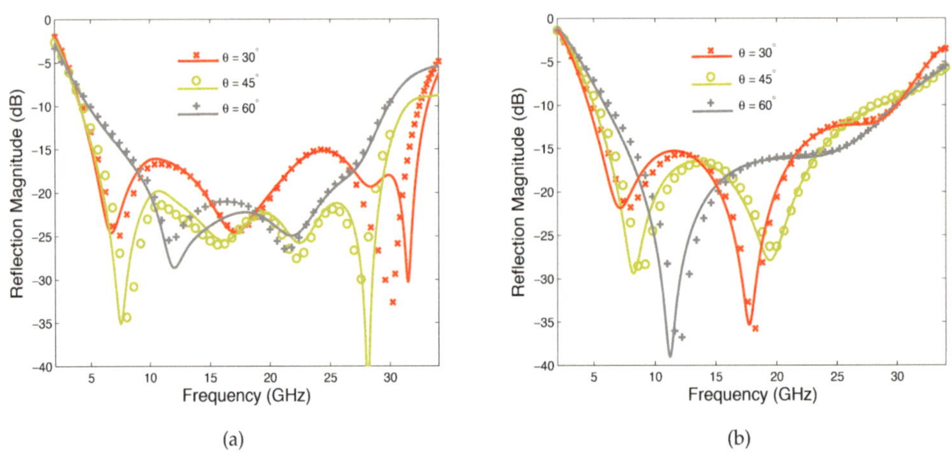

Figure 21. Comparison of analytical (solid lines) and full-wave HFSS results (crosses, circles, and plus signs) of the reflection coefficient for the three-layer HIS absorber excited by a TM-polarized plane wave at oblique angles of incidence θ: (a) with vias. (b) without vias.

3.2.3. *Three-layer mushroom HIS structure with thin resistive patches*

Figure 20 shows a three-layered mushroom HIS absorber, which consists of three resistive patch arrays separated by dielectric slabs perforated with metallic vias, and backed by a ground plane. The parameters of the dielectric slabs together with the dimensions and sheet resistivities of the square patches are given in the caption of Fig. 20. Figs. 21(a) and 21(b) show the reflection magnitude curves for different incidence angles.

Referring to Fig. 21(a), it can be observed that the analytical results obtained using the nonlocal homogenization model described in Section 2.3 agree very well with the HFSS results. As stated before, the absorption level increases for increasing angles of incidence due to increase in the interaction with vias. By comparing Fig. 21(a) with Fig. 21(b) one can clearly see significant improvements in the absorption bandwidth for the structure with vias for increasing angles of incidence. For example, for $\theta = 60^o$, the 20 dB absorption bandwidth of the structure with vias covering the frequency band from 9.93 GHz to 24.93 GHz shows a 38 % increase in the bandwidth when comparing to the structure with no vias, having the bandwidth from 9.01 GHz to 14.6 GHz. Also, it should be noted that the HFSS results shown in Fig. 21(a) for $\theta = 45^o$ and 60° have been obtained up to 30 GHz only (due to lack of convergence at higher frequencies).

4. Conclusion

Generalized additional boundary conditions are derived for the interface of two uniaxial wire media with thin resistive sheet at the junction. Based on these conditions a nonlocal homogenization model has been proposed to characterize the reflection properties of the multilayered mushroom HIS structures with thin resistive patches. The homogenization model has been applied to four different structures and the results are validated using the full-wave numerical simulations. An analytical design procedure for selecting resistivities of the patch arrays has been presented. The prospect of using the multilayer mushroom structures as absorbers has been studied. It has been shown that the presence of vias in fact enhances the absorption band and increases the absorption level for increasing angles of incidence for the obliquely incident TM-polarized plane wave.

Author details

Alexander B. Yakovlev[1], Yashwanth R. Padooru[1],
George W. Hanson[2] and Chandra S.R. Kaipa[1]

1 The University of Mississippi, USA
2 University of Wisconsin-Milwaukee, USA

References

[1] J. van Kranendonk & J. E. Sipe. (1977). *Foundtions of the macroscopic electromagnetic theory of dielectric media*, chapter 5 in *Progress in Optics* XV, E. Wolf, (Ed.), New York: North-Holland, 1977.

[2] V. Agranovich & V. Ginzburg. (1966). *Spatial Dispersion in Crystal Optics and the Theory of Excitons*, NewYork: Wiley-Interscience, 1996.

[3] J. D. Jackson. (1998). *Classical Electrodynamics*, Sect. 6.6, NewYork: Wiley, 1998.

[4] D. R. Smith, W. J. Padilla, D. C. Vier, S. C. Nemar-Nasset, & S. Schultz. (2000). A composite medium with simultaneously negative permeability and permittivity, *Phys. Rev. Lett.*, Vol. 84, p. 4184, May 2000.

[5] L. Landau & E. Lifchitz. (1984). *"Electrodynamics of continuous media," in Course of Theoretical Physics*, 2nd ed., Burlington, MA: Butterworth-Heinemann, 1984, Vol. 8.

[6] P. A. Belov, Y. Hao, & S. Sudhakaran. (2006). Subwavelength microwave imaging in photonic crystals, *Phys. Rev. B*, Vol. 73, 2006, 033108 (1-4).

[7] M. G. Silveirinha, P. A. Belov, & C. R. Simovski. (2007). Subwavelength imaging at infrared frequencies using an array of metallic nanorods, *Phys. Rev. B*, Vol. 75, 2007, 035108 (1-12).

[8] P. A. Belov, R. Marques, S. I. Maslovski, I. S. Nefedov, M. G. Silveirinha, C. R. Simovski, & S. A. Tretyakov. (2003). Strong spatial dispersion in wire media in the very large wavelength limit, *Phys. Rev. B*, Vol. 67, 2003, 113103 (1-4).

[9] M. G. Silveirinha, C. A. Fernandes, & J. R. Costa. (2008). Electromagnetic characterization of textured surfaces formed by metallic pins, *IEEE Trans. Antennas Propag.*, Vol. 56, No. 2, pp. 405-415, Feb. 2008.

[10] M. G. Silveirinha, C. A. Fernandes, & J. R. Costa. (2008). Additional boundary condition for a wire medium connected to a metallic surface, *New J. Phys.*, Vol. 10, 2008, 053011 (1-17).

[11] M. G. Silveirinha. (2006). Additional boundary condition for the wire medium, *IEEE Trans. Antennas Propag.*, Vol. 54, pp. 1766-1780, 2006.

[12] M. G. Silveirinha. (2009). Broadband negative refraction with a crossed wire mesh, *Phys. Rev. B*, Vol. 79, 2009, 153109 (1-4).

[13] S. I. Maslovski, T. A. Morgado, M. G. Silveirinha, C. S. R. Kaipa, & A. B. Yakovlev. (2010). Generalized additional boundary conditions for wire media, *New J. Phys.*, Vol. 12, 2010, 113047 (1-19).

[14] O. Luukkonen, M. G. Silveirinha, A. B. Yakovlev, C. R. Simovski, I. S. Nefedov, & S. A. Tretyakov. (2009). Effects of spatial dispersion on reflection from mushroom-type artificial impedance surfaces, *IEEE Trans. Microw. Theory Tech.*, Vol. 57, No. 11, pp. 2692-2699, Nov. 2009.

[15] A. B. Yakovlev, M. G. Silveirinha, O. Luukkonen, C. R. Simovski, I. S. Nefedov, & S. A. Tretyakov. (2009). Characterization of surface-wave and leaky-wave propagation on wire-medium slabs and mushroom structures based on local and nonlocal homogenization models, *IEEE Trans. Microw. Theory Tech.*, Vol. 57, No. 11, pp. 2700-2714, Nov. 2009.

[16] A. B. Yakovlev, Y. R. Padooru, G. W. Hanson, A. Mafi, & S. Karbasi. (2011). A generalized additional boundary condition for mushroom-type and bed-of-nails-type wire media, *IEEE Trans. Microw. Theory Tech.*, Vol. 59, No. 3, pp. 527-532, Mar. 2011.

[17] C. S. R. Kaipa, A. B. Yakovlev, & M. G. Silveirinha. (2011). Characterization of negative refraction with multilayered mushroom-type metamaterials at microwaves, *J. Appl. Phys.*, Vol. 109, 2011, 044901 (1-10).

[18] C. S. R. Kaipa, A. B. Yakovlev, S. I. Maslovski, & M. G. Silveirinha. (2011). Indefinite dielectric response and all-angle negative refraction from a structure formed by deeply-subwavelength inclusions, *Phys. Rev. B*, Vol. 84, 2011, 165135 (1-7).

[19] C. S. R. Kaipa, A. B. Yakovlev, S. I. Maslovski, & M. G. Silveirinha. (2011). Mushroom-type high-impedance surface with loaded vias: homogenization model and ultra-thin design, *IEEE Antennas Wireless Propag. Lett.*, Vol. 10, pp. 1503-1506, Dec. 2011.

[20] C. S. R. Kaipa, A. B. Yakovlev, F. Medina, F. Mesa, C. A. M. Butler, & A. P. Hibbins. (2010). Circuit modeling of the transmissivity of stacked two-dimensional metallic meshes, *Opt. Express*, Vol. 18, No. 13, 2010, 13309 (1-12).

[21] D. Sievenpiper, L. Zhang, R. Broas, N. Alexopoulos, & E. Yablonovitch. (1999). High-impedance electromagnetic surfaces with a forbidden frequency band, *IEEE Trans. Microw. Theory Tech.*, Vol. 47, No. 11, pp. 2059-2074, Nov. 1999.

[22] G. W. Hanson. (2008). Dyadic Greens functions and guided surface waves on graphen, *J. Appl. Phys.*, Vol. 103, 2008, 064302 (1-8).

[23] S. Tretyakov. (2003). *Analytical Modeling in Applied Electromagnetics*, Norwood, MA: Artech House, 2003.

[24] O. Luukkonen, C. Simovski, G. Grant, G. Goussetis, D. Lioubtchenko, A. Raisanen, & S. Tretyakov. (2008). Simple and accurate analytical model of planar grids and high-impedance surfaces comprising metal strips and patches, *IEEE Trans. Antennas Propag.*, Vol. 56, No. 6, pp. 1624-1632, Jun. 2008.

[25] Y. R. Padooru, A. B. Yakovlev, C. S. R. Kaipa, F. Medina, & F. Mesa. (2011). Circuit modeling of multi-band high-impedance surface absorbers in the microwave regime, *Phys. Rev. B*, Vol. 84, No. 3, 2011, 035108 (1-11).

[26] Y. R. Padooru, A. B. Yakovlev, C. S. R. Kaipa, G. W. Hanson, F. Medina, F. Mesa, & A. W. Glisson. (2012). New absorbing boundary conditions and homogenization model for multilayered mushroom-type materials: Applications to wideband absorbers, *IEEE Trans. Antennas Propag.*, Vol. 60, No. 12, pp. 5727-5742, Dec. 2012.

[27] L. J. Du Toit & J. H. Cloete. (1996). Electric screen Jauman absorber design algorithms, *IEEE Trans. Microw. Theory Tech.*, Vol. 44, pt. 1, pp. 2238-2245, 1996.

[28] S. Chakravarty, R. Mittra, & N. R. Williams. (2002). Application of the microgenetic algorithm (MGA) to the design of broad-band microwave absorbers using multiple frequency selective surface screens buried in dielectrics, *IEEE Trans. Antennas Propag.*, Vol. 50, pp. 248-296, 2002.

[29] W. J. Lee, J. W. Lee, & C. G. Kim. (2008). Characteristics of an electromagnetic wave absorbing composite structure with a conducting polymer electromagnetic bandgap (EBG) in the X-band, *Composites Science and Technology*, Vol. 68, pp. 2485-2489, 2008.

[30] S. A. Tretyakov, & S. I. Maslovski. (2003). Thin absorbing structures for all incidence angles based on the use of a high-impedance surface, *Microw. Opt. Technol. Lett.*, Vol. 38, pp. 153-157, 2003.

[31] High Frequency Structure Simulator (HFSS) ver. 12.0, Ansoft Corporation, PA, 2010. [Online]. Available: http://ansoft.com

Chapter 11

Progressive Failure Analysis of Glass/Epoxy Composites at Low Temperatures

Mohammad Amin Torabizadeh and Abdolhossein Fereidoon

Additional information is available at the end of the chapter

http://dx.doi.org/10.5772/55093

1. Introduction

Fiber reinforced composite materials have been increasingly used as structural members in many structures such as airplane. The advantages of these materials are derived from their high strength, stiffness and damping together with low specific weight. Low temperature mechanical properties of glass fiber-reinforced epoxy have to be assessed, because of composite materials are subjected to low temperatures in service. Experimental or analytical investigation on the tensile failure behavior of glass/epoxy laminated composite with/or without stress concentration subjected to thermo-mechanical static loadings at low temperatures has not been done yet. In the present work, a model was developed to perform the progressive failure analysis of quasi isotropic composite plates at low temperatures. The initial failure load is calculated by means of an elastic stress analysis. The load is increased step by step. For each given load, the stresses are evaluated and the appropriate failure criterion is applied to inspect for possible failure. For the failed element, material properties are modified according to the failure mode using a non-zero stiffness degradation factor. Then, the modified Newton–Raphson iteration is carried out until convergence is reached. This analysis is repeated for each load increment until the final failure occurs and the ultimate strength is determined. The present method yields results in a reasonable agreement with the experimental data at room temperature and -60 ºC. The effect of low temperature on the failure mechanism of the plates was also determined.

2. Materials and mechanical testing

2.1. Material properties

Unidirectional glass fibers have been used in this investigation as reinforcing material, while epoxy resin has been utilized as the matrix material. Hand lay-up method was used to fabricate thin laminates with epoxy resin ML-506 with hardener HA11. Test specimens were cut from laminates according to relevant standard codes. All specimens had a constant cross section with tabs bonded to the ends. The fiber volume fraction of the composites was 55%.

2.2. Mechanical test equipments

The specimens are tested from beginning to the failure mode both at room and low temperatures. All tests were conducted under displacement control condition using Instron 5582 machine adaptable for cryogenic service by an environmental chamber. The displacement rate was 2 mm/min. Environmental chamber has the ability to cool down the temperature to -196 °C by evaporating liquid cryogenic Nitrogen. During the tests, a pressurizing device is used to control the cooling time from room temperature to -20 °C and -60 °C and maintain an evaporating pressure of 152 kPa. In order to reach thermal equilibrium in mechanical parts of machine one hour prior to first time of testing, temperature of environmental chamber was kept at desired low temperatures. Since composite materials require more time than metal materials to reach thermal equilibrium states in low temperatures, specimens stayed in a constant low temperature for at least an hour prior to start loading. Environmental chamber was equipped with a fan to perform a uniform temperature. A digital thermometer was mounted inside of environmental chamber to monitor active temperature during the tests. Experimental set up for mechanical tests at different temperature is shown in Figure 1.

Figure 1. Experimental setup for mechanical tests at both room and low temperatures

3. Material characterization

In order to apply progressive damage modeling to a laminated composite, mechanical properties of unidirectional laminates should be assessed. For this purpose test specimens was fabricated for tensile, compressive and shear loading in longitudinal and transverse direction. In this section, for each case, specimen dimensions and experimental results are reported. Also some master curves are illustrated to predict mechanical properties in the assumed temperature range. Finally all of mechanical properties are assessed by a regression function.

3.1. Tensile test

The unidirectional glass fiber-reinforced epoxy which composed of ten plies was used in this test; the total thickness of the laminate is 2 *mm*. Unidirectional tensile specimens were cut out of the laminates in both fiber and matrix directions (longitudinal and transverse) according to ASTM D3039 [1]. Woven glass/epoxy tabs with tapered ends were locally bonded on each side of the specimens. These tabs allow a smooth load transfer from the grip to the specimen especially for low temperature test. The geometries of the specimens for longitudinal and transverse tensile tests are shown in Figure 2.

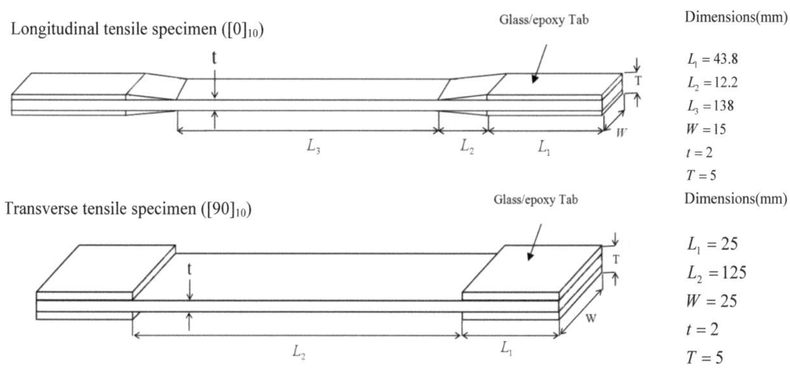

Figure 2. Geometry of the specimens for longitudinal and transverse tensile tests at different temperatures

The purpose of the static tensile tests was to determine the low temperature effects on the tensile strength, Young modulus and ultimate strain of unidirectional laminates in longitudinal and transverse directions. Figure 3 show the typical stress-strain behavior of unidirectional laminate under tensile loads in longitudinal direction at room temperature, -20 °C and -60 °C. The laminate exhibits a linear elastic behavior until breakage and the slope of the stress-strain curve increases as the temperature decreases. On the other hand, by decreasing temperature, strain to failure decreases slightly. The strain to failure reduction was very small (from 0.037, at room temperature, to 0.032 at -60 °C). However, the strength and stiffness increase significantly as temperature decreases. The average value of tensile strength increases from 700.11 *MPa*, at room temperature, to 784.98 *MPa*, at -60 °C, whereas the Young modulus also increases from 23.05 *GPa* at room temperature, to 28.65 *GPa* at -60 °C. Some authors eported increasing of tensile strength and modulus for a UD glass-epoxy composite at cryogenic temperatures [10], [38]. Consequently, it was speculated that the brittleness of the fibers had a major influence on the stiffness and strength increase of UD composites at low temperatures. Especially the brittleness of the fibers rapidly increased within a temperature range from RT to -60 °C which was indicated by Ref [10].

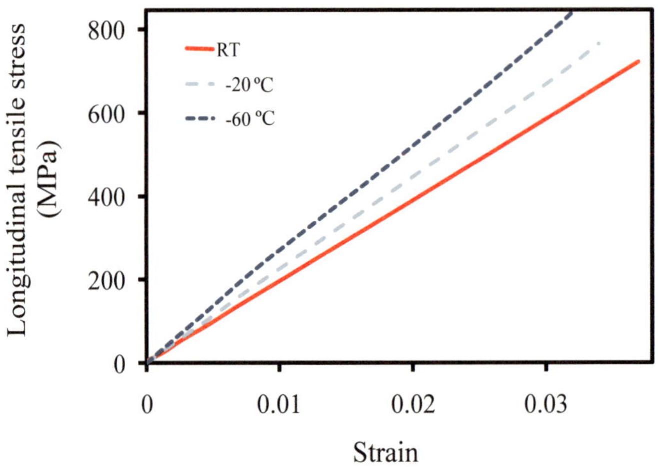

Figure 3. Typical stress-strain behavior of unidirectional laminate under tensile loads in longitudinal direction at room temperature, -20 °C and -60 °C

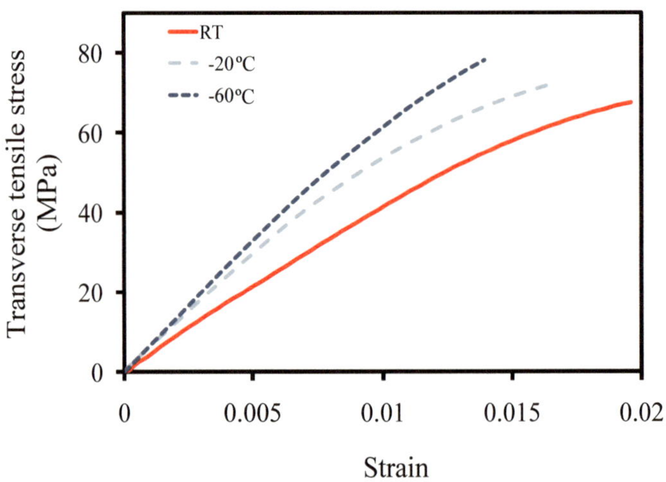

Figure 4. Typical stress-strain behavior of unidirectional laminate under tensile loads in transverse direction at room temperature, -20 °C and -60 °C

Figure 4 show the load-deflection curve of unidirectional laminates under tensile loads in transverse direction. The behavior is the same as described in the previous tests. The only difference is that the stress-strain curve in this case showed insignificant nonlinearity before reaching the maximum stress, which is due to the nature of the plasticity of epoxy. It was

http://dx.doi.org/10.5772/55093

observed because the composite becomes more brittle in low temperatures, this nonlinearity behavior of matrix decreased by decreasing temperature in Figure 4 and Ref. [37]. Typical results was also reported that both transverse tensile strength and stiffness increased by decreasing temperature from RT for glass-epoxy composites [39].

Failure regions of the composite specimens at different temperatures in longitudinal and transverse directions under tensile loading are shown in Figure 5 and Figure 6 respectively. It is found that in the case of longitudinal tension, the failure mechanism changes with temperature while no significant change was observed on the transverse direction.

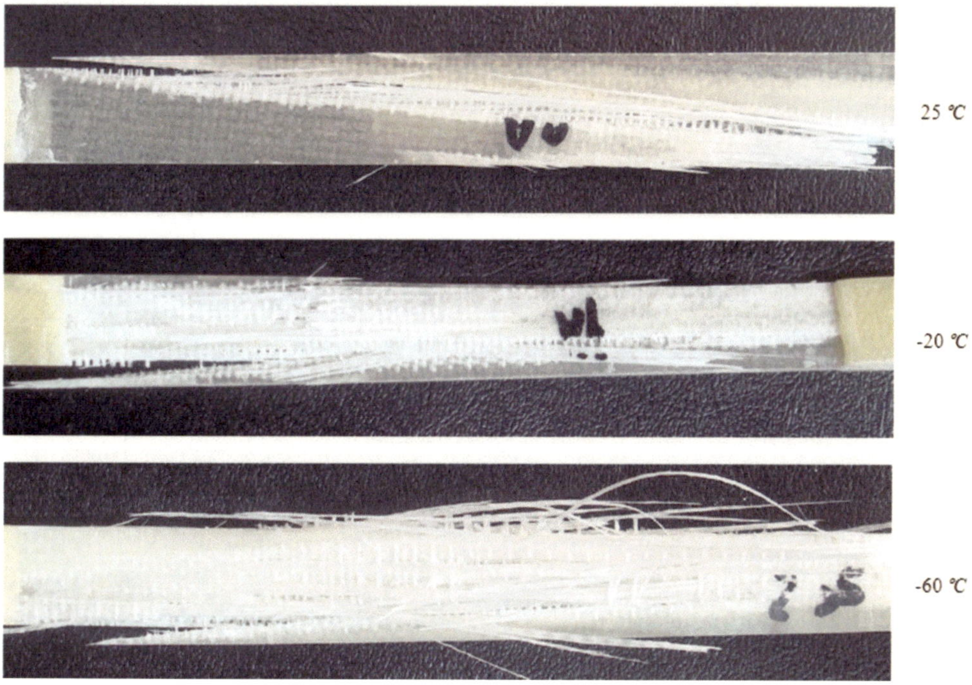

Figure 5. Failure regions of the composite specimens at different temperatures in longitudinal direction under tensile loading

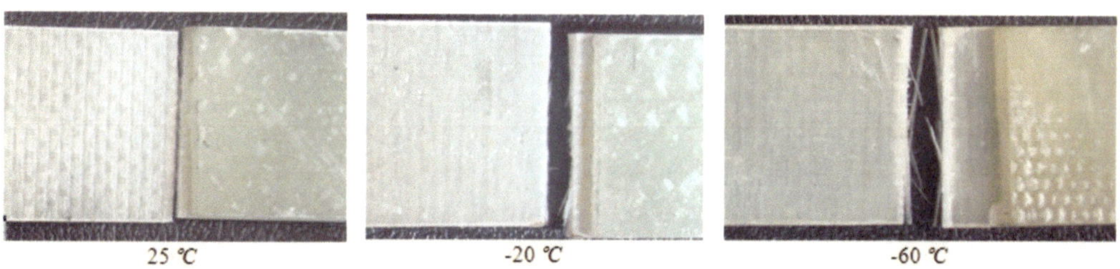

Figure 6. Failure regions of the composite specimens at different temperatures in transverse direction under tensile loading

For the longitudinal direction, from a visual inspection of the damage surfaces, a significant change could be observed in the fracture surface at room temperature to -60 ºC. There is a small amount of tab de-bonding near the gauge area for room temperature tested specimens. Also, fiber breakage which is more visible for low temperature case was occurred. At low temperature, due to weak interface between the fiber and matrix, fibers lose their bonds to the matrix. Moreover, matrix cracking and fiber pull-out were also observed. At shown in Figure 5, damage area was more limited at RT while by decreasing temperature to -60 °C; the damage was extend further and covered the entire gauge region. In the transverse tension loading, net matrix failure observed near the gauge area. The major difference between room temperature and low temperatures tests in this case is that in low temperatures (Figure 6) small fiber-matrix deboning is observed which is not obvious in room temperature specimens. Also, bigger failure region of matrix is obvious at low temperatures. There is no significant change in failure mechanism for specimens at -20 ºC and -60 ºC.

3.2. Compression test

The specimens used for compression test had the same material as those of the tensile test except that these specimens were made of fifteen plies so that the total thickness is 3 *mm*. Unidirectional compression specimens were cut in the longitudinal and transverse directions for longitudinal and transverse compression tests and also a fixture is used for the tests according to ASTM D3410 [2]. The geometries of the specimens are shown in Fig. 2. Woven glass/epoxy tabs were locally bonded on each side of the specimens to reduce the gripping effects. Also, a thin Teflon sheet, as a weakly bonded area, was inserted on each of the contacting surfaces between the specimens and the tabs to reduce the gripping effects in fiber compression specimens.

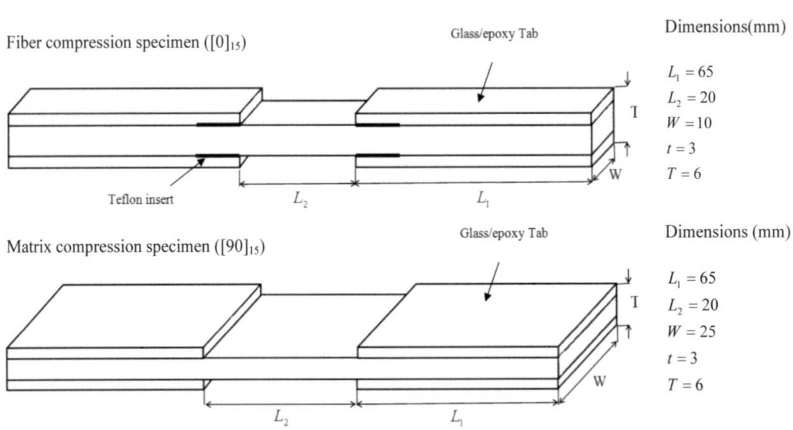

Figure 7. Geometry of the specimens for longitudinal and transverse compressive tests at different temperatures

In order to determine the effect of low temperature service on the strength and strain to failure of composite specimens in longitudinal and transverse directions, a set of experimental tests were performed. For this reason, a special fixture recommended in ASTM D3410 was used to prevent possible buckling. Figure 8 illustrate typical stress-strain behavior of composite laminate under compressive loading in longitudinal direction at various temperatures. As shown, the stress-strain relation is slightly nonlinear and final failure occurs catastrophically in all cases. The main specific property of longitudinal compressive stress-strain curves is that against other nonlinear stress-strain curves of UD composite in longitudinal or transverse directions, these curves exhibit positive curvature. This may be due to interface bonding between matrix and fibers. Imperfect fiber-matrix bonding is a manufacturing defect that is created during curing process. This defect influences the role of the matrix that supports the fibers when the composite is under compression [28], [29]. From stress-strain curves, it seems that by increasing compression in longitudinal direction, this bonding becomes grater and the role of matrix against microbuckling of fibers is more highlighted. As shown in the figure, significant non-linear deformation was often observed before the maximum load [30], [31], and this behavior was associated to the plastic deformation of the polymeric matrix. Compressive strength in longitudinal direction increases significantly by decreasing temperature (from 570.37 *MPa* at room temperature, to 731.94 *MPa* at -60 °C).

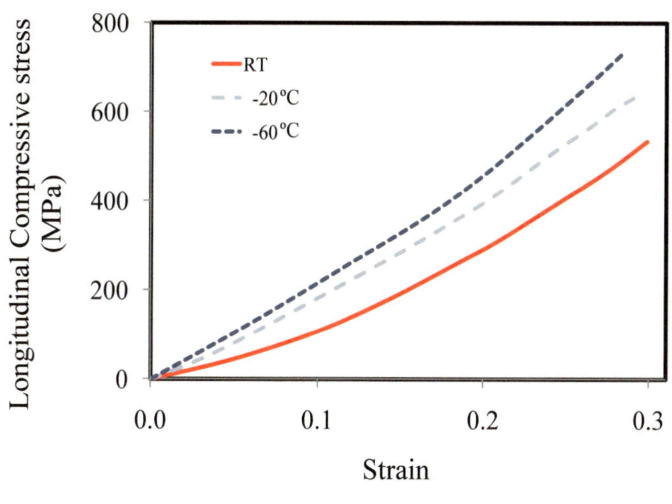

Figure 8. Typical stress-strain behavior of unidirectional laminate under compressive loads in longitudinal direction at room temperature, -20 °C and -60 °C

Figure 9 depict stress-strain curves of unidirectional composite under transverse compressive loading at different temperatures. Again significant increase in strength of composite was observed by decreasing temperature (from 122.12 MPa at room temperature, to 186.22 MPa at -60 °C). Typical behavior was reported by [32] for glass-epoxy composites. But in this case against other directions of loading, by decreasing temperature, transverse compressive stiffness was decreased and strain to failure was increased significantly. The reason of the observed phenomenon in this type of loading is unknown and it is interested to investigate more about this behavior in future works.

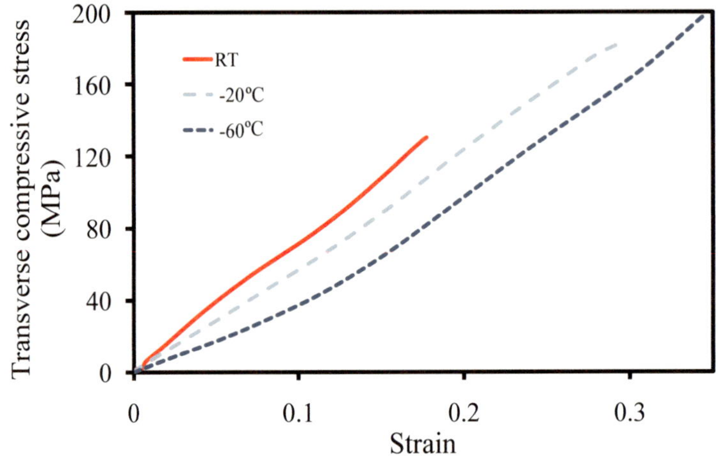

Figure 9. Typical stress-strain behavior of unidirectional laminate under compressive loads in transverse direction at room temperature, -20 °C and -60 °C

Figure 10 and Figure 11 show failed specimens under compressive loading in longitudinal and transverse directions at different temperatures, respectively.

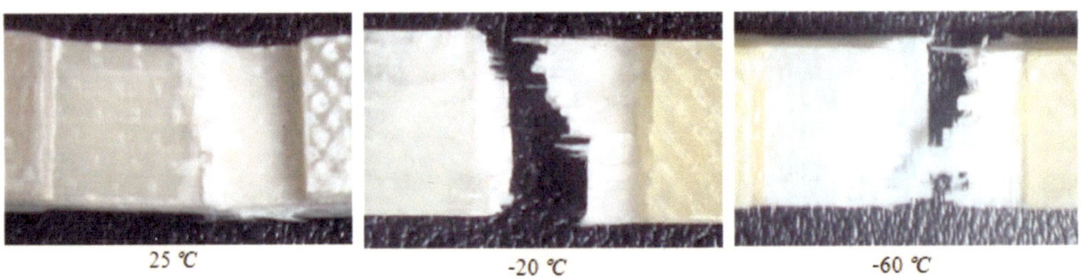

Figure 10. Failure regions of the composite specimens at different temperatures in longitudinal direction under compressive loading

Figure 11. Failure regions of the composite specimens at different temperatures in transverse direction under compressive loading

The main failure mode of specimens in this case is micro buckling of fibers in shear or extensional mode. Also, in this case, there is a small amount of tab deboning near the gauge area for room temperature specimens. The change in the mechanical response with decreasing temperature is associated with a change in failure modes. As shown in Figure 11, failure occurred at the middle of specimen under micro-buckling failure mode. The difference is that, in low temperature cases, more fiber pull-out was occurred. Also, extensive deboning between fibers and the matrix was also observed because of weaker fiber and matrix bonding at low temperatures. In this case, only at low temperatures failed specimens fractured in two separate parts. In the case of compression failure in the transverse direction at low temperature, the failure area is much more than the room temperature specimens. Also at room temperature, specimen is not fractured in two separate parts. This is because of major effects of low temperature on mechanical properties of epoxy matrix of composites. In this case, there is no significant change in failure mechanism for specimens at -20 ºC and -60 ºC. It is found the role played by two dominant damage mechanisms (decohesion at the interface and shear band formation in the matrix) in controlling the composite strength. On the other hand, if decohesion is inhibited, failure took place by the development of shear bands in the matrix which propagated through the microstructure at angle of ±45° with respect to the plane perpendicular to the compression axis [32]. The compressive strength was slightly higher than the matrix strength under compression due to the additional strengthening provided by the fibers. Parametric studies showed that other factors (such as the matrix friction angle, the interface fracture energy and the thermo-elastic residual stresses) exerted a secondary influence on the compressive strength of PMC under transverse compression [32]. But the matrix is more susceptible to the formation of shear bands. The angle formed between the failure plane and in-plane loading direction is slightly above 45° and typical values reported are in the range 50-56° [40]-[42].

3.3. In-plane shear test

In the following, method of static experiments for characterizing material properties under in-plane shear loading is illustrated. Among numerous testing methods, the three-rail shear test method, described by the ASTM A4255 [3], is a fairly reliable technique. This method is modified by Shokrieh [28]. He developed new specimen by inserting notches into the edges of specimen. The reason behind the insertion of notches at the locations of stress concentrations is to replace a very sharp crack with a blunt crack with much lower stress concentration factor. To increase the stability of the specimen during the test, a $[0,90]_s$ configuration is selected in this test. The specimen and the dimension are shown in Figure 12.

Electric strain gauges were used to measure the strains of the specimens during the shear tests. The strain gauges are attached to the specimens in the direction of 45°. The strain gauges and lead wires were manufactured by TML Corporation for low temperature ranges. However, since the changing in temperature would also affect the strain in strain gauges, Wheatstone half bridge circuit has been used to reduce this unfavorable effect.

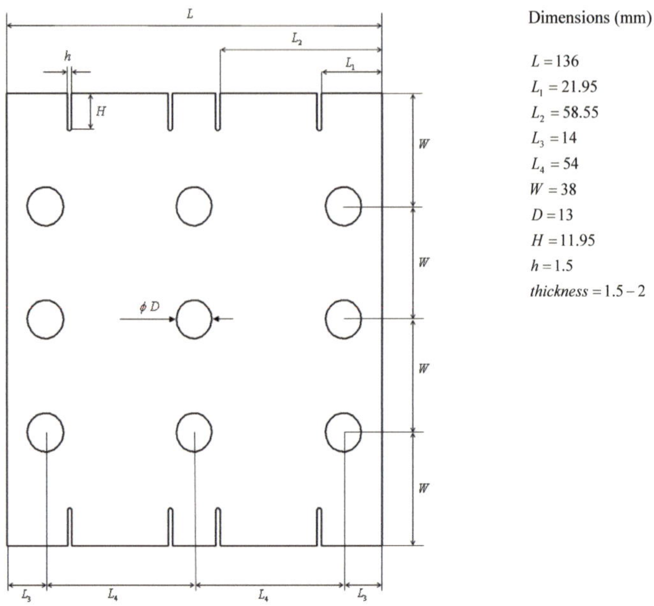

Figure 12. Geometry of the specimens for in-plane shear tests at different temperatures

Figure 13 show the stress-strain behavior of modified specimen under in-plane shear loadings at room temperature, -20 °C and -60 °C. As shown, this behavior is highly nonlinear for all temperature tests. Also, as temperature decreases to -60 °C from the room temperature, both the in-plane shear modulus and strength increased significantly.

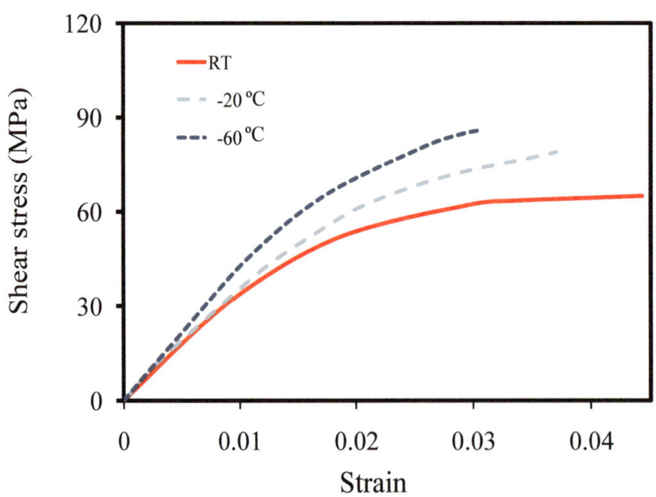

Figure 13. Typical stress-strain behavior of unidirectional laminate under in-plane shear loading at room temperature, -20 °C and -60 °C

For metallic materials, the linear region of the stress-strain curve corresponds to elastic deformation and the nonlinear region to plastic deformation. However, when working with non-crystalline substances (such as epoxy), the glass transition temperature (T_g) determines whether a material can exhibit viscous, or non-linear, behavior. Below the glass transition temperature a non-crystalline material is considered an amorphous solid, above it a rubbery solid and, as the temperature increases, a viscous liquid [33]. A non-crystalline material such as a polymer can only exhibit viscous behavior above T_g; below T_g, the material will fail before it plastically deforms. All of the tests in the current study were conducted well below T_g for the epoxy resin (≈157 °C). This indicates that the nonlinear regions of the shear stress-shear strain curves generated by the tests at room and low temperatures are not the result of viscous, or plastic, deformation [37]. This nonlinear behavior is thought to be the result of micro-crack accumulation throughout the matrix. The extensive nonlinear region displayed in the room temperature stress-strain curves would correspond to a high density of micro-cracks throughout the material (Figure 13). The more limited nonlinear region in the curves generated from the tests done at low temperatures imply that there is a progressive decrease in micro-crack accumulation within the specimen with decreasing temperature. Reduction the nonlinear region of stress-strain curves at low temperatures also confirmed by a suitable candidate model to study the nonlinear behavior of composites under shear loading which was proposed by Hahn and Tsai [28]. Parameter of material nonlinearity decreased from 1.76e^{-8} MPa^{-3} at room temperature to 1.33e^{-8} MPa^{-3} at -60 ºC. Figure 14 show tested specimens and failure regions under in-plane shear loading at different temperatures. As shown in the figure, more damaged area was observed with decreasing temperature from room temperature to -60 ºC.

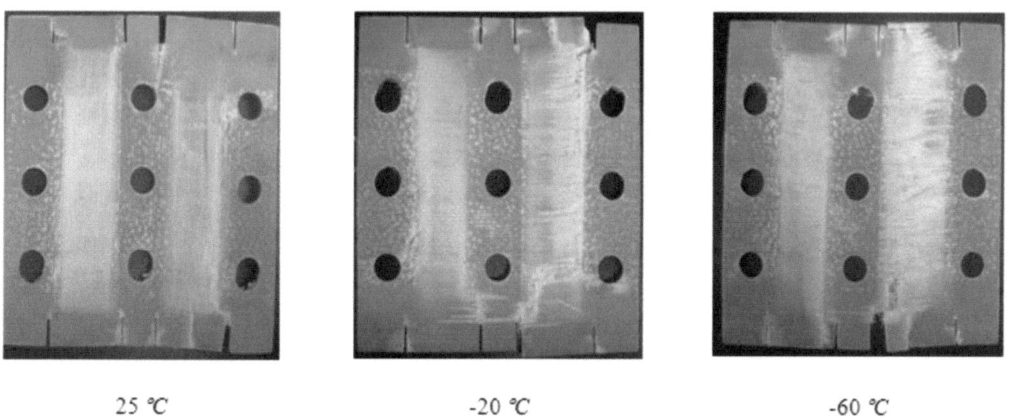

Figure 14. Tested specimens under shear loading at different temperatures

3.4. Master curves for mechanical properties

Figure 15 show the experimental values of tensile and compressive strengths for unidirectional laminates at different temperatures in longitudinal and transverse directions with their calculated standard deviations.

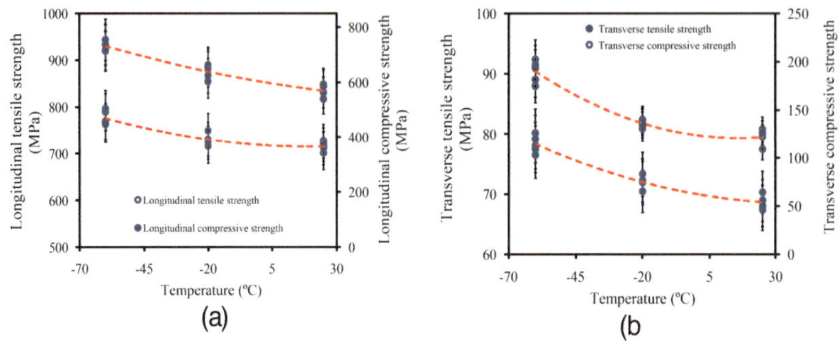

Figure 15. Experimental values of the tensile and compressive strengths for unidirectional laminates at different temperature in a) longitudinal and b) transverse directions

In each case in the above figures, by a polynomial curve fitting to experimental data, the increasing trend of strengths by decreasing the temperature is shown. However, the increasing rate of compressive strength by decreasing temperature is more for that of the tensile strength in both longitudinal and transverse directions. Figure 16 show experimental values of tensile modulus for unidirectional laminates at different temperatures in longitudinal and transverse directions. Both longitudinal and transverse tensile modulus increased by decreasing temperature to -60 ºC.

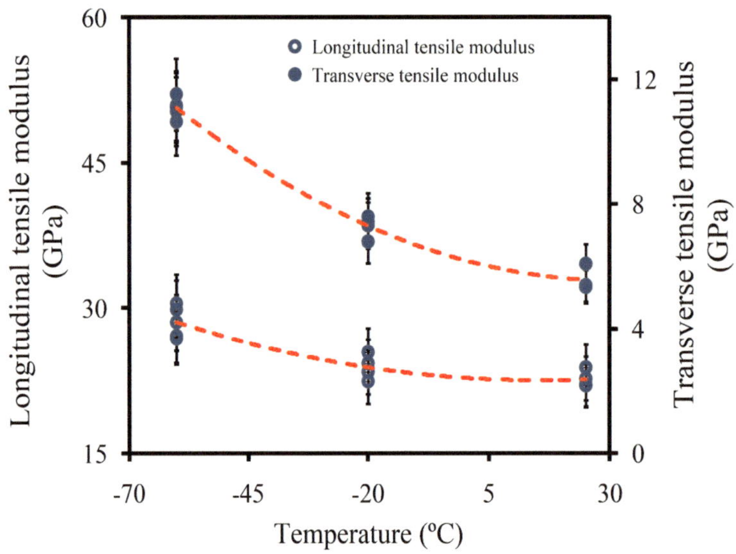

Figure 16. Experimental values of compressive strength for unidirectional laminates at different temperature in longitudinal and transverse directions

Decreasing temperature, in this case, has lower effect on the final failure mode of the compression tests of the unidirectional plies in longitudinal and transverse directions at various temperatures in comparison with the previous one. Experimental magnitudes of shear strength and modulus at different temperatures are compared in Figure 17. In the case of shear loading, low temperature has major effect on mechanical properties of composites.

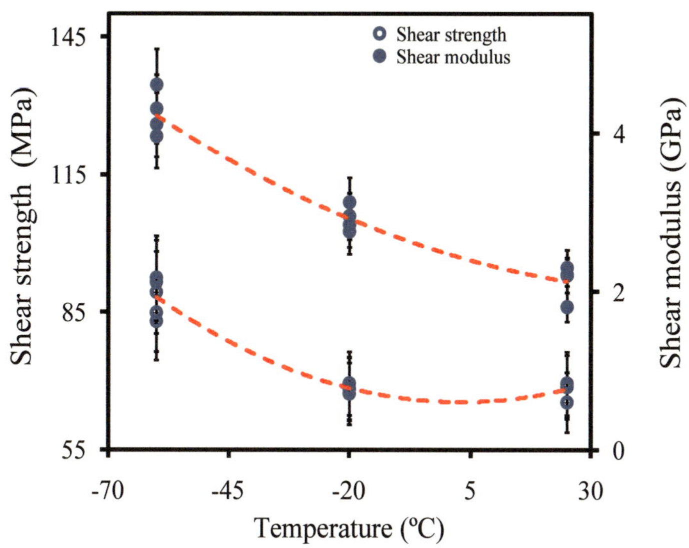

Figure 17. Experimental values of shear strength and modulus for cross ply laminates at different temperatures

Temperature effects on mechanical properties of UD composites which are plotted in Figure 15 to Figure 17 can be assessed also using a regression function defined by:

$$M(T)=\alpha T^{2}-\beta T+\gamma$$

where, M and T are the mechanical property and temperature, respectively. Also α, β and γ are the material constants. The values for α, β, γ and R (the correlation coefficient) are given in Table 1.

Material constants	***α***	***β***	***γ***	***R***
Longitudinal compressive strength	0.0093	1.5706	604.05	0.9575
Transverse compressive strength	0.0121	0.3784	124.04	0.9678
Longitudinal tensile strength	0.0098	0.3808	718.52	0.9642
Transverse tensile strength	0.0010	0.0789	70.390	0.9622
In-plane shear strength	0.0058	0.0341	65.339	0.9682
Longitudinal modulus	0.0011	0.0326	22.800	0.9666
Transverse modulus	0.0007	0.0420	6.1913	0.9920
In-plane shear modulus	0.0002	0.0184	2.477	0.9769

Table 1. Material constants and correlation coefficients -60 °C ≤ T ≤ 23 °C

4. Specimen geometry

Quasi-isotropic lay-up ($[0/\pm45/90]_{2s}$) was used in this study for tensile tests at room temperature and -60 °C. For this reason, thin laminate composed of ten plies of reinforcement with epoxy resin were fabricated with considered configuration, giving a laminate approximately 2 mm in thickness. For laminate with stress concentration, a central hole was made by a machine. Woven glass/epoxy tabs with tapered ends were locally bonded on each side of the specimens. These tabs allow a smooth load transfer from the grip to the specimen especially for low temperature test. All specimens had a constant cross section with tabs bonded to the ends. The geometry of the specimen with stress concentration for tensile tests is shown in Figure 18.

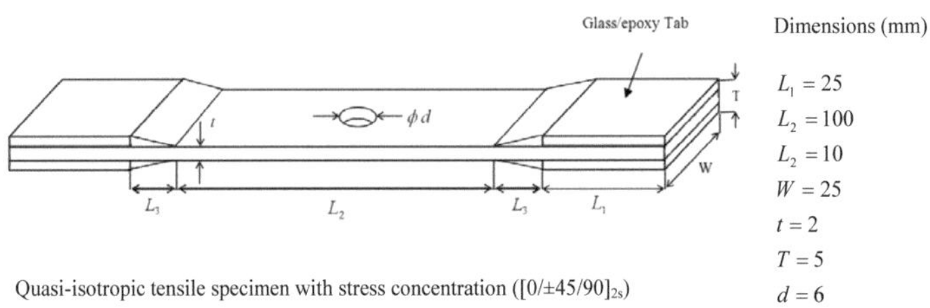

Figure 18. Geometry of the specimen with stress concentration for tensile test at room temperature and -60°C

5. Progressive Damage Modeling (PDM)

5.1. Stress analysis

The first component of the PDM is finite element stress analysis. Consider a composite plate with/without stress concentration (Figure 18). The plate has width w = 25 mm, length l =170 mm, thickness t = 2 mm and central hole D = 6 mm (if applicable). The plate is a laminated composite with quasi-isotropic ply orientation. A two dimensional macro code by APDL of ANSYS [33] is developed to perform finite element analysis. In this paper, the 8-node layered element SHELL 99 is adopted to model the laminates, which allows up to 250 different material layers in the thickness direction in each element without much increase of counting time. Mechanical properties of a unidirectional laminate at room temperature and -60°C, which is used as initial values in FEM and listed in Table 2, were tested by the present authors [34].

In this table, the script x refers to the fiber direction, and y refers to direction perpendicular to the fiber direction. In this study, the stress resultants are defined as follow:

$$\sigma_{Total} = \sigma_M + \sigma_T \tag{1}$$

Mechanical properties	23 °C	-60 °C
Longitudinal modulus *Ex* (*GPa*)	19.94	28.65
Transverse modulus *Ey* (*GPa*)	5.83	11.03
Shear modulus *Gxy* (*GPa*)	2.11	4.21
Longitudinal tensile strength *Xt* (*MPa*)	700.11	784.98
Longitudinal compression strength *Xc* (*MPa*)	570.37	731.94
Transverse tensile strength *Yt* (*MPa*)	69.67	75.2
Transverse compression strength *Yc* (*MPa*)	122.12	186.22
Shear strength *S* (*MPa*)	68.89	91.22

Table 2. Mechanical properties of GFRP at room and low temperatures

where σ_M and σ_T are the mechanical and thermal stresses respectively. Thermal stresses are due to decreasing temperature from room temperature to -60 °C. These stress resultants will be calculated by the following relation:

$$\sigma_T = \begin{Bmatrix} \sigma_x{}^T \\ \sigma_y{}^T \\ \sigma_{xy}{}^T \end{Bmatrix} = \begin{bmatrix} \overline{Q}_{11} & \overline{Q}_{12} & \overline{Q}_{16} \\ \overline{Q}_{21} & \overline{Q}_{22} & \overline{Q}_{26} \\ \overline{Q}_{61} & \overline{Q}_{62} & \overline{Q}_{66} \end{bmatrix} \begin{Bmatrix} \varepsilon_x{}^T \\ \varepsilon_y{}^T \\ \gamma_{xy}{}^T \end{Bmatrix} \tag{2}$$

In the above equation, $[\overline{Q}_{ij}]$ is transformed reduced stiffness matrix for a laminated composite and $\{\varepsilon_{ij}^T\}$ is thermal strain vector which is defined as follow:

$$\begin{Bmatrix} \varepsilon_x{}^T \\ \varepsilon_y{}^T \\ \gamma_{xy}{}^T \end{Bmatrix} = \Delta T \begin{Bmatrix} \alpha_x \\ \alpha_y \\ \alpha_{xy} \end{Bmatrix} \tag{3}$$

where ΔT and $\{\alpha_{ij}\}$ are temperature difference from room temperature and coefficients of thermal expansion for an angle ply laminate, respectively. $\{\alpha_{ij}\}$ can be given in terms of the coefficients of thermal expansion for a unidirectional laminate as:

$$\begin{Bmatrix} \alpha_x \\ \alpha_y \\ \alpha_{xy}/2 \end{Bmatrix} = [T]^{-1} \begin{Bmatrix} \alpha_1 \\ \alpha_2 \\ 0 \end{Bmatrix} \tag{4}$$

and $[T]$ is a transformation matrix for an angle ply laminate [35].

5.2. Failure criterion

The second part of PDM is failure analysis. By using finite element results, at layer level in each element, stiffness reduction is carried out considering five types of damages: fiber and matrix in tension and compression and fiber-matrix shearing modes. To detect them, a set of two dimensional stress based failure criterion is selected. The following Hashin criteria [28] are used to detect five different failure modes. The first two of the failure modes are catastrophic and the others are not.

Condition	Failure criteria	Failure relation
$\left(\frac{\sigma_{xx}}{X_t}\right)^2+\left(\frac{\sigma_{xy}}{S}\right)^2=e_{F^+}^2 \quad \begin{cases} e_{F^+}\geq 1 & Fail \\ e_{F^+}<1 & Safe \end{cases}$	Fiber tensile failure	$\sigma_{xx} \succ 0$
$\left(\frac{\sigma_{xx}}{X_c}\right)=e_{F^-} \quad \begin{cases} e_{F^-}\geq 1 & Fail \\ e_{F^-}<1 & Safe \end{cases}$	Fiber compressive failure	$\sigma_{xx} \prec 0$
$\left(\frac{\sigma_{yy}}{Y_t}\right)^2+\left(\frac{\sigma_{xy}}{S}\right)^2=e_{M^+}^2 \quad \begin{cases} e_{M^+}\geq 1 & Fail \\ e_{M^+}<1 & Safe \end{cases}$	Matrix tensile failure	$\sigma_{yy} \succ 0$
$\left(\frac{\sigma_{yy}}{Y_c}\right)^2+\left(\frac{\sigma_{xy}}{S}\right)^2=e_{M^-}^2 \quad \begin{cases} e_{M^-}\geq 1 & Fail \\ e_{M^-}<1 & Safe \end{cases}$	Matrix compressive failure	$\sigma_{yy} \prec 0$
$\left(\frac{\sigma_{xx}}{X_c}\right)^2+\left(\frac{\sigma_{xy}}{S}\right)^2=e_{FM}^2 \quad \begin{cases} e_{FM}\geq 1 & Fail \\ e_{FM}<1 & Safe \end{cases}$	Fiber-Matrix shearing failure	$\sigma_{xx} \prec 0$

Table 3. Hashin failure criterian

5.3. Material properties degradation rules

The last component of PDM is material properties degradation. As failure occurs in a unidirectional ply of a laminate, material properties of that failed ply are changed by a set of sudden material property degradation rules. In the present method after failure occurrence in a ply of the laminate, instead of inducing real crack, the failed region of the unidirectional ply is replaced by an intact ply of lower material properties. A complete set of sudden material property degradation rules for all various failure modes of a unidirectional ply under a uniaxial static stress is explained in the following. The rules must be carefully applied to avoid numerical instabilities during computation by the computer program.

- Fiber tension failure

Fiber tension failure mode of a ply is a catastrophic mode of failure and when it occurs, the failed material cannot sustain any type or combination of stresses. Thus, all material properties of the failed ply are reduced, as follows:

$$\left[E_x, E_y, G_{xy}, \upsilon_{xy}, \upsilon_{yx}\right] \rightarrow \left[\lambda_{cdr}E_x, \lambda_{cdr}E_y, \lambda_{cdr}G_{xy}, \lambda_{cdr}\upsilon_{xy}, \lambda_{cdr}\upsilon_{yx}\right] \tag{5}$$

$$\left[X_t, Y_t, X_c, Y_c, S\right] \rightarrow \left[\lambda_{cdr}X_t, \lambda_{cdr}Y_t, \lambda_{cdr}X_c, \lambda_{cdr}Y_c, \lambda_{cdr}S\right] \tag{6}$$

where λ_{cdr} is coefficient of degradation rules. Extensive comparative studies are carried out to study the effect of λ_{cdr}, which indicates that λ_{cdr} would greatly influence the strength prediction and failure mechanism in the progressive damage model. After a careful comparative study, λ_{cdr}=0.001 is applied in the current model.

- Fiber compression failure

Fiber compression failure mode of a unidirectional ply is a catastrophic mode of failure and when it occurs, the failed material cannot sustain any type or combination of stresses. Thus, all material properties of the failed ply are reduced. Equations 7 and 8 show this degradation rule.

$$\left[E_x, E_y, G_{xy}, \upsilon_{xy}, \upsilon_{yx}\right] \rightarrow \left[\lambda_{cdr}E_x, \lambda_{cdr}E_y, \lambda_{cdr}G_{xy}, \lambda_{cdr}\upsilon_{xy}, \lambda_{cdr}\upsilon_{yx}\right] \tag{7}$$

$$\left[X_t, Y_t, X_c, Y_c, S\right] \rightarrow \left[\lambda_{cdr}X_t, \lambda_{cdr}Y_t, \lambda_{cdr}X_c, \lambda_{cdr}Y_c, \lambda_{cdr}S\right] \tag{8}$$

As mentioned, these two modes of failure are catastrophic, therefore if it occurs, the other modes of failure do not need to also be verified.

- Matrix tension failure

In matrix tension failure mode of a ply, that is not catastrophic failure, only matrix direction affected, therefore other material properties are left unchanged. (Eq. 9 and 10)

$$\left[E_x, E_y, G_{xy}, \upsilon_{xy}, \upsilon_{yx}\right] \rightarrow \left[E_x, \lambda_{cdr}E_y, G_{xy}, \upsilon_{xy}, \lambda_{cdr}\upsilon_{yx}\right] \tag{9}$$

$$\left[X_t, Y_t, X_c, Y_c, S\right] \rightarrow \left[X_t, \lambda_{cdr}Y_t, X_c, Y_c, S\right] \tag{10}$$

- Matrix compression failure

Matrix compression failure mode results in the same type of damage to the composite ply as the matrix tension failure mode. This mode of failure is not catastrophic; therefore, other material properties are left unchanged:

$$\left[E_x, E_y, G_{xy}, \upsilon_{xy}, \upsilon_{yx}\right] \rightarrow \left[E_x, \lambda_{cdr}E_y, G_{xy}, \upsilon_{xy}, \lambda_{cdr}\upsilon_{yx}\right] \tag{11}$$

$$[X_t, Y_t, X_c, Y_c, S] \rightarrow [X_t, Y_t, X_c, \lambda_{cdr} Y_c, S] \tag{12}$$

- Fiber-Matrix shear failure

In fiber-matrix shearing failure modes of a ply, the material can still carry load in the fiber and matrix directions, but in-plane shear stress can no longer be carried. This is modeled by reducing the in-plane shear material properties of the failed ply, as follows:

$$[E_x, E_y, G_{xy}, \upsilon_{xy}, \upsilon_{yx}] \rightarrow [E_x, E_y, \lambda_{cdr} G_{xy}, \lambda_{cdr} \upsilon_{xy}, \lambda_{cdr} \upsilon_{yx}] \tag{13}$$

$$[X_t, Y_t, X_c, Y_c, S] \rightarrow [X_t, Y_t, X_c, Y_c, \lambda_{cdr} S] \tag{14}$$

The PDM is an integration of the three important components: stress analysis, failure analysis and material property degradation. The model is capable of simulating the first and final failure load of composite laminates with arbitrary geometry and stacking sequence under tensile static loading at room temperature and -60 °C.

A computer program, the algorithm of which is shown in Figure 19, is established to analyze the failure mechanism of composite plates at low temperatures using APDL of ANSYS. All material properties are set to initial values which are experimentally evaluated by present authors [34]. The initial failure load is calculated by means of an elastic stress analysis. The load is increased step by step. For each given load, the stresses at each integration point are evaluated and the appropriate failure criterion is applied to inspect for possible failure. At the point with failure, the material properties are modified according to the failure mode using a non-zero stiffness degradation factor. Then, the modified Newton–Raphson iteration is carried out until convergence is reached. The convergence tolerance is assumed to be 0.001. This analysis is repeated for each load increment until the final failure occurs and the ultimate strength is determined.

Theoretically, the smaller load increment between successive steps, the more accurate analysis result can be achieved. However, a reasonable load increment should be prescribed to avoid too much analysis time and to ensure accuracy. After sensitivity analysis on load increment, 1 KN is applied in the current model.

6. Results and discussion

The specimens are tested under static tensile loading at room and low temperatures. In each case (with or without stress concentration) at room temperature four specimens and at -60 ºC five coupons were tested to show statistic scatter of experiments. By statistical evaluation (mean values and standard deviation) reliability of results were appraised. The experimental setup for low temperature tests using an environmental chamber is shown in Figure 20.

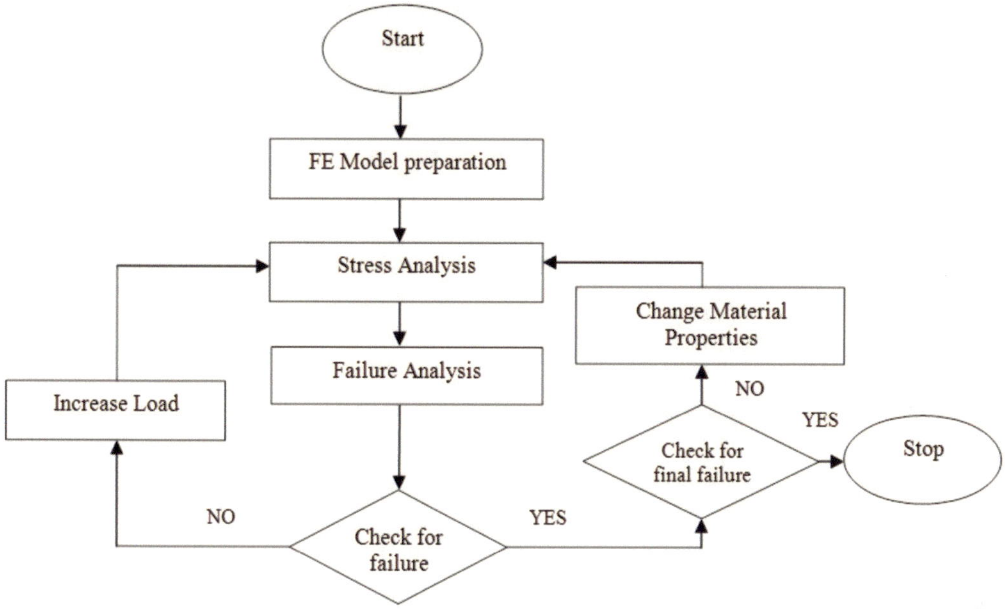

Figure 19. The algorithm of progressive damage modeling

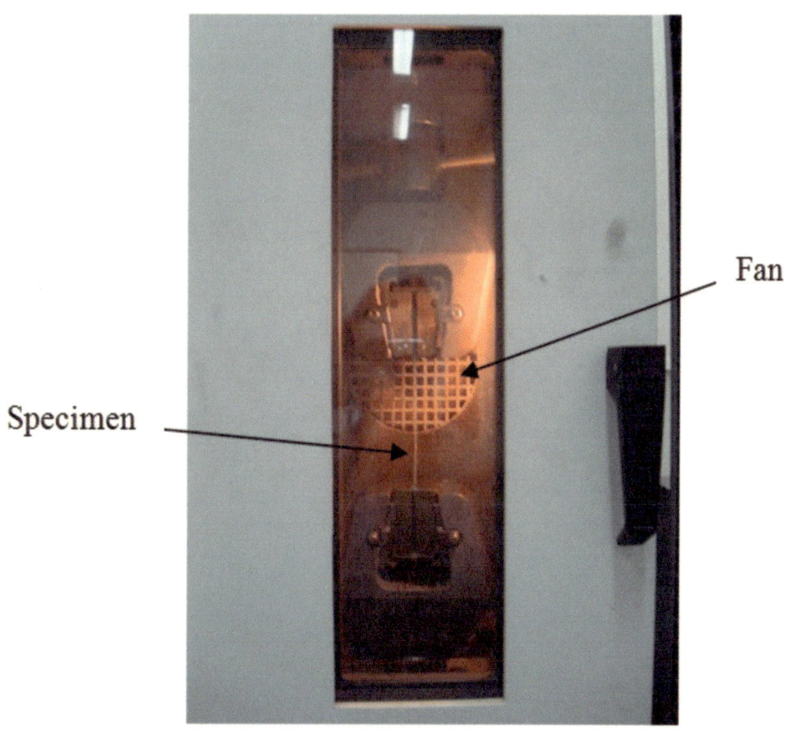

Figure 20. Environmental chamber for low temperature tests

The average tensile properties such as first ply failure (FPF) load, final failure (FF) load and ultimate strain to failure (USF) for quasi-isotropic laminates with/without stress concentration determined based on results of stress–strain curves from experimental tests and numerical analysis are summarized in Table 4.

	With stress concentration				Without stress concentration			
	Analytical		Experimental		Analytical		Experimental	
	RT	-60ºC	RT	-60ºC	RT	-60ºC	RT	-60ºC
FPF (KN)	2.045	6.28	1.8	5.75	3.12	5.62	2.15	4.52
FF (KN)	13.72	18.27	11.96	16.31	15.58	18.53	13.86	17.37
USF	0.041	0.036	0.040	0.037	0.056	0.058	0.055	0.057

Table 4. Average test results on glass/epoxy laminated composite at room temperature and -60 °C

Figure 21 and Figure 22 show the failure process predicted by the model at room temperature and -60 ºC, respectively. At the FPF load, a mainly obvious damage around the hole of plate is matrix cracking (see Figure 21a and Figure 22a). By increasing the load, other failure modes are also occurred (see Figure 21b and Figure 22b). At the final failure load, the plate breaks along the transverse direction through the central hole edge, the same as noticed in the experimental tests. In this load, the mainly failure mode is fiber breakage (see Figure 21c and Figure 22c). As shown in the following figures, the failure regions of specimens at -60 ºC at each step is much more than room temperature.

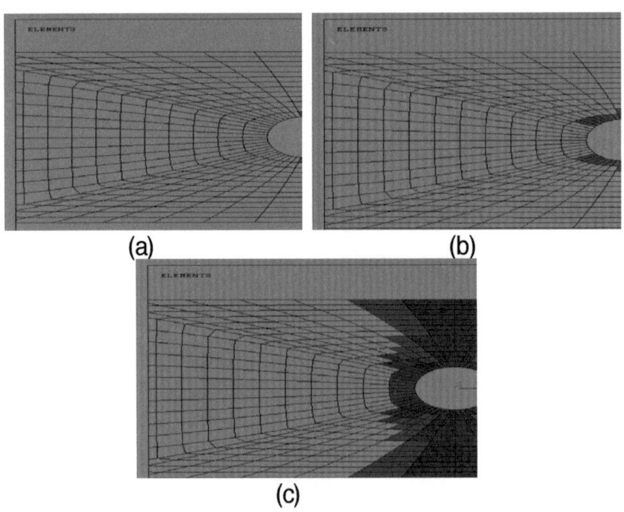

Figure 21. Failure process of plate with stress concentration at room temperature: (a) *F=2.04 kN* (FPF), (b) *F=8 kN*, (c) *F=13.72 kN* (FF)

http://dx.doi.org/10.5772/55093

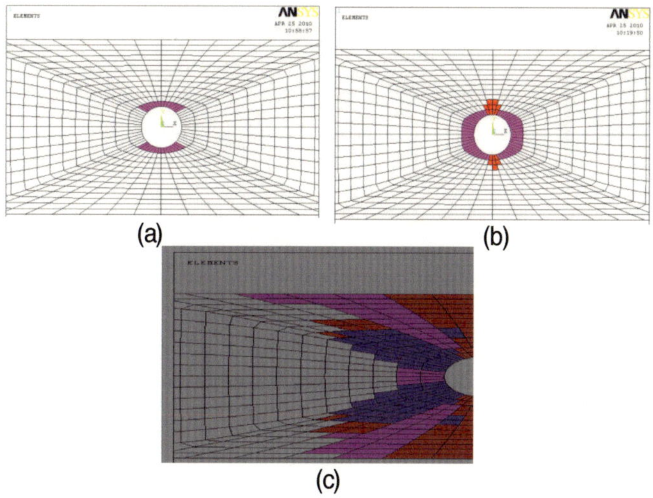

Figure 22. Failure process of plate with stress concentration at -60 °C: (a) *F=6.28 kN* (FPF), (b) *F=10 kN*, (c) *F=18.27 kN* (FF)

In all cases, major failure mode was fiber breakage, matrix cracking and fiber-matrix shearing which are shown in the figures by red, pink and blue paints respectively. Other failure modes are also occurred in the final failure but can not be shown in the figures.

Figure 23 illustrates mean values of tensile strength for quasi-isotropic laminate at room temperature and -60 °C with and without stress concentration. This figure also compares experimental results with those obtained from the present finite element model. Results show that strength of laminate increased by decreasing temperature. This is because of change of micromechanical properties of composites at low temperature.

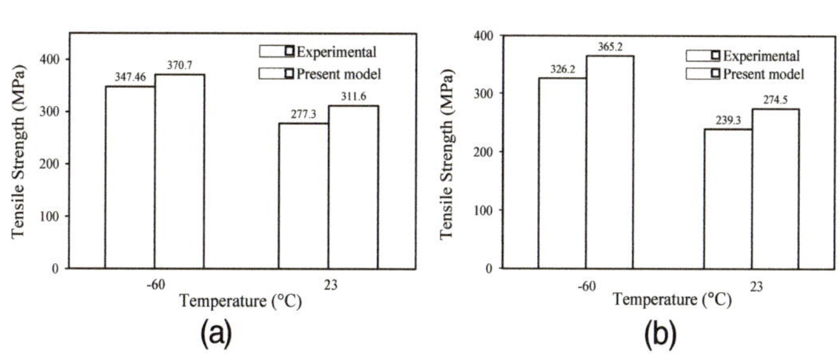

Figure 23. Tensile strength of laminates at different temperatures (a) without (b) with stress concentration

Figure 24 shows typical stress-strain curve for the laminate with stress concentration based on experimental results at room temperature and -60 °C.

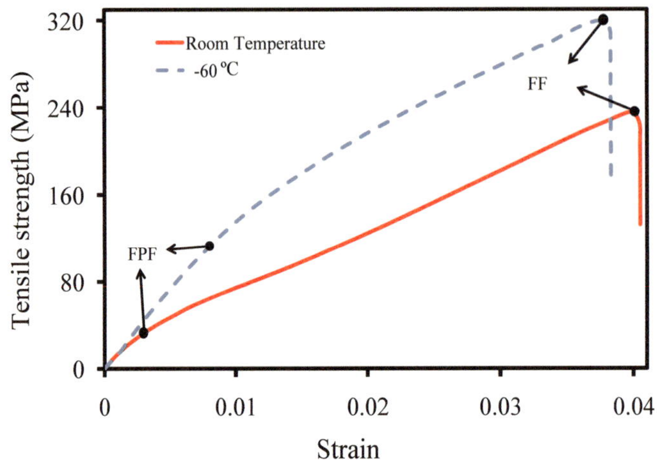

Figure 24. Typical stress-strain curve for quasi-isotropic laminate with central hole at room temperature and -60°C

Failure mechanism of tested specimens with central hole at room temperature and -60 °C are different. Figure 25 shows failed specimens at two different temperatures. From a visual inspection, there is a small amount of tab de-bonding near the gage area for both two cases with more fiber pull-out for low temperature specimen. At low temperature, because of the interface between fiber and matrix are much weaker and the fiber debond from the matrix, synchronous with fiber breakage, matrix cracking and a few fiber-matrix shearing were occurred. However, the mainly failure mode for all cases is fiber breakage and matrix cracking.

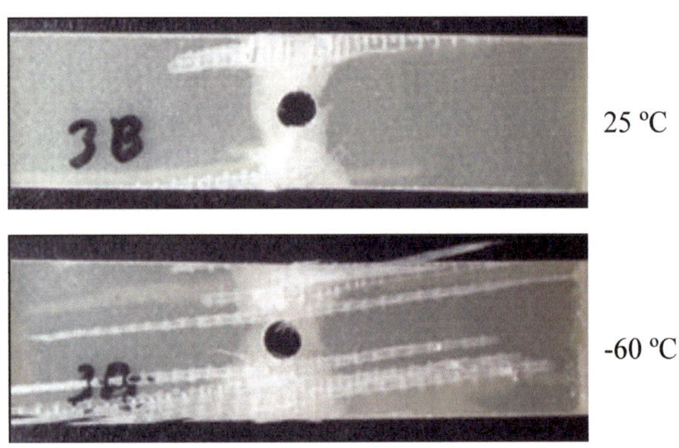

Figure 25. Failure regions of glass–fiber reinforced epoxy composites at room temperature and -60 °C

7. Conclusion

Tensile failure behavior of glass/epoxy laminated composite subjected to thermo-mechanical loadings at low temperatures with/without stress concentration was investigated experimentally and numerically. A finite element code was utilized to model the progressive failure analysis of quasi-isotropic composite plates at low temperatures under static loading. For each given load step, the stresses at each integration point are evaluated and the appropriate failure criterion is applied to inspect for possible failure by using Hashin failure criteria. At the point with failure, the material properties are modified according to the failure mode using a non-zero stiffness degradation factor. In case of failure detection, because of nonlinear behavior, the modified Newton–Raphson method was carried out until convergence is reached. This analysis is repeated for each load increment until the final failure occurs and the ultimate strength is determined. Based on the results of the present study, the following conclusions can be drawn:

1. The stress-strain behavior of UD laminate under tensile loads in longitudinal direction was linear elastic until breakage and the slope of the stress-strain curve and the strength increased about 12% as the temperature decreased to -60 °C. On the other hand, by decreasing temperature, strain to failure decreased slightly about 10%. However, the laminate under transverse tensile loading exhibited insignificant nonlinearity (especially at room temperature) before reaching the maximum stress, which is due to the nature of the plasticity of epoxy.

2. The UD laminate showed a nonlinear elastic relation between stress and strain under compressive loads in both directions until final failure occurred catastrophically. The compressive strength increased by decreasing temperature (for longitudinal direction 28% and transverse direction 50%) while the strain to failure decreased. But in transverse direction, strain to failure increased by decreasing temperature and against other nonlinear curves, positive curvature was observed.

3. The UD laminate under in-plane shear loading behaves highly nonlinear for all temperature tests until final failure. Again, in this case, both shear strength and stiffness increased by decreasing temperature in about 32% and 70% respectively. Also nonlinear region decreased by decreasing temperature due to increasing brittleness of epoxy matrix.

4. Failure type of UD laminates under various loadings was affected by low temperature. It was found that, by decreasing temperature a small amount of tab deboning occurred near the gauge area of specimens in longitudinal tensile loading. Also, because of the interface between fiber and matrix was much weaker at low temperature, fibers deboned the matrix in all test cases. Therefore, it may be concluded that the low temperature affects the micro mechanisms of damage in composite specimen.

5. From general master curves and illustrated regression function, mechanical properties of unidirectional glass fiber polymeric composites at temperature range of -60 °C to 23 °C can be evaluated.

6. The stress-strain behavior of laminate under tensile loads was linear elastic until first ply failure (FPF). After this, the behavior of laminate was nonlinear until final failure occurred. This trend was observed for laminated composite with/without stress concentration at both temperatures.

7. The slope of the stress-strain curve and the strength of laminate increased as the temperature decreased to -60 °C. On the other hand, by decreasing temperature, strain to failure decreased slightly. So, in spite of improvement in strength and stiffness of composites under static loading at low temperatures in comparison with room temperature, their strain to failure under these environmental conditions becomes weaker.

8. The failure mode of laminated composite at low temperature changes from matrix cracking at FPF to mixed mode failure (fiber breakage, fiber matrix shearing and matrix cracking) at final failure load.

9. Failure type of laminates under various loadings was affected by low temperature. It was found that, by decreasing temperature a small amount of tab deboning occurred near the gage area with more fiber pull-out. Also, due to weakness of the interface between fiber and matrix at low temperature, fiber debones the matrix. Therefore, it may be concluded that the lower temperature affects the micro mechanisms of damage.

10. Good agreement was achieved between results from experimental and analytical calculation at room temperature and -60 °C. This agreement also showed the validity of model.

Author details

Mohammad Amin Torabizadeh[1*] and Abdolhossein Fereidoon[2]

*Address all correspondence to: Torabizadeh@yahoo.com

1 University of Applied Science and Technology, Tehran, Iran

2 University of Semnan, Semnan, Iran

References

[1] Astm, D. D 3039M-95a, Standard test method for tensile properties of polymer matrix composite materials, (1997).

[2] Astm, D. D 3410M-95, Standard test method for compressive properties of polymer matrix composite materials with unsupported gage section by shear loading, (1997).

[3] Astm, D. D 4255M-83, Standard guide for testing in-plane shear properties of composite laminates, (1994).

[4] Schutz, J. S. Properties of composite materials for cryogenic application, Cryogenics. 38, 3-12 ((1998).

[5] Sun, C. T. and A. Wanki Jun, Compressive strength of unidirectional fiber composites with matrix non-linearity, Composite Science and Technology, (1994). , 52, 577-587.

[6] Bechel, V. T, & Kim, R. Y. Damage trends in cryogenically cycled carbon/polymer composites, Composite Science and Technology. 64, 1773-1784 ((2004).

[7] Kim, R. Y, & Donaldson, S. L. Experimental and analytical studies on the damage initiation in composite laminates at cryogenic temperature, Composite Structure. 76, 62-66 ((2006).

[8] Ifju, P, Myers, D, & Schulz, W. Residual stress and thermal expansion of graphite epoxy laminates subjected to cryogenic temperatures, Composite Science and Technology. 66, 2449-2455 ((2006).

[9] Rupnowski, P, Gentz, M, & Kumosa, M. Mechanical response of a unidirectional graphite fiber/polyimide composite as a function of temperature, Composite Science and Technology. 66, 1045-1055 ((2006).

[10] Kim, M. G, Kang, S. G, Kim, C. G, & Kong, C. W. Tensile response of graphite/epoxy composite at low temperatures, Composite Structures. 79(1), 84-89 ((2007).

[11] Takeda, T, Shindo, Y, & Narita, F. Three-dimensional thermoelastic analysis of cracked plain weave glass/epoxy composites at cryogenic temperatures, Composite Science and Technology. 64, 2353-2362 ((2004).

[12] Shindo, Y, Horiguchi, K, Wang, R, & Kudo, H. Double cantilever beam measurement and finite element analysis of cryogenic Mode I interlaminar fracture toughness of glass-cloth/epoxy laminates, Journal of Engineering Materials and Technology. 123, 191-197 ((2001).

[13] Melcher, R. J, & Johnson, W. S. Mode I fracture toughness of an adhesively bonded composite-composite joint in a cryogenic environment, Composite Science and Technology. 67(3-4), 501-6 ((2007).

[14] Shindo, Y, Inamoto, A, & Narita, F. Characterization of Mode I fatigue crack growth in GFRP woven laminates at low temperatures, Acta Materialia. 53, 1389-1396 ((2005).

[15] Shindo, Y, Inamoto, A, Narita, F, & Horiguchi, K. Mode I fatigue delamination growth in CFRP woven laminates at low temperatures, Engineering Fracture Mechanics. 73, 2080-2090 ((2006).

[16] Kumagai, S, Shindo, Y, & Inamoto, A. Tension-tension fatigue behavior of GFRP woven laminates at low temperatures, Cryogenics. 45, 123-128 ((2005).

[17] Shindo, Y, Takano, S, Horiguchi, K, & Sato, T. Cryogenic fatigue behavior of plain weave glass/epoxy composite laminates under tension-tension cycling, Cryogenics. 46, 794-798 ((2006).

[18] Labeas, G, Belesis, S, & Stamatelos, D. Interaction of damage failure and post-buckling behavior of composite plates with cut-outs by progressive damage modeling, Composites: Part B. 39(2), 304-15 ((2008).

[19] Liu, X, & Wang, G. Progressive failure analysis of bonded composite repairs, Composite Structures. 81, 331-340 ((2007).

[20] Zhao, Q, Hoa, S. V, & Ouellette, P. Progressive failure of triaxial woven fabric (TWF) composites with open holes, Composite Structures. 65, 419-431 ((2004).

[21] Takeda, T, Takano, S, Shindo, Y, & Nurita, F. Deformation and progressive failure behavior of woven-fabric-reinforced glass/epoxy composite laminates under tensile loading at cryogenic temperatures, Composite Science and Technology. 65, 1691-1702 ((2005).

[22] Shindo, Y, Takano, S, Narita, F, & Horiguchi, K. Tensile and damage behavior of plain weave glass/epoxy composites at cryogenic temperatures, Fusion Engineering and Design. 81, 2479-2483 ((2006).

[23] Akhras, G, & Li, W. C. Progressive failure analysis of thick composite plates using spline finite strip method, Composite Structures. 79, 34-43 ((2007).

[24] Shokrieh, M. M, Torabizadeh, M. A, & Fereidoon, A. Progressive failure analysis of composite plates, Proceedings of 8th Iranian aerospace society conference. Esfahan, Iran ((2009).

[25] ANSYSVer. 10, Canonsburg (PA): SAS IP; ((2005).

[26] Shokrieh, M. M, Torabizadeh, M. A, & Fereidoon, A. An investigation on damage of quasi-isotropic laminated composite, Proceedings of 18th annual international conference on mechanical engineering (In Persian), Tehran, Iran ((2010).

[27] Kaw, A. W. Mechanics of composite materials, Taylor and Francis Group, LLC ((2006).

[28] Shokrieh, M. M. Progressive fatigue damage modeling of composite materials, Ph.D. Thesis, McGill University ((1996).

[29] Wilczynski, A. P. Longitudinal compressive strength of a unidirectional fibrous composite, Composite Science and Technology, (1992)., 45, 37-41.

[30] Vogler, T. J, & Kyriakides, S. Inelastic behavior of an AS4/PEEK composite under combined transverse compression and shear. Part I: Experiments, International Journal of Plastics, (1999)., 15, 783-806.

[31] Hsiao, H. M, & Daniel, I. M. Strain rate behavior of composite materials, Composites: Part B, (1998)., 29, 521-533.

[32] Gonzalez, C, & Llorca, J. Mechanical behavior of unidirectional fiber-reinforced polymers under transverse compression: Microscopic mechanisms and modeling, Composite Science and Technology, (2007)., 67, 2795-2806.

[33] ANSYSVer. 10, Canonsburg (PA): SAS IP; (2005).

[34] Shokrieh, M. M, Torabizadeh, M. A, & Fereidoon, A. An investigation on damage of quasi-isotropic laminated composite," Proceedings of 18th annual international conference on mechanical engineering (In Persian), Tehran, Iran, ISME (2010).

[35] Kaw, A. K. Mechanics of composite materials, Taylor and Francis Group, LLC, 978-0-84931-343-1(2006).

[36] Callister, J. R. W.D., Materials Science and Engineering: An Introduction, Third edition, John Wiley & Sons, New York, (1994)., 480.

[37] Nettles, A. T, & Biss, E. J. Low temperature mechanical testing of carbon-fiber/epoxy-resin composite materials, NASA Technical Paper, 3663, (1996).

[38] Gong, M, Wang, X. F, & Zhao, J. H. Experimental study on mechanical behavior of laminates at low temperatures, Cryogenics, (2007)., 47, 1-7.

[39] Walsh, R. P, Colskey, J. D, & Reed, R. P. Low temperature properties of a unidirectionally reinforced epoxy fiberglass composite, Cryogenics, (1995)., 35, 723-725.

[40] Pinho, S. T, Iannucci, L, & Robinson, P. Physically-based failure models and criteria for laminated fiber-reinforced composites with emphasis on fiber-kinking. Part I: development, Composites: Part A, (2006)., 37, 63-73.

[41] Puck, A, & Schurmann, H. Failure analysis of FRP laminates by means of physically based phenomenological models, Composite Science and Technology, (2002). , 62, 1633-1662.

[42] Aragones, D. Fracture micromechanisms in C/epoxy composites under transverse compression, Master thesis, Universidad Politecnica de Madrid, (2007).

Chapter 12

Materials Assessment and Process Characterization for Lead-Free Soldering

Chien-Yi Huang, Chen-Liang Ku, Hao-Chun Hsieh,
Ming-Shu Li, Chia-Hsien Lee and Cheng-I Chang

Additional information is available at the end of the chapter

http://dx.doi.org/10.5772/51392

1. Introduction

Trends in the packaging of semiconductors are towards miniaturization and high functionality. Increased functional density and reduced I/O (Input/Output) spacing are the market trends. In the assembly process with surface mount technology (SMT), a solder paste consisting of solder particles and flux is deposited onto the PCB's bonding pads through stencil printing. The electronic components are then placed onto the printed circuit board (PCB) by a pick-and-placement machine. During reflow soldering, the solder paste melts and solidifies to form reliable solder joints. With the continuously increasing demands of electronic appliances, energy consumption and the use of hazardous substances, such as heavy metals and chemical ingredients, have become of serious concern. Traditionally used tin-lead solder is known to be harmful to human life and not compatible with an eco-environment. Environmental protection has become an international issue, and on July 1, 2006, the. European Union (EU) implemented a directive restricting hazardous substances (RoHS). One lead-free solder candidate, SAC305 (Sn96.5/Ag3.0/Cu0.5), has been widely used as a replacement for traditional tin-lead solder (Sn63/Pb37). Industry reports indicate that approximately 50–70% of soldering defects are attributed to the solder paste printing process. A sufficient amount of paste and the desired reflow soldering profile are mandatory for reliable interconnections. Solderability analysis is one of the major approaches in characterizing PCB assembly, quality control and reliability.

First, the procedure of solder paste evaluation is illustrated. In the lead-free process, the flux and solder alloy play important roles in the effectiveness of electrical resistance and the strength of the solder joints. Test items include wettability, spread, solder balls, slump, tackiness, viscosity, copper minor, silver chromate, assembly test and reliability tests. This study demonstrates the use of measuring tools combined with statistical methods to investigate the effects of the laser cutting taper angle and speed on the stencil quality for the fine pitch solder paste printing process. The stencil quality is defined by the amount of stainless steel residue after laser cutting and the roundness of the stencil aperture for a circular aperture design. The solder paste inspection data is analyzed to investigate the influences of electro-polishing and the stencil area ratio on the transfer ratio during solder paste deposition. The aim of this study is to determine the optimal parameters for stencil fabrication and the corresponding performance in solder paste deposition. The temperature profiles on the critical locations of the PCB during reflow soldering are also important to achieve the desired solder joint quality. This study establishes a procedure to predict the temperature profile based on given information about the PCB design and component loading. Critical factors, such as the conveyor speed and temperature settings of adjacent zones, that can influence the heating process are identified and investigated. A regression model and artificial neural network are constructed to more accurately predict the temperature profile. Results of this study will help improve the efficiency of the temperature setting process, especially in the pilot run stage. Finally, the analytical techniques used for soldering properties, including visual inspection, side-view microscopy, x-ray inspection and dye staining analysis, are also discussed. The crack size and percentage are classified according to the crack area. In this study, the solderability analysis is carried out to determine whether any cracks in the solder joints occur in the CPU and RAM of the ball grid arrays (BGAs) components. It may cause the failure symptom of no display to appear on an ultrabook while the power is on.

2. Stencil evaluation of ultra-fine pitch solder paste printing process

The surface mount technology (SMT) is commonly used in electronics industry to assemble the components onto the PCB. It (SMT) enables the production with relatively low cost. Stencil printing is known to significantly influence the process quality in Surface Mount Technology (SMT). For ultra-fine pitch applications, solder paste printing acts as one of the most critical steps in the SMT assembly process. A review of the literature indicates that solder paste printing is responsible for around 60% of the total process defects. One of the key factors for the performance of solder paste deposition is the stencil fabrication quality. Other important factors include the stencil aperture design, paste printing parameters and stainless steel raw material. The incoming inspection of the stencil typically includes a frame appearance examination and tension force measurement. Inspection of individual apertures under a microscope is not feasible due to the large size of the stencil. Related research also shows that stencil aperture and wall roughness impact the solder paste deposition quality. Laser cutting is widely used due to its advantageous performance with respect to cost and aperture quality. While electroformed stencils are typically considered for the assembly of 01005 passives, evaluations of the laser cut stencil manufacturing process through solder paste printing optimization are lacking. The stencil manufacturing process generally includes three stages, namely, screen stretching, laser cutting and electropolishing/deburring (Fig. 1).

http://dx.doi.org/10.5772/51392

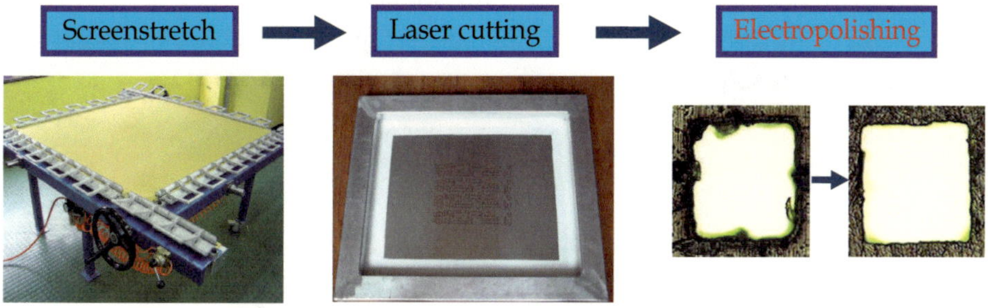

Figure 1. Stencil manufacturing process

Measuring tools combined with statistical methods were used to investigate the effects of the laser cutting taper angle and speed on the stencil quality. The stencil quality is defined by the amount of stainless steel residue after laser cutting and the roundness of the stencil aperture for a circular aperture design. The solder paste inspection data was analyzed to investigate the influence of electropolishing and stencil area ratio on the transfer ratio during solder paste deposition. Weighted ranking was adopted to indicate the variation in solder paste deposits for 01005 and 0.4/0.3mm pitch CSP pads due to the stencil fabrication process. The aim was to determine the optimal parameters for stencil fabrication and corresponding performance in solder paste deposition.

The laser machine and materials used for stencil aperture cutting and stencil materials were evaluated. Experiments included: (1) material properties assessment – to test the durability of the screen; (2) stencil quality inspection – to evaluate the occurrence of clogging, aperture roundness and taper angle; (3) electropolishing process inspection – to determine the optimal process parameters for laser cutting for various taper angles; and (4) printing process assessment – to evaluate the stencil printing performance for 01005/0201 RLC and 0.4/0.3mm pitch CSP applications.

2.1. Experiments

2.1.1. Material properties assessment

The durability of various raw materials against IPA cleaning solvent was evaluated. The test samples were 736 mm square and 0.1 mm thick. The stencils (without aperture) were cleaned by the cleaning machine. After cleaning for 90 min, the samples were dried for 10 min. As the cleaning time is 10 min in the process currently in use, 90 min of cleaning was equivalent to 9 times the cleaning cycle. The criteria for acceptance were readings above -0.22 mm at the stencil center and above -0.22 mm on the other locations.

2.1.2. Stencil quality inspection

The two-factor three-level full factorial experimental design was used. The factors considered were taper angle and laser cutting speed. Details of the experimental design are shown in Table 1. The taper angle (Fig. 2.) was determined by Equation (1). Quality characteristics included the occurrence of clogging and aperture roundness, as in Fig. 3. Clogging was defined as the percentage of aperture area covered by the steel scrap, while roundness was determined by image measurement software. The stencil aperture design is shown in Fig. 4.

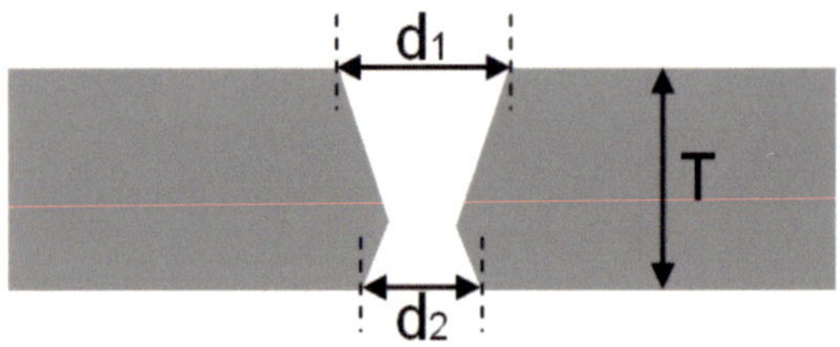

Figure 2. Taper angle

$$\alpha = \tan^{-1}((d1-d2)/2*T) \quad (1)$$

The d1 and d2 are the aperture sizes at the entrance side and exit side, respectively; T is the stencil thickness.

Factors	Levels		
Taper Angle	2°	4°	7°
Speed (mm/s)	12	10	8

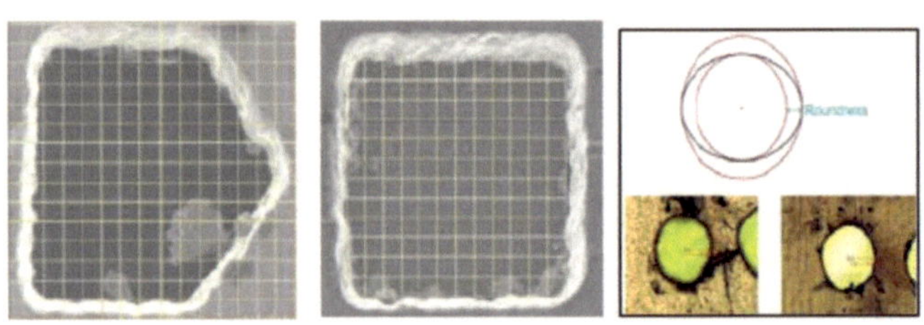

Table 1. Laser cutting experimental design

2.1.3. Electropolishing process

The anodic polarization chart (Fig. 5) shows the relationship between the applied voltage and the anode current density. When the voltage reached the polishing region (DE), further

http://dx.doi.org/10.5772/51392

Factors	Levels		
Taper Angle	2°	4°	7°
Speed (mm/s)	12	10	8

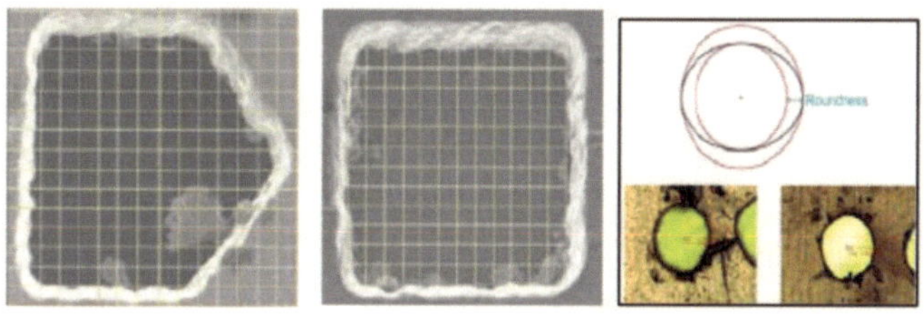

Figure 3. Clogging and aperture roundness

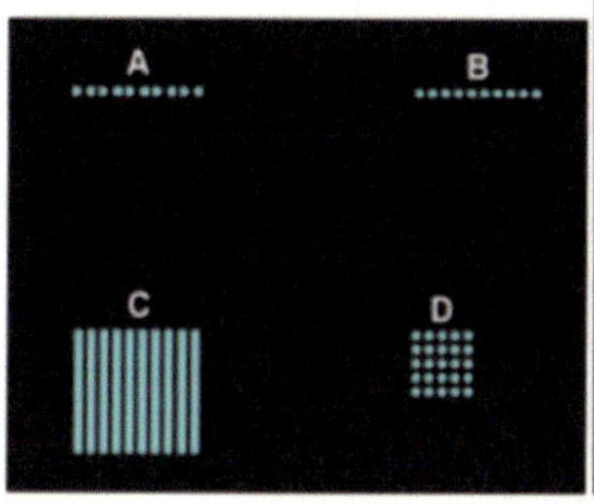

Area	Component pad	Scale (mm)	Inspection Item
A	01005 Home Plate	Side:0.186 Space:0.129mm	Clogging
B	Square	0.2mmx0.2mm	Clogging
C	Connecter	0.15mmx2.65mm	Clogging
D	BGA Pitch 0.3	Diameter:0.2mm	Roundness

Figure 4. Stencil design for stencil cutting evaluation

increases in voltage did not have a significant effect. The stencil aperture logging scenario considered three polishing times.

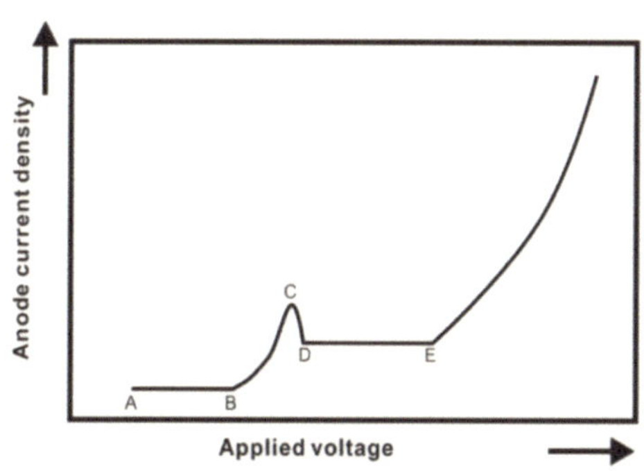

Figure 5. Anodic polarization chart

2.1.4. *Printing process assessment*

Three stencil thicknesses (0.08mm, 0.1mm, 0.12mm), three post treatment methods (CP: chemical polishing, EP: electropolishing, NP: no polishing) and three taper angles (2°, 4°, 7°) were considered. Details of the experimental design are shown in Table 2. Ten stencils in total were used for the printing experiments. The squeegee printing angle remained at 60°. Stencils were cleaned after every printing. The separation velocity was 0.5 mm/sec, the printing pressure was 12 Kg and the printing velocity was 40 mm/sec. The sample size was forty, i.e., there were forty identical apertures on the stencil for each treatment. The printing was duplicated 15 times. The transfer ratio of printing was used as quality performance.

Stencil thickness	0.08				0.1					0.12
Machine	Machine A								Machine B	
Taper Angle	2°	4°	7°	4°	2°	4°	7°	4°	4°	4°
Polishing	EP	EP	EP	CP	EP	EP	EP	CP	NP	NP
Stencil ID	A	B	C	D	E	F	G	H	I	J

Table 2. Experimental design for printing performance evaluation

The test board was a bare board without a solder mask in order to minimize the potential printing variation. It is not reasonable to compare the transfer ratio when evaluating the printing performance of stencils with different thicknesses. While the amount of solder paste printed is a major factor influencing the solder joint reliability, a thick stencil with a smaller transfer ratio may still have a greater amount of solder paste deposition. Therefore, a 0.1 mm stencil was used as the standard thickness (Fig. 6). The printing performance was normalized and evaluated through the steps below.

The equation used to calculate the transfer ratio is shown below:

$$\text{volume transfer ratio} = \frac{\text{solder paste volume}}{\text{stencil thickness} * \text{aperture area}} \tag{2}$$

After ten printings, the average and standard deviation are calculated.

The stencil with the highest average printing volume is assigned a score of 10, and the second highest is assigned a score of 9….

The performance of printing variation is considered for an extra score. The top three stencils with the least printing variation (standard deviation) are assigned a score of 3, 2, 1, respectively. The results of printing for a 0.3 mm BGA are shown below.

Stencil	F	I	D	A	C	J	H	E	G	B
Mean	10	9	8	7	6	5	4	3	2	1
Standard deviation		3				1			2	
Total	10	12	8	7	6	6	4	3	4	1

Table 3. The results of printing for a 0.3 mm BGA

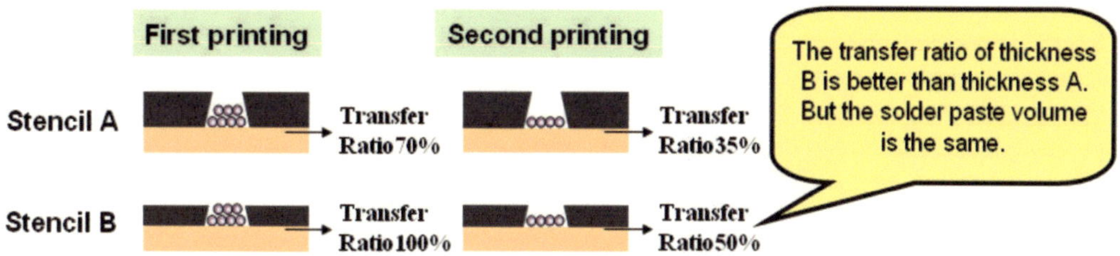

Figure 6. Plot showing difference in transfer ratio for various stencil thicknesses even with similar solder volume deposition

2.2. Results

2.2.1. Material properties assessment

After a total of 30 hours IPA cleaning (180 cleaning cycles), separation occurred to one of screen model. It showed that the test method can be an effective evaluation of the durability of raw materials.

2.2.2. Stencil quality inspection

Results show that a taper angle of 2° and a laser speed of 8 mm/s resulted in minimal stencil aperture clogging (Fig. 7(a)). A slower laser speed was preferred for better aperture roundness (Fig. 7(b)).

2.2.3. Electropolishing process

Results showed that clogging was minimized by increasing the polishing time (Fig. 8). The residue was found to be carbon and oxide through scanning electron microscope/energy dispersive X-ray spectroscopy (SEM/EDS) analysis (Fig. 9). Therefore, the electropolishing process did not seem to be effective in removing the non-metal contamination and improving solder deposition. Also, chloride was detected and this may have introduced additional concerns.

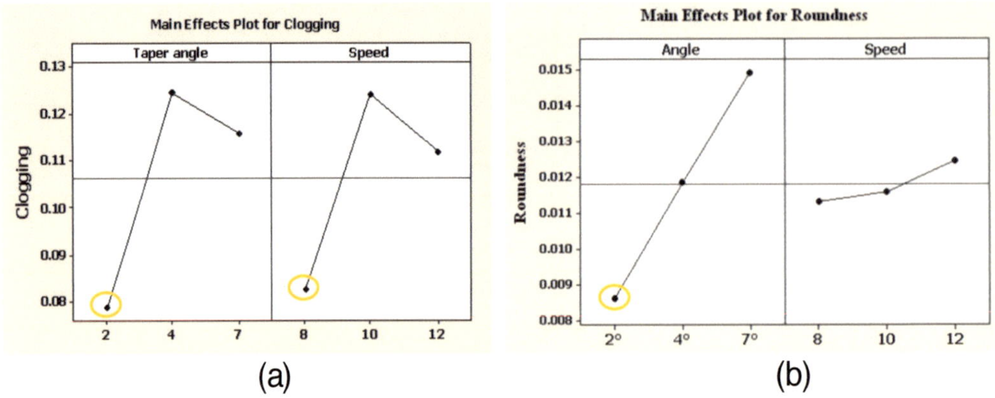

Figure 7. Main effects of taper angle and laser speed on (a) clogging and (b) aperture roundness

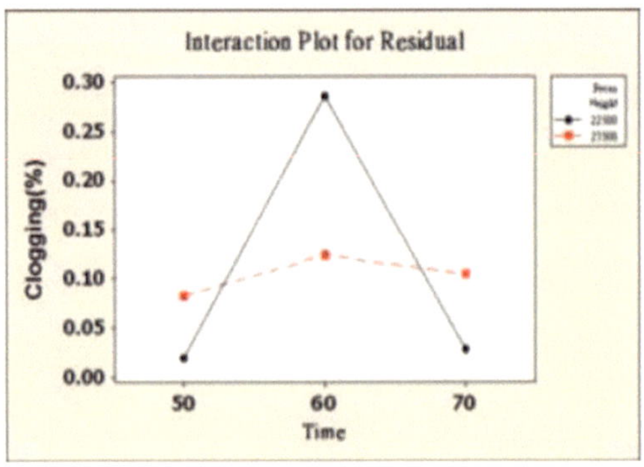

Figure 8. Main effect of polishing time on aperture clogging

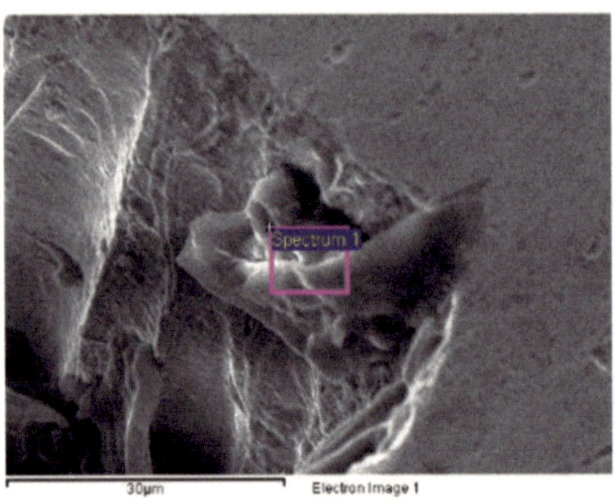

Element	Weight%	Atomic%
C K	52.48	62.64
O K	37.08	33.22
Na K	2.79	1.74
Cl K	2.94	1.19
Fe K	4.72	1.21
Totals	100.00	

Figure 9. Residue compound SEM/EDS analysis

http://dx.doi.org/10.5772/51392

2.2.4. Printing process assessment

The effect of the stencil aperture shape was evaluated. For both 0.08 mm and 0.1 mm stencils, the circular aperture design was superior to the home plate shape aperture design for 01005 solder paste printing with respect to the transfer ratio (Figs. 10 and 11). For the 0.08 mm stencil, a chemical polished stencil was superior to an electropolished stencil. However, for the 0.1 mm stencil, the electropolished stencil and the stencil with no polishing performed better than the chemical polished stencil. This result was consistent with SEM/EDS analysis in that post treatment did not remove steel scraps and improve solder deposition.

The solder paste transfer ratio scores for various components are shown in Table 4. The transfer ratio for the 0201, 0.5mm BGA and 0.4mm connector were all above 100%. The stencil performance was evaluated for miniaturization applications. The components with a transfer ratio above 100% were excluded. The top three scenarios with the highest scores were (1) the electropolished stencil with a 0.1 mm thickness and a 40° taper angle; (2) the stencil with no polishing, 0.1 mm thick and a 2° taper angle; and (3) the stencil with no polishing stencil, 0.12 mm thick and a 2° taper angle.

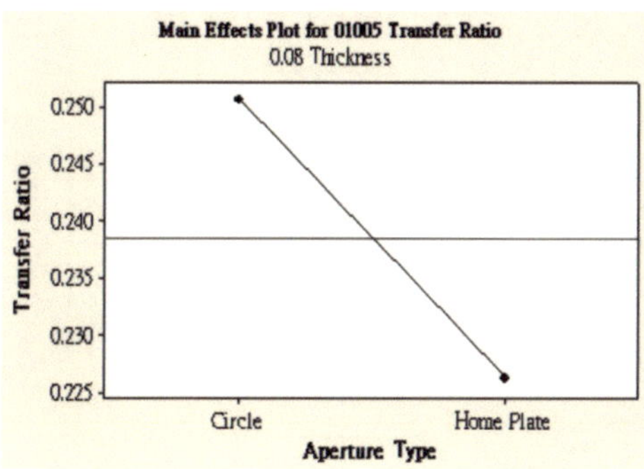

Figure 10. Influence of aperture shape on transfer ratio for stencil with thickness 0.08 mm

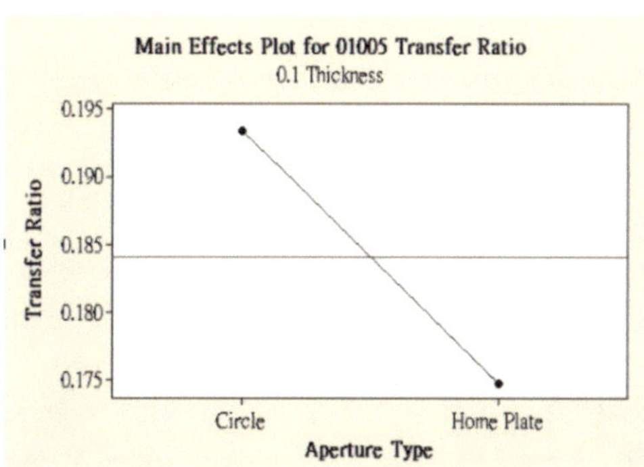

Figure 11. Influence of aperture shape on transfer ratio with thickness 0.1 mm

Also, the 0.1 mm stencil performed better than the 0.08 mm stencil with respect to the transfer ratio. The stencil with no polishing performed better than the electropolished or chemical polished stencils.

The correlations between the transfer ratio and area ratio were investigated for the stencils with three different post treatment technologies (Figs. 12 and 13). No correlation was found for either the electropolished or chemical polished stencils. The process variation of laser machine A was more severe than that of machine B. While the industry standard IPC-7525 (IPC-7525 3.2.1.1, 2000) suggests an area ratio above 0.66, stencils with an area ratio of 0.5 still achieved a 60% transfer ratio for machine B. Also, as regards stencil manufacturing, stencils with no polishing performed better than stencils that were either electropolished or chemical polished. Significant variations in the stencil manufacturing machine were observed, which may have affected the effectiveness of the stencil evaluation. The area ratio could be used to verify the process variations of stencil manufacturing in future stencil quality assessment studies. The development of printing capability contributes to the miniaturization of electronic products with greater functionality.

Stencil ID	F	I	J	G	A	E	D	C	H	B
01005	12.5	10.5	9	4	8	3	6	5	1	2
BGA0.4	10	10	10	6	4	3	5	4	8	1
BGA0.3	10	12	6	4	7	3	8	6	4	1
CN 0.3P	9	11	10	6	4.5	7	4	5.5	3	1
CN 0.3V	10	6	8	10	5	11	2	3.5	3.5	1
Total	51.5	49.5	43	30	28.5	27	25	24	19.5	6

Table 4. Solder paste transfer ratio score for various components

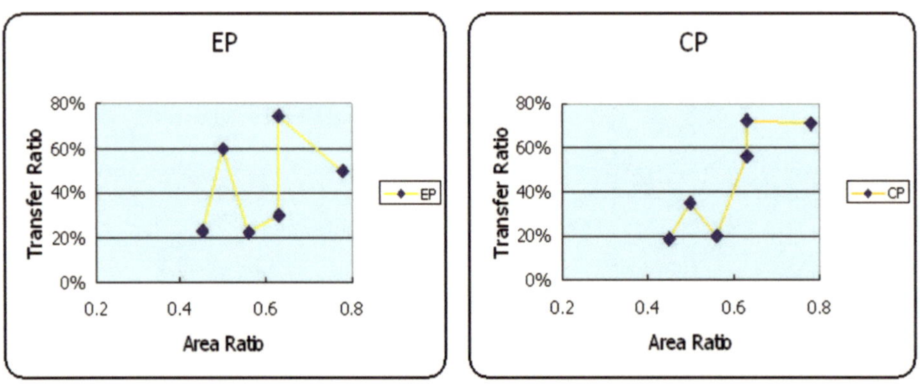

Figure 12. Correlation between the transfer ratio and area ratio for laser machine A

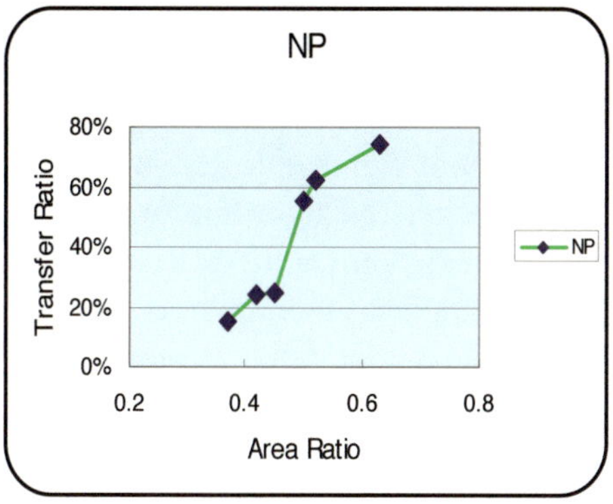

Figure 13. Correlation between the transfer ratio and area ratio for laser machine B

3. Characterization of reflow temperature profile

During reflow soldering, the solder paste melts and solidifies to form reliable solder joints. In the reflow soldering process, the PCBs are transported by the conveyor and go through the various heating zones. Prior to mass production, the process engineer has to ensure that the desired temperature is achieved by using a profile board with the same characteristics as the real product to be assembled. This is critical to ensure that the desired solder joint is formed, while at the same time taking into consideration the heat resistibility of the PCB and the electronic components. The trial and error approach is commonly used, which involves adjusting the temperature setups at various heating zones and the conveyor speed to achieve the temperature profiles on important locations of the PCB so that all comply with the process specifications [Su 1997]. The process to determine the appropriate temperature setups is time consuming as it includes the time to measure the PCB temperature, time for the thermal tracker to cool down to room temperature and the time for the reflow chamber to reach a stabilized temperature. Therefore, the schedule for production could be detained. Characterization of the reflow heating process helps to reduce setup time and the number of trials while ensuring the quality of the solder interconnections.

Temperature simulation using finite difference (FD) and computational fluid dynamics (CFD) modeling tools can achieve a high degree of accuracy, but they are complex to generate and analysis times are unacceptably long. Whalley (2004) developed a less complex approach using simplified representations of both the product and the process. Sarva and Conway (1998) abandoned mathematical modeling and utilized the concept of What-If to forecast temperature profiles. This approach, however, is appropriate primarily at the phase of product design and process planning. Su (1997) proposed a knowledge-based approach to design and implement a profile identification decision support system. Inferences through an expert system usually apply only to certain categories of products. Gao (2008) offered a convenient method to develop an oven recipe for a specific soldering profile. The method was devised to quickly achieve the proper profile shape, but the accuracy of the profiling was constrained, especially for complicated products in real applications.

Our research considered real PCBs of server applications, usually having six to eight copper layers, but sometimes up to a total of eighteen layers, which are loaded with complicated components with a large heat capacity. The algorithm suggested by Gao (2008) was used in this study as a baseline model for prediction. Data (real temperature setups) of a total of twelve server models collected from production environments were then employed to make the needed modifications to the baseline model. A 'correction value' for the baseline model was the 'response', and the temperature setting of the heating zone, conveyor speed and the total PCBA heat capacity were the independent variables. Stepwise regression analysis (SRA) was used to effectively express the relationship between the response variable and the independent variables (for each heating zone).

3.1. Preliminary experiments

Preliminary experiments were conducted to assess the stability of the reflow oven under study and the influence of the heating zone setup on the resulting temperature profile. The effect of the PCB carrier was also considered. The three preliminary experiments considered a total of five heating zone setups (Table 5). The oven used in these experiments had nine heating zones and natural cooling (Fig. 14). The test vehicle was a bare board with eight stacking layers (Fig. 15). The length, width and thickness were 241 mm, 335 mm and 1 mm, respectively. The thickness of the copper layers on the PCB surface was 0.5 oz, and 1 oz for the inner layers. Six thermal couples were attached at four corners on the top side and center locations of both the top and bottom sides. The fixture used in this study is shown in Fig. 16.

Heating zone		Z1	Z2	Z3	Z4	Z5	Z6	Z7	Z8	Z9
The 1st preliminary experiment		180	180	180	180	180	180	180	180	180
The 2nd preliminary experiment		180	180	180	200	200	200	180	180	180
The 3rd preliminary experiment	Temp setting (1)	180	180	180	180	200	200	240	240	240
	Temp setting (2)	180	180	180	180	200	200	250	245	240
	Temp setting (3)	180	180	180	180	200	200	260	240	220

Table 5. Heating zone setups considered in the preliminary experiments

In the first experiment, the temperature setting for all heating zones was maintained at 180℃. The measured preheating and cooling temperature slopes for the PCB samples with a carrier were smaller than those for PCB samples without a carrier. The peak temperature (T_p) during the reflow zone for the PCB samples with a carrier was also lower than that for PCB samples without a carrier and with less variation (ΔT). The above mentioned observations showed the inconsistency in heating phenomena due to the carrier's heat capacity. Secondly, the measured

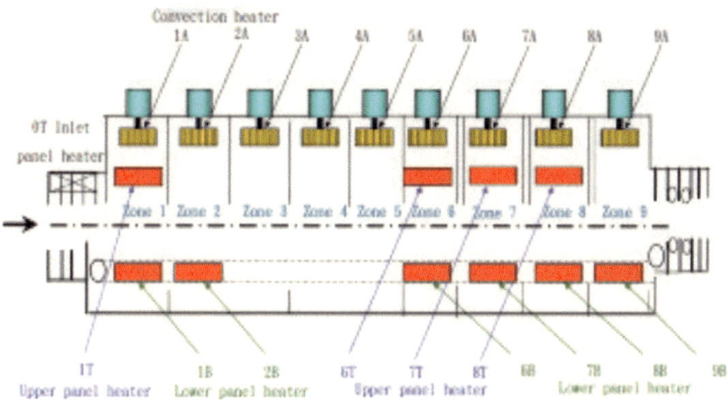

Figure 14. Reflow oven used in the preliminary experiments

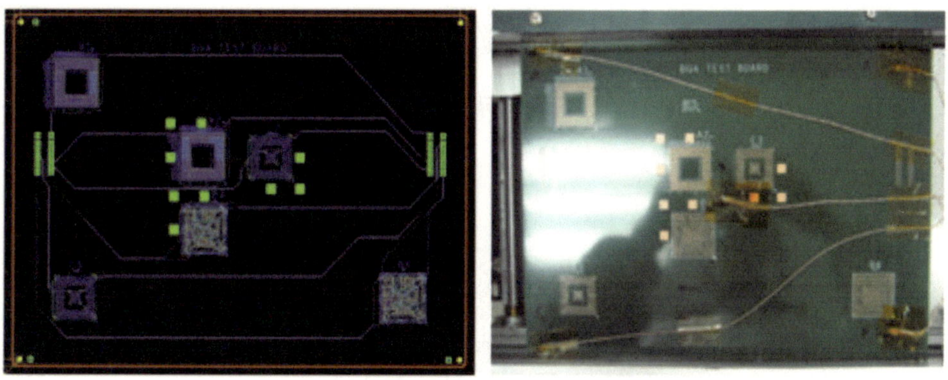

Figure 15. PCB used in the preliminary experiments

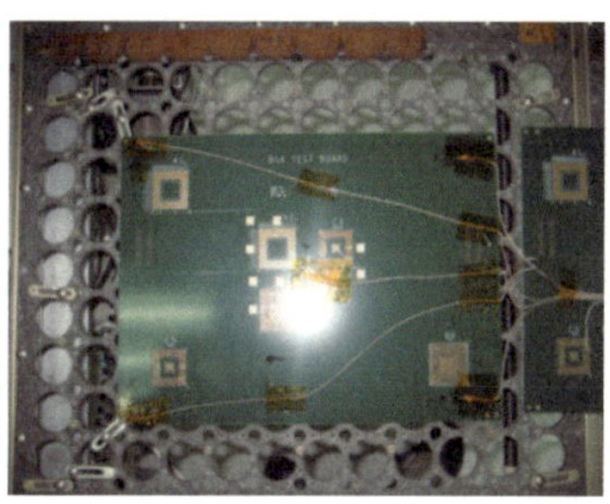

Figure 16. Fixture used in this study

temperatures near the adjustable side of the conveyor were higher than those near the fixed side of the conveyor. This was because the adjustable side of the conveyor was close to the center of the heater and the heating source was more stable and effective (Fig. 17). In the second experiment, the temperature setting at Z4~Z6 was 200℃ and all the other zones maintained at

180℃. The results showed that the PCB temperature also increased by approximately 20℃ when the temperature setting at the heating zone increased by 20℃. This indicated that the heating zone was stabilized and that the heat transfer was effective. The third experiment, in reference to Cho (2008) (Fig. 18), considered three temperature settings: (1) reflow heating zones Z7-Z9 maintained at 240℃; (2) reflow zone Z7 at 250℃ and the remaining zones (Z8 and Z9) decreasing 5℃ progressively; and (3) reflow zone Z7 at 260℃ and the remaining zones decreasing 20℃ progressively. The purpose was to investigate the effect of the temperature setting at the reflow zones on the reflow time (time above 217℃) and the peak temperature. Temperature setting (2) resulted in a higher reflow peak temperature, a longer reflow time and the least variation in peak temperature for the various PCB locations (Table 6), i.e., the reflow temperature setting decreasing 5℃ progressively resulted in the desired stable (hat type) temperature profile.

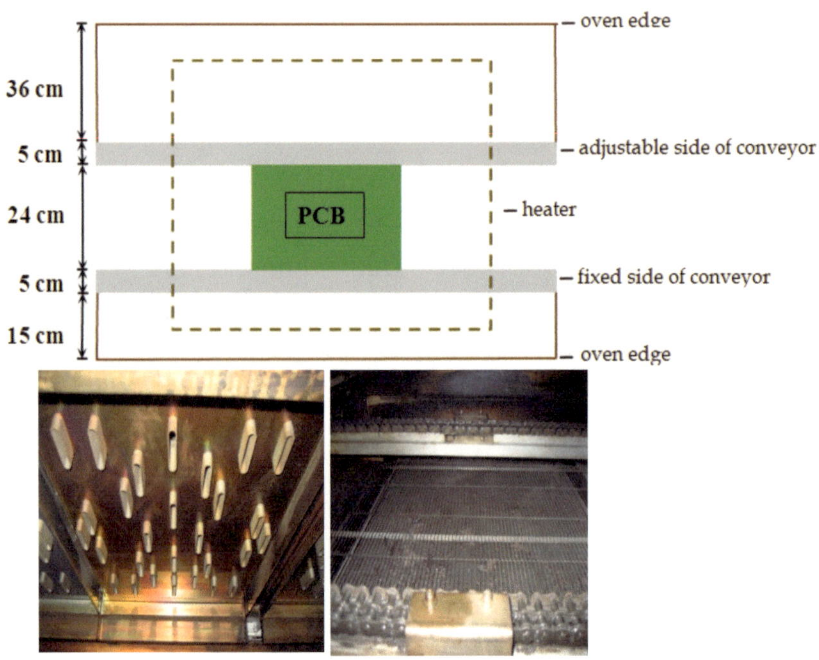

Figure 17. Plot showing location of heater and the PCB

3.2. Reflow process specification

The specifications used in this study are shown in Fig. 19 and described below: (1) the temperature slope increase from 50℃ to 120℃ during the preheat stage should be less than 3℃/s to prevent damaging the PCB/component due to thermal shock; (2) a slow temperature rise at the soak stage reduces temperature variation across the PCB before the reflow soldering commences. A temperature rise from 150℃ to 180℃ should be within 60s to 120s; (3) the time above soldering material SAC305's melting point 217℃ should be within 40s to 90s; and (4)

http://dx.doi.org/10.5772/51392

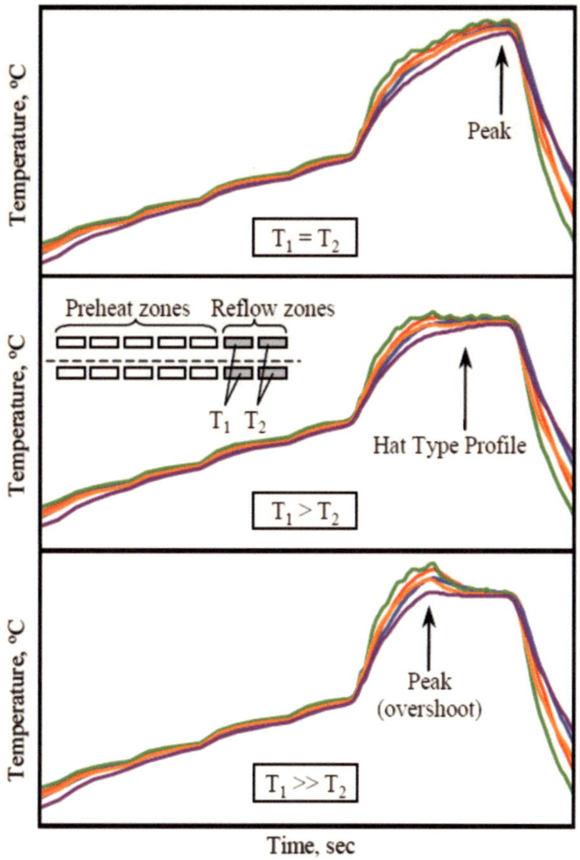

Figure 18. Temperature setting at reflow zone from Cho (2008)

	Temperature setting (1)		Temperature setting (2)		Temperature setting (3)	
	without carrier	with carrier	without carrier	with carrier	without carrier	with carrier
Average Tp(℃)	227	222	230	224	228	220
Reflow time (s)	60	41	71	52	68	44
△T(℃)	6.3	7.8	5.7	6	6.5	10.4

Table 6. Results of the third experiment

rapid cooling is preferred to achieve the desired solder joint metallurgy. The temperature slope from 210℃ to 170℃ should not exceed 3℃/s to reduce thermal shock.

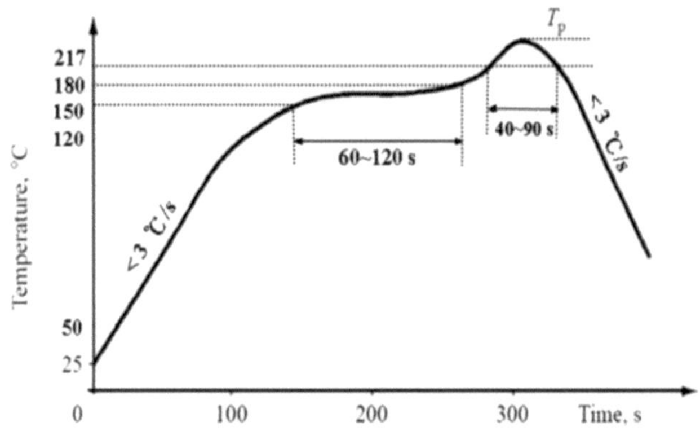

Figure 19. Reflow temperature profile process specifications

3.3. Basic model suggested temperature settings

This study started with calculations of the suggested temperature settings based on the reflow profile prediction algorithm proposed by Gao (2008). The results were compared with the real temperature settings in the manufacturing process so that the measured PCB temperature complied with the process specifications. The result of a sample product is shown in Table 7. This task was then duplicated for a total of twelve server products.

heating zone	Z1	Z2	Z3	Z4	Z5	Z6	Z7	Z8	Z9	Z10	Z11
basic algorithm	70	125	158	174	185	191	221	248	260	248	216
real settings	120	170	180	180	185	185	240	260	260	185	145
difference	50	45	22	6	0	-6	19	12	0	-63	-71

Table 7. Basic algorithm suggested temperature settings versus real settings

3.4. Constructing the function of the correction value through stepwise regression analysis

The temperature settings of twelve server products used in real manufacturing environments were collected. It is noted that the settings were determined by experienced process engineers through trial and error. This was time consuming and iterative modification efforts were required to ensure that the temperatures at critical locations of the PCB were within process specifications. This was essential before the mass-production commenced. Table 7 and Fig. 20 show the basic algorithm suggested temperature settings compared to the real settings. Real temperature settings exhibited longer soak times (in a temperature range of 150℃~180℃) and reflow times (above 217℃). Apparently, PCBs with a higher layer count and loaded with complicated electronic components were more suitable. In contrast, the basic model was only appropriate for the experimental bare board and/or low-end products. Therefore, the aim of this study was to determine the 'correction value' used to modify the suggested settings proposed by the basic algorithm and feasible for high-end products, such as a high-layer count server.

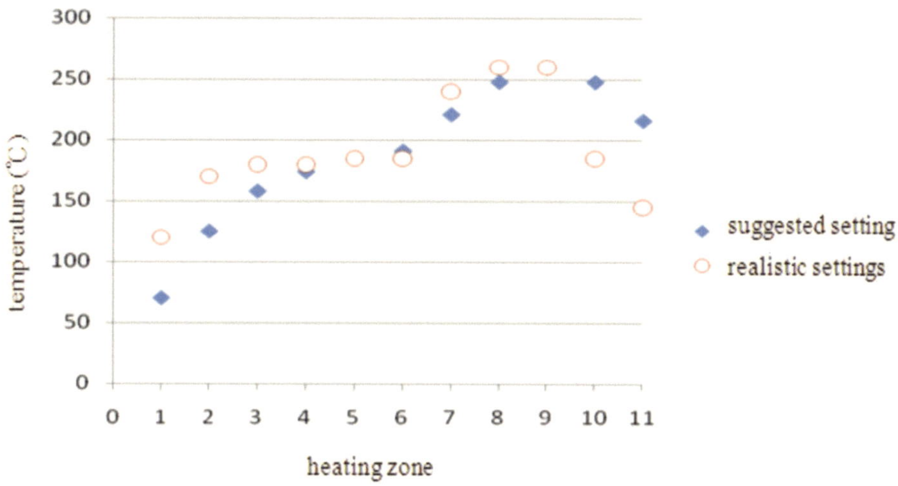

Figure 20. Basic algorithm suggested temperature settings compared to the real settings

The correction value is considered as a function, f(x), of the product and process-related factors, namely, the total PCB assembly's (PCBA's) heat capacity, $H_{total}(m)$; the temperature setting at zone i, x_1; the temperature setting at the previous zone (zone i-1), x_2; the temperature setting at the following zone (zone i+1), x_3; and the conveyor speed, x_4. The PCBA consisted of FR4 (flame retardant-4) material, copper layer and components. The total PCBA heat capacity (J/k) is the product of the materials' mass (g) and corresponding specific heat capacities (J/g-k) (Equation 3). In this study, the mass of PCBA was measured with a digital balance. While CAD Gerber file provided information on the total volume of copper, the mass of copper was determined by multiplying the copper density (1.9 g/cm3). The components mass could thus be determined by deducting the mass of PCBA by the masses of copper and components.

$$H_{total}(m) = m_1c_1 + m_2c_2 + m_3c_3 \tag{3}$$

where m1, m2 and m3 are the masses of FR4, copper and components of a given product; c_1, c_2 and c_3 are 0.6 J/g-k, 0.38 J/g-k and 0.1 J/g-k, respectively, which are the specific heat capacities of FR4, copper and components, respectively.

The function of the correction value was constructed through the data of eleven server products. Data of the other server product was then used for model verification. SRA integrates the forward regression and the backward regression. First, the forward regression brings in the item with the most prediction ability; the backward regression then removes the item(s) that are not significant (expected explanation ability) from the model. The test criteria, partial

F value and type I error, α, used in this study were 4 and 0.15, respectively. The cross validation continued until all the items in the complete function were tested. Statistics software Minitab was used to analyze data of the eleven server products and construct the correction value function for each heating zone. The tenth heating zone, Z10, was used for illustration (Table 8). The complete function included first power items, second power items and two way interaction items (Equation 4). The twenty items were tested for their significance, and the results showed that only item $x_1{}^2$ was brought into the function model.

$$f(x) = \beta_0 + \beta_1 g(m) + \beta_2 X_1 + \beta_3 X_2 + \beta_4 X_3 + \beta_5 X_4 + \beta_6 H(m)^2 + \beta_7 X_1^2 + \beta_8 X_2^2 + \beta_9 X_3^2 + \beta_{10} X_4^2 + \beta_{11} H(m) X_1 + \beta_{12} H(m) X_2 + \beta_{13} H(m) X_3 + \beta_{14} H(m) X_4 + \beta_{15} X_1 X_2 + \beta_{17} X_2 X_3 + \beta_{19} X_2 X_4 + \beta_{20} X_3 X_4 \quad (4)$$

product	difference (℃)	total heat capacity (J / k)	Z10 set temp (℃)	Z9 set temp (℃)	Z11 set temp (℃)	conveyor speed (cm/min)
1	0	286.20	260	260	180	80
2	-13	485.54	240	265	160	75
3	-21	411.48	240	275	160	75
4	16	386.06	265	270	180	82
5	-7	401.34	250	270	180	75
6	-3	242.58	250	260	180	87
7	-2	316.65	250	260	150	70
8	0	191.45	260	255	180	80
9	-63	864.21	185	260	145	68
10	-62	403.69	190	265	145	80
11	11	219.52	260	260	180	73

Table 8. Data to construct correction value function for Z10

The correction value functions for all heating zones determined by SRA are shown in Table 9. Firstly, the setting at Z1 was at 120℃ for an efficient temperature rise and to prevent thermal shock to the PCBA during the preheat stage. The setting at Z9 was in the range of 260℃~275℃ for the desired wetting performance and to limit the exposure of the PCBA in the elevated temperature environment during the reflow stage. Secondly, the effectiveness of the regression functions were determined by the index of model explanation ability, R-sq(adj), and model prediction ability index, R-sq(pred). The R-sq(pred) of the prediction function for the Z11 temperature setting was only 13%, and thus considered an invalid model. For heating zones Z4-Z8, no item was brought into the function with α equal to 0.15. On the other hand, prediction functions for the Z2, Z3 and Z10 temperature settings were valid. Both prediction functions for Z2 and Z3 included the interaction item of 'total heat capacity', $H_{total}(m)$, and 'temperature setting'. The reason for this was as follows. Z2 and Z3 corresponded to the preheat stage of the temperature profile. PCBA at this stage had a significant heat absorption which resulted in a rapid rise in temperature. The efficiency of the heat absorption and the rise in temperature depended on the heat capacity of the PCBA; therefore, the interaction item (between 'heat capacity' and 'temperature setting') appeared as a critical item. Also, the prediction function for Z10 included only the 'temperature setting', x_1. This was because Z9-Z11 corresponded to the reflow stage of the temperature profile. The purpose of Z10 was not to raise the PCBA temperature, but to extend the reflow time duration to ensure the complete wetting of the soldering material onto the bonding pads. The PCBA's heat absorption and rise in temperature were slow; therefore, 'total heat capacity' did not play a critical role in the prediction function. Another server product was used for the verification test. The PCB temperatures were measured with the settings suggested by the baseline model and the SRA model (Table 10). The resulting temperature profiles were then compared to the process specifications. As shown in Table 11 and Fig. 21, the temperature profile resulting from the settings suggested that the SRA model was close to the target.

Heating zone	Function model	R^2(adj.)	R^2(pred.)
Z1	temperature setting at 120℃	-	-
Z2	$y=-37.75+0.0023H(m)x_1-0.00457H(m)x_3+0.00739H(m)x_4$	96%	91%
Z3	$y=-27.04-0.00235H(m)x_3+0.00645H(m)x_4$	78%	61%
Z4~ Z8	no item brought into the function	-	-
Z9	setting in the range of 260℃~275℃	-	-
Z10	$y=-155.7+0.00241x_1^2$	86%	77%
Z11	invalid model	40%	13%

Table 9. Correction value functions by SRA

Heating zone	Z1	Z2	Z3	Z4	Z5	Z6	Z7	Z8	Z9	Z10	Z11
baseline model	92	135	164	182	194	205	230	252	265	252	243
SRA model	(120)	162	167	182	194	205	230	252	265	250	243
ANN model	(120)	180	183	185	187	192	238	250	265	255	245

Table 10. PCB temperatures measured with settings suggested by the baseline model and the SRA model

Temperature profile	Temp rising slope (℃/s) 50℃~120℃	Soak time (s) 150℃~180℃	Peak temp (℃)	Reflow time (time above 217℃) (sec)	Cooling rate 210℃~170 ℃(℃/s)
spec requirement	0~3	60~120	230~25	40~90	0~3
baseline model	1.38	40	234	45	1.06
SRA model	1.53	45	235	46	1.07
ANN model	1.85	72	240	62	1.51

Table 11. Temperature profiles versus process specs

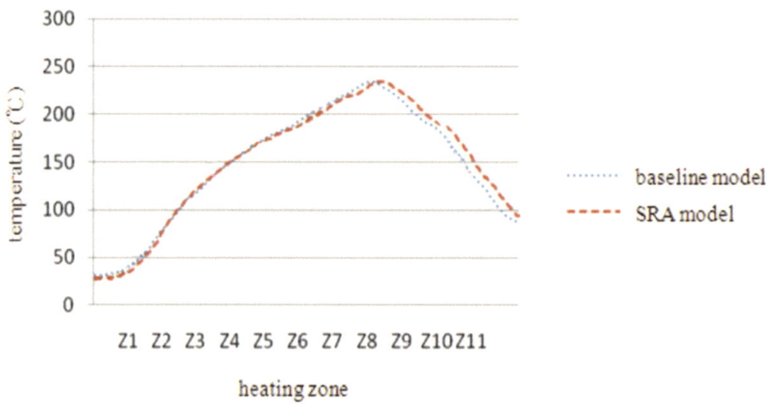

Figure 21. Temperature profiles versus process specs

3.5. Summary

This research characterized the IR-based reflow oven and calculated the desired temperature setting through a baseline algorithm proposed in reference to twelve server samples. Correction values were as suggested by the SRA model. The results of the SRA analysis indicated that the regression functions of heating zones Z2, Z3 and Z10 were valid models with significant prediction ability. Functions of Z2 and Z3 included the item of interaction between 'total heat capacity' and 'temperature setting', while functions of Z10 included only 'temperature setting'. The results of the confirmation tests indicated that the temperature profile resulting from the settings suggested that the SRA model was close to the target.

4. Solderability analysis

Solderability analysis is one of the most important characterizations for PCB assembly (PCBA), quality control and reliability (Huang et al., 2009 and 2011). The analytical techniques used for

http://dx.doi.org/10.5772/51392

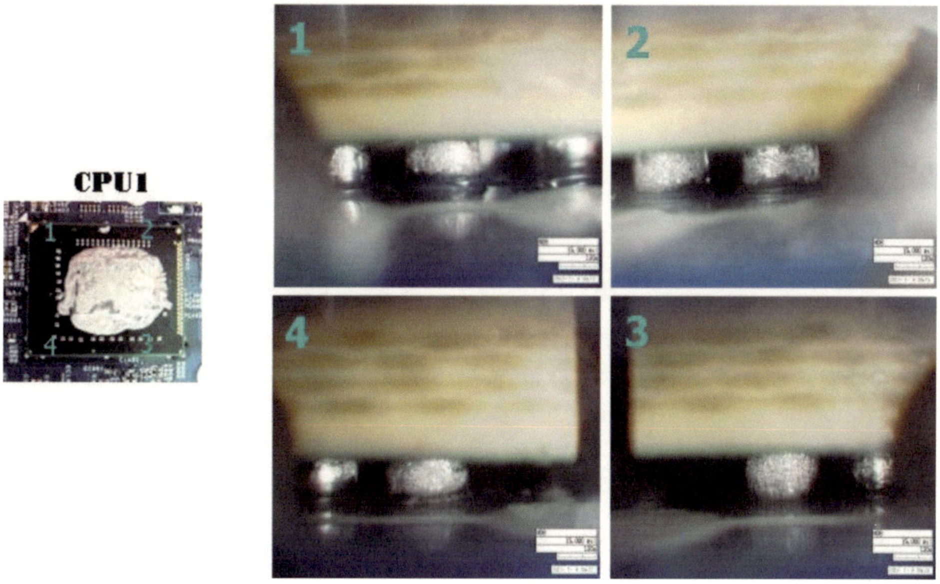

Figure 22. Side-view inspection results from the CPU1 BGA

soldering properties, including visual inspection, side-view microscopy, x-ray inspection and dye staining analysis, were conducted (Huang et al., 2011 and Castello et al., 2006). The crack size percentage was classified according to the crack area. In this case, the solderability analysis was carried out to determine whether any cracks in the solder joints occurred in the CPU and RAM BGAs. The one CPU and eight RAM chips were examined to clarify the suspect ones. It may cause the failure symptom of no display to appear on an ultrabook while the power is on.

4.1. Side-view inspection

Side-view is mainly executed to observe the surface mount components, especially the exterior row BGA solder joints in detail, non-destructively, such as foreign matter, solder joint cracks, BGA shift, missing balls and soldering related defects.

However, for an interior analysis or inspection of BGAs or PCBA components, side-view microscopy has blind spots and should be combined with other analysis methods, such as x-ray, dye staining and cross section, for further confirmation and judgment. In this study, from the side view inspection results, no abnormalities were found for the CPU1 and RAM1~8 BGAs. An illustration of the CPU1 inspection results is shown in Fig. 22. The corresponding and complementary analyses were characterized and are presented in the next sections.

4.2. X-ray inspection

X-ray is mainly executed to examine soldering related defects for BGA, CSP and QFN, etc. X-ray inspection is regarded as a non-destructive method and is used when a high proportion of BGA solder joints is hidden under the IC package. As the internal material of BGA package shows different levels of absorbency, x-rays can determine the solder joint related defects, for example, solder bridges, missing balls, solder voids, insufficient solder or excessive solder volume. The presence and location of solder joint defects under these types of IC package may be determined by conventional 2-D X-ray in a production environment. X-ray inspection has become a common tool for soldering characterization analysis performed to verify product quality (Harrison et al., 2001).

In this study, from the x-ray inspection results, no soldering defects (short, missing balls, etc.) were observed for the CPU1 and RAM1~8 BGAs. An illustration of the CPU1 inspection results is shown in Fig. 23. Voids may exist in the solder joints and make surface mount area arrays risky. According to IPC-A-610E (IPC-A-610E 8.3.12.4, 2010), the acceptance criterion for voids in solder joints is 25% or less voiding of the ball x-ray image area. In this case, void percentages were all within the inspection criterion (<25% ball x-ray image area). While some critical defects in BGA solder joints were not readily detectable by the side-view microscope and/or the x-ray inspections, dye staining analysis was carried out for more accurate soldering analysis.

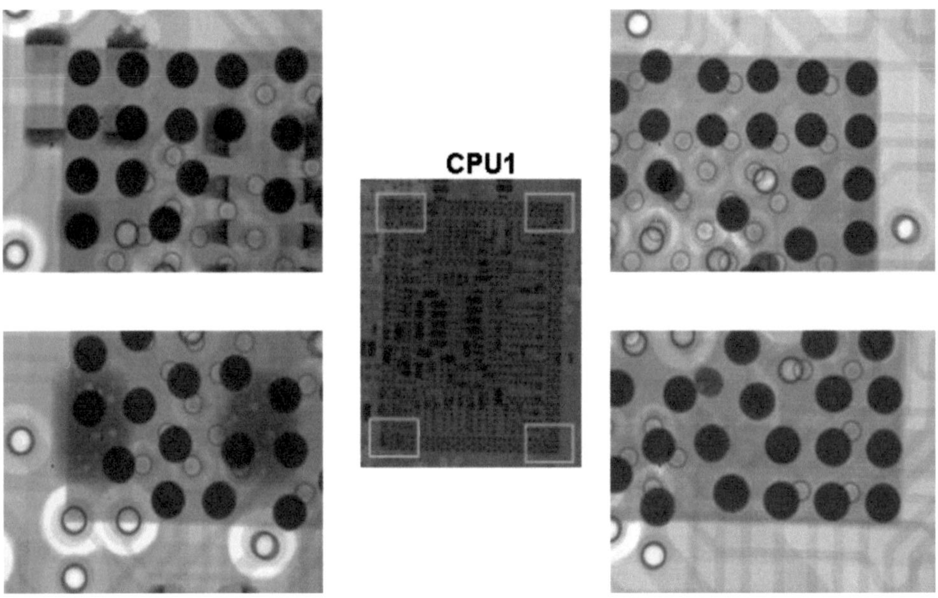

Figure 23. X-ray inspection results from the CPU1 BGA

4.3. Dye staining analysis

The dye staining analysis is commonly applied to verify whether solder joint cracks have occurred under BGA packages (Liu et al., 2010 and Lau et al., 2004). The dye staining technology, which is a destructive test method used to reveal defects on the solder balls, relies on a liquid dye that penetrates existing micro cracks or under open solder balls. After the dyeing process, a heating process is used to dry the dye followed by prying off the BGA. The presence of the dye on solder balls shows the location of the failure within the solder joint and the degree of crack propagation. It is a commonplace method applied to grid array soldering problems (Huang et al., 2011).

Since RD can not identify the failure locations, dye staining analysis is performed to determine the possible causes for the no display issue when the ultrabook power is on. If the failure locations can be identified, cross section analysis is suggested to obtain more failure symptom information from other aspects.

In this study, from the dye staining analysis, solder joint cracks were found in the CPU1 (corners 2 and 3) and RAM2 (corner 4) BGAs (Figs. 24 and 25). For CPU1, the failure mode was located between PCB pad/PCB laminate (Type 5) and the crack size percentage was between 1-25%. For RAM2, the failure mode was also Type 5 and the crack size percentage was 26-50%. Cracks are one of the most important reasons for the failure symptom showing no display for the integrated chips and board in an ultrabook product.

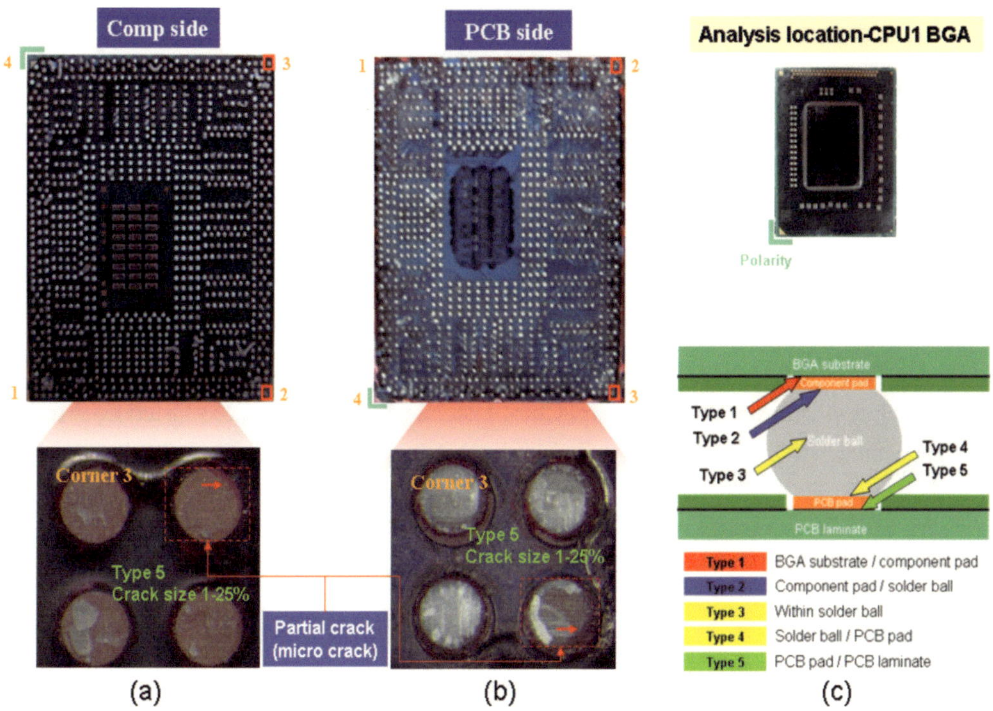

Figure 24. Dye staining analysis results from the CPU1 BGA (a) component and (b) PCB side (corner 3), (c) failure mode

4.4. Summary

In this study, visual inspection, side-view microscopy, x-ray inspection and dye staining analysis were successfully used to characterize the soldering quality associated with microelectronics assembly. Side-view microscopy and x-ray inspection were used for a preliminary screening for failure symptoms. Consequently, dye staining analysis was used to identify the

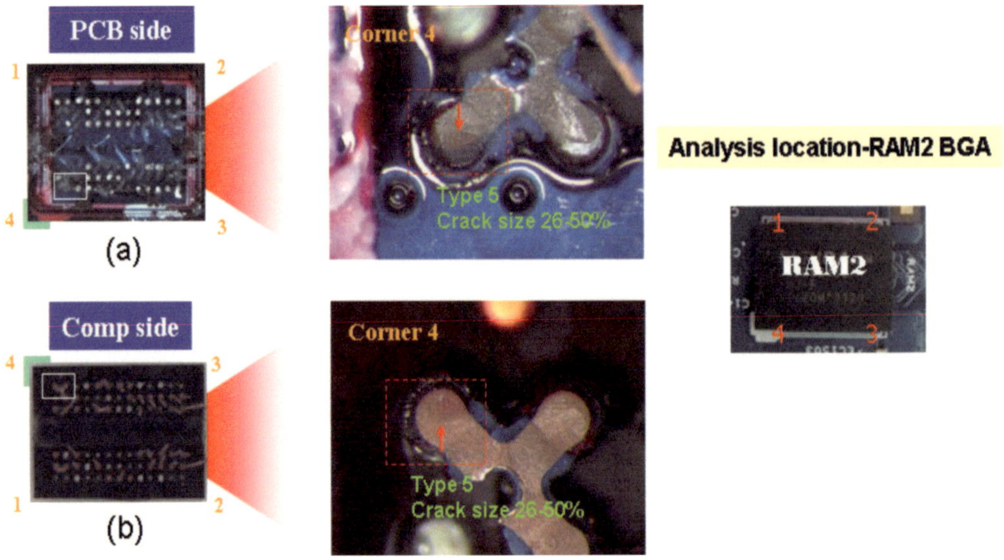

Figure 25. Dye staining analysis results from the RAM2 BGA (a) PCB and (b) component side (corner 4)

failure mode and crack size accurately. This method facilitated determining the failure causes for most soldering problems.

Author details

Chien-Yi Huang[1], Chen-Liang Ku[2], Hao-Chun Hsieh[2], Ming-Shu Li[2], Chia-Hsien Lee[2] and Cheng-I Chang[2]

1 Department of Industrial Engineering and Management, National Taipei University of Technology, Taipei, Taiwan, R.O.C.

2 Process Technology Enabling & Materials Characterization Div., Global Operations, Wistron Corporation, Hsinchu, Taiwan, R.O.C.

References

[1] Castello, T., Rooney, D. & Shangguan, D. (2006). Failure analysis techniques for lead-free solder joints. *Soldering & Surface Mount Technology*, Vol. 18, No. 4, pp. 21-27.

[2] Clark, D. & Rohrer, D. (2003). Specification and qualification of machine vision technology in the SMT process, *Technical Paper, GSI Lumonics.* pp.1-8.

[3] Gangue, Z. & Wang, C. (2009). Shape and fatigue life prediction of chip resistor solder joints, *International Conference on Electronic Packaging Technology & High Density Packaging (ICEPT-HDP).*

[4] Gao, J.G., Wu, Y. P., Ding H. & Wan, N.H. (2008). Thermal profiling: a reflow process based on the heating factor. *Soldering & Surface Mount Technology,* Vol. 20, No. 4, pp. 20-27.

[5] Harrison, M.R., Vincent, J.H., Steen, H.A.H. (2001). Lead-free reflow soldering for electronics assembly. *Soldering & Surface Mount Technology,* Vol. 13 Iss: 3, pp. 21–38.

[6] Huang, C.Y., Li, M.S., Ku, C.L., Hsieh, H.C. & Li, K.C. (2009). Chemical characterization of failures and process materials for microelectronics assembly. *Microelectronics International,* Vol. 26, No. 3, pp. 41-48.

[7] Huang, C.Y., Li, M.S., Huang, S.Y., Chang, C.I. & Huang, M.H. (2011). *Material Characterization and Failure Analysis for Microelectronics Assembly Processes*. Wide Spectra of Quality Control, InTech publication, ISBN 978-953-307-683-6, Chapter 26, pp. 509-532.

[8] Huang, C.Y., Lin, C.T., & Tsai, H.L. (2004). Study of Process Capability for The Stencil Printing in PCB Assembly. *Fifth Asia-Pacific Industrial Engineering and Management Systems Conference*, Gold Coast, Australia.

[9] IPC-7525 3.2.1.1 (2000). Area Ratio/Aspect Ratio, pp. 3.

[10] IPC-A-610E 8.3.12.4 (2010). Surface Mount Area Array – Voids, pp. 8-89.

[11] Jianbiao, P., Tonkay, G.L., Storer, R.H., Sallade, R.M. & Leandri, D.J. (2004). Critical variables of solder paste stencil printing for micro-BGA and fine-pitch QFP. *IEEE Transactions on Electronics Packaging Manufacturing,* Vol. 27, No. 2.

[12] Lau, J., Shangguan, D., Castello, T., Horsley, R., Smetana, J., Hoo, N., Dauksher, W., Love, D., Menis, I. & Sullivan, B. (2004). Failure analysis of lead-free solder joints for high-density packages. *Soldering & Surface Mount Technology,* Vol. 16, No. 2, pp. 69-76.

[13] Liu, F., Meng, G. & Zhao, M. (2010). Experimental investigation on the failure of lead-free solder joints under drop impact. *Soldering & Surface Mount Technology,* Vol. 22, No. 3, pp. 36-41.

[14] Rumelhart, D.E. & McClelland, J.L. (1986). *Parallel Distributed Processing: Explorations in the Micro-structure of Cognition*. Cambridge: MIT Press, Vol. 1.

[15] Sarvar, F. & Conway, P.P. (1998). A modeling tool for the thermal optimization of the reflow soldering of printed circuit assemblies. *Finite Elements in Analysis and Design,* Vol. 30, pp. 47-63

[16] Su, Y.Y., Srihari, K. & Emerson, C.R. (1997). A profile identification system for surface mount printed circuit board assembly. *Computers and Industrial Engineering,* Vol. 33, pp.377-380.

[17] Tsai, T.N. (2008). Modeling and optimization of stencil printing operations A comparison study. *Computers & Industrial Engineering,* Vol. 54, pp. 374–389.

[18] Whalley, D.C. (2004). A simplified reflow soldering process model. *Journal of Materials Processing Technology,* Vol. 150, pp. 134-144.

[19] Yeh, I.C. (2009). Taguchi method based on neural networks and cross validation methodology. *Journal of Quality,* Vol. 14, No. 4.

Chapter 13

Tensile Mechanical Properties and Failure Behaviors of Friction Stir Processing (FSP) Modified Mg-Al-Zn and Dual-Phase Mg-Li-Al-Zn Alloys

Chung-Wei Yang

Additional information is available at the end of the chapter

http://dx.doi.org/10.5772/54313

1. Introduction

With the gradually depletion of natural resources, especially for the petrochemical fuel in the Earth, and the global warming problems, the energy-saving, increasing fuel-efficiency and eco-technologies have become worldwide interested subjects of nowadays. Recently, alternative green energies, such as solar energy/cells, fuel cells, Li-ion batteries, biomass fuel and algae fuel, have been developed and used as eco-friendly energies. In addition, materials choosen to give high performance in use with minimized impact to the environment during manufacture and delivery is another important concept to take into consideration. For the purpose of reducing fuel consumption, therefore, lightweight metallic materials are generally applied to replace traditional ferrous structural metals in the transportation vehicles

The definition of "Light metals" has typically been given to metallic materials with which density less than 4.5 g cm^{-3}. On this basis, the alkali metals for lithium (Li, 0.54 g cm^{-3}), sodium (Na, 0.97 g cm^{-3}), potassium (K, 0.86 g cm^{-3}), rubidium (Rb, 1.53 g cm^{-3}), cesium (Cs, 1.9 g cm^{-3}), and alkaline earth metals for berylium (Be, 1.85 g cm^{-3}), magnesium (Mg, 1.74 g cm^{-3}), calcium (Ca, 1.55 g cm^{-3}), strontium (Sr, 2.63 g cm^{-3}), barium (Ba, 3.51 g cm^{-3}) can be thought of light metals. In addition, the core group of metals, such as aluminum (Al, 2.7 g cm^{-3}) and titanium (Ti, 4.5 g cm^{-3}), are also classed as light metals. Considering for the engineering applications of these metals in structural components, Mg, Al, Ti metals and their alloys are the generally recognized as light metals in inductrial applications with respect to other engineering alloys because the specific strength (i.e., a strength-to-density ratio) is an important factor to be concerned with the weight reduction. The advantages of decreased density become important in engineering design when mechanical properties of stiffness and resistance to buckling are involved. Comparative stiffness for equal weights of a similar beam increase in the ratios 1:2.9:8.2:18.9 for steel, Ti, Al and Mg, respectively.

Lightweight magnesium (Mg), which has the lowest density (1.74 g cm^{-3}) of any metallic constructional material in inductrial applications [1,2]. It has many advantages such as high specific strength, high specific stiffness, well recyclability and radiation absorption of electromagnetic waves. It also provides high damping capacities, high thermal and electrical conductivities [1,3]. These effects combine to make Mg alloy be an efficient heat sink with good noise and vibration reduction properties. Researches and developments of Mg and its alloys are expanded during the past decade. They are now widely used in consumer electronic industries, biomedical and sports applications. There are numerous applications within the automobiles and aerospace [4,5] of nowadays for the purpose of reducing vehicle weight and fuel consumption [1,6,7]. However, the more widespread commercial applications of Mg alloys are limited due to its poor formability at room temperature arising from its hexagonal close-packed (hcp) crystal structure. Therefore, it is important to develop the Mg alloys having excellent mechanical properties, especially for applications at elevated temperatures. Additionally, the welded structure is inevitably used in the engineering application of wrought Mg alloys. The welding of Mg alloys can be achieved by using TIG, MIG, resistance spot welding (RSW), laser beam welding [8,9] and vacuum electron beam welding [10] processes. But the Mg-based alloys are generally of poor weldability because of their exceptional thermal and electrical properties and oxidation characteristics. In 1991, a novel friction stir welding (FSW) process was invented by The Welding Institute (TWI) of UK [11]. FSW is a solid-state joining technique and the joining temperature is lower then that in the melting welding process. It uses a high speed rotating tool consisting of a pin and tool shoulder to apply frictional heat and severe plastic deformation of the base metal and to produce a reliable metallurgical joint. Compared with the traditional melting welding techniques, FSW has many advantages [12-14] and it is quite suitable for the welding of Mg alloys and other lightweight aluminum (Al) and titanium (Ti) alloys [15-19]. In recent years, numbers of researches have been conducted to evaluate the feasibility of FSW process for Mg aloys. Moreover, the joining of dissimilar metals of Mg, Al, Cu and Ti can also be easily achieved by the FSW process [20-24].

The present chapter focuses on investigating the modification effect of friction stir process (FSP) on the tensile mechanical properties and failure behaviors of Mg-Al-Zn (AZ-series) and Mg-Li-Al-Zn and LAZ-series alloys. Section 2 will give a overview for the recently researches and applications of the these Mg-based alloys and the FSW-joining/FSP-modification techniques. In section 3, extruded Mg-6Al-1Zn (AZ61) alloys with full-annealed (AZ61-O) and FSP-modified (AZ61-FSP) specimens were used for evaluating the strain rate and texture effects on the tensile properties and failure behaviors. In section 4, we focused on investigating tensile mechanical properties and failure behaviors of the α/β-dual-phase Mg-10Li-2Al-1Zn (LAZ1021) extruded alloy. A significant ductile-to-brittle transition phenomenon was confirmed for the dual-phase Mg-Li alloy. Since the phase composition and crystallographic structure vary significantly according to Li content, it is important to investigate the mechanical properties, as well as deformation and failure behaviors of Mg-Li alloys with performing the FSP modification. The aim in section 5 also focuses on improving tensile mechanical properties and evaluating failure behaviors of compositional-stabilized dual-phase LAZ1021 extruded alloy through the modification of FSP technique.

2. Overview of magnesium-based alloys and friction stir process

2.1. Lightweight magnesium (Mg) alloys

Mg alloys can generally be divided into casting alloys and wrought alloys. The casting Mg alloys include die casting, sand casting and thixomolding alloys, and wrought Mg alloys are available as bars, forgings, extrusion sheets and plates. Because of the difficulty of cold working Mg alloys, the castings are more prevalent product (account for about 85-90%) of Mg alloys usage than wrought products. Die casting is one of the most effective fabrication methods and has been extensively used to produce Mg components. Mg alloys are designated by a combination of two capital letters followed by two numbers. The first two letters indicate the two major alloying elements in the alloy. The following two numbers stand for the approximate amounts (in weight percentage, wt.%) for the first and second alloying elements, respectively. The Mg alloy designation of alloying elements is indicated according to the following code: A-aluminum, B-bismuch, C-copper, D-cadmium, E-rare earth elements, F-iron, H-thorium, K-zirconium, L-lithium, M-manganese, N-nickel, P-lead, Q-silver, R-chromium, S-silicon, T-tin, W-yttrium, Y-antimony and Z-zinc. For example, the alloy AZ91contains approximately 9 wt.% Al and 1 wt.% Zn. The present section will give an overview of the generally used Mg-Al based, Mg-Zn based alloys, Mg-Li based alloys and rare earth elements (RE)-containing Mg alloys. In addition, topics of Mg alloys applied as orthopaedic, dental implants and surface treating methods for corrosion resistance are also introduced in the end of this section.

2.1.1. Mg-Al based alloys

The Mg-Al alloy system are the most popularly used Mg alloys. Aluminum (Al) is alloyed with Mg because it increases strength, castability and corrosion resistance. The addition of Al also provides solid solution strengthening by increasing the critical resolved shear stress for slip along the basal planes of Mg-Al alloys. Fig. 1(a) shows the equilibrium phase diagram of a binary Mg-Al alloy system. Since Al has a maximum solubility in Mg of 12.7 wt.% at 437°C and its solubility decreases to about 2 wt.% at room temperature, it is expected that this system can be strengthened by precipitation hardening when the addition of Al in amount greater than 6 wt.%. However, a coarse, incoherent precipitate of the γ-$Mg_{17}Al_{12}$ phase intermetallic compound that discontinuously precipitated around grain boundaries without the formation of GP zones or intermediate metastable phase [25]. Since Mg alloys undergo creep mainly through the grain boundary sliding (GBS), the γ-$Mg_{17}Al_{12}$ phase, which has a melting point of about 460°C, can not serve to pin grain boundaries at elevated temperatures, and then the creep resistance of Mg-Al alloys is reduced.

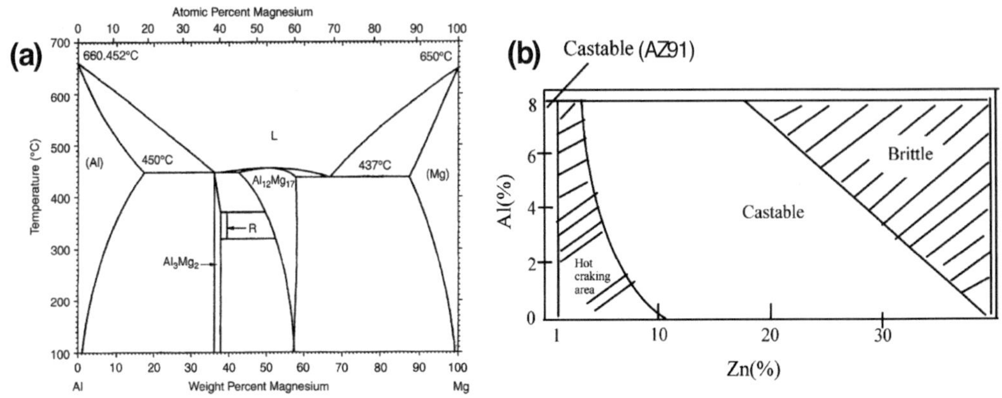

Figure 1. (a) Mg-Al equilibrium phase diagram, and (b) castability of the Mg-Al-Zn alloy system [26].

Zinc (Zn) is the second most important alloying elements, and the Mg-Al-Zn alloys system, which is known as the AZ-series, is the one most widely used for castings. Generally, Mg alloys are limited to a total Al and Zn content of less than 10 wt.% (the narrow region in Fig. 1(b) [26]) for obtaining good castability and weldability, whereas the hot cracking is occurred and the ductility of Mg is significantly reduced with the formation of γ-phase $Mg_{17}Al_{12}$. The castable domain of high-Zn content Mg alloys is obtained with controlling suitable Al content and the Zn/Al ratio, as shown in Fig. 1(b). When Zn is added to the Al-containing Mg alloys, it can help to refine the precipitates. It increases the strength by a combination of solid solution strengthening and precipitation hardening because Zn is approximately three times more effective than Al in increasing the yield strength for alloys that are in the solution treated and quenched condition. Zn also assists to overcome the harmful corrosive effect of iron (Fe) and nickel (Ni). For the casting Mg alloys, AZ91D is the most widely used die casting Mg alloy, and other commonly used alloys of this type are AZ81 and AZ63 alloys. AZ91D alloy offers a good room temperature mechanical properties, corrosion resistance and die castibility, however, it is unsuitable for use at temperatures higher than 120°C due to its poor creep resistance. The Mg-Al and Mg-Al-Zn alloys with the addition of manganese (Mn) element are known as the AM-series alloys, such as AM60B and AM50A. They display better ductility and fracture resistant property than AZ91 alloy, and these alloys are often used for cast wheels on automobiles. Moreover, adding sufficient Mn element can help to improve the corrosion resistance by removing Fe and other heavy-metal impurties from the alloy.

Wrought Mg alloys are much less applied than castings. A relatively limited number of the AZ-series (for AZ31, AZ61 and AZ80) and ZK-series (for ZK60) alloys are used for most extrusion and forging Mg products. AZ31B is the most widely used alloy in plates and sheets. Since the room temperature formability of Mg alloys is limited, heavy deformation must be performed at above 250°C. In addition, even if wrought Mg alloys are strengthened by a combination of cold working, grain refinement and solid solution strengthening, they still relatively limited use due to the strength is much lower than medium-strength Al-Mg-Si based alloys and the softening at elevated temperatures. Thus, the attempts to obtain high-strength wrought Mg alloys are made and some alloys have been developed [27].

2.1.2. Mg-Zn based alloys

Mg-Zn alloys respond to age hardening, which is associated with two intermediate precipitates of β_1'-$MgZn_2$ phase (rods, coherent) and β_2'-$MgZn_2$ phase (discs, semicoherent). The final equilibrium precipitate is β-Mg_2Zn_3 phase (incoherent). Since AZ91D alloy is unsuitable for applications at elevated temperatures due to its decrease in strength and poor creep resistance [28,29], Mg-Zn-Al alloys (designated as the ZA-series) with a high Zn content is developed as a promising alloy system, which is able to improve the above requirements [29-31]. Fig. 2(a) shows the ternary phase diagram of Mg-rich Mg-Zn-Al alloy, the ternary intermetallic compounds of τ-phase ($Mg_{32}(Al,Zn)_{49}$), Mg-rich φ-phase ($Mg_5Zn_2Al_2$) and the binary intermetallic compound of ε-phase (MgZn) are found in the Mg-Zn-Al system. In addition, a ternary icosahedral quasi-crystalline phase, denoted as the Q-phase, is also identified in the ZA-series Mg alloys [30,32]. Zhang et al. indicated that the change of microstructral constituent of alloys is dominated by both the content of Zn, Al elements and the Zn/Al ratio [32]. As shown in Fig. 2(b), ZA-series alloys with a high Zn/Al ratio and a low Al content fall into the ε-phase, and alloys with an intermediate Zn/Al ratio and an intermediate Al content display the τ-phase. Those alloys with a low Zn/Al ratio and a high Al content are dominated by the icosahedral quasi-crystalline Q-phase with an approximate composition of $Mg_9Zn_4Al_3$, and such phase shows a short-range order of the crystalline state. Althought the ZA-series alloys show a lower elongation than the AZ91, however, it is noted that the creep resistance of ZA-series alloys is significanly higher than the AZ91 alloy because no γ-$Mg_{17}Al_{12}$ phase is formed. Moreover, the elevated temperatures creep resistance property of the τ-type and the Q-type ZA-series alloys are superior to that of the ε-type alloys [32].

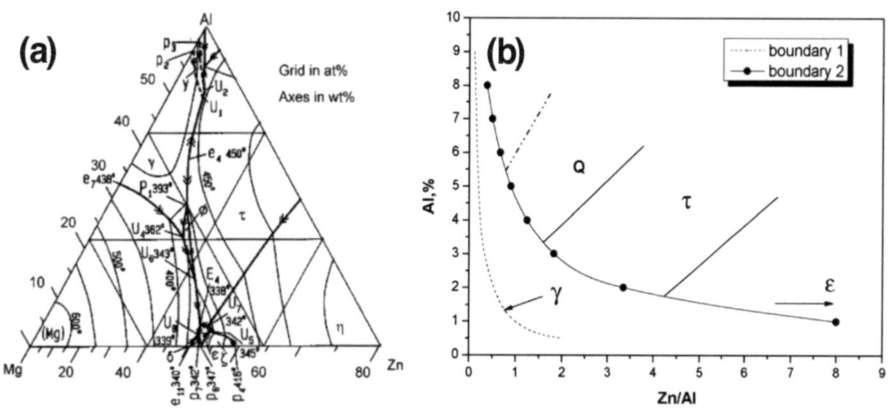

Figure 2. (a) Liquidus projection of the Mg-rich Mg-Zn-Al ternary phase diagram, and (b) schematic illustration of microstructural constituent as a function of the Zn/Al ratio and the Al content [32]. Boundary 1 and 2 are the two boundaries of the hot cracking area in Fig. 1(b).

The Mg-Zn alloys with the addition of zirconium (Zr) element are known as the ZK-series alloys, such as ZK51 and ZK61 casting alloys. Zr is added to Mg to reduce the harmful effects of Fe and Si elements on corrosion performance. Zr is used in casting alloys for grain refinement, and the strength of ZK-series alloys is higher than AZ-series alloys. However, they are not widely used due to they are vulnerable to microporosity during casting, and they cannot be welded for their high Zn content.

2.1.3. Mg-Zn-RE alloys

For aerospace and emerging automotive applications, Mg alloys with high strength, wear and creep resistance are required for engine blocks, steering wheels and transmission cases. Since the formation of low-melting $Mg_{17}Al_{12}$ eutectic at grain boundaries sharply degrades creep resistance, the creep resistance of cast Mg-Al-based alloys will be improved when the amount of discontinuous $Mg_{17}Al_{12}$ precipitates in grain boundary regions is reduced by adding minor elements. For example, the addition of Si can form the a new stable Mg_2Si precipitate that increase the creep strength by suppress grain boundary sliding at elevated temperatures. Reports indicated that Mg-Al-based die-cast alloys containing Ca, Sr or Si elements are now available for automotive applications [33-36]. In addition, rare earth elements (RE) containing Mg alloys exhibit improved corrosion resistance, excellent high-temprature strength and creep resistance. Therefore, the aim of recent researches is to improve the properties and broaden the structural applications of Mg alloys by alloying with RE elements. The RE elements have been shown to have a wide range of benefits for Mg alloys. It is recognized that the addition of RE elements, such as yttrium (Y), lanthanum (La), neodymium (Nd) or cerium (Ce), can improve the high temperature mechanical properties and creep strength of Mg-Al and Mg-Zn based alloys [37-46]. The Mg-Y-Nd alloys for WE54 and WE43, Mg-2.5wt.%Ag-2.0wt.%RE (QE22), Mg-RE-Zn alloys for EZ33 and EZ41 are also creep-resistant alloys which can be used for aerospace applications [47-49].

Recently, high strength Mg-Zn-RE (the alloying RE elements are Y or Gd) alloys with a long period stacking ordered (LPSO) phase were developed. The LPSO phase leads to the excellent mechanical properties. The 0.2% proof yield strength of RE-containing Mg-Zn alloys can exceed 600 MPa when they are produced by rapid solidification process [50], and a value of over 470 MPa has been obtained for the 0.2% proof yield strength in an Mg-1.8Gd-1.8Y-0.7Zn-0.2Zr alloy produced by conventional hot extrusion [51]. It is recognized that the Mg-Zn-Y alloys consisting α-Mg and LPSO phases have higher strength [52,53] and superior heat resistance properties as compared to conventional Mg alloys, and the plastic deformation behavior of $Mg_{97}Zn_1Y_2$ alloy has been investigated [54,55]. In the Mg-Zn-Gd alloys, LPSO phase precipitates from α-Mg supersaturated solid solution with annealing at high temperature results in strengthening of the $Mg_{97}Zn_1Gd_2$ alloy [56]. $Mg_{96.5}Zn_1Gd_{2.5}$ alloys with an LPSO structure exhibited high tensile yield strength (345 MPa) and large elongation (about 7%) due to the refinement of α-Mg grains and the high dispersion of a hard LPSO structure phase [57]. Although Mg-Zn-RE alloys with alloy-reinforcing LPSO phase precipitates display superior mechanical properties, this two-phase structure negatively affects corrosion resistance between α-Mg and LPSO phases. The corrosion resistance of Mg-Zn-Y alloy can be improved with the addition of Al element, and it is expected to obtain oxide layer reforming in Mg alloys and improves corrosion resistance without changing the structure [58]. However, this Mg alloys system still require a larger quantity of adding RE elements, and the drawback is that RE elements are obtained from few sources and relatively expansive for a widespread commercial application.

2.1.4. Mg-Li based alloys

Mg displays the lowest density (1.74 g cm^{-3}) of any constructional metal, and lithium (Li) element (0.53 g cm^{-3}) containing Mg-Li alloys show lower density (1.4-1.6 g cm^{-3}) than commonly used AZ-series and ZK-series alloys (the density is about 1.8-1.9 g cm^{-3}). The density of Mg-Li alloys can even lower to about 1.3 g cm^{-3} when the addition of Li content is higher than 40 wt.% [59]. The Mg-Li alloys system has attracted attention as a basis for ultra-lightweight materials, which can be a good candidate for making components for aerospace vehicles [60], such as the skin of fuselage, wings and landing frame. Adding Li to Mg alloys can transform the hcp structure to a body-centered cubic (bcc) structure, substantially increasing the ductility of the Mg-Li alloys and further reducing its density [59]. According to the equilibrium Mg-Li phase diagram as shown in Fig. 3 [61], Li has a high solid solubility in Mg. When Li content is less than 5 wt.%, only Mg-rich α-phase (hcp) exists. Mg alloying with the addition of about 5-11 wt.% Li content exhibits a dual-phase eutectic crystal structure, which consists of a Mg-rich α-phase and a Li-rich β-phase (bcc). A bcc single-phase Mg-Li alloy is obtained if Li content of more than 11 wt.% is added. Mg-Li alloys have excellent formability and better vibration resistance than commonly used wrought AZ-series alloys [59,62,63]. The superplasticities of Mg-Li based alloys were also investigated [64-68]. But Mg-Li alloys exhibit low mechanical strength and they are not very useful in engineering applications. Recent studies investigated the effects of cold working, addition of alloying elements, aging and precipitation hardening on improving mechanicals of Mg-Li alloys [69-78]. Al and Zn are the commonly used elements, and Mg-Li-Al (LA-series), Mg-Li-Zn (LZ-series) and Mg-Li-Al-Zn (LAZ-series) are generally applied Mg-Li alloys. It is recognized that the precipitation of θ'-phase ($MgLi_2X$, X = Al or Zn) in the β-phase results in the aging hardening effect of Mg-Li-(Al,Zn) alloys. The over aging is attributed to the precipitationof the α and θ-MgLi(Al,Zn) phases [74].

In orthopaedic and dental fields, commonly used stainless steel (e.g., ASTM F-138/139a, type 316L SS), cobalt-chromium-based alloys (e.g., ASTM F-799: Co-28Cr-6Mo, hot forged, ASTM F-562: Co-35Ni-20Cr-10Mo, cold worked and aged) and titanium-based alloys (Ti6Al4V-ELI, ASTM F-136) play an essential role as metallic biomaterials to assist with the replacement of hard tissues. However, limitations of these commonly used metallic biomaterials are the possible release of toxic metallic ions [79-81] and the stress shielding effects [82] resulted from their much higher elastic moduli than human bone tissues. Besides the applications in automotive, aerospace and electronics industry fields, Mg-based alloys have been identified as potential lightweight metallic implants in the orthopaedic field due to their high specific strength and much lower elastic modulus [83-85]. The elastic modulus of Mg alloys (about 41-45 GPa) is closer to the natural human bone (about 3-20 GPa) and much lower than commonly used Ti alloys (about 110 GPa) [85]. In addition, the fracture toughness of Mg alloys (about 15-40 MPa $m^{1/2}$) is higher than dense HA bulks. *In vivo* studies also indicated that Mg, which is a basic element in the growth of new bone tissue, metabolism and an essential element of the enzyme system in human body [86-89], are gradually applied as biodegradable, load bearing orthopaedic implants [84]. Therefore, several Mg alloy systems, such as commonly used Mg-Al-Zn series (e.g., AZ31B and AZ91D), Mg-Ca, Mg-Zn, Mg-Zr, Mg-4wt.%Y-3wt.%RE (WE43, RE is a mixture consisting of Nd, Ce and Dy) and Mg-4wt.%Li-4wt.%Y-2wt.%RE (LAE442, RE is a mixture consisting of Ce, La and Nd) alloys [1,83-85,90-94], have been investigated and developed for using in biomedical applications in recent years. However, Mg alloys have high electrochemical activity, and the major drawback of Mg alloys in many applications is their low corrosion resistance, especially in an electrolytic aqueous environment of the human body. Thus, biomedical usage of Mg alloys is not widespread.

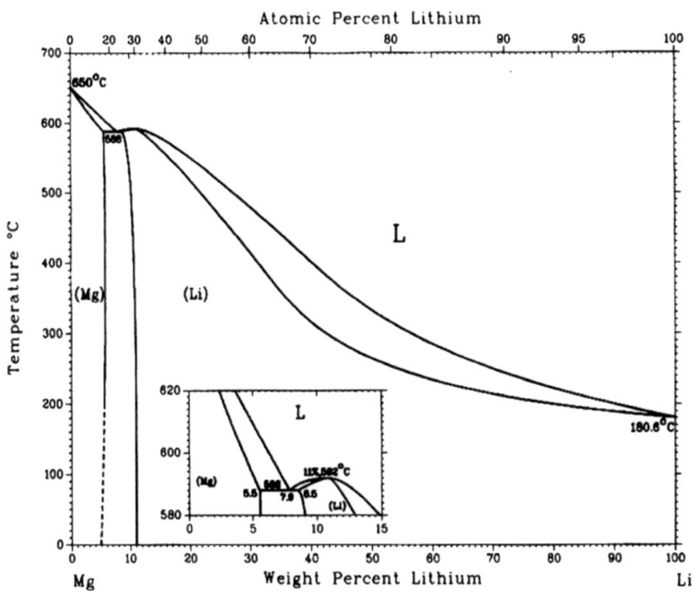

Figure 3. Mg-Li equilibrium phase diagram [61].

Mg alloys are susceptible to corrosion. An appropriate alloying composition can improve their mechanical properties and corrosion resistance. Mg alloys with moderate corrosion resistance and improved mechanical properties can be obtained by adding 2-10 wt.% Al with trace addition of Zn and Mn elements. The corrosion resistance and mechanical properties can also be further enhanced with adding a small amount of RE elements, such as Y, La, Nd, Ce, Zr and etc. [84]. Surface treatments, including chemical conversion coatings, anodizing, electrochemical plating, electroless nickel plating and etc. [95-97], are commonly used methods to enhance the corrosion resistance. Among these various surface treatments, the chemical conversion of Mg alloys is easier to perform for improving corrosion resistance, and chromate conversion coatings have attracted much interest due to the simplicity of the coating process and the good corrosion resistance that offers [98]. However, environmental toxicity issues of the hexavalent chromium (Cr^{6+}) compounds are increasingly restricting the use of the chromate bath [99,100], and Cr is also a toxic element to the human body. Thus, a Cr^{6+}-free conversion coating process is developed for the surface treatment of Mg alloys. Relative studies indicate an environmentally clean method can be achieved for synthesizing a chemical conversion coating on Mg alloys. Lin et al. [101,102] reported that an Mg,Al-hydrotalcite ($Mg_6Al_2(OH)_{16}CO_3{\bullet}4H_2O$) layer was developed on die cast AZ91D in an HCO_3^-/CO_3^{2-} aqueous solution to protect the alloy against corrosion. The hydrotalcite shows a crystalline structure, which has Mg,Al-layered double hydroxides $(Mg_6Al_2(OH)_{16})^{2+}$ intercalated by interlayers $(CO_3\ 4H_2O)^{2-}$ [103]. It is demonstrated that the corrosion rate of Mg,Al-hydrotalcite coated specimens is evidently lower than that of the AZ91D substrate in a chloride solution [104].

2.2. Friction stir welding/processing (FSW/FSP) techniques

Mg alloys can be gas welded with an oxyacetylene torch and required careful fluxing to minimize oxidation. This welding process is quite difficult, and extensive corrosion of welds is occurred when the flux is incomletely removed by the applied cleaning methods. Thus, virtually the welding of Mg alloys has been done using inert gas shielded tungsten arc welding (TIG) or consumable metal electrode arc welding (MIG) processes. In these processes, a continuously fed Mg alloy wire acts as electrode for maintaining the arc while the Ar gas shield prevents oxidation of the welds. Resistance spot welding (RSW) is applied for joining steels, Al, Mg alloys in the automotive industries of nowadays [104-108]. But RSW is relatively not suitable for joining Al or Mg alloys because of its high electrical current requirements and the inconsistent quality of final welds [109-111]. Increasing interest in using Mg alloys in aerospace and automobiles is requiring more attention to be given to alternative joining methods. Some success are achieved with the laser beam and the vacuum electron beam welding processes [8-10].

Friction stir welding (FSW) is a relatively new solid-state joining technique, which was invented by The Welding Institute (TWI) of United Kingdom in 1991 [11]. FSW can be considered as the most important development in metal joining in last decade. This joining technique is versatile, energy efficiency and environment friendly without using any cover gas and flux. It can be used to join high-strength aerospace aluminum alloys, magnesium alloys and other metallic alloys that are hard to weld by conventional fusion welding. Compared with the traditional melting welding, FSW has many advantages, and it does not need the consumable metallic wire and the protection atmosphere. FSW is quite suitable for the welding of Mg alloys because it can mitigate the susceptibility to hot cracking, compositional segregation of alloying elements and precipitation of divorced intermetallic particles. Numbers of researches have been conducted to evaluate the feasibility of FSW process for Mg alloys.

Fig. 4(a) schematically illustrates the FSW process. A non-consumable rotating tool with a specially designed pin and shoulder is plunged into the abutting edges of sheets or plates to be joined and transversed along the welding direction (WD). The plane normal of the WD, normal direction (ND) and transverse direction (TD) are denoted in Fig 4(a). The advancing side (AS) means that the tool rotating direction is the same as the WD, and the retreating side (RS) is the inverse direction to the WD. Additionally, friction stir processing (FSP) has been developed as a thermo-mechanical microstructural modification technique of metallic materials based on the basic principles of FSW [112-114]. FSW/FSP is emerging as an effective solid-state joining/processing technique. For both FSW/FSP processes, the frictional heat is generated by the friction between the high-speed rotating tool and the workpiece. The localized heating softens of the base metals around the pin, and it causes severe plastic deformation to produce a strong metallurgical joint. This region is usually referred to as the nugget zone (or called the stir zone, SZ). The contribution of intense plastic deformation and high-temperature exposure within the SZ during FSW/FSP result in generation of a dynamically recrystallized fine-grained microstructure, development of texture, precipitate dissolution and coarsening within the SZ [114-119]. In addition, roughly positions of the thermo-mechanically affected zone (TMAZ) and the heat-affected zone (HAZ) can also be identified as indicated in Fig. 4(a) based on the microstructural characterization of grains and precipitates. The final grain size and microstructural evolution is dependent on the FSW/FSP parameters [120-122]. Many studies have been reported on the microstructural evolutions, grain refining effect, texture effect, dynamic recrystallization and mechanical properties of various Mg alloys after the FSW/FSP process [116-119,123-129].

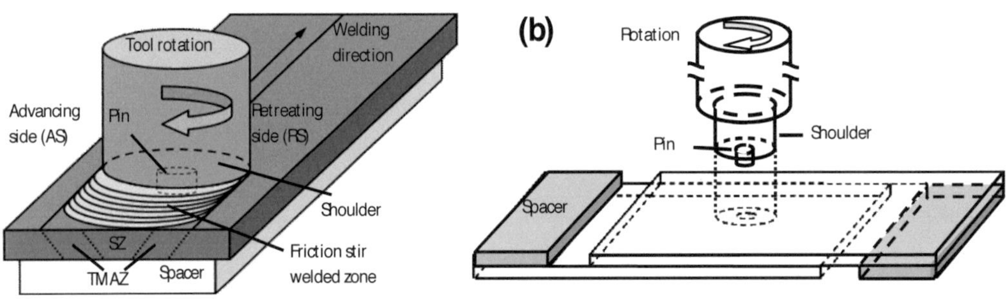

Figure 4. Schematic illustrations of (a) friction stir welding (FSW); (b) friction stir spot welding (FSSW).

Fig. 4(b) shows another joining type of the friction stir spot welding (FSSW), which is a derivative process of the FSW, has been developed as a widespread technique and successfully applied for producing lap-joints [130-135]. Compared with the traditional RSW process, FSSW can avoid severe heating and cooling cycles induced during welding process. Furthermore, the HAZ and residual stresses associated with the welds are relatively small [136]. Therefore, FSSW as well as FSW is now a simple and indispensable process to acquire better joining strength and vibration fracture resistance for the lightweight structural metals [130,131].

3. Tensile properties and failure behaviors of FSP-modified Mg-6Al-1Zn alloys

The effects of varing deformation temperature and strain rate on the microstructural features, tensile properties, deformation and failure behaviors of AZ61 Mg alloy with FSP modification are discussed in this section. The base metal is a 3 mm-thick AZ61-F extruded sheets. The sheets are full-annealed at 410°C for 20 hours, furnace cooled, and then machined into rectangular specimens with dimensions of 100 mm (*l*) × 30 mm (*w*). The full-annealed specimens are denoted by "AZ61-O" in the following. During the FSP, a stirring pin of 6-mm diameter and 2-mm depth, protruding from a rotating rod of 20-mm diameter, is plunged into the rectangular AZ61-O specimens. The rotation speed is set at 1200 rpm, and the downward push pressure is controlled at about 22 MPa. With a tilting angle of 1.5°, the stirring pin moved along the center line of the specimens at a traverse speed of about 1 mm s^{-1}. These specimens will be designated as "AZ61-FSP", and the plane normal of the processed direction (PD), normal direction (ND) and transverse direction (TD) are denoted in Fig. 5(a). The AZ61-O and AZ61-FSP specimens are machined into testing samples with dimensions of 50 mm (*l*) × 15 mm (*w*). Fig. 5(a) schematically illustrates the orientation of AZ61-FSP tensile specimens, and Fig. 5(b) shows the dimensions of the tensile specimens, which gage length is completely within the stir zone (SZ).

Uniaxial tensile tests, which are conducted parallel to the PD, are performed with different initial strain rates of 8.83×10^{-3} s^{-1}, 1.67×10^{-3} s^{-1} and 8.83×10^{-4} s^{-1} at various deformation temperatures of 200°C, 225°C and 250°C. The specimens that failed are examined using an optical microscope (OM) to observe the failure sub-surfaces on the TD plane. The micro-Vickers hardness test across the cross-section of AZ61-FSP specimen is applied using a Vickers indenter (Hv) with a 100 g load for 10 s dwell time. Each datum is the average of three tests.

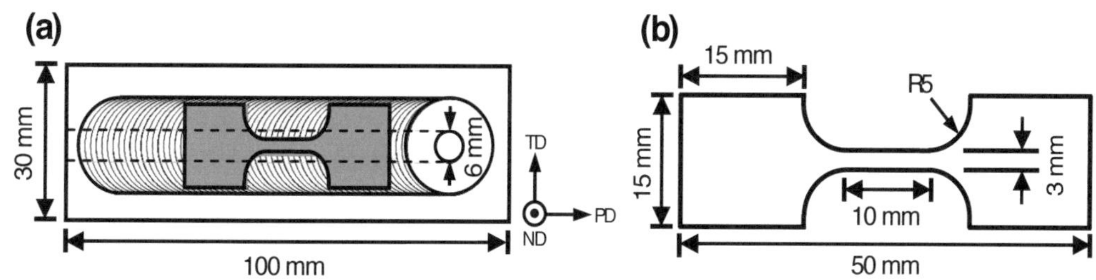

Figure 5. Schematic illustrations of (a) the orientation of AZ61-FSP tensile specimen, (b) the tensile specimen dimension.

Fig. 6(a) shows the microstructural feature of full-annealed AZ61-O specimens. It displays equiaxed grains with an average grain size of about 19.3 ± 2.2 μm. After the FSP modification, the average grain size within the SZ of AZ61-FSP specimens is significantly refined to about 8.2 ± 2.5 μm, as shown in Fig. 6(b). It is recognized that the dynamic recrystallization effect (DRX) governs the grain refining effect during the FSP [115]. Fig. 7 displays the microhardness (Hv) profiles along the cross-section (i.e. on the plane with a plane normal parallel to the PD) of AZ61-FSP specimen. The dash line is the micrograph represents the Vickers indenter testing area, which is located at 1 mm depth from the surface. The result shows that the microhardness within SZ (average value of Hv82.7) is increased and significantly higher than the average level of the non-stir zone (the base metal region, average value of Hv74.8).

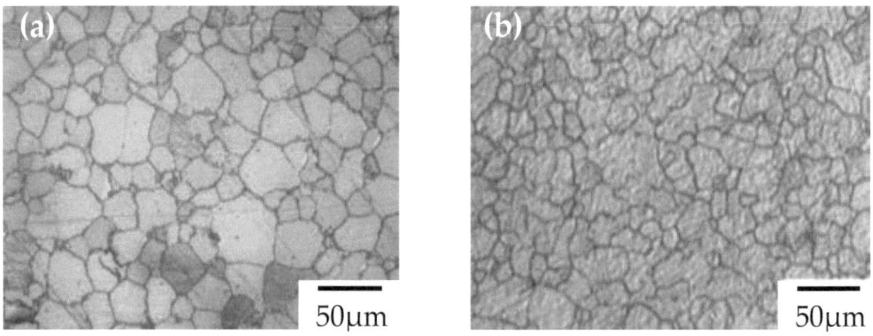

Figure 6. Optical micrograph of the (a) AZ61-O, and AZ61-FSP specimens.

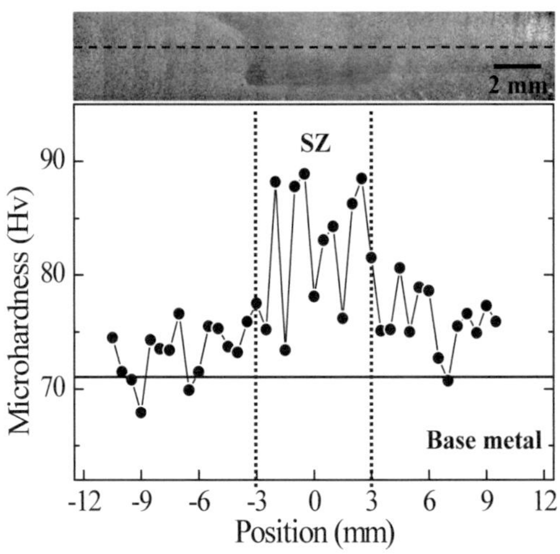

Figure 7. Microhardness (Hv) of the AZ61-FSP. The indentations are made with a spacing of 0.5 mm along the parallel dash line.

3.1. Deformation temperature effect

Fig. 8 shows the stress-strain curves of the AZ61-O and AZ61-FSP specimens obtained from the uniaxial tensile tests under an initial strain rate of 1.67×10^{-3} s^{-1} at various deformation temperatures. It can be seen that the yield strength (YS) and ultimate tensile strength (UTS) are decreased with increasing deformation temperatures for both AZ61-O and AZ61-FSP specimens. It is noted that the total elongation (TE) is significantly increased with increasing deformation temperatures for the AZ61-FSP.

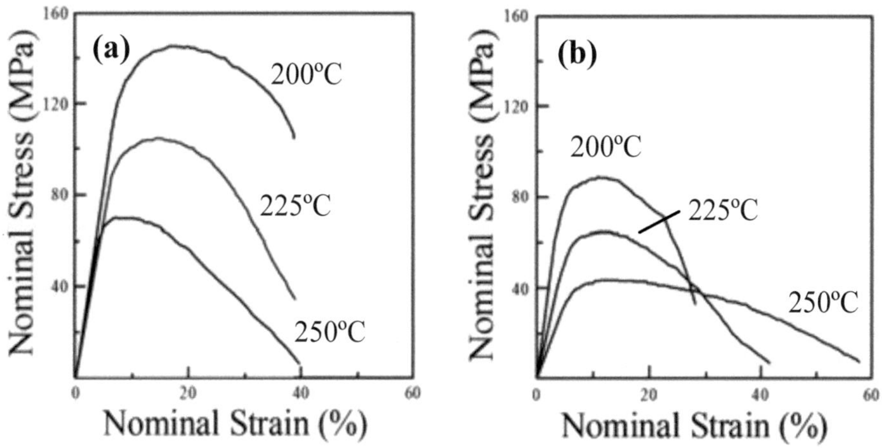

Figure 8. Stress-strain curves of the (a) AZ61-O, and (b) AZ61-FSP specimens with an initial strain rate of $1.67 \times 10^{-3}\ s^{-1}$ at 200, 225 and 250°C.

In addition, the obvious serration of stress-strain curves is observed beyond the UTS, especially for the AZ61-FSP specimens at higher deformation temperatures. This phenomenon is resulted from the dynamic recrystallization (DRX) during the tensile deformation [137,138]. Several mechanisms have been proposed for the DRX process in Mg and its alloys, such as discontinuous dynamic recrystallization (DDRX), continuous dynamic recrystallization (CDRX) and geometric dynamic recrystallization (GDRX). A CDRX process is a recovery process and proceeds by continuous absorption of dislocations in sub-grain boundaries which eventally results in the formation of high angle grain boundaries and new grains. Alternatively, DDRX, which is characterized by nucleation of new grains at original high-angle grain boundaries and nucleus growth by high-angle boundary migration [139], has been recognized as an operative mechanism for the DRX process in the SZ of FSW/FSP aluminum and magnesium alloys. Referring to the failure sub-surfaces as shown in Fig. 9(a) and 9(b), the ultra-fine new grains (as those indicated by arrows in Fig. 9(b)) nucleated and growth at the grain boundaries can be significantly recognized as the occurrence of DRX effect for AZ61-O at higher deformation temperature. From the failure sub-surfaces of AZ61-FSP as shown in Fig. 9(c) and 9(d), we can see fine-grains, which are nucleated and growth from the DRX effect, with an average grain size of about 9 μm are formed at 200-250°C. It is recognized that the DRX effect is activated at lower temperature and the driving force of DRX is much higher for the AZ61-FSP modified specimens than the AZ61-O.

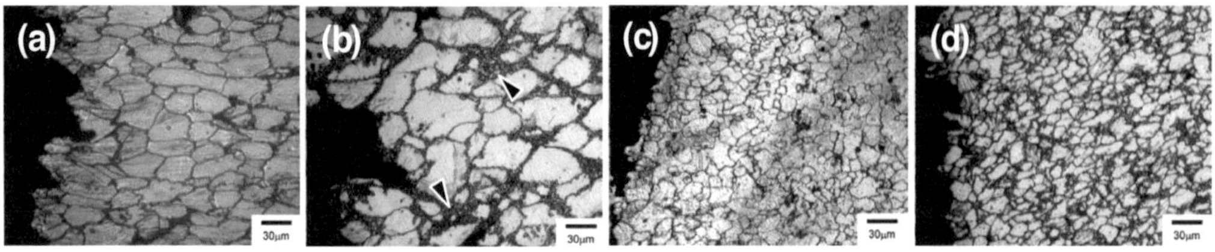

Figure 9. Failure sub-surfaces of AZ61-O specimen tensile tested at (a) 200°C, (b) 250°C, and AZ61-FSP specimen at (c) 200°C, (d) 250°C with an initial strain rate of 1.67 × 10^{-3} s^{-1}.

3.2. Strain rate effect

Fig. 10 shows the stress-strain curves of AZ61-O and AZ61-FSP under different initial strain rates tested at 250°C. The YS and UTS are decreased, but the TE is increased with decreasing the initial strain rate. The AZ61-FSP specimens display higher TE value than AZ61-O. A significant serration of stress-strain curves is also observed beyond the UTS, especially for the AZ61-FSP specimens at lower initial strain rate.

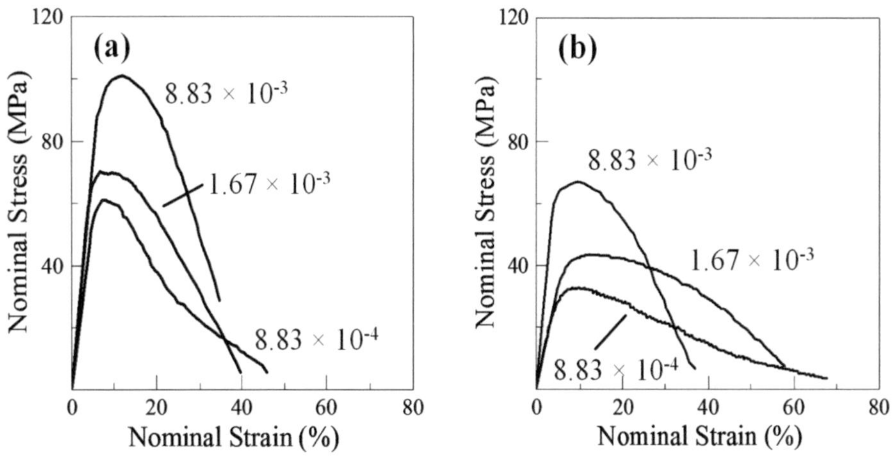

Figure 10. Stress-strain curves of the (a) AZ61-O, and (b) AZ61-FSP specimens with different initial strain rates of 8.83 × 10^{-3} s^{-1}, 1.67 × 10^{-3} s^{-1} and 8.83 × 10^{-4} s^{-1} at 250°C.

Fig. 11 shows the failure sub-surfaces of AZ61-O and AZ61-FSP specimens tested at 250°C for initial strain rates of 8.33 × 10^{-3} s^{-1} and 8.33 × 10^{-4} s^{-1} (the failure sub-surfaces of AZ61-O and AZ61-FSP for the strain rate 1.67 × 10^{-3} s^{-1} are referred to Fig. 9(b) and 9(d), respectively). We can see that the DRX effect is occurred for both of AZ61-O (new fine-grains nucleated and growth at the grain boundaries as encircled in Fig. 11(a)) and AZ61-FSP specimens at 250°C. The DRX effect during tensile deformation is significantly occurred at relatively lower strain rate (8.33 × 10^{-4} s^{-1}) because the recrystallized fine-grains are observed with an average grain size of about 5 μm all over the microstructure, as shown in Fig. 11(b) and 11(d). Considering the AZ61-FSP specimens, the calculation of strain hardening exponent (*n*-value) can help to clarify the deformation mechanism of FSP-modified specimens during the DRX of tensile deformation. Fig. 12(a) shows different n-values with respect to various strain rates and deformation temperatures. The modulus E is calculated by Eq. [1], and T is for deformation temperatures in Kelvin.

$$E = 4.3 \times 10^4 \cdot [1 - 5.3 \times 10^{-4} \cdot (T - 300)] \tag{1}$$

Based on the calculation of n-value and the average grain size, the deformation mechanism of AZ61-FSP can be deduced according to the deformation mechanism map (DMM) as shown in Fig. 12(b). It is demonstrated that the grain boundary sliding (GBS) is a dominant deformation mechanism for the DRX effect during the tensile deformation performed at a lower strain rate (8.33×10^{-4} s^{-1}) and higher deformation temperatures (at 225 and 250°C) for the AZ61-FSP specimens.

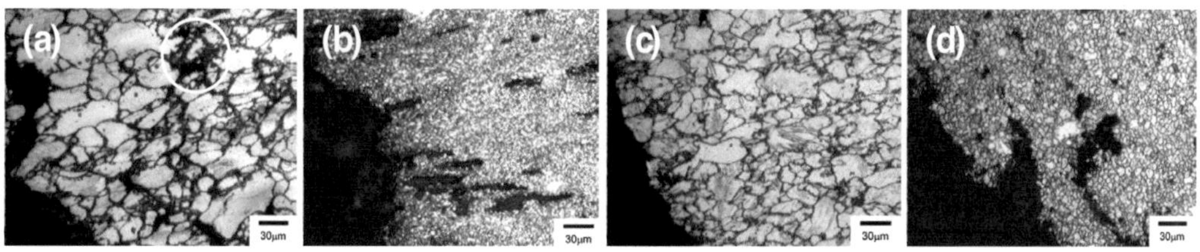

Figure 11. Failure sub-surfaces of AZ61-O specimen tensile tested with an initial strain rate of (a) 8.33×10^{-3} s^{-1}, (b) 8.33×10^{-4} s^{-1}, and AZ61-FSP specimen with (c) 8.33×10^{-3} s^{-1}, (d) 8.33×10^{-4} s^{-1} at 250°C.

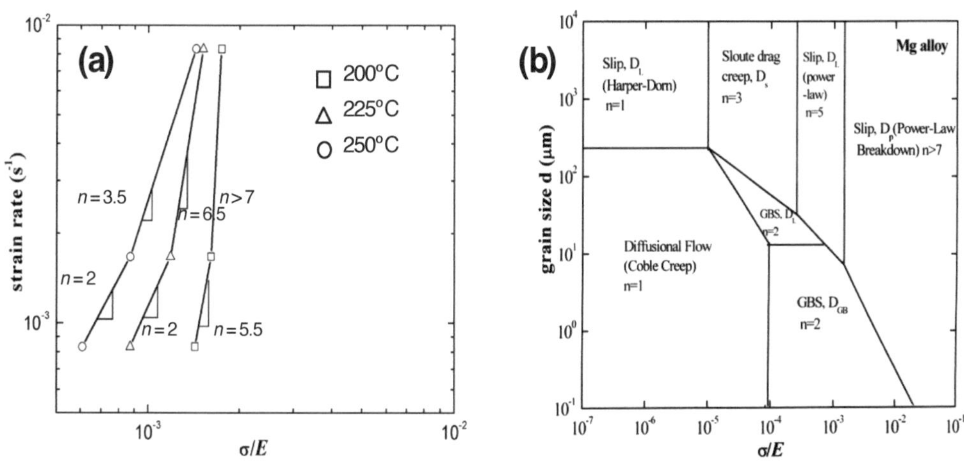

Figure 12. (a) The relationship between strain hardening exponent (n-value) and deformation temperatures for the AZ61-FSP at 0.2% strain, and (b) the deformation mechanism map of Mg alloy at 250°C [140].

4. Tensile ductile-to-brittle transition behavior of the dual-phase Mg-Li-Al-Zn alloy

Since the phase composition and crystallographic structure vary significantly according to Li content, it is important to investigate the mechanical properties, as well as deformation and failure behaviors of Mg-Li alloys. The temperature dependence of fracturing which changes from ductile at higher temperatures to brittle at low temperatures (i.e. the ductile-to-brittle transition temperature, DBTT) is well established and occurs below room temperature for bcc metals [141-143]. The ductile-to-brittle transition has been studied in hcp Zn, Mg alloys and β-phase bcc Mg-Li alloy using the tensile and Charpy impact tests [75,144,145]. The Zr-based alloy with a α/β-type dual-phase microstructure also shows ductile-to-brittle transition failure behavior [146]. Considering the variations in microstructure and mechanical properties with Li addition, it is likely that the α/β-type dual-phase Mg-Li alloys may exhibit a transition temperature at which fracturing changes from ductile to brittle [147]. But the DBTT is seldom discussed vis-à-vis Mg alloys. This section is to investigate tensile properties at various testing temperatures and study the ductile-to-brittle transition failures for the α/β-type Mg-Li-Al-Zn alloy.

The base metal is 3 mm-thick as-extruded Mg-10.3Li-2.4Al-0.7Zn sheets (LAZ1021). Fig. 13(a) shows the 3-D optical microstructure of the as-extruded LAZ1021 alloy, which is composed of a dual-phase crystal structure. The light gray and dark gray regions correspond to the Mg-rich α-phase and the Li-rich β-phase, respectively. The α-phase is surrounded by the β-phase, and the volume fraction of α-phase is about 30% (vol.%). The micro-Vickers hardness test showed the average microhardness of α-phase is Hv64.3, while that of β-phase is Hv54.0. The tensile specimen dimensions is the same as shown in Fig. 5(b). Tensile tests, which are conducted parallel to the extruded direction (ED), have an initial strain rate of $1.67 \times 10^{-3}\ s^{-1}$. The tensile tests are performed at 25, 25, 50, 100, 150, 200 and 250°C using isothermal heating equipment. The samples that failed are examined using a SEM to observe the fracture surfaces, while the failure sub-surfaces on the TD plane are examined by an OM.

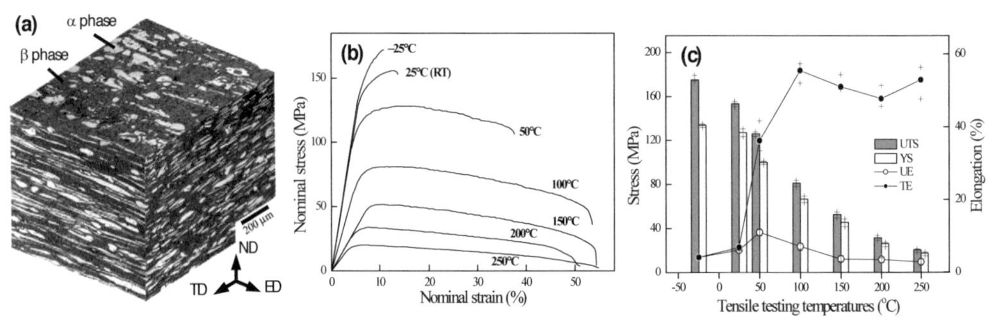

Figure 13. (a) Optical micrograph of as-extruded LAZ1021 base metal at the ND, ED and TD planes, (b) stress-strain curves, and (c) tensile testing results of the LAZ1021 alloy at various temperatures.

Fig. 13(b) shows the stress-strain curves of the α/β-type LAZ1021 alloy obtained from the uniaxial tensile tests conducted with a strain rate of 1.67×10^{-3} s^{-1} at various temperatures. To describe and compare the tensile properties conveniently, Fig. 13(c) displays the effect of testing temperatures on the yield strength (YS), ultimate tensile strength (UTS), uniform elongation (UE) and total elongation (TE) of the α/β dual-phase LAZ1021 alloy. The cross markers denote the highest and the lowest values (representing the data extremes) of the measured tensile strength and elongation. It can be seen that the YS and UTS decreased with increasing deformation temperatures. The difference between YS and UTS is very small for temperatures higher than 150°C. It is worth noting that the elongation (especially for the TE) significantly increased at deformation temperatures higher than 100°C, and the ductile-to-brittle transition temperature (DBTT) can be pinpointed at about 50°C for the α/β-type LAZ1021 alloy.

Fig. 14 shows the fracture surfaces and sub-surfaces (on the TD plane) within the gauge length of deformed tensile specimens for a strain rate of 1.67×10^{-3} s^{-1} at certain specific deformation temperatures. Fig. 14(a) shows a brittle fracture surface, which displays a lot of cleavage features, generally observed at temperatures lower than room temperature (RT). The brittle fracturing is reduced with increasing testing temperatures, and Fig. 14(b) shows a mixed fracture surface of quasi-cleavage with tear ridges and dimpled ruptures at 50°C. Ductile failure with obvious dimpled ruptures appears with increasing frequency at temperatures higher than 100°C, as shown in Fig. 14(c) and 14(d). As seen from the fracture sub-surfaces, significant brittle fractures occurred with a mixed transgranular cleavage within α-phase and grain boundary separation fracture at β-phase, as denoted in Fig. 14(e). Cleavage fracturing is still occurred within the α-phase at a deformation temperature of 50°C, as shown in Fig. 14(f). However, some other α-phase is clearly elongated along the tensile direction without fracture, while decohesion-induced cracking is observed at the α/β-phase interface. Previous study indicated that cavities form easily after decohesion between the second-phase and the matrix [148]. We found that cavities appeared and the cracking at the α/β-phase interface became more evident with increasing the deformation temperature, as denoted in Fig. 14(g). Since many of the cavities are fairly enlarged, as shown in Fig. 14(h), it is reasonable to suggest that failure occurs ultimately through cavity coalescence and interlinkage. Although the ductility of Mg-Li alloy is improved, the TE being restricted to about 55-60% can be considered as resulting from the serious cavitation and the interlinkage of cavities when specimens deformed at high temperatures. Based on the fractography, it is likely that brittle and ductile fractures are related to variations in the crystal structure and the change of slip systems.

When a metal commences to deform plastically, slip begins when the shearing stress on the slip plane in the slip direction reaches a critical resolved shear stress (CRSS). According to Schmid's law, if the tension axis is normal or parallel to the slip plane, the resolved shear stress is zero. Slip will not occur for these orientations since there is no shear stress on the slip plane. For the present used α/β-type LAZ1021 extruded alloy, the preferred orientation of Mg-rich α-phase is (0002) with a plane normal in the TD. Since the tensile direction is parallel to the basal plane, basal slip as well as the deformation twin will barely be present during tensile tests. Brittle fracturing without significant twinning crosses through the Mg-rich α-phase as shown in Fig. 14(a) and 14(e). Cleavage fracturing of the α-phase generally occurs as the plastic deformation proceeds at temperatures lower than RT. In addition, brittle failure behavior can also be found in the intergranular fractures, which go along the grain boundaries within the bcc Li-rich β-phase, as denoted in Fig. 14(e). These phenomena cause the lack of ductility and low elongation of the α/β-type Mg-Li-Al-Zn alloy at deformation temperatures lower than RT.

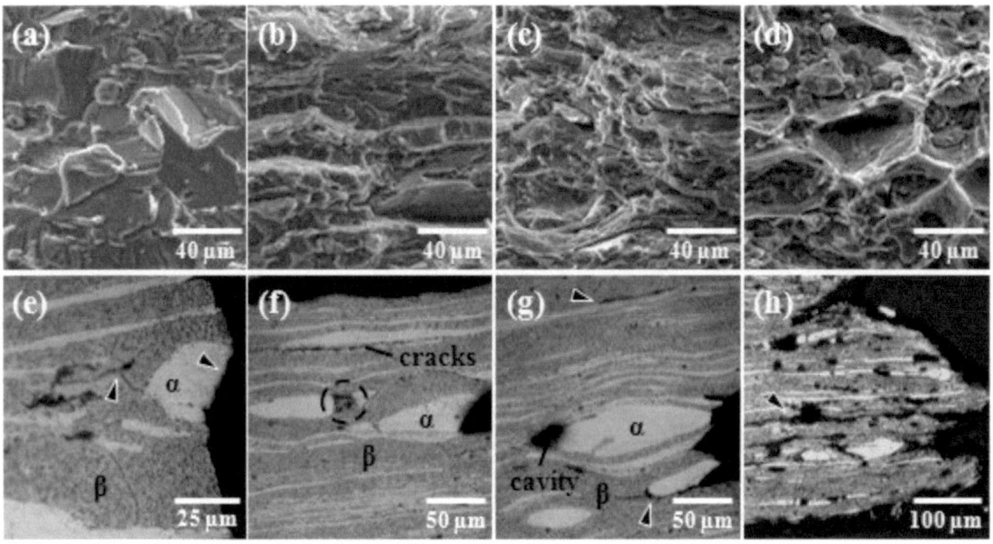

Figure 14. Tensile fracture surface of samples tested at (a) 25°C, (b) 50°C, (c) 100°C, (d) 250°C and failure sub-surface (at the TD plane) of samples tested at (e) 25°C, (f) 50°C, (g) 100°C and (h) 200°C.

Greater non-basal slips, including prismatic and pyramidal slip, occur in situations with a reduced c/a ratio for hcp metals (compared with the ideal value of 1.634) at elevated temperatures or when the concentration of the solid solution changes [149]. For Mg-Li-based alloys, it has been reported that the addition of Li to form a solid solution in Mg matrix will decrease the length of the c-axis and thus lead to a reduction in the c/a ratio. Research results suggest that this helps to suppress basal slip, and encourages prismatic slip. The variation in stacking fault energy should also be considered for the Mg-Li alloys. Agnew et al. [150] indicated that Li additions may lower the non-basal stacking fault energy for the glissile dislocations and increase the stability of the glissile configuration. Since the pyramidal slip mode offers five independent slip systems, it provides an explanation for the improved ductility of Mg-Li alloy. The changes in failure behavior can be attributed to variations in the c/a ratio, slip systems and stacking fault energy of the α/β-type Mg-Li alloy.

At higher testing temperatures, we can see that the elongation significantly increased and the fraction of cleavage fracturing decreased at higher than 100°C. Since the prismatic slip of the hcp α-phase is a thermally activated process and tends to be dominant above RT, lattice twisting of hcp α-phase will occur with a large number of prismatic slips; this in turn results in an increase in the mobility of basal slips during tension. Although cleavage fracturing is still likely to occur within some α-phase, most of the α-phase can be easily elongated, as shown in Fig. 14(f) and 14(g). The ductility of LAZ1021 alloy is significantly increased and the ductile-to-brittle transition occurs between about 50 and 100°C. Attendant with the improvement in ductility, α/β-phase interfacial cracking, which is indicated in Fig. 14(f), begins during the ductile-to-brittle transitional interval. At testing temperatures above 100°C, cavities and α/β-phase interfacial cracking become more obvious, as shown in Fig. 14(g) and 14(h). The cavitation and interfacial cracking may result from the elongation difference between α-phase and β-phase. Referring to the stress-strain curves displayed in Fig. 13(b), the flow curves for 50-200°C exhibited serrations while specimens elongated beyond the UTS. Although prismatic slip and basal slip of the α-phase are more active at high temperatures, deformation occurred much more easily in the bcc β-phase because of more slip systems. Thus, the DRX effect of the β-phase can be seen with refined grains as shown in Fig. 15(a). The ductile failure is due entirely to dimpled ruptures (Fig. 14(c) and 14(d), Fig. 15(b)) and resulted from the detrimental coalescence and interlinkage of cavities (Fig. 14(h)) as plane slips, DRX and α/β-phase interfacial decohesion became more prevalent.

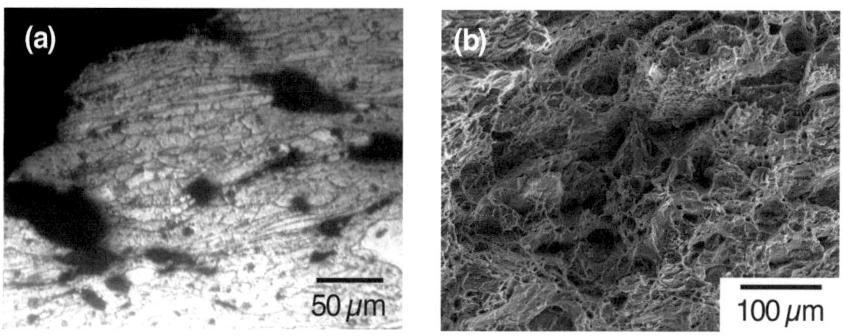

Figure 15. (a) The tensile failure sub-surface, and (b) the fracture surface of samples tested at 100°C.

5. Microstructural refining effect on mechanical properties of FSP-modified dual-phase Mg-Li-Al-Zn alloy

The aim of this section focuses on evaluating microstructural evolution, tensile mechanical properties and failure behaviors of the α/β-dual-phase LAZ1021 extruded alloy through the modification of FSP technique. A rotating rod of 20-mm diameter with a stirring pin of 6-mm diameter and 2-mm depth is used for the FSP modification of LAZ1021. The rotation speed is set at 2500 rpm, and the downward push pressure is controlled at about 15 MPa. With a tilting angle of 1.5°, the stirring pin moved along the center line of the specimens at a traverse speed of about 2.5 mm s^{-1}. These specimens are designated as "LAZ1021-FSP". The phase composition and crystal structure of as-extruded LAZ1021 alloy and LAZ1021-FSP specimens are identified by X-ray diffractometry (XRD), using CuKα radiation at 30 kV, 20 mA with a scan speed of 1° (2θ) min^{-1}. XRD analysis for the texture of the LAZ1021-FSP is obtained from the stir zone (SZ) region. The micro-Vickers hardness test across the cross-section of LAZ1021-FSP specimen is applied using a Vickers indenter (Hv) with a 50 g load for 10 s dwell time. Each datum is the average of three tests. The dimension of tensile specimens, plane normal of PD, ND and TD planes are the same as defined in Fig. 5. Uniaxial tensile tests are conducted parallel to the PD with an initial strain rate of 1.67×10^{-3} s^{-1} at RT. The failure sub-surfaces on the TD plane and the fracture surfaces are also examined by OM and SEM, respectively.

Fig. 16 shows the cross-sectional features of LAZ1021-FSP specimen. We can see that the microstructure is significantly changed after the FSP modification. The bottom SZ region (SZ2) displays an obvious plastic flow with a finer microstructure than the upper SZ region (SZ1), as shown in Fig. 16(a). Compared with the base metal (Fig. 16(b)), the grain size of Mg-rich α-phase and Li-rich β-phase for the LAZ1021-FSP are refined and redistributed with an obvious texture within the SZ, as shown in Fig. 16(c) and 16(d). In addition, it is noted that the Mg-rich α-phase is likely dissolved in the matrix during FSP, and then the α-phase (as indicated by arrows in Fig. 16(c)) re-precipitated not only at the grain boundaries, but also within the refined grains of the β-phase, especially occurred in the SZ. The volume fraction of α-phase is reduced to about 11-15% (vol.%) after FSP modification.

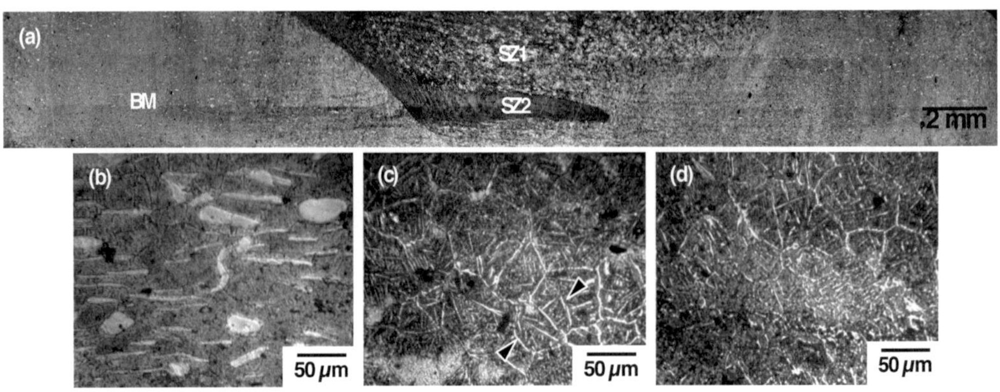

Figure 16. (a) The cross-sectional microstructure (on the PD plane) of the LAZ1021-FSP specimen. Microstructural features of (b) the base metal (BM), the stir zone of (c) SZ1 and (d) SZ2 denoted in (a).

Fig. 17(a) shows the XRD patterns, which display typical diffraction peaks of Mg-rich α-phase and Li-rich β-phase, of the LAZ1021 extruded alloy on the ND, ED and TD planes. The prismatic planes of $(11\bar{2}0)$ and $(10\bar{1}0)$ are the preferred orientations of hcp α-phase on the ND and ED planes, respectively. Some crystal planes of the α-phase are also observed, though the peak intensities are low. Compared with the standard powder diffraction of Mg (JCPDS 35-0821), the (0002) basal plane, which displays a stronger peak intensity, can be recognized as the preferred orientation of the TD plane. For the hcp α-phase, therefore, the basal plane is parallel to the ED, and its plane normal (the *c*-axis of hcp crystal structure) is perpendicular to the TD plane. As for the bcc β-phase, the (200) and (110) planes are identified as the preferred orientations of the ND plane and of both the ED and TD planes, respectively. Apart from the peaks of α and β-phases, relatively weak peaks are also observed in Fig. 17(a), and these peaks belong to the AlLi intermetallic compound (JCPDS 71-0362). Fig. 17(b) shows the XRD patterns obtained from SZ regions on the ND, PD and TD planes of the LAZ1021-FSP specimen, respectively. Diffraction peaks of the AlLi intermetallic compound is reduced, and it is recognized that the AlLi compound is dissolved after the FSP. It is noted that the peaks intensity of the hcp α-phase is significantly reduced and the texture of extruded LAZ1021 alloy is changed after the FSP, especially for the ND plane of LAZ1021-FSP specimens. We can see that the preferred orientation of the hcp α-phase is varied from the prismatic plane $(11\bar{2}0)$ to the pyramidal plane $(10\bar{1}1)$, and the bcc β-phase is varied from (200) to (110) on the ND plane. The reduction of diffraction peaks intensity for the α-phase can be resulted from the decrease in the volume fraction of α-phase. The (110) plane of the bcc β-phase can be identified as the preferred orientation for the LAZ1021-FSP modified specimens, as shown in Fig. 17(b).

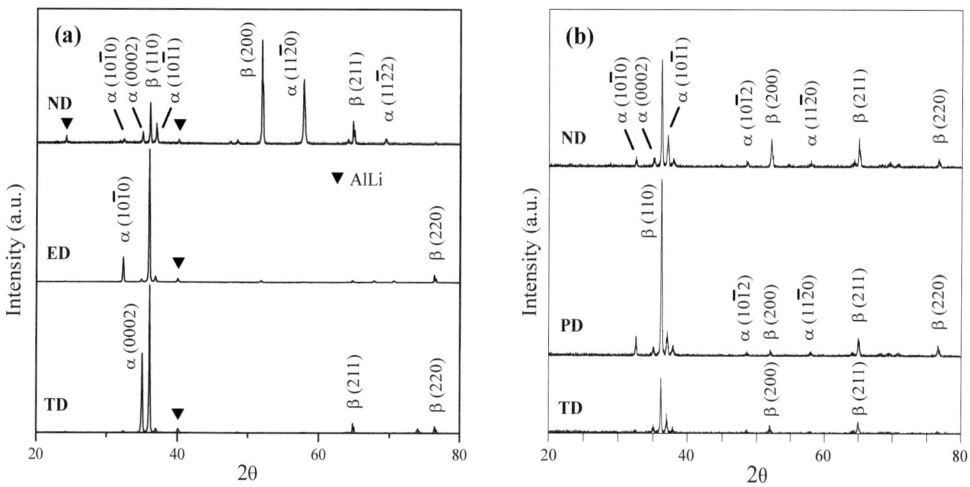

Figure 17. X-ray diffraction patterns of (a) the as-extruded LAZ1021, and (b) the LAZ1021-FSP specimen.

Fig. 18(a) displays the microhardness (Hv) profiles along the cross-section of LAZ1021-FSP specimen. The dash line is the micrograph represents the Vickers indenter testing area, which is located at 1 mm depth from the surface. The result shows that the microhardness within SZ (average value of Hv72.5) is significantly increased and higher than the the base metal region. Fig. 18(b) displays the stress-strain curves of the as-extruded LAZ1021 alloy and the LAZ1021-FSP specimen. After FSP modification, the tensile strength is increased to about 180 MPa, but the elongation is significantly decreased for the LAZ1021-FSP specimen. The total elongation (TE) of LAZ1021 alloy is reduced from about 10% to about 3% after FSP. The variation of microhardness and tensile mechanical properties can be resulted from the microstructural features and preferred orientations (Figs. 16 and 17) are significantly varied with performing FSP modification. Fig. 19 shows the fracture surface and failure sub-surface of the LAZ1021-FSP specimen. Compared with the failure morphologies of as-extruded LAZ1021 alloy (Fig. 14(a) and 14(e)), the LAZ1021-FSP specimen displays an intergranular fracture feature, as shown in Fig. 19(a). Since the α-phase is precipitated at grain boundaries of the β-phase after FSP, it is recognized that the intergranular fracture can be resulted from the α/β-interface decohesion and the cracks propagation along the grain boundaries of β-phase, as shown in Fig. 19(b).

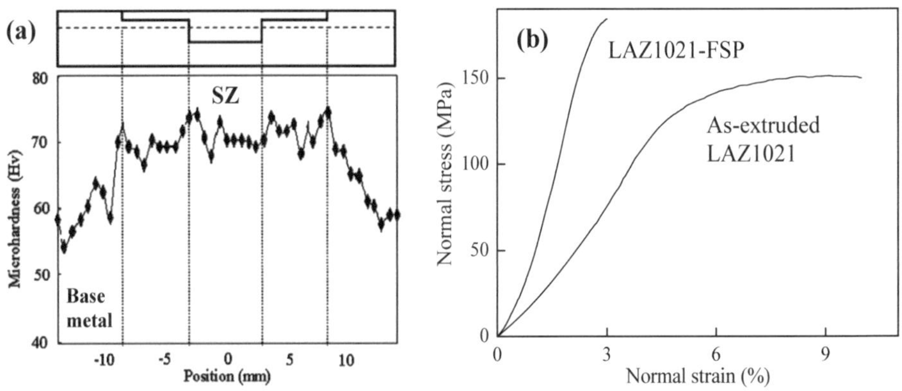

Figure 18. (a) Microhardness (Hv) of the LAZ1021-FSP. The indentations are made with a spacing of 0.5 mm along the parallel dash line. (b) The stress-strain curves of the as-extruded LAZ1021 alloy and the LAZ1021-FSP specimen.

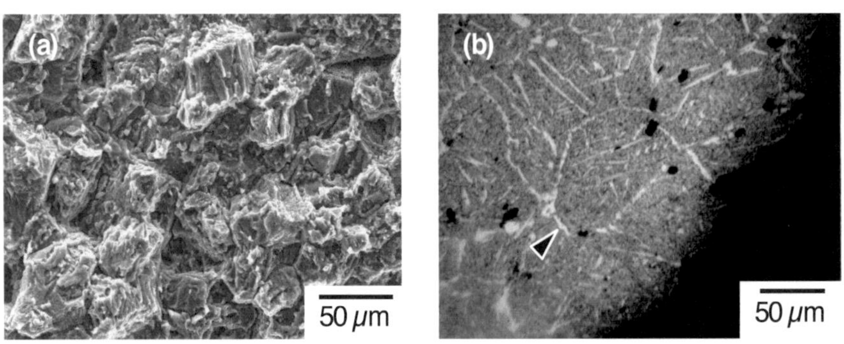

Figure 19. (a) Fracture surface, and (b) failure sub-surface on the TD plane of the LAZ1021-FSP specimen.

6. Conclusion

The evolution of microstructural features, tensile mechanical properties and failure behaviors of full-annealed Mg-6Al-1Zn (AZ61-O) and as-extruded α/β-dual-phase Mg-10Li-2Al-1Zn (LAZ1021) alloys with applying a microstructural modification of the friction stir processing (FSP) are evaluated in this chapter. In summary, the grain size of AZ61-O and LAZ1021 alloys is significantly refined after the FSP modification. It is noted that the preferred orientations of hcp Mg-rich α-phase and bcc Li-rich β-phase is changed for the LAZ1021-FSP specimens. The volume fraction of α-phase is reduced after FSP, and the formation of a specific texture of the LAZ1021-FSP is resulted from the re-precipitation of α-phase within the grains and at the grain boundaries of β-phase in the stir zone (SZ). After the FSP, the microhardness within SZ is significantly improved for both of the AZ61-O and LAZ1021 alloys.

Tensile strength is decreased and total elongation is increased with decreasing the initial strain rate for both of the AZ61-O and AZ61-FSP. The serration of tensile stress-strain curves for AZ61-FSP specimens are resulted from the dynamic recrystallization (DRX) effect. The DRX effect is activated at lower temperature and the driving force of DRX is much higher for the AZ61-FSP modified specimens than the AZ61-O. Based on the evaluation of strain hardening exponent (n-value) and average grain size, it is demonstrated that the grain boundary sliding (GBS) is a dominant deformation mechanism for the DRX during the tensile deformation performed at a lower strain rate and higher deformation temperatures for the AZ61-FSP specimens.

For α/β-dual-phase LAZ1021 alloy, the tensile strength is increased, but the total elongation is significantly decreased after the FSP modification. LAZ1021-FSP displays an intergranular fracture, which is occurred from the α/β-interface decohesion and crack propagation along the grain boundaries of β-phase. A significant ductile-to-brittle transition effect is confirmed for the α/β-dual-phase Mg-Li-Al-Zn alloy. At temperatures lower than 25°C, fracturing occurs by transgranular cleavage of the α-phase and intergranular fracture at grain boundaries of the β-phase. At deformation temperatures higher than 100°C, the dominant failure behaviors are interfacial cracking between α/β interface and ductile dimpled ruptures of the β-phase. The ductile-to-brittle transition occurred at temperatures between 25 and 100°C, and the ductile-to-brittle transition temperature (DBTT) can be pinpointed at around 50°C for the α/β-dual-phase LAZ1021 alloy. The dimpled rupture for specimens with higher elongation is resulted from the cavitation due to the α/β interfacial cavities coalescence and interlinkage. Failures with dimpled ruptures occurred from the cavitation is more significant with increasing deformation temperatures.

Acknowledgements

This study was financially supported by the National Science Council of Taiwan (Contract No. NSC 100-2221-E-150-037 and NSC 101-2221-E-150-028) for which we are grateful.

Author details

Chung-Wei Yang

Department of Materials Science and Engineering, National Formosa University, Huwei, Yunlin, Taiwan

References

[1] Mordike BL, Ebert T. Magnesium properties-applications-potential. Materials Science and Engineering A 2001;302(1):37-45.

[2] Eliezer D, Aghion E, Froes FH. Magnesium science technology and applications. Advanced Performance Materials 1998;5:201-202.

[3] Deetz J. The use of wrought magnesium in bicycles. JOM 2005;57(5):50-53.

[4] Plomear IJ. Magnesium alloys and applications. Materials Science and Technology 1994;10(1):1-16.

[5] Aghion E, Bronfin B. Magnesium alloys development towards the 21th century. Materials Science Forum 2000;350:19-28.

[6] VanFleteren R. Magnesium for automotive applications. Advanced Materials and Processes 1996;149(5):33-34.

[7] Aghion E, Bronfin B, Eliezer D. The role of the magnesium industry in protecting the environment. Journal of Materials Processing Technology 2001;117:381-385.

[8] Wang HY, Li ZJ. Investigation of laser beam welding process of AZ61 magnesium-based alloy. Acta Metallurgica Sinica 2006;19(4):287-294.

[9] Padmanaban G, Balasubramanian V. Optimization of laser beam welding process parameters to attain maximum tensile strength in AZ31B magnesium alloy. Optics & Laser Technology 2010;42(8):1253-1260.

[10] Luo Y, Ye H, Du C, Xu H. Influence of focusing thermal effect upon AZ91D magnesium alloy weld during vacum electron beam welding. Vacuum 2012;86(9):1262-1267.

[11] Thomas WM, Nicholas ED, Needham JC, Murch MG, Templesmith P, Dawes CJ. G. B. Patent Application, No. 9125978.8, 1991.

[12] Mahoney MW, Rhodes CG, Flintoff JG, Bingle WH, Spurling RA. Properties of friction-stir-welded 7075-T651 aluminum. Metallurgical and Materials Transactions A – Physical Metallurgy and Materials Science 1998;29(7):1955-1964.

[13] Squillace A, De Fenzo A, Giorleo G, Bellucci F. A comparison between FSW and TIG welding techniques: modification of microstructure and pitting corrosion resistance

in AA 2024-T3 butt joints. Journal of Materials Processing Technology 2004;152(1): 97-105.

[14] Padmanaban G, Balasubramanian V. Metallurgical characterization of pulsed current gas tungsten arc, friction stir and laser beam welded AZ31B magnesium alloy joints. Materials Chemistry and Physics 2011;125(3):686-697.

[15] Wang X, Wang K. Microstructure and properties of friction stir butt-welded AZ31 magnesium alloy. Materials Science and Engineering A 2006;431:114-117.

[16] Xie GM, Ma ZY, Geng L, Chen RS. Microstructural evolution and mechanical properties of friction stir welded Mg-Zn-Y-Zr alloy. Materials Science and Engineering A 2007;471:63-68.

[17] Dobriyal RP, Dhindaw BK, Muthukumaran S, Mukherjee SK. Microstructure and properties of friction stir butt-welded AE42 magnesium alloy. Materials Science and Engineering A 2008;477(1-2):243-249.

[18] Cerri E, Leo P. Warm and room temperature deformation of friction stir welded thin aluminum sheets. Materials & Design 2010;31(3):1392-1402.

[19] Fratini L, Micari F, Buffa G, Ruisi VF. A new fixture for FSW processes of titanium alloys. CIRP Annals – Manufacturing Technology 2010;59(1):271-274.

[20] Somasekharan AC, Murr LE. Microstructures in friction-stir welded dissimilar magnesium alloys and magnesium alloys to 6061-T6 aluminum alloy. Materials Characterization 2004;52(1):49-64.

[21] Yan Y, Zhang DT, Qiu C, Zhang W. Dissimilar friction stir welding between 5052 aluminum alloy and AZ31 magnesium alloy. Transactions of Nonferrous Metals Society of China 2010;20(2):s619-s623.

[22] Xue P, Ni DR, Wang D, Xiao BL, Ma ZY. Effect of friction stir welding parameters on the microstructure and mechanical properties of the dissimilar Al-Cu joints. Materials Science and Engineering A 2011;528(13-14):4683-4689.

[23] Anouma M, Nakata K. Effect of alloying elements on interface microstructure of Mg-Al-Zn magnesium alloys and titanium joint by friction stir welding. Materials Science and Engineering B 2009;161(1-3):46-49.

[24] Anouma M, Nakata K. Dissimilar metal joining of ZK60 magnesium alloy and titanium by friction stir welding. Materials Science and Engineering B 2012;177(7):543-548.

[25] Clark JB. Age hardening in a Mg-9wt.%Al alloy. Acta Metallurgica 1968;16(2): 141-152.

[26] Foerster GS. Proceeding of the IMA 33rd Annual Meeting, May 23-25, 1976, Montreal, Quebec, Canada, p.35-39.

[27] Hono K, Mendis CL, Sasaki TT, Oh-ishi K. Towards the development of heat-treatable high-strength wrought Mg alloys. Scripta Materialia 2010;63:710-715.

[28] Miller WK. Creep of die cast AZ91 magnesium at room temperature and low stress. Metallurgical and Materials Transactions A – Physical Metallurgy and Materials Science 1991;22(4):873-877.

[29] Luo A, Pekguleryuz MO. Cast magnesium alloys for elevated temperature applications. Journal of Materials Science 1994;29(20):5259-5271.

[30] Zhang J, Li ZS, Guo ZX, Pan FS. Solidification microstructural constituent and its crystallographic morphology of permanent-mold-cast Mg-Zn-Al alloys. Transactions of Nonferrous Metals Society of China 2006;16:452-458.

[31] Xiao W, Jia S, Wang J, Wu Y, Wang L. Effects of cerium on the microstructure and mechanucal properties of Mg-20Zn-8Al alloy. Materials Science and Engineering A 2008;474:317-322.

[32] Zhang J, Guo ZX, Pan F, Li Z, Luo X. Effect of composition on the microstructure and mechanical properties of Mg-Zn-Al alloys. Materials Science and Engineering A 2007;456(1-2):43-51.

[33] Pekguleryuz MO, Baril E. Creep resistant magnesium diecasting alloys based on alkaline earth elements. Materials Transactions 2001;42(7):1258-1367.

[34] Luo AA, Powell BR, Balogh MP. Creep and microstructure of magnesium-aluminum-calcium based alloys. Metallurgical and Materials Transactions A – Physical Metallurgy and Materials Science 2002;33(3):567-574.

[35] Luo AA. Recent magnesium alloy development for elevated temperature applications. International Materials Reviews 2004;49(1):13-30.

[36] Saddock ND, Suzuki A, Jones JW, Pollock TM. Grain-scale creep processes in Mg-Al-Ca base alloys: Implications for alloy design. Scripta Materialia 2010;63:692-697.

[37] Lu YZ, Wang QD, Zeng XQ. Effects of rare earths on the microstructure, properties and fracture behavior of Mg-Al alloys. Materials Science and Engineering A 2000;278(1-2):66-76.

[38] Zhou HT, Zhang XQ, Ding WJ, Ma CJ, Zhu YP. Effect of La and Nd on microstructures and mechanical properties of AZ61 wrought magnesium alloy. Transactions of Nonferrous Metals Society of China 2004;14(1):67-70.

[39] Wang MX, Zhou H, Wang L. Effect of yttrium and cerium addition on microstructure and mechanical properties of AM50 magnesium alloy. Journal of Rare Earths 2007;25(2):233-237.

[40] Zhang J, Wang J, Qiu X, Zhang D, Tian Z, Niu X, Tang D, Meng J. Effect of Nd on the microstructure, mechanical properties and corrosion behavior of die-cast Mg-4Al-based alloy. Journal of Alloys and Compounds 2008;464(1-2):556-564.

[41] Pan FS, Chen MB, Wang JF, Peng J, Tang AT. Effect of yttrium addition on microstructure and mechanical properties of as-extruded AZ31 magnesium alloys. Transactions of Nonferrous Metals Society of China 2008;18:s1-s6.

[42] Zhang J, Liu K, Fang D, Qiu X, Tang D, Meng J. Microstructure, tensile properties and creep behavior of high pressure die-cast Mg-4Al-4RE-0.4Mn (RE=La, Ce) alloys. Journal of Materials Science 2009;44(8):2046-2054.

[43] Kaveh MA, Afshin M, Farzad K. The effect of different rare earth elements content on microstructure, mechanical and wear behavior of Mg-Al-Zn alloy. Materials Science and Engineering A 2010;527(7-8):2027-2035.

[44] Tong G, Liu H, Liu Y. Effect of rare earth additions on microstructure and mechanical properties of AZ91 magnesium alloys. Transactions of Nonferrous Metals Society of China 2010;20:s336-s340.

[45] Zou H. Effect of microstructure on creep behavior Mg-5%Zn-2%Al (-2%Y) alloy. Transactions of Nonferrous Metals Society of China 2008;18(3):580-587.

[46] Xiao W, Jia S, Wang L, Wu Y, Wang L. The microstructure and mechanical properties of Mg-Zn-Al-RE alloys. Journal of Alloys and Compounds 2009;480(2):L33-L36.

[47] Wei LY, Dunlop GL, Westengen H. Age hardening and precipitation in a cast magnesium-rare-earth alloy. Journal of Materials Science 1996;31(2):387-397.

[48] Nie JF, Muddle BC. Characterization of strengthening precipitate phases in a Mg-Y-Nd alloy. Acta Materialia 2000;48(8):1691-1703.

[49] Smola B., Stulikova I., Pelcova J., Mordike L. Magnesium Alloys and Their Applications. In: Kainer K.U. (ed.) Wiley-VCH Verlag, Weinheim; 2003, p43.

[50] Kawamura Y, Hayashi K, Inoue A, Masumoto T. Rapid solidified powder metallurgy $Mg_{97}Zn_1Y_2$ alloys with excellent tensile yield strength above 600 MPa. Materials Transactions 2001;42(7):1172-1176.

[51] Homma T, Kunito N, Kamado S. Fabrication of extraordinary high-strength magnesium alloy by hot extrusion. Scripta Materialia 2009;61(6):644-647.

[52] Hagihara K, Kinoshita A, Sugino Y, Yamasaki M, Kawamura Y, Yasuda HY, Umakoshi Y. Effect of long-period stacking ordered phase on mechanical properties of $Mg_{97}Zn_1Y_2$ extruded alloy. Acta Materialia 2010;58(19):6282-6293.

[53] Yamasaki M, Hashimoto K, Hagihara K, Kawamura Y. Effect of multimodal microstructure evolution on mechanical properties of Mg-Zn-Y extruded alloy. Acta Materialia 2011;59(9):3646-3658.

[54] Matsuda M, Ii S, Kawamura Y, Ikuhara Y, Nishida M. Interaction between long period stacking order phase and deformation twin in rapidly solidified $Mg_{97}Zn_1Y_2$ alloy. Materials Science and Engineering A 2004;386(1-2):447-452.

[55] Hagihara K, Kinoshita A, Sugino Y, Yamasaki M, Kawamura Y, Yasuda HY, Umakoshi Y. Plastic deformation behavior of $Mg_{97}Zn_1Y_2$ extruded alloys. Transactions of Nonferrous Metals Society of China 2010;20(7):1259-1268.

[56] Yamasaki M, Sasaki M, Nishijima M, Hiraga K, Kawamura Y. Formation of 14H long period stacking ordered structure and profuse stacking faults in Mg-Zn-Gd alloys during isothermal aging at high temperature. Acta Materialia 2007;55(20):6798-6805.

[57] Yamasaki M, Anan T, Yoshimoto S, Kawamura Y. Mechanical properties of warm-extruded Mg-Zn-Gd alloy with coherent 14H long perodic staking ordered structure precipitate. Scripta Materialia 2005;53(7):799-803.

[58] Yamasaki M, Izumi H, Kawamura Y, Habazaki H. Corrosion and passivation behavior of Mg-Zn-Y-Al alloys prepared by cooling-rate controlled solidification. Applied Surface Science 2011;257(19):8258-8267.

[59] Haferkamp H, Boehm R, Holzkamp U, Jaschik C, Kaese V, Niemeyer M. Alloy development, processing and applications in magnesium lithium alloys. Materials Transactions 2001;42:1160-1166.

[60] Wang JY, Hong WP, Hsu PC, Tan L. Microstructures and mechanical behavior of processed Mg-Li-Zn alloy. Materials Science Forum 2003;419-422:165-170.

[61] Massalski T, Okamoto H, Subramanian P, Kacprzak L. Binary alloy phase diagrams. United States of America: ASM International; 2001, p.2445.

[62] Matsuzawa K, Koshihara T, Kojima Y. Age-hardening and mechanical properties of Mg-Li-Al alloys. Journal of Japan Institute of Light Metals 1989;39(1):45-51.

[63] Song JM, Wen TX, Wang JY. Vibration fracture properties of a lightweight Mg-Li-Zn alloy. Scripta Materialia 2007;56:529-532.

[64] Metenier P, Gonzalez-Doncel G, Ruano OA, Wolfenstine J, Sherby DD. Superplastic behavior of a fine-grained two-phase Mg-9wt.%Li alloy. Materials Science and Engineering A 1990;125(2):195-202

[65] Sivakesavam O, Prasad YVRK. Characteristics of superplasticity domain in the processing map for hot working of as-cast Mg-11.5Li-1.5Al alloy. Materials Science and Engineering A 2002;323(1-2):270-277.

[66] Dong SL, Imai T, Lim SW, Kanetake N, Saito N. Superplasticity evaluation in an Mg-8.5Li alloy. Journal of Materials Science 2007;42(13):5296-5298.

[67] Cao FR, Ding H, Li YL, Zhou G, Cui JZ. Superplasticity, dynamic grain growth and deformation mechanism in ultra-light two-phase magnesium-lithium alloys. Materials Science and Engineering A 2010;527:2335-2341.

[68] Liu X, Du G, Wu, R, Niu Z, Zhang M. Deformation and microstructure evolution of a strain rate superplastic Mg-Li-Zn alloy. Journal of Alloys and Compounds 2011;509(39):9558-9561.

[69] Clark J, Sturkey L. The age-hardening mechanism in magnesium-lithium-zinc alloys. The Japan Institute of Metals 1957;86:272-276.

[70] Alamo A, Banchik D. Precipitation phenomena in the Mg-31at% Li-1 at% Al alloy. Journal of Materials Science 1980;15:222-229.

[71] Hatta H, Ramesh C, Kamado S, Kojima Y. Heat treatment characteristics and mechanical properties of superlight Mg-Li-Al alloys. Journal of Japan Institute of Light Metals 1997;47(4):202-207.

[72] Saito N, Mabuchi M, Nakanishi M, Kubota K, Higashi K. The aging behavior and the mechanical properties of the Mg-Li-Al-Cu alloy. Scripta Materialia 1997;36:551-555.

[73] Takuda H, Matsusaka H. Tensile properties of a few Mg-Li-Zn alloy thin sheets. Journal of Materials Science 2002;37:51-57.

[74] Yamamoto A, Ashida T, Kouta Y, Kim KB, Fukumoto S, Tsubakino H. Precipitation in Mg-(4-13)%Li-(4-5)%Zn ternary alloys. Materials Transactions 2003;44(4):619-624.

[75] Song GS, Staiger M, Kral M. Some new characteristics of the strengthening phase in β-phase magnesium-lithium alloys containing aluminum and beryllium. Materials Science and Engineering A 2004;371:371-376.

[76] Hsu CC, Wang JY, Lee S. Room temperature aging characteristic of MgLiAlZn alloy. Materials Transactions 2008;49(11):2728-2731.

[77] Chiu CH, Wu HY, Wang JY, Lee S. Microstructure and mechanical behavior of LZ91 Mg alloy processed by rolling and heat treatments. Journal of Alloys and Compounds 2008;460:246-252.

[78] Wang PC, Lin HC, Lin KM, Yeh MT, Lin CY. A study of aging treatment on the Mg-10Li-0.5Zn alloy. Materials Transactions 2009;50(9):2259-2263.

[79] Puleo DA, Huh WW. Acute toxicity of metal ions in cultures of osteogenic cells derived from bone marrow stromal cells. Journal of Applied Biomaterials 1995;6:109-116

[80] Jacobs JJ, Skipor AK, Patterson LM, Hallab NJ, Paprosky WG, Black J, Galante JO. Metal release in patients who have had a primary total hip arthroplasty. A prospective, controlled, longitudinal study. Journal of Bone Joint Surgury 1998;80:1447-1458.

[81] Lhotka C, Szekeres T, Steffan I, Zhuber K, Zweymuller K. Four-year study of cobalt and chromium blood levels in patients managed with two different metal-on-metal total hip replacements. Journal of Orthopaedic Research 2003;21:189-195.

[82] Nagels J, Stokdijk M, Rozing PM. Stress shielding and bone resorption in shoulder arthroplasty. Journal of Shoulder and Elbow Surgery 2003;12:35-39.

[83] Witte F, Fischer J, Nellesen J, Crostack HA, Kaese V, Pischd A. In vitro and in vivo corrosion measurements of magnesium alloys. Biomaterials 2006;27:1013-1018.

[84] Witte F, Kaese V, Switzer H, Maeyer-Lindenberg A, Wirth CJ, Windhag H. In vivo corrosion of four magnesium alloys and the associated bone response. Biomaterials 2005;26:3557-3563.

[85] Staiger MP, Pietak AM, Huadmai J, Dias G. Magnesium and its alloys as orthopedic biomaterials: a review. Biomaterials 2006;27:1728-1734.

[86] Okuma T. Magnesium and bone strength. Nutrition 2001;17:679-680.

[87] Vormann J. Magnesium: nutrition and metabolism. Molecular Aspects of Medicine 2003;24:27-37.

[88] Revell PA, Damien E, Zhang XS, Evans P, Howlett CR. The effect of magnesium ions on bone bonding to hydroxyapatite. Key Engineering Materials 2004;254-256:447-450.

[89] Zreiqat H, Howlett CR, Zannettino A, Evans P, Schulze-Tanzil G, Knabe C, et al. Mechanisms of magnesium-simulated adhesion of osteoblastic cells to commonly used orthopaedic implants. Journal of Biomedical Materials Research 2002;62:175-184.

[90] Stroganov GB, Savitsky E, Mikhailovich T, Nina M, Terekhova V, Fedorovna V et al. (1972). Magnesium-base alloys for use in bone surgery. US Patent No. 3,687,135.

[91] Li Z, Gu X, Lou S, Zheng Y. The development of binary Mg-Ca alloys for use as biodegradable materials with bone. Biomaterials 2008;29:1329-1344.

[92] Gao L, Zhang C, Zhang M, Huang, X, Sheng N. The corrosion of a novel Mg-11Li-3Al-0.5RE alloy in alkaline NaCl solution. Journal of Alloys and Compounds 2009;468: 285-289.

[93] Zhang S, Li J, Song Y, Zhao C, Zhang X, Xie C, Zhang Y, Tao H, He Y, Jiang Y, Bian Y. In vitro degradation, hemolysis and MC3T3-E1 cell adhesion of biodegradable Mg-Zn alloy. Materials Science and Engineering C 2009;29:1907-1912.

[94] Tsai MH, Chen MS, Lin LH, Lin MH, Wu CZ, Ou KL, Yu CH. Effect of heat treatment on the microstructures and damping properties of biomedical Mg-Zr alloy. Journal of Alloys and Compounds 2011;509:813-819.

[95] Gray JE, Luan B. Protective coatings on magnesium and its alloys – a critical review. Journal of Alloys and Compounds 2002;336:88-113.

[96] Shi Z, Song G, Atrens A. Influence of anodizing current on the corrosion resistance of anodized AZ91D magnesium alloy. Corrosion Science 2006;48:1939-1959.

[97] Yu BL, Uan JY. Sacrificial Mg film anode for cathodic protection of die cast Mg-9wt.%Al-1wt.%Zn alloy in NaCl aqueous solution. Scripta Materialia 2006;54:1253-1257.

[98] Ono S, Asami K, Masuko N. Mechanism of chemical conversion coating film growth on magnesium and magnesium alloys. Materials Transactions 2001;42:1225-1231.

[99] Elsentriecy HH, Azumi K, Konno H. Improvement in stannate chemical conversion coatings on AZ91D magnesium alloy using the potentiostatic technique. Electrochimica Acta 2007;53(2):1006-1012.

[100] Lin CS, Lee CY, Li WC, Chen YS. Fang GN. Formation of phosphate/permanganate conversion coating on AZ31 magnesium alloy. Journal of the Electrochemical Society 2006;153:B90-B96.

[101] Lin JK, Hsia CL, Uan JY. Characterization of Mg,Al-hydrotalcite conversion film on Mg alloy and Cl- and CO_3^{2-} anion-exchangeability of the film in a corrosive environment. Scripta Materialia 2007;56:527-530.

[102] Lin JK, Uan JY. Formation of Mg,Al-hydrotalcite conversion coating on

[103] Mg alloy in aqueous HCO_2^-/CO_3^{2-} and corresponding protection against corrosion by the coating. Corrosion Science 2009;51:1181-1188.

[104] Miyata S, Okada A. Synthesis of hydrotalcite-like compounds and their physico-chemical properties. Clays and Clay Minerals 1977;25:14-18.

[105] Qiu R, Iwamoto C, Satonaka S. Interfacial microstructure and strength of steel/aluminum alloy joints welded by resistance spot welding with cover plate. Journal of Materials Processing Technoligy 2009;209(8):4186-4193.

[106] Ambroziak A, Korzeniowski M. Using resistance spot welding for joining aluminum elements in automotive. Archives of Civil and Mechanical Engineering 2010;10(1): 5-13.

[107] Hayat F. The effect of the welding current on heat input, nugget geometry, and the mechanical and fractural properties of resistance spot welding on Mg/Al dissimilar materials. Materials & Design 2011;32(4):2476-2484.

[108] Xu W, Chen DL, Liu L, Mori H, Zhou Y. Microstructure and mechanical properties of weld-bonded and resistance spot welded magnesium-to-steel dissimilar joints. Materials Science and Engineering A 2012;537(1):11-24.

[109] Kishore Babu N, Brauser S, Rethmeier M, Cross CE. Characterizaton of microstructure and deformation behaviour of resistance spot welded AZ31 magnesium alloy. Materials Science and Engineering A 2012;549(15):149-156.

[110] Thornton PH, Krause AR, Davies RG. Aluminum spot weld. Welding Journal 1996;75(3):101s-108s.

[111] Fukumoto S, Lum I, Biro E, Boomer DR Zhou Y. Effects of electrode degradation on electrode life in resistance spot welding of aluminum alloy 5182. Welding Journal 2003;82(11):307s-312s.

[112] Lang B, Sun DQ, Li GZ, Zhu BQ. Electrode degradation in resistance spot welding of magnesium alloy. ISIJ International 2009;49(11):1744-1748.

[113] Mishra RS, Mahoney MW, McFadden SX, Mara NA, Mukherjee AK. High strain rate superplasticity in a friction stir processed 7075 Al alloy. Scripta Materialia 2000;42(2): 163-168.

[114] Mishra RS, Mahoney MW. Friction stir processing: A new grain refinement technique to achieve high strain rate superplasticity in commercial alloys. Materials Science Forum 2001;357-359:507-514.

[115] Mishra RS, Ma ZA. Friction stir welding and processing. Materials Science and Engineering R 2005;50:1-78.

[116] Jata KV, Semiatin SL. Continuous dynamic recrystallization during friction stir welding of high strength aluminum alloys. Scripta Materialia 2000;43(8):743-749.

[117] Chang CI, Lee CJ, Huang JC. Relationship between grain size and Zener-Holloman parameter during friction stir processing in AZ31 Mg alloys. Scripta Materialia 2004;51(6):509-514.

[118] Woo W, Choo H, Prime MB, Feng Z, Clausen B. Micristructure, texture and residual stress in a friction-stir-processed AZ31B magnesium alloy. Acta Materialia 2008;56(8): 1701-1711.

[119] Yu S, Chen X, Huang Z, Liu Y. Microstructure and mechanical properties of friction stir welding of AZ31B magnesium alloy added with cerium. Journal of Rare Earths 2010;28(2):316-320.

[120] Chowdhury SH, Chen DL, Bhole SD, Cao X, Wanjara P. Friction stir welded AZ31 magnesium alloy: microstructure, texture, and tensile properties. Metallurgical and Materials Transactions A – Physical Metallurgy and Materials Science 2012; DOI: 10.1007/s11661-012-1382-3.

[121] Hirata T, Oguri T, Hagino H, Tanaka T, Chung SW, Takigawa Y, Higashi K. Influence of friction stir welding parameters on grain size and formability in 5083 aluminum alloy. Materials Science and Engineering 2007;456:344-349.

[122] Cao X, Jahazi M. Effect of welding speed on the quality of friction stir welded butt joints of a magnesium alloy. Materials & Design 2009;30(6):2033-2042.

[123] Yang J, Wang D, Xiao BL, Ni DR, Ma ZY. Effects of rotation rates on microstructure, mechanical properties and fracture behavior of friction stir-welded (FSW) AZ31 magnesium alloy. Metallurgical and Materials Transactions A – Physical Metallurgy and Materials Science 2012;DOI:10.1007/s11661-012-1373-4.

[124] Park SHC, Sato YS, Kokawa H. Effect of micro-texture on fracture location in friction stir weld of Mg alloy AZ61 during tensile test. Scripta Materialia 2003;49(2):161-166.

[125] Park SHC, Sato YS, Kokawa H. Basal plane texture and flow pattern in friction stir weld of a magnesium alloy. Metallurgical and Materials Transactions A – Physical Metallurgy and Materials Science 2003;34(4):987-994.

[126] Woo W, Choo H, Brown DW, Liaw PK, Feng Z. Texture variation and its influence on the tensile behavior of a friction-stir processed magnesium alloy. Scripta Materialia 2006;54:1859-1864.

[127] Feng AH, Ma ZY. Enhanced mechanical properties of Mg-Al-Zn cast alloy via friction stir processing. Scripta Materialia 2007;56:397-400.

[128] Afrin N, Chen DL, Cao X, Jahazi M. Microstructure and tensile properties of friction stir welded AZ31B magnesium alloy. Materials Science and Engineering A 2008;472(1-2):179-186.

[129] Zhang DT, Suzuki M, Maruyama K. Study on the texture of a friction stir welded Mg-Al-Ca alloy. Acta Metallurgica Sinica 2009;48(2):335-340.

[130] Mansoor B, Ghosh AK. Microstructure and tensile behavior of a friction stir processed magnesium alloy. Acta Mateialia 2012;60(13-14):5079-5088.

[131] Lathabai S, Painter MJ, Cantin GMD, TyagiVK. Friction spot joining of an extruded Al-Mg-Si alloy. Scripta Materialia 2006;55(10):899-902.

[132] Chen YC, Nakata K. Friction stir lap joining aluminum and magnesium alloys. Scripta Materialia 2008;58(6):433-436.

[133] Yang CW, Hung FY, Lui TS, Chen LH, Juo JY. Weibull statistics for evaluating failure behaviors and joining reliability of friction stir spot welded 5052 aluminum alloy. Materials Transactions 2009;50(1):145-151.

[134] Yin YH, Ikuta A, North TH. Microstructural features and mechanical properties of AM60 and AZ31 friction stir spot welds. Materials & Design 2010;31(10):4764-4776.

[135] Chen YC, Gholinia A, Prangnell PB. Interface structure and in bonding and abrasion circle friction stir spot welding: A novel approach for rapid welding aluminum alloy to steel automotive sheets. Materials Chemistry and Physics 2012;134(1):459-463.

[136] Sun YF, Fujii H, Takaki N, Okitsu Y. Microstructure and mechanical properties of mild steel joints prepared by a flat friction stir spot welding technique. Materials & Design 2012;37:384-392.

[137] Lin PC, Lin SH, Pan J, Pan T, Nicholson JM, Garman MA. SAE Technical Paper No. 2004-01-1330, Society of Automotive Engineering, Warrendale, PA; 2004.

[138] Kaibyshev R, Stidikov O. Dynamic recrystallization of magnesium at ambient temperature. Zeitschrift fuer Metallkunde 1994;85:738-743.

[139] McQueen HJ, Imbert CAC. Dynamic recrystallization: plastic enhancing structural development. Journal of Alloys and Compounds 2004;378:35-43.

[140] Humphreys FJ, Hotherly M. Recrystallization and Related Annealing Phenomena. New York: Pergamon Press; 1995.

[141] Watanabe H, Mukai T, Kohzu M, Tanabe S, Higashi K. Effect of temperature and grain size on the dominant diffusion process for superplastic flow in an AZ61 magnesium alloy. Acta Materialia 1999;47:3753-3758.

[142] Lynch SP, Muddle BC, Pasang T. Ductile-to-brittle fracture transitions in 8091 Al-Li alloys. Acta Materialia 2001;49(15):2863-2874.

[143] Argon AS. Mechanics and physics of brittle to ductile transitions in fracture. Journal of Engineering Materials Technology, Transactions of the ASME 2001;123(1):1-11.

[144] Crocker AG, Flewitt PEJ, Smith GE. Computational modelling of fracture in polycrystalline materials. International Materials Reviews 2005;50(2):99-124.

[145] Russell AM, Chumbley LS, Gantovnik VB, Xu K, Tian Y, Laabs FC. Anomalously high impact fracture toughness in BCC Mg-Li between 4.2K and 77K. Scripta Materialia 1998;39:1663-1667.

[146] Counts WA, Friák M, Raabe D, Neugebauer J. Using ab initio calculations in designing bcc Mg-Li alloys for ultra-lightweight applications. Acta Materialia 2009;57(1): 69-76.

[147] Hughes GM, Flewitt PEJ. Temperature dependence of mechanical properties of zinc and Zircaloy measured using miniaturized disc tests. Materials Science and Technology 2008;24(5):567-574.

[148] Yang CW, Lui TS, Chen LH, Hung HE. Tensile mechanical properties and failure behaviors with the ductile-to-brittle transition of the α+β-type Mg-Li-Al-Zn alloy. Scripta Materialia 2009;61:1141-1144.

[149] Bae DH, Ghosh AK. Cavity formation and early growth in a superplastic Al-Mg aloy. Acta Materialia 2002;50(3): 511-523.

[150] Drozd Z, Trojanová Z, Kúdela S. Deformation behavior of Mg-Li-Al alloys. Journal of Alloys Compounds 2004;378(1-2):192-195.

[151] Agnew SR, Horton JA, Yoo MH. Transmission electron microscopy investigation of <c+a> dislocations in Mg and α-solid solution Mg-Li alloys. Matallurgical and Materials Transactions A – Physical Metallurgy and Materials Science 2002;33(3):851-858.

Chapter 14

Photothermal Techniques in Material Characterization

Marios Nestoros

Additional information is available at the end of the chapter

http://dx.doi.org/10.5772/56039

1. Introduction

The 21st century is characterized by a rapid increase in technological applications which involve the design and development of materials with tailored properties that manifest and take advantage of scientific developments from the region of modern physics. In medicine and biology there is a need of new materials that are biocompatible and exhibit similar biological behavior to that of several types of human tissue. In structural engineering there are new approaches for the production of concrete materials that can withstand tension, can self-repair and self-monitor structural damage. Organic semiconductors and devices based on nanomaterials unfold new dimensions in optoelectronics, information technology and renewable energy applications.

The field of photothermal physics and the related characterization techniques is becoming increasingly important with the advance in the design and production of "new materials" (composite materials/heterostructures). Several photothermal techniques present significant advantages over traditional material characterization techniques, mainly due to their non-contact and non-destructive nature. Photothermal techniques offer high sensitivity relative to the thermal and optoelectronic properties of materials although in most cases the extraction of the aforementioned parameters is not direct and involves much computational effort. In addition in several cases the above techniques can be used to monitor in real time industrial or in-lab processes.

In this chapter the concept of thermal and electronic wave methodology will be initially introduced, followed by information on a brief description of experimental setups and recent advances in the field. The chapter will focus on the presentation of examples from the application of some photothermal techniques in the characterization of composite materials and thin metal films used as gas sensors.

2. Thermal and electronic waves

The optical absorption of an intensity modulated beam of light results in the diffusion of the generated heat through the sample according to Fourier's law. Due to the periodic nature of the excitations, one expects a periodic temperature response. The combination of Fourier's Law and energy conservation provides the conduction heat transfer equation [1]:

$$\nabla^2 T(r,t) - \frac{1}{D_T}\frac{\partial}{\partial t}T(r,t) = -\frac{1}{\kappa}Q(r,t) \tag{1}$$

In the above equation the temperature **T** is a function of position **r** and time **t**. The temperature depends on the thermal diffusivity $\mathbf{D_T}$ and on the thermal conductivity $\boldsymbol{\kappa}$ of the material. The heat source term **Q**(r,t) is measured in W·m^{-3} and is modulated at an angular frequency $\boldsymbol{\omega = 2\pi f}$. In the case of one dimensional heat flux along the z axis the heat source term takes the form

$$Q(x,t) = \frac{1}{2}\cdot I_0 \cdot (1-R)\cdot \alpha \cdot e^{-\alpha\cdot z}(1+e^{i\cdot\omega\cdot t}) \tag{2}$$

Assuming one dimensional heat flow along **z** direction, introducing eq.(2) into eq.(1) and taking the temporal Fourier transform of eq.(1) one gets the following differential equation for the ac temperature field $\mathbf{\Delta T}$ in the material

$$\frac{d^2\Delta T(z,\omega)}{dz^2} - \sigma_T^2(\omega)\Delta T(\omega) = -\frac{\alpha\cdot I_0\cdot(1-R)}{k}e^{-\alpha\cdot z} \tag{3}$$

In the above equations$\boldsymbol{\alpha}$ and **R**, are respectively the optical absorption coefficient and reflectivity at the excitation wavelength and $\mathbf{I_0}$ is the incident optical intensity (in W/m^2).

A fundamental quantity for the thermal waves [1] which are the solutions of eq.(3) is the thermal wavenumber $\boldsymbol{\sigma_T}$ [1]defined as:

$$\sigma_T = \sqrt{\frac{i\omega}{D}} = (1+i)\sqrt{\frac{\omega}{2D}} \equiv \frac{(1+i)}{\mu} \tag{4}$$

The quantity $\boldsymbol{\mu_T}$ is the thermal diffusion length defined as:

$$\mu_T = \sqrt{\frac{2D_T}{\omega}} = \sqrt{\frac{D_T}{\pi\cdot f}} \tag{5}$$

http://dx.doi.org/10.5772/56039

At high modulation frequencies (**f**) the thermal diffusion length μ_T is small, so the thermal waves are restricted near the surface of the sample, while at low modulation frequencies the thermal wave centroid lies deeper in the material. The ac temperature field $\Delta\mathbf{T}$ has an amplitude and phase (relative to the excitation) that can be measured with different photothermal configurations and reveal the thermal properties of the material under investigation.

In semiconducting materials, if the incident photon energy (**hν**) is above the energy gap ($\mathbf{E_g}$), the optical excitation results in the generation of free electron and hole populations that contribute to the conductivity of the material. On a picosecond time scale electrons and holes become thermalized with the lattice through phonon interaction, releasing the excess energy h ν -E_g, and lie on the bottom of the conduction band (electrons) and top of the valence band (holes). Then, electrons and holes diffuse for some time, known as carrier lifetime $\boldsymbol{\tau}$, before they recombine through non-radiative recombination processes, giving thermal energy h ν to the lattice. Following an approach similar to the one described above for thermal waves, one can formulate an equation for the plasma (electron –hole) diffusion equation:

$$\frac{d^2\Delta N(z,\omega)}{dz^2} - \sigma_N{}^2 \cdot \Delta N(z,\omega) = -\frac{I_0 \cdot (1-R)}{D_N} \alpha \cdot e^{-\alpha \cdot z} \tag{6}$$

Where $\Delta\mathbf{N}$ is the ac component of the plasma concentration in m^{-3}, $\mathbf{D_N}$ is the electronic diffusivity of the material and $\boldsymbol{\sigma}_{\mathbf{N}}$ is the electronic wave number defined as

$$\sigma_N = \sqrt{\frac{1 + i \cdot \omega \cdot \tau}{D \cdot \tau}} \tag{7}$$

The electronic diffusion length is defined as

$$\mu_N \equiv \frac{1}{\mathrm{Re}(\sigma_N)} = \sqrt{\frac{2 \cdot D_N \cdot \tau}{1 + \sqrt{1 + \omega^2 \cdot \tau^2}}} \tag{8}$$

At high modulation frequencies $\omega\tau \gg 1$ the expression for the electronic diffusion length takes a similar form to that for the thermal diffusion length and depends on the modulation frequency, $\mu_N=(D_N/\pi f)^{1/2}$. At small modulation frequencies the electronic diffusion length is constant, $\mu_N=(D_N\cdot\boldsymbol{\tau})^{1/2}$, and the plasma wave is in phase with the optical excitation.

The dimensionality of the diffusion problem (thermal/plasma) depends on the relation of the diffusion length (thermal, electronic) to the spot sizes of the pump beam and the probe beam. The assumption of one dimensional propagation for the thermal/electronic wave is a valid as long as the thermal/electronic diffusion length is smaller than the spot size of the beam.

Here we have to mention that in the case of semiconductors the heat source [2,3] included in the thermal diffusion equation has to be modified in order to include the heat released by the thermalization of free carriers which is proportional to h ν-E_g as well as the one due to non-radiative recombination which is proportional to E_g/τ. Hence the thermal diffusion equation takes the form

$$\frac{d^2\Delta T(z,\omega)}{dz^2}+\frac{E_g}{\kappa\cdot\tau}\Delta N(z,\omega)-\sigma_T{}^2\cdot\Delta T(z,\omega)=-\frac{(h\nu-E_g)\cdot\alpha\cdot I_0\cdot(1-R)\cdot e^{-\alpha\cdot z}}{\kappa\cdot h\nu} \tag{9}$$

This means that the system of thermal and plasma diffusion equations is coupled and one should first seek the solution of the plasma equation and then the thermal one.

The appropriate boundary conditions [2,3] have to be applied depending on the sample structure (two layer, multilayer) in order to get the solutions of the ac fields ΔT(z, ω) and ΔN(z, ω). The boundary conditions involve the continuity of heat flux and temperature at the interfaces, as well as the continuity of carrier concentration at the interfaces and the recombination of free carriers at the surfaces and interfaces.

3. Instrumentation and setup of some photothermal techniques

3.1. Photomodulated Thermoreflectance (PMTR)

Photomodulated Thermoreflectance (PMTR) has been used extensively for the characterization of a variety of implanted semiconductors and layered structures [3-9]. Concerning applications in electronic materials one can mention among others: ion implantation monitoring, annealing kinetics of defects [5,6], etch monitoring and carrier lifetime evaluation [6,7], thin metal film measurements [4]. The technique is based on the detection of local temperature excursions on the surface of the material under investigation induced by an intensity modulated light beam (pump) via the monitoring of the surface reflectance with the aid of another light beam (probe). The non-contact and non-destructive character of the technique in addition to the high sensitivity to thermal inhomogeneities, make PMTR an attractive material evaluation technique. The technological trend towards smaller structures in the field of microelectronics creates the need for characterization techniques with improved spatial resolution. Such a technique is PMTR-microscopy (micro PMTR) [10], a variant of PMTR which combines the advantages of the PMTR technique mentioned above, with the various functions of a microscope offering the ability of viewing the sample under test while characterizing it. In addition it offers an improved spatial resolution as compared to the conventional PMTR technique. A conventional PMTR setup (Figure 1) as the one used for the experiments described in the applications discussed in this chapter, is described below.

An Ar^+ laser operating at the main lines of 488 nm and 514.5 nm is used as the excitation source providing the pump beam with an intensity that is modulated by an acousto-optic modulator. The pump beam (typical power 60 mW) is then focused onto the sample's surface at a nearly normal incidence to a spot size that can be varied. The reflectivity changes of the surface due to

http://dx.doi.org/10.5772/56039

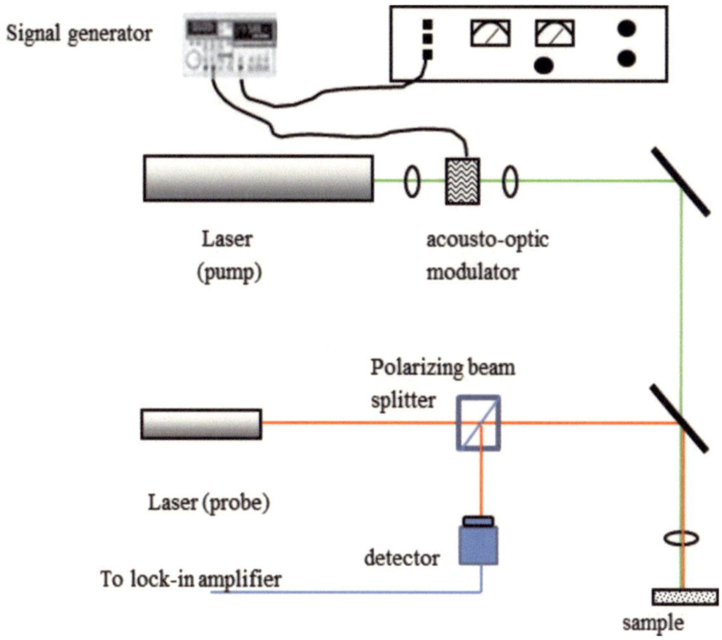

Figure 1. PMTR setup

the photo-induced thermal and or plasma waves are probed by a He–Ne Laser operating at 632.8 nm and at a power of a few mW. The probe beam is focused onto the sample collinearly to the pump beam. The reflected probe beam is diverted with the aid of a polarizing beam splitter to a silicon photodiode, filtered by a 632.8nm band pass interference filter. In order to reject any signal contributions arising from the thermoelastic deformation of the sample surface, the detector is deliberately underfilled, and is operated in the sum mode. The output of the detector is monitored with a fast lock-in amplifier, synchronous with the Ar^+ modulation of various frequencies. The in-phase and quadrature components of the signal are stored in the computer for subsequent analysis and display. The PMTR signal $\mathbf{S_{PMTR}}$ is proportional to the reflectivity change $\mathbf{\Delta R}$ of the sample's surface (z=0) and can be expressed as

$$S_{PMTR}(\omega) \equiv \frac{\Delta R}{R} = \frac{1}{R}\frac{\partial R}{\partial T} \cdot \Delta \mathrm{T}(z=0,\omega) + \frac{1}{R}\frac{\partial R}{\partial N} \cdot \Delta N(z=0,\omega) \tag{10}$$

3.2. Photothermal Radiometry (PTR)

Photothermal radiometry [9-11] relies on the monitoring of modulated blackbody radiation emitted from a material that is optically excited by an intensity modulated monochromatic light beam. Following partial or total absorption of the incident radiation, a portion of it is converted into heat. The subsequent diffusion of heat is dictated by thermal transport properties like thermal diffusivity and thermal conductivity. In semiconducting samples the photoexcited carriers behave as Planck radiators that contribute to the black body emission.

In the case of good quality semiconductors the contribution of free carriers is the main component of the PTR signal which has an amplitude that is proportional to free carrier's lifetime. As a result photothermal techniques allow the determination of optical, electronic as well as thermophysical properties of materials even in cases of efficient light scattering materials, which prohibit their characterization through conventional optical techniques. Photothermal radiometry has been used to characterize a variety of materials and thin films ranging from semiconducting films [11,12,14] to nanotube/nanoparticle composites [15,16] and biomaterials [17]. The PTR experimental setup (shown in Figure 2) consists of an excitation laser source (DPSS laser emitting at 512 nm), an a acousto-optic cell that chops periodically the laser intensity, an MCT (Mercury, Cadmium- Telluride) detector with detection in the 2 μm-12 μm wavelength range, a preamplifier, a phase sensitive detection system (lock in amplifier) and a couple of off axis Ag coated paraboloid mirrors to gather and collimate the black body radiation emitted from the excited sample in the sensitive detector area. The excitation beam with incident power illuminates the sample surface almost at normal incidence. As the modulation frequency is varied (typical frequency span 10^0-10^6 Hz) the signal amplitude and phase is recorded. In order to calculate the instrumental transfer function with which we have to correct the experimental data, a frequency scan of a thick and opaque sample (steel disc) is usually performed. The PTR signal S_{PTR} can be expressed in most of the cases as an integral of the temperature field and the plasma field (in the case of semiconductors) over the thickness of the sample multiplied by appropriate coefficients C_1 and C_2 [14].

$$S_{PTR}(\omega) = C_1 \cdot \int_0^l \Delta T(z,\omega) \cdot dz + C_2 \cdot \int_0^l \Delta N(z,\omega) \cdot dz \qquad (11)$$

4. Applications of photothermal techniques in materials characterization

4.1. Analysis of Hydrogen adsorption and desorption on palladium thin films with photomodulated thermoreflectance measurements

The importance of hydrogen gas technology is growing due to the many industrial applications as well as the needs for environmentally clean energy resources. Nevertheless hydrogen gas has some disadvantages concerning its storage and its flammability. For these reasons there is a need for hydrogen sensors with high sensitivity and fast response. In addition the reaction kinetics of hydrogen gas with the surface of the sensor is quite interesting from the viewpoint of physical chemistry and materials science. In this paragraph the kinetics of hydrogen adsorption and desorption on palladium thin films was studied [18] via photomodulated thermorefelectance (PMTR) measurements. The analysis of the experimental data is based on a Langmuirian isothermal model which supports a dissociative adsorption of hydrogen on palladium followed by molecular desorption. The values of the adsorption and desorption rate constants were determined and their values are discussed. The response and recovery times

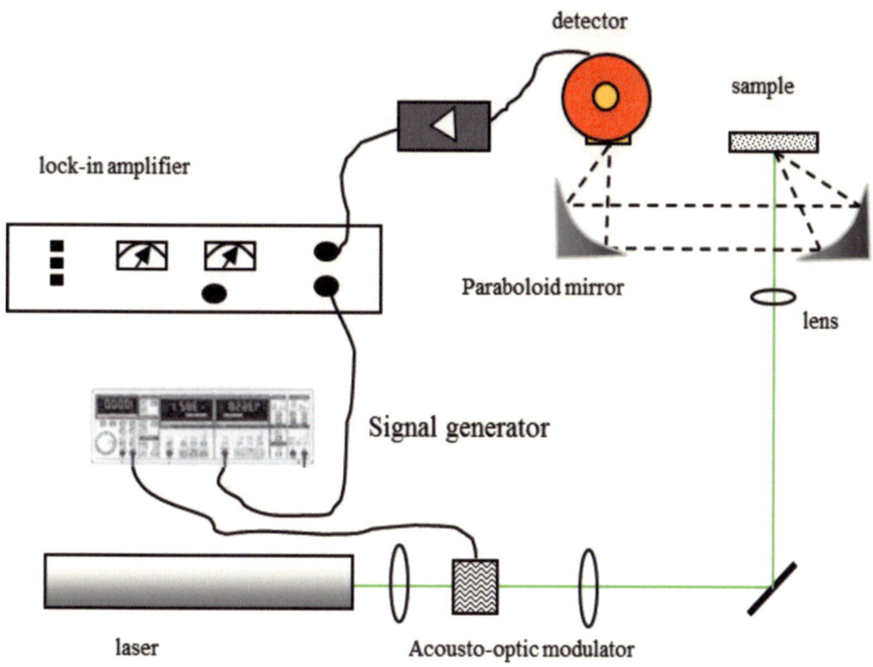

Figure 2. PTR setup

of the sensor were measured and their dependence on hydrogen concentration is also explored and discussed.

4.1.1. Sample details and experimental methodology

The sensor consists of a palladium thin film, with a thickness of 6 nm, evaporated on silicon substrate, with an intermediate layer of silicon oxide. In hydrogen atmosphere the PMTR signal from such sensors alters due to the adsorption of hydrogen onto the palladium film and the formation of the palladium hydride. The formation of the PdH_X modifies the electronic structure of the palladium metal and also changes the Pd lattice constant. These modifications finally change the refractive index and the reflectance of the Pd film surface [19,20].The experimental setup has been discussed earlier in this chapter. The excitation beam (Ar^+ laser operating at 488 nm) power was set at 100 mW and the diameter of the spot size on the samples was 50 μm. The reflectivity changes of the surface were monitored by a He-Ne Laser operating at 632.8 nm and a power of 0.5 mW. The probe beam was focused onto the sample to a spot size of 35 μm, collinearly to the pump beam. The modulation frequency value of the pump beam was set at 1 kHz since at this frequency the signal was stable, with a good signal to noise ratio.

The thin film sensor was placed in a cylindrical cell (radius of 4.7 cm and length of 15.6 cm) with a gas input and output system as well as a temperature controller. High accuracy flowmeters were used in order to supply hydrogen and nitrogen in the cylinder at the desired ratio and at a constant flow rate of 150 mL/min. The temperature was raised up to 100°C, and the samples were exposed to different hydrogen concentrations (from 25 to 150 ppb), in nitrogen atmosphere.

4.1.2. Reaction kinetics and the PMTR signal

The differential equations describing the adsorption process result from the kinetics of the chemical reaction and the application of the Langmuirian model [20,21].

In the case of the molecular adsorption $S^* + H_2(gas) \underset{k_d}{\overset{k_a}{\rightleftharpoons}} S - H_2$ and considering that the rate constant of adsorption $\mathbf{k_a}$ is many orders of magnitude bigger than the rate constant of desorption $\mathbf{k_d}$ the time evolution of the coverage is

$$\theta_P(t) = 1 - \exp\left[-k_a P(t - t_0)\right] \tag{12}$$

The partial pressure of hydrogen H_2 is $\mathbf{P}$ and $\mathbf{t_0}$ the time at which hydrogen is deployed to the chamber, i.e. $\theta(t_0)=0$.

In the case of dissociative adsorption $2S^* + H_2(gas) \underset{k_d}{\overset{k_a}{\rightleftharpoons}} 2S - H$ the solution of the governing differential equation is

$$\theta_P(t) = \frac{k_a P \tanh\left(\sqrt{k_a k_d P}\,(t - t_0)\right)}{\sqrt{k_a k_d P} + k_a P \tanh\left(\sqrt{k_a k_d P}\,(t - t_0)\right)} \tag{13}$$

After a Taylor expansion of the hyperbolic tangents around $\sqrt{k_a k_d P} \approx 0$ up to first order, the solution is

$$\theta_P(t) = \frac{k_a P(t - t_0)}{1 + k_a P(t - t_0)} \tag{14}$$

The processes of adsorption and desorption happen simultaneously and the negative sign of the desorption term shows that the two processes are competitive. The terms which describe the desorption from the Pd surface and the adsorption of hydrogen at the Pd-SiO_2 interface were not taken into account. This is because the diffusion of hydrogen adsorbed atoms and their adsorption to interface happen so rapidly, that the amount of the adsorbed hydrogen at the interface is constantly in equilibrium with that at the surface [22]. According to Lundström et al. [22], when the hydrogen partial pressure is lower than 200 Pa, the response signal of the hydrogen sensor $\Delta\xi$ is proportional to the surface coverage θ. At a partial pressure P, the signal's time dependence is given by:

$$\Delta\xi = \alpha\,\theta_P(t) \tag{15}$$

Hence the PMTR signal evolution, presented in Fig. 3, is given by the equation:

$$\xi(t) = \frac{\Delta\xi_s}{\theta_s}\,\theta_P(t) + \xi_0 \tag{16}$$

where $\Delta\xi_s$ is the maximum change of the photothermal signal, which happens at $t\rightarrow\infty$ and is defined in Figure 3(b). It's obvious that, θ_s and $\Delta\xi_s$ are slightly larger than θ_f and $\Delta\xi_f$, because the former is defined at $t\rightarrow\infty$ while the latter is defined at t_f. In the steady state ($d\theta/dt=0$) the coverage at saturation θ_s for molecular and dissociative adsorption is given respectively by:

$$\begin{aligned} \theta_s\big|_{mol} &= \frac{k_a P}{k_d + k_a P} && \text{(a)} \\ \theta_s\big|_{diss} &= \frac{\sqrt{k_a P}}{\sqrt{k_d} + \sqrt{k_a P}} && \text{(b)} \end{aligned} \tag{17}$$

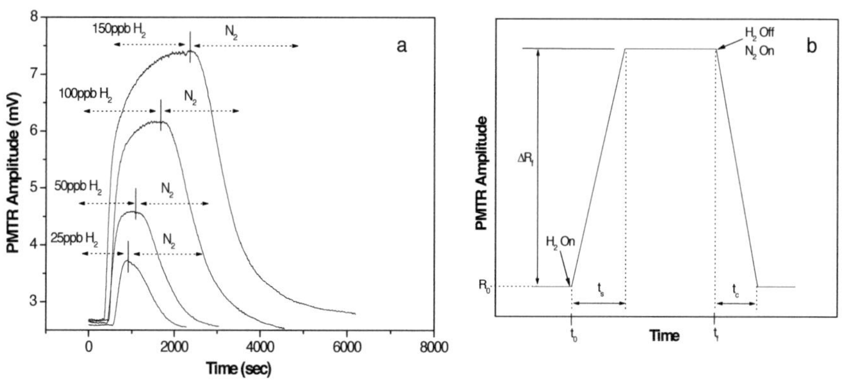

Figure 3. (a) PMTR signal response during adsorption and desorption of different H_2 concentrations, at 100°C, (b) definition of ΔR_f, t_s and t_c. Demetriou C., Nestoros M. and Christofides C. Appl. Phys. A 2008; 92 (3), 651-658

The equilibrium constants K_{mol} and K_{diss} for the cases of molecular adsorption and dissociative adsorption respectively are:

$$K_{mol} = {k_a}/{k_d} \quad \text{(a)}$$
$$K_{diss} = \sqrt{{k_a}/{k_d}} = \sqrt{K_{mol}} \quad \text{(b)} \qquad (18)$$

4.1.3. Experimental results and discussion

The sensor was exposed to various concentrations of hydrogen in nitrogen environment, and afterwards only to nitrogen gas. One can observe a quick rise upon the introduction of H_2 followed by a slower one up to the saturation point. The saturation point occurs at longer times with increasing concentrations of hydrogen (Figure 3(a)). Once the nitrogen gas was introduced following hydrogen saturation in the chamber, the PMTR signal decreased and returned to equilibrium. It is important to note that the photothermal signal decreases faster at low hydrogen concentration. In addition, according to Figure 3(a), the PMTR signal change is higher for larger amounts of hydrogen in the chamber. This is consistent with eqs.(18a) and (18b). The experimental data were fitted using the Langmuirian model defined from eqs. (12)-(14) as well as eqs. (15) and (16). As result the rate constants of adsorption k_a, desorption k_d where extracted. The equilibrium constant K_{diss} and the surface coverage at saturation θs were then calculated. As shown in Figure 4(a) and (b) the experimental results are following the model describing molecular adsorption in the early part of the signal rise, while at later times there is a significant deviation. As it can be seen from Figure 4(c) and (d), the theoretical model describing dissociative adsorption is in very good agreement with the experimental data, indicating that the process of hydrogen adsorption on palladium is dissociative as mentioned in bibliography [21,23]. The extracted values of k_a and k_d are presented in Tables 1 and 2 (Table 2 for k_d =0). One can observe a decrease of the values of k_a and k_d with increasing hydrogen concentration. Furthermore the rate constant of desorption k_d is several orders of magnitude smaller than the k_a constant which supports the decision of ignoring the desorption process for both types of molecular and dissociative adsorptions. The decrease of the adsorption rate constant k_a as the hydrogen partial pressure increases could be the result of site to site interactions upon high coverage. For this reason the saturation point occurs at longer times for higher hydrogen concentration.

P (ppb)	k_a ($Pa^{-1}s^{-1}$)	$\sigma(k_a)$ ($Pa^{-1}s^{-1}$)	k_d (s^{-1})	$\sigma(k_d)$ (s^{-1})	α (mV)	σ(α) (mV)
25	3.457	±0.555	1.382×10^{-3}	$\pm1.260\times10^{-3}$	1.781	±0.187
50	2.220	±0.148	1.281×10^{-3}	$\pm0.312\times10^{-3}$	2.702	±0.094
100	0.717	±0.011	5.930×10^{-4}	$\pm0.342\times10^{-4}$	4.596	±0.033
150	0.448	±0.004	5.799×10^{-7}	$\pm55.030\times10^{-7}$	5.186	±0.002

Table 1. Parameters obtained by fitting the dissociative adsorption model

Using the values of k_a and k_d which were obtained by fitting to the dissociative adsorption equation, the equilibrium constant K_{diss} and the surface coverage at saturation θ_{diss} were

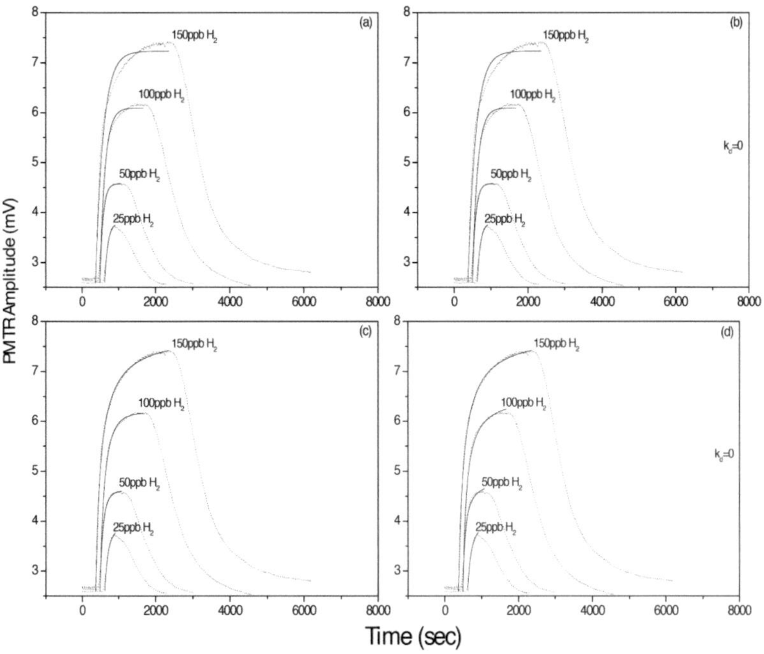

Figure 4. Molecular adsorption: (a) and (b), dissociative adsorption: (c) and (d), and fits with solid line. Demetriou C., Nestoros M. and Christofides C. Appl. Phys. A 2008; 92 (3), 651-658

calculated. The calculated values of K_{diss} are presented in Table 3 and are of the same order of magnitude for the different values of hydrogen partial pressure. At 150 ppb the equilibrium constant K_{diss} is an order of magnitude greater than the ones calculated for the lower concentrations. This is due to the small value of the rate constant of desorption k_d at 150 ppb.

P (ppb)	k_a ($Pa^{-1}s^{-1}$)	Δk_a ($Pa^{-1}s^{-1}$)	α (mV)	σ(α) (mV)	θ_f (%)
25	4.221	±0.202	1.568	±0.028	70.16
50	3.200	±0.086	2.269	±0.013	83.75
100	0.995	±0.010	3.956	±0.007	88.48
150	0.449	±0.002	5.185	±0.005	90.64

Table 2. Parameters obtained by fitting dissociative adsorption model and neglecting any desorption process

The calculated value of the equilibrium constant has also large errors. This is due to the large statistical errors of the rate constants of adsorption and mainly of the rate constants of desorption as shown in Table 1. Since the mechanism of desorption is negligible during adsorption process, it is not possible to determine the rate constant of desorption and consequently the equilibrium constant, taking into account only the adsorption process. The values

P (ppb)	θ_f (%)	$K_{diss}(Pa^{-1/2})$	$\sigma(K_{diss})$ $(Pa^{-1/2})$	θ_s (%)
25	61.77	50.016	±23.150	71.44
50	70.33	41.628	±5.256	74.64
100	76.15	34.778	±1.038	77.67
150	90.62	879.380	±4170.413	99.08

Table 3. Calculations of K_{diss} and θ_s using the values of k_a and k_d from Table 1

of the palladium surface coverage θ_f, at time t_f, presented in Tables 2 and 3 are larger for higher hydrogen concentrations. The values of θ_f were calculated using the following relationship $\theta_f = \Delta\zeta_f/\alpha$.The quantity α was obtained by fitting experimental data to theory and $\Delta\xi_f$.

A similar process was followed in order to study the desorption process (nitrogen enters the chamber). The equations were solved once more letting the partial pressure to be zero. The solutions for molecular and dissociative adsorption are given by eqs. (19a) and (19b), respectively.

$$\theta_P(t) = \theta_f \exp\left[-k_d\left(t - t_f\right)\right] \quad (a)$$
$$\theta_P(t) = \frac{\theta_f}{1 + k_d\theta_f\left(t - t_f\right)} \quad (b) \tag{19}$$

In Figure 5 (a) and (b) the experimental data are fitted using the above equations with the theory of molecular and dissociative desorption respectively. Note that by fitting the equation of molecular desorption to the experimental data one gets much better results. Table 4 presents the resulting values of the rate constant of desorption k_d for the different hydrogen concentrations. As shown in Table 4, the rate constant k_d is of the same order of magnitude and varies slightly with hydrogen concentration. This contradicts the fact that the recovery time of the PMTR signal is larger for higher hydrogen concentrations, as will be shown later. Using the data of Figure 3(a), the response time t_s and the recovery time t_c of the hydrogen sensor were determined. The response time is defined as the time needed for the PMTR signal to reach saturation, i.e. for the system to reach equilibrium. The time needed for the system to recover from saturation to its initial state is defined as the recovery time.

P (ppb)	k_d (s^{-1})	σk_d (s^{-1})
25	2.592×10^{-3}	$\pm0.012\times10^{-3}$
50	1.887×10^{-3}	$\pm0.007\times10^{-3}$
100	1.475×10^{-3}	$\pm0.002\times10^{-3}$
150	1.463×10^{-3}	$\pm0.002\times10^{-3}$

Table 4. Desorption rate constants

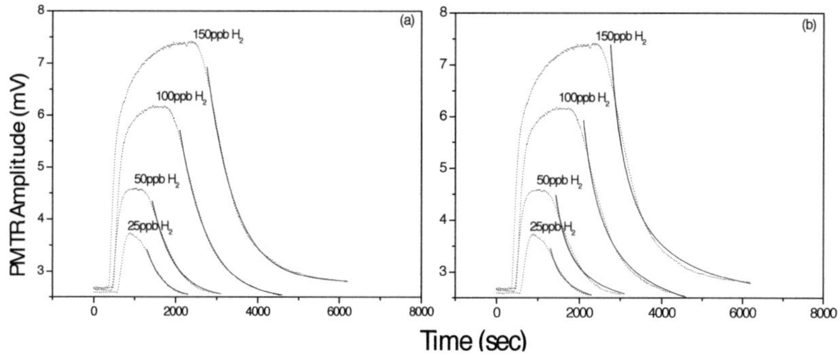

Figure 5. Molecular desorption (a) and dissociative desorption (b), data and fit. Demetriou C., Nestoros M. and Christofides C. Appl. Phys. A 2008; 92 (3), 651-658

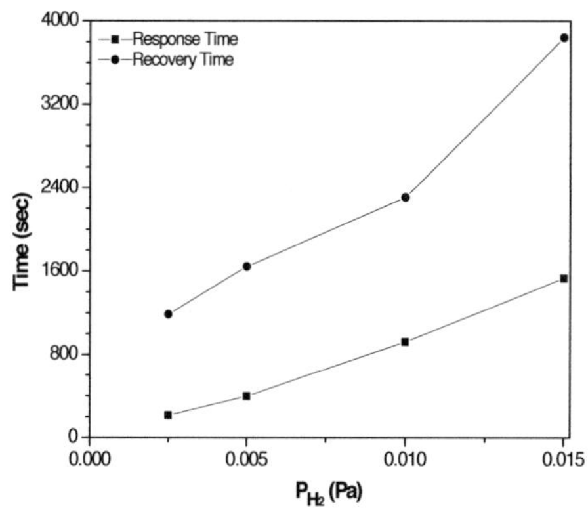

Figure 6. Response (lower curve) and recovery (upper) times versus hydrogen partial pressure. Demetriou C., Nestoros M. and Christofides C. Appl. Phys. A 2008; 92 (3), 651-658

As shown in Figure 6, the response and the recovery times increase by introducing larger concentrations of hydrogen gas into the cell. As mentioned before, the increase of t_c for higher hydrogen partial pressure contradicts the results that occur by studying the time progress of the desorption process. The apparent contradiction of recovery time increase for higher partial pressures of hydrogen may be due to hydrogen bonding between adjacent adsorbate protons or even molecules (i.e. before dissociation occurs on the Pd surface). This would effectively account for the dependence of the rate constant on pressure and its decrease with increasing coverage. Moreover, it should be pointed out that the response and the recovery times are determined by the experimental conditions. Previous experimental work has shown that the sensor's response time is decreased if the hydrogen gas flow rate into the cell is increased. In addition, it has been proved that by introducing into the cell an inert gas, in order to remove hydrogen, the recovery time decreases drastically. In the experiment presented [18] in this chapter nitrogen gas was introduced into the measurement cell. The presence of nitrogen gas in the cell helps the hydrogen desorption process in the following mechanism: At saturation point, the hydrogen concentration in gas phase (above the palladium film) is in balance with the adsorbed hydrogen concentration. When nitrogen is introduced into the cell, it drifts the hydrogen molecules which are in gas phase, thus disturbing the system's equilibrium. In order for the system establish its balance again, hydrogen is desorbed from the palladium film and returns to gas phase. For bigger gas flow rates in the chamber the recovery of the signal from the saturation point to the initial value occurs faster. The measurements presented in Figure 3(a) were carried out under constant flow rate of gases of 150 mL/min, hence the comparison between the resulted values of ts and tc is possible. It should be noted that the sorption rate constants k_a and k_d, have an exponential dependence on the absolute temperature T of the form $K \sim \exp(-E_A/R\,T)$ where E_A is the activation energy of adsorption (E_A^{ads}) or desorption (E_A^{des}). Therefore, at higher temperatures the adsorption and desorption rate constants are increased resulting to the reduction of the response and the recovery time. It is also important to note that the Pd-H (hydride) phase transition does not occur in this system. According to [24] for a 6.5 nm palladium film, which is close to the thickness of our film (6 nm) no transition has been observed for hydrogen concentrations lower to 1%. Thus, for hydrogen concentrations of a few ppb no phase transition can occur. The possibility of a Pd-H (hydride) phase transition is considered to be very small, since the hydrogen concentrations used are extremely low and the sample thickness is very small. Some previous work done [25,26] from other researchers investigating the above phase transition does not present data for palladium films with thickness below 6 nm.

4.2. Photothermal radiometry applied on polymer-nanotube blends

4.2.1. Introduction

Polymer-based materials with electro-active functionalities have attracted significant attention due to their potential technological applications in many fields ranging from optoelectronics (solar cells, LED's, sensors, thin film transistors), to informatics (optical data storage) and biomaterials (artificial muscles). Carbazole-containing polymers present a lot of interest due to their potential applications as photoconductors, solar cells, gas sensors and photorefractive or charge transporting materials [27-28]. Carbazolyl groups easily form relatively stable radical cations (holes), present comparatively high charge carrier mobility and exhibit high thermal and photochemical stability. Furthermore polymers with fluorinated groups present either in the main backbone or in the side chain, exhibit some outstanding properties, including high thermal stability, chemical inertness, low dielectric constant and dissipation factors, low water absorptivity, and good resistance to surface properties [29-30]. In addition the ability of these materials to stabilize carbon nanotubes in organic solvents via non-covalent (physical) adsorption, renders them promising for use in carbon nanotube (CNT)-based nanotechnology. In general the introduction of carbon nanotubes (CNT's) enhances the mechanical stability of polymers and modifies the thermal and optoelectronic properties since CNT's act as electron acceptors. Although the mechanical properties of composites improve with the addition of CNT's, the electronic behavior is often optimum at low concentrations of nanotubes usually close to the percolation threshold.

Much work has been done concerning the modelling and measurement of the mechanical, optical and electrical properties of the composites while less work has been done concerning the evaluation of their thermal properties. Nevertheless the knowledge of thermal properties (thermal effusivity and conductivity) is of great interest in applications (biological implants, electronic packaging) where thermal management is important or in thermoelectric power applications. The polymers have low values of thermal conductivity (a representative value is 0.2 W/m K) which depends on the degree of crystallinity of the polymer among other factors. Since carbon nanotubes (CNT's) single wall (SWCNT's) or multi-wall (MWCNT's) have very high thermal conductivity (2800-6000 W/m K) it is expected that their introduction to the polymer will greatly enhance the effective thermal conductivity. Although there is a large scatter in the experimental results [31] (some of them are contradictory) the polymer-CNT composites still present quite low values of thermal conductivity. This is attributed to the phonon mismatch at the polymer/CNT interface resulting in high interface thermal resistance. The work [32] described in this paragraph employs a non-contact and non-destructive technique, photothermal radiometry (PTR) to study heat transport in a novel type of polymer loaded with SWCNT's at various concentrations. A two layer photothermal model is used to analyze thermal transport in the composite film which is deposited on quartz substrate. Furthermore, optical measurements are used to extract the absorption coefficient of the thin films at the excitation beam wavelength and hence eliminate one of the unknown parameters in the theoretical model. A simultaneous fitting of the experimental signal amplitude and phase is compared to the calculated signal in order to extract the thermal properties of the thin composite films.

4.2.2. Materials

A series of well-defined diblock copolymers consisting of 2-(N-carbazolyl)ethyl methacrylate (CbzEMA) and 2,2,3,3,4,4,4-heptafluorobutyl methacrylate (HFBMA) ($CbzEMA_x$-*b*-$HFBMA_y$) was synthesized by Reversible Addition-Fragmentation chain Transfer (RAFT) controlled radical polymerization [33]. The RAFT process allows for the preparation of functional polymeric materials characterized by pre-determined molecular weights, narrow molecular weight distributions and pre-defined architectures [34,35]. The polymer films were generated by spin coating on 0.9 cm x 0.9 cm quartz substrates of a dilute THF solution of $CbzEMA_{52}$-*b*-$HFBMA_{69}$ and SWCNTs at room temperature. The single wall carbon nanotubes (SWCNT's) used, where 4-5nm (diameter) x 500-1500nm (length) bundles, with 80-90 % carbonaceous purity from Sigma-Aldrich.The first series of samples consisted of three samples (A1,A2,A3) with varying thicknesses and constant % weight of SWCNT's. The thickness of the A series samples measured by a mechanical profilometer varied in a big extend across the surface of the sample. Later on a second series of samples (B0,B1,B2,B3) with different % weight of SWCNT's was fabricated. The spin coater angular velocity was increased at 100 rev/s in order to achieve better thickness homogeneity across the sample, something which was verified by the profilometer measurements. The thickness (with a tolerance of 10 nm) and % weight of SWCNT's are shown in Table 5. The PTR experimental setup was described earlier. The excitation beam (512 nm) with incident power of 45 mW and dimensions of approximately 2mm illuminated almost at normal incidence the center of the sample surface (composite film). As the modulation frequency was varied in the 10^2-10^5 Hz the signal amplitude and phase was recorded. The reproducibility of the PTR signal measurements was verified within less than 1%. In order to calculate the instrumental transfer function with which we have to correct the experimental data a frequency scan of a thick and opaque sample (steel disc) was performed.

sample	film thickness	SWCNT % weight
A1	4,62 μm	4.11
A2	6,16 μm	4.11
A3	8 μm	4.11
B0	35 nm	0
B1	30 nm	1.64
B2	40 nm	4.76
B3	62 nm	9.09

Table 5. Sample characteristics

4.2.3. Two layer photothermal model and simulations

A two layer model was developed in order to simulate the ac temperature field in the samples.

$$\frac{d^2\Delta T_f(z,\omega)}{dz^2} - \sigma^2_{T,f} \cdot \Delta T_f(z,\omega) = -\frac{\alpha_f \cdot I_0 \cdot (1-R) \cdot e^{-\alpha_f \cdot z}}{\kappa_f} \text{ for } 0 \le z \le l \tag{20}$$

$$\frac{d^2\Delta T_s(z,\omega)}{dz^2} - \sigma^2_{T,s} \cdot \Delta T_s(z,\omega) = -\frac{\alpha_s \cdot I_0 \cdot (1-R_2) \cdot e^{-\alpha_s \cdot (z-d)}}{\kappa_s} \text{ for } l \le z \le d+l \tag{21}$$

The illuminated surface of the polymer-nanotube blend is at z=0 while the interface between the blend and the substrate at z=l. The rear surface of the quartz glass is at z=l+d. The ac temperature ΔT is a function of depth z, measured from the surface of the sample and the angular modulation frequency ω=2πf. The subscripts s and f stand for the film and substrate respectively. Using the boundary conditions below, concerning the continuity of heat flux and temperature at the interfaces, the solutions for ΔT(z) in the film and substrate can be evaluated at different values of the modulation frequency.

http://dx.doi.org/10.5772/56039

$$\begin{aligned}
&-[-\kappa_f \frac{dT_f}{dz}|_{z=0}] = 0 \\
&(-\kappa_s \frac{dT_s}{dz}\Big|_{z=l}) - (-\kappa_f \frac{dT_f}{dz}\Big|_{z=l}) = 0 \\
&T_f(z=l) = T_s(z=l) \\
&-(-\kappa_s \frac{dT_s}{dz})\Big|_{z=l+d} = 0
\end{aligned} \tag{22}$$

The modulated blackbody radiation emission $\mathbf{S_{PTR}}$ of the sample is proportional to the weighted depth integral of the temperature field in the sample and it can be separated in two terms, the contribution of the composite film and the contribution of the substrate, as presented in eq. (23) [17] below. The weighting is related to the infrared absorption coefficient $\boldsymbol{\beta_{ir}}$ of the sample over the infrared detection range (2-12 μm).

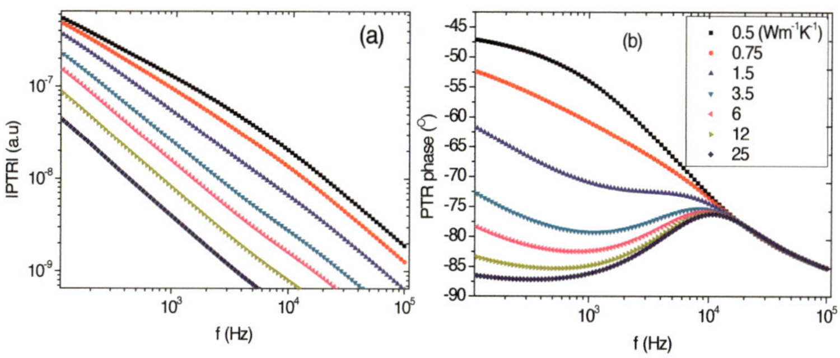

Figure 7. Dependence of photothermal signal amplitude (a) and phase (b) on the thermal conductivity (κ_f) of the film

$$S_{PTR}(\omega) \propto [\beta_{ir,f} \cdot \int_0^l \Delta T_f(z,\omega) \cdot \exp(-\beta_{ir,f} \cdot z) \cdot dz + \beta_{ir,s} \cdot \int_l^{l+d} \Delta T_s(z,\omega) \cdot \exp(-\beta_{ir,s} \cdot z) \cdot dz] \tag{23}$$

A series of simulations was performed in order to examine the influence of the film properties (absorption coefficient α, thickness, thermal conductivity and diffusivity) on the signal amplitude and phase.

The values of the optical and thermal properties of the quartz substrate used in the simulations and calculations used for fitting the theoretical results to the experimental data are: D_s=1.4x10^{-6} m^2/s, κ_s=1.5x10^{-6} W/(m K), α_s=30 m^{-1}, β_{irs}=1x10^7 m^{-1}. The thermal and properties of the quartz substrate were taken from literature while the absorption coefficient at the excitation beam was calculated from the transmission measurements. The IR absorption coefficient of the quartz substrate was taken to be practically infinite. This simplification was backed from the experiment since the PTR signal was dropping from a high value (mV) to practically noise level when the sample orientation was reversed (beam entering the sample from the quartz substrate and then meeting the thin film). This means that the quartz substrate is opaque to the IR emission of the polymer-nanotube blend film. In addition the substrate was assumed to be transparent at the excitation beam wavelength 512 nm). Concerning the thermal properties of the composite film typical values for polymers were used for the simulations, as well as for starting values of the fitting process. The simulations show that the contribution of the quartz substrate to the overall PTR signal (second term of eq.(23)) can be ignored since the quartz substrate has a very low absorption coefficient at the excitation wavelength. Nevertheless the substrate presence in the theoretical model is necessary since it defines the correct boundary conditions at the interface. The effect of thermal conductivity on the PTR signal as function of modulation frequency is shown in Figure 7. An increase of the thermal conductivity of the film allows the generated heat to diffuse away more efficiently and hence the PTR signal amplitude decreases as the temperature field in the film weakens. The signal phase in the low frequency regime reduces with the increase of thermal conductivity, leading to the formation of an extremum at a frequency defined by the thermal diffusivity of the film.

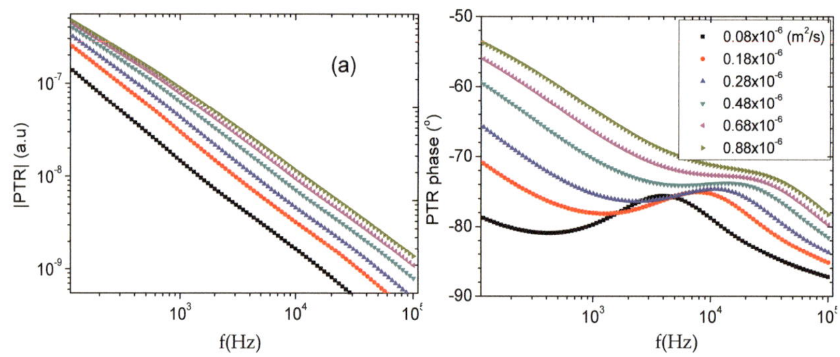

Figure 8. Dependence of photothermal signal amplitude (a) and phase (b) on the thermal diffusivity (D_f) of the film

An increase in thermal diffusivity results in an increase of the thermal diffusion length causing the shift of the phase extremum to higher modulation frequencies as seen in Figure 8, as well as an overall increase of the signal phase. The signal amplitude increases with increasing thermal diffusivity although at a first sight one would expect the opposite since high values of thermal diffusivity would mean that heat is carried out faster from the region generated. One should not forget though that the modulated IR emission of the sample is proportional to the integral of the temperature field over the thickness of the sample. Since a higher diffusivity results in a longer thermal diffusion length a thicker part of the sample will contribute to the IR emission leading to stronger signal amplitude.

A series of transmission and reflection measurements was performed in the 200 nm -3000 nm in order to help evaluate the absorption coefficient of the samples and reveal its dependence on the SWCNT content. Although the thicknesses of the samples are quite different the optical transmission measurements presented in Figure 9(a) indicate a decrease of transmission with the increase of concentration of SWCNT's. In addition it is clear that the polymers absorb

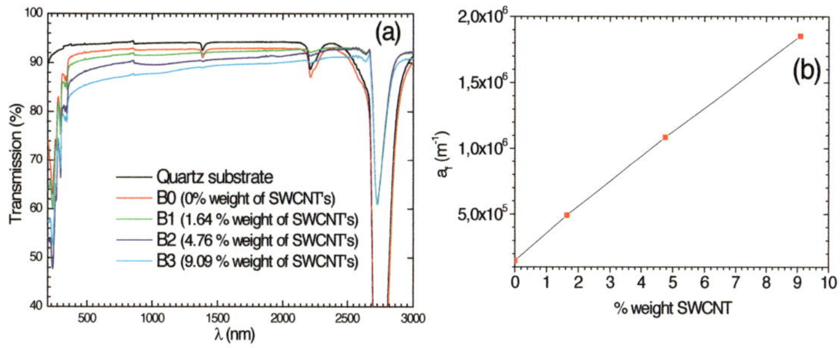

Figure 9. Transmission measurements (a) and calculated value of absorption coefficient (b) at excitation wavelength as a function of SWCNT concentration.

significantly in the UV range. The transmission deep in the 2650 nm-2900 nm is associated with the OH- bonds. The optical absorption coefficient, presented in Figure 9(b), at the excitation beam wavelength (512 nm) was calculated from the transmission and reflection spectra of the samples. Since the thickness tolerance was 10 nm the samples were assumed to have an average thickness of 42 nm. The PTR signal amplitude and phase, for A series samples, as a function of modulation frequency are shown in Figure 10. In the low frequency (<10^3 Hz) range the signal amplitude differentiates well for the three samples. Since the % weight of SWCNTS's is the same for the three samples one can assume that the optical and thermal properties of the films should be very similar, ignoring non-homogeneities due to the fabrication process, hence the signal differences should be attributed to the composite film thickness variation. More specifically the signal amplitude decreases as the composite film thickness increases.

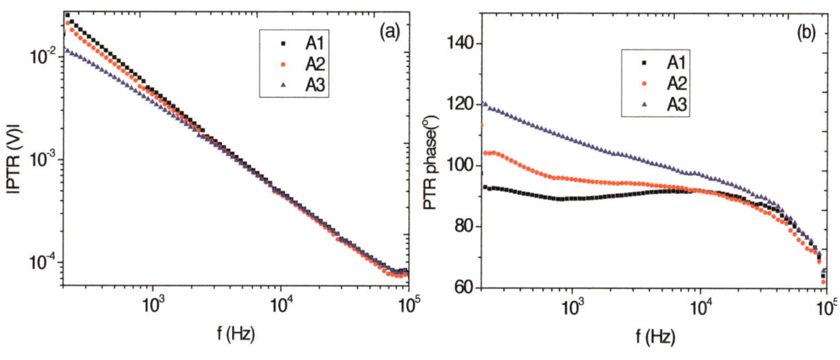

Figure 10. Experimental PTR signal amplitude (a) and phase (b) for the A series of samples

The phase channel of the signal (Figure 10(b)) can differentiate between the three samples up to a higher frequency (10^4 Hz) range and is increasing and changing slope as the composite film thickness is increasing. The decrease of the signal amplitude has to do with the fact that the contribution of the film to the signal depends on the IR emission of the film, which is proportional to the integral of the temperature field over the thickness of the sample multiplied by the factor exp ($-\beta_{ir}.z$) which results in a lower signal for thicker films. The extraction of the optical and thermal properties for the composite films was accomplished by comparing (fitting) the theoretical signal to the experimental data, with the aid of a MATLAB code.

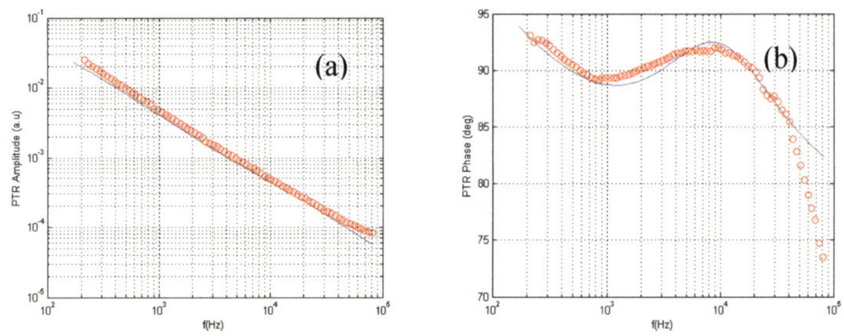

Figure 11. Theoretical signal amplitude (a) and phase (b) fit (solid lines) to the experimental signal amplitude and phase for sample A1

	A1	A2	A3
D (m^2/s)	$0.19x10^{-6}$	$0.33x10^{-6}$	$0.59x10^{-6}$
κ (W/m K)	2.8	1.45	0.98
β_{ir} (m^{-1})	$4.4x10^5$	$4.7x10^5$	$4.7x10^5$
α (m^{-1})	$1.4x10^6$	$1.4x10^6$	$1.4x10^6$

Table 6. Extracted thermal and optical parameters for the A series samples

Both channels of experimental information i.e signal amplitude and signal phase were used in order to achieve higher accuracy. The fitting results for samples A1, A2 are shown in Figures 11 and 12 respectively while the extracted values of the thermal and optical properties of the samples are tabulated in Table 6. The theoretical curves fit quite well the experimental data, except at the edges of the frequency spectrum. This discrepancy could be due to the simplified theoretical model which ignores multiple reflections in the thin film. According to Arjona et.al [36] this is more significant for phase channel in the higher modulation frequencies while for the amplitude channel is more significant for the lower modulation frequencies. The calculated thermal parameters are within the range of values reported in the literature [30] for polymer carbon-nanotube composites. The second series (B series) of samples consists of very thin composite layers (nm range) and is more interesting since in this group we have a variation of the concentration of SWCNT's. The increase of the experimental signal amplitude among the B series samples as observed in Figure 13 could be attributed to the increased optical absorption coefficient and thermal conductivity as a function of the % weight of SWCNT's which increases from B1 to B3. The samples of B series have different concentration of SWCNTS's ranging from 0-9 % weight. Due to the increase of the angular speed of the spin coater the thicknesses of the films were now in the nm range and are actually very similar within the experimental tolerance which is 10 nm. The observed differences have then to be attributed to the change of the optical

http://dx.doi.org/10.5772/56039

absorption coefficient and the thermophysical parameters (thermal conductivity and diffusivity) due to the variation of the concentration of the CNT's. Nevertheless the fact that the relative uncertainty in the film thickness is quite high creates an additional obstacle for the fitting process.

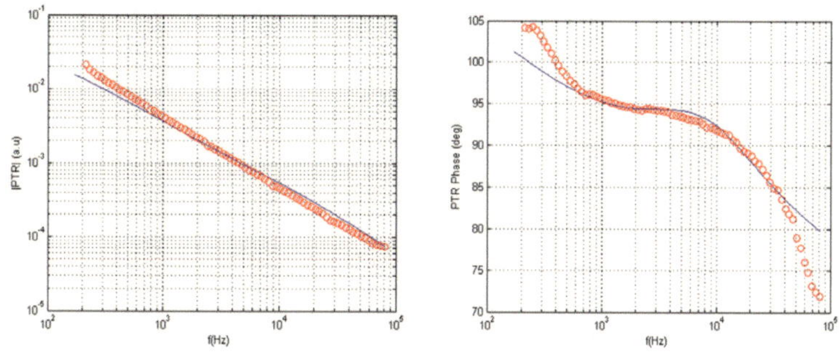

Figure 12. Theoretical signal amplitude (a) and phase (b) fit (solid lines) to the experimental signal amplitude and phase for sample A

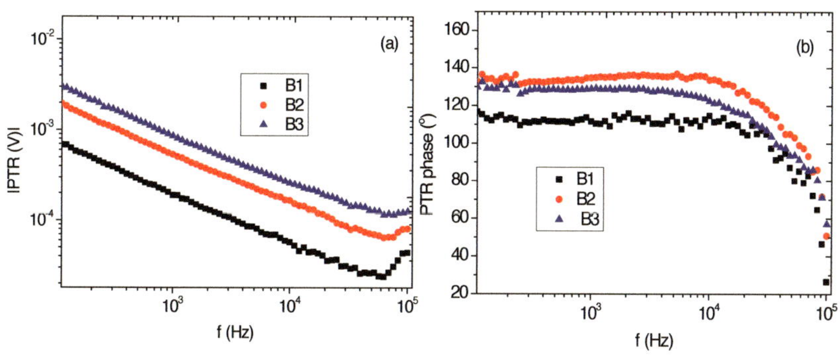

Figure 13. Experimental PTR signal amplitude (a) and phase (b) for the B series of samples

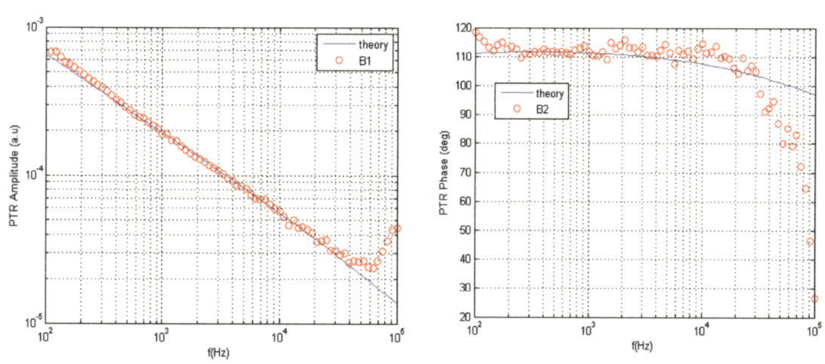

Figure 14. Theoretical signal amplitude (a) and phase (b) fit (solid lines) to the experimental signal amplitude and phase for sample B1

In general an increase in thermal conductivity which is expected with the addition of CNT's should be accompanied with a reduction of the signal amplitude in contrast with the experimental results shown in Figure 13. Nevertheless the simulations indicate that for thinner (~20 nm-B series) the amplitude variation with the increase of thermal conductivity is much smaller as compared to the case of thicker films (~μm range-A series). In addition the simulations show that the signal amplitude is affected more by the optical absorption coefficient. This means that the small decrease in amplitude due to the increase of thermal conductivity is masked by a much bigger amplitude increase due to the increase of the absorption coefficient at the excitation wavelength. In the case of sample B1 the fitted values were κ_f=2.0 W/mK and D_f=1.06x10^{-7} m^2/s.The fitting for samples B1and B2 was not attempted due to the high uncertainty in the film thickness which was about the same size as the measured value of the thickness.

5. Conclusion

In this chapter we introduced the physics of thermal and electronic waves that underpin photothermal phenomena and described two basic photothermal techniques, namely Photo-Modulated Thermoreflectance (PMTR) and Photothermal Radiometry (PTR). The abilities of these experimental techniques were explored in the study of thin metal film-gas interaction kinetics and the optical and thermal characterization of composite materials.

In the first application it was verified that that hydrogen adsorbs to palladium dissociatively while desorption is characterized by a molecular reaction. As a consequence of the high sensitivity of the PMTR sensor, the determination of the rate constants of adsorption k_a and desorption k_d for very small H_2 concentrations was possible. In addition the analysis of the time evolution of the PMTR signal showed that the values of k_d were several orders of magnitude smaller than the ones resulting for the rate constant of adsorption k_a. Thus, in order to study the adsorption process, the competitive mechanism of desorption should not be taken into account. For the adsorption mechanism it was observed that the resulting values of the adsorption rate constant k_a decrease with the increase of H_2 concentration. These results also justify the increase of the sensor's response time for larger H_2 concentrations.

In the second application it was shown that Photothermal Radiometry (PTR) is able to characterize composite materials like polymer-carbon nanotube blends. The photothermal signal (amplitude and phase) is very sensitive to the change of SWCNT's concentration as well as to the polymer film thickness. With the aid of a two layer photothermal model it was possible to extract thermal and optical properties of the composite materials.

Acknowledgements

I gratefully acknowledge the contribution and collaboration of Prof. C. Christofides and Prof. A. Othonos, directors of the Photonics and Optoelectronics Laboratory at the University of

Cyprus where the photothermal experiments took place. I would like to express my thanks to Dr. T. Krasia and Ms. I. Savva from the Department of Mechanical and Manufacturing Engineering, University of Cyprus for the fabrication of the polymer-nanotube blend samples as well as to Dr. N. C. Papanicolaou from the Department of Computer Science, University of Nicosia for his contribution in the modelling and simulations of the photothermal response of the polymer-carbon nanotube blends.

Author details

Marios Nestoros

Department of Electrical & Computer Engineering, University of Nicosia, Nicosia, Cyprus

References

[1] Mandelis A. Diffusion-Wave Fields, Mathematical Methods and Green Functions. New York: Springer; 2001.

[2] Fournier D., Boccara C., Skumanich A. and Amer N.M. Photothermal investigation of transport in semiconductors: Theory and experiment. Journal of Applied Physics 1986; 59(3) 787-795.

[3] Ikari T., Salnick A. and Mandelis A. Theoretical and experimental aspects of three-dimensional infrared photothermal radiometry of semiconductors. Journal of Applied Physics 1999; 85(10), 7392 – 7397.

[4] Rosengwaig A., Opsal J., Smith W. L. and Willenborg D. L. Detection of thermal waves through optical reflectance. Applied Physics Letters 1985;46(11) http://dx.doi.org/10.1063/1.95794.

[5] Wurm S., Alpern P., Savignac D and Kakoschke R. Modulated optical reflectance measurements on amorphous silicon layers and detection of residual defects. Applied Physics A 1988; 47(2) 147-155.

[6] Christofides C. Photomodulated Thermoreflectance of Implanted Wafers. In Christofides C., Ghibaudo G. (eds.) Effect of Disorder and Defects in Ion-Implanted Semiconductors: Optical and Photothermal Characterization, Semiconductors and Semimetals, Vol. 46. New York: Academic Press; 1997. p 115-147.

[7] Fotsing J. L. N., Hoffmeyer M., Chotikaprakhan S., Dietzel D., Pelzl J., Bein B. K., Cerqueira F., Macedo F., and Ferreira J. A. Laser modulated optical reflectance of thin semiconductor films on glass. Review of Scientific Instruments 2003;74(1) 873-876.

[8] Lavega-Sanchez A., Salazar A., Ocariz A., Pottier L., Gomez E., Villar L.M, Macho E., Thermal diffusivity measurements in porous ceramics by photothermal methods. Applied Physics A 1997; 65 (1) 15-22.

[9] Schmidt A. J., Cheaito R., and Chiesa M. Characterization of thin metal films via frequency-domain thermoreflectance. Journal of Applied Physics 2010; 107(2), 024908.

[10] Ikari T., Roger J. P, and Fournier D. Photothermal microscopy of silicon epitaxial layer on silicon substrate with depletion region at the interface. Review of Scientific Instruments 2003;74(1) 553-556.

[11] Sheard S., Somekh M., and Hiller T. Non-contacting determination of carrier lifetime and surface recombination velocity using photothermal radiometry. Material Science and Engineering: B 1990; 5(2) 101-105.

[12] Othonos A., Christofides C. and Mandelis A. Photothermal radiometric investigation of implanted silicon: The influence of dose and thermal annealing. Applied Physics Letters 1996; 69(6) 821-823.

[13] Ravi J., Lu Y., Longuemart S., Paoloni S., Pfeiffer H.,Thoen J., and Glorieux C. Optothermal depth profiling by neural network infrared radiometry signal recognition. Journal of Applied Physics 2005; 97(1) http://dx.doi.org/10.1063/1.1821635

[14] Christofides C., Nestoros M. and Othonos. A Photothermal Radiometric Study of Implanted Semiconductors. In: Mandelis A. and Hess P. (eds.) Progress in Photothermal and Photoacoustic Science and Technology, Volume IV: Semiconductors and Electronic Materials. Bellingham; SPIE; 2000. p110-142.

[15] Depriester M., Hus P., Delenclos S., and Sahraoui H. A. Study of thermal parameters' temperature dependence in solids using photothermal radiometry. Review of Scientific Instruments 2007; 78(3), 109-113.

[16] Macedo F., Vaz1 F., Torrelli M., Faria R. T., Cavaleiro A., Barradas N. P., Alves E., Junge K. H. and Bein B. K. TiO2 coatings with Au nanoparticles analysed by photothermal IR radiometry. Journal of Physics D: Applied Physics 2012; doi: 10.1088/0022-3727/45/10/105301

[17] Hellen A., Matvienko A., Mandelis A., Finer Y., and Amaechi B. T. Optothermophysical properties of demineralized human dental enamel determined using photothermally generated diffuse photon density and thermal-wave fields. Applied Optics 2010;49, (36) 6938-6951.

[18] Demetriou C., Nestoros M. and Christofides C. Photothermal hydrogen sensor: the technique, experimental process, and physicochemical analysis. Applied Physics A 2008; 92 (3), 651-658.

[19] Schlapbach L., Züttel A., Gröning P., Gröning O. and Aebi P. Hydrogen for novel materials and devices. Applied Physics A 2001;72(2), 245-253.

[20] Kalli K., Othonos A. and Christofides C. Characterization of reflectivity inversion, α- and β-phase transitions and nanostructure formation in hydrogen activated thin Pd films on silicon based substrates. Journal of Applied Physics 2002; 91(6) 3829 – 3840.

[21] Mandelis A. and Christofides C. Physics, Chemistry and Technology of Solid State Gas Sensor Devices, Vol. 125 in Chemical Analysis. New York: Wiley Interscience; 1993.

[22] Lundström I. Why bother about gas-sensitive field-effect devices? Sensors and Actuators A 1996; 56(1-2),75-82.

[23] Lundström I., Shivaraman M.S, and Svensson C. Chemical reactions on palladium surfaces studied with Pd-MOS structures. Surface Science 1977;64 497-519.

[24] Wang C., Mandelis A. and Garcia J.A. Pd/PVDF Thin Film Hydrogen Sensor System based on Photopyroelectric Purely-Thermal-Wave Interference. Sensors Actuators B 1999; 60 228 – 237.

[25] Christofides C., and Mandelis, A. Operating characteristics and comparison of photopyroelectric and piezoelectric sensors for trace hydrogen gas detection. I. Development of a new photopyroelectric sensor. Journal of Applied Physics (1989); 66(9) 3986-3992.

[26] Christofides C., Mandelis A., Rawski J. and Rehm S. Photopyroelectric detection of hydrogen/oxygen mixtures. Review of Scientific Instruments 1993; 64(12) 3563- 3571.

[27] Gerad M., Chaubey A. and Malhotra B. Application of conducting polymers to biosensors Biosensors & Bioelectronics 2002; 17(5) 345–359.

[28] Brar A. S. and Gandhi S. Microstructure determination of 9-ethyl-3-hydroxymethylcarbazolyl acrylate/methacrylonitrile using two-dimensional NMR spectroscopy. Journal of Molecular Structure, 2007; 832(1-3) 26–37.

[29] Zheng-hong L. and H. Teng-yun H. Synthesis and characterization of poly(dimethylsiloxane)-block-poly(2,2,3,3,4,4,4-heptafluorobutyl methacrylate) diblock copolymers with low surface energy prepared by atom transfer radical polymerization. Reactive & Functional Polymers, 2008, 68(5) 931–942.

[30] Wang QY., Zhang QH., Zhan XL. and Chen FQ., Low Surface Energy Polymer with Fluorinated Side Group. Progress in Chemistry 2009; 21(10) 2183-2187.

[31] Han Z., Fina A. Thermal conductivity of carbon nanotubes and their polymer nanocomposites: A review. Progress in Polymer Science 2011; 36(7) 914-944.

[32] Nestoros M., Savva I., Krasia T., Papanicolaou N. C., Othonos A. and Christofides C. Thermal and optical characterization of methacrylate-based polymers with embedded carbon nanotubes, 9th International Conference on Nanosciences & Nanotechnologies, 3-6 July, 2012, Thessaloniki, Book of Abstracts p. 134.

[33] Savva I., Demetriou M., Othonos. A, Turcu R., Popa A., Macavei S. and Krasia-Christoforou T. Well-defined fluoro- and carbazole-containing diblock copolymers: syn-

thesis, characterization and immobilization onto Au-coated silicon surfaces. RSC Advances 2012; 2(10) 8741-8751.

[34] Sun G., Cheng C. and Wooley K.L. Reversible Addition-Fragmentation Chain Transfer (RAFT) Polymerization of 4-Vinylbenzaldehyde. Macromolecules 2007; 40(4) 793-795.

[35] Konkolewicz D., Hawkett B. S., Weale A. G., Perrier S. Raft Polymerization Kinetics: How long are the Cross-Terminating Oligomers? Journal of Polymer Science: Part A: Polymer Chemistry 2009; 47(14) 3455-3466.

[36] Arjona-Zambrano M. A., Miranda-Ordonez J., Esquivel-Medina R. A, Peñuñuri F., Martinez P. and Gil-Alvarado J. J. Effect of the multiple reflections of a light beam on the thermal wave field of a sample of finite thickness. Journal of Applied Physics 2012; 111(9) http://dx.doi.org/10.1063/1.4717115

Section 3

Soldering

Chapter 15

Laser Soldering

Ezeonu Stella Ogochukwu

Additional information is available at the end of the chapter

http://dx.doi.org/10.5772/51527

1. Introduction

The trends toward miniaturization of electronic devices and the use of expensive, temperature –sensitive components used in the telecom equipment industry have led to the demand for new, highly controllable selective laser soldering technology [1, 2]. Secondly, modern high-density electronic and electro-optic subassemblies usually include thermally-sensitive components as well as complex three-dimensional (3-D) circuit geometries that cannot be soldered using conventional wave or reflow soldering techniques [6]. The laser is especially used for soldering of the temperature-sensitive assemblies and boards and it is also used for joining of assemblies with high thermal capacity that could not be conventionally soldered in the reflow process [2].

1.1. Laser soldering

Laser soldering is a technique where a precisely focused laser beam provides controlled heating of the solder alloy leading to a fast and non-destructive of an electrical joint. The process uses a controlled laser beam to transfer energy to a soldering location where the absorbed energy heats the solder until it reaches its melting temperature leading to the soldering of the contact and this completely eliminates any mechanical contact [3].

Laser soldering uses solder, which in liquid state wet the materials to be joined and provide mechanically and electrically stable connections when solidified. The ease of controlling the shape and location of the heating area leads to reliable solder joints with minimal component heating and is well suitable for high density packaging. Since it is only the solder joint region is heated during this type of soldering process without damaging the electronic components or the board, laser soldering process is of more benefit to interconnections using solders with a wide range of melting temperatures when compared with convectional infrared reflow process [4].The energy for melting of solder is applied by mean of laser beam. The use of laser technology provides precise heating, preventing the high sensitive components from being subjected to thermal stresses while allowing soldering of such sensitive components at high temperatures [2, 3]. Using focusing optics allows soldering in small spaces and also of fine pitch components and the availability of motorized optics optimizes the focal point for each joint. [3].

Laser soldering has been applied in the production of electronic assemblies and in the soldering of electronics components to circuit boards [5-8]. The biomedical sector is also experimenting with the laser soldering as a means of joining tissues without sutures. Its use in this sector is driven by the desire for minimally invasive surgical joining procedure in which the surrounding tissues is unaffected [9]. A lot of work has been carried out on its biomedical applications [10-13].

1.2. The laser soldering system

The laser soldering system consist of laser generator, fiber optic module, focusing optics, computer vision module with camera, illuminator and the motion module, the X-Y positioning stage with the servo control system shown in figure 1

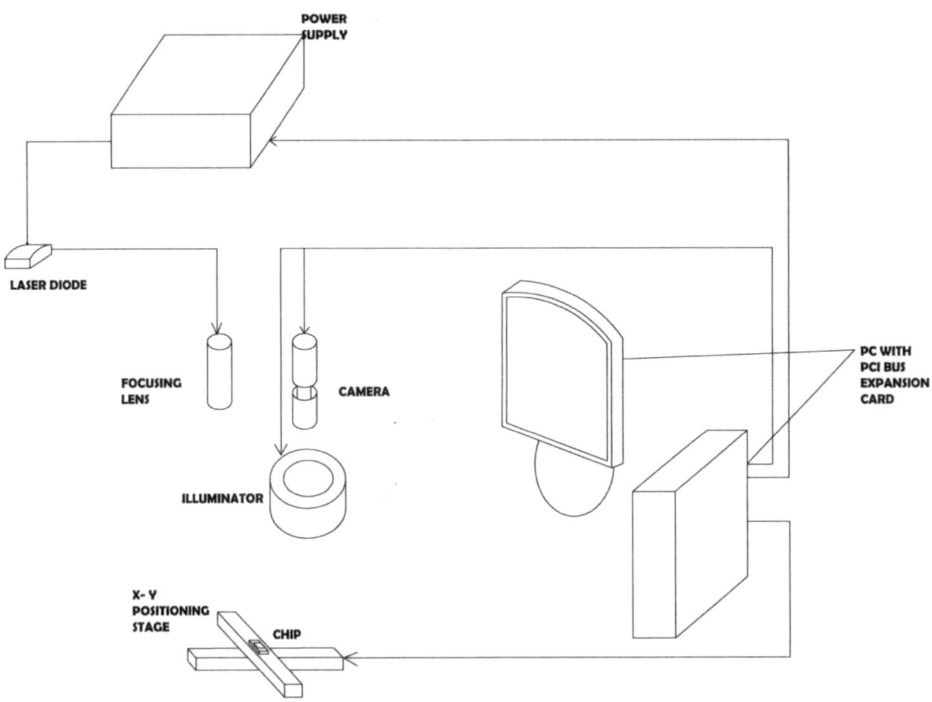

Figure 1. The block diagram of a laser soldering system

The laser beam is generated from the laser diode and modulated. With the optical system the laser beam is focused accurately on the solder joint since the laser beam is guided by a flexible

http://dx.doi.org/10.5772/51527

fiber optic cable. The required temperature at the solder joint is generated by absorbing of the heat irradiation. The application of energy can be controlled precisely. This procedure is suitable both for reflow soldering with soldering paste and soldering with solder wire.

The output of the laser is delivered by the fiber optic module, consisting of input coupling optics mounted on an adjustable mount, a connectorized armor jacketed fiber cable, and the output coupling optics, which collimates and focuses the beam toward the target region.

Due to the precision and controllability offered by laser, the devices are normally mated to automated precision X-Y positioning tables. The devices are placed into soldering position by a servo-operated XY table. This X-Y table is used to accurately position and move the workpiece under a fixed laser beam and this actually provides precise, contactless soldering. A peripheral component interconnect-bus card in the computer controls the X-Y stage [14]. During soldering, the movement of the laser spots on a given surface mount device (SMD) is taken care of by a computer controlled positioning mirrors, called galvanometers and these galvanometers are used to control the path of the laser. These galvanometers are programmed for each component type, and paths are stored in computer interface.

The solder material needed can be provided by wire feeding system or can be applied by solder paste.

The high level of automation of the laser soldering system results in a very repetitive process. For the process stability to be increased a closed loop temperature control of the solder joint is necessary and this can be done by using a pyrometer. This pyrometer serves as a reliable quality control during the soldering of miniaturized components in temperature sensitive surroundings [15]. The pyrometer continually monitors the surface temperature every 10ms during the optimal control process and this ensures a repeatable and reliable process [16]. This pyrometer is integrated in the processing head and aligned in the optical path of the laser beam.

The CCD camera could be integrated to directly view the laser spot using accessories that allow co-axial real time viewing of the laser spot. When there is a deviation in the temperature (either high or low) of the emitted heat waves, a feedback control circuit would help in adapting the laser output in less than few seconds.

In contrast to other conventional soldering techniques, laser soldering offers a lot of advantages. They include [6, 17]:

- Contactless, locally limited application of energy
- Temporally and spatially well-controlled energy input,
- Low thermal stress
- It reduces intermetallic compound formation, due to rapid joint formation which results high quality joint.
- It is also has low maintenance
- Very flexible and easily adapted
- It involves a fine grain size (due to rapid cooling), resulting in better fatigue properties.

The limitation of laser soldering lies with its extreme non-equilibrium nature. For a given laser soldering tasks, every joint have its own individual thermal mass and reflectivity and therefore demands a precisely defined laser impulse. Any slight deviation such as bent lead or a slight change in the amount of solder paste may lead to open or destroyed joint [18]. Other limitations of laser soldering include; limited ability to solder large areas, high relative price of the laser and additional cost for safety requirements.

1.3. Process parameters

Process parameters such as laser power, process time, and geometry of the laser beam can easily be programmed, permitting for consistent soldering results. According to [3], some variables such as the pad geometry or the size of the crown which can affect the formation of the solder joint interfere with the laser soldering process. Of importance is the ratio between the crown size and through hole. If the ratio is not optimal- if the hole is too big or too small with regard to the pin diameter, problems may occur in the reflow of tin on components sides. If the pin is either too long or too short, problems of poor adherence may occur both on the pin and pad.

In addition to physical and geometrical variables of the printed circuits and components, it is essential to take into consideration the soldering parameters applied during the process. The thermal profile control, assigned to each joint, permits the optimization of these soldering parameters particularly where the pads are connected to ground planes, the circuit has a number of layers or when the components to be soldered have a large thermal mass [3].

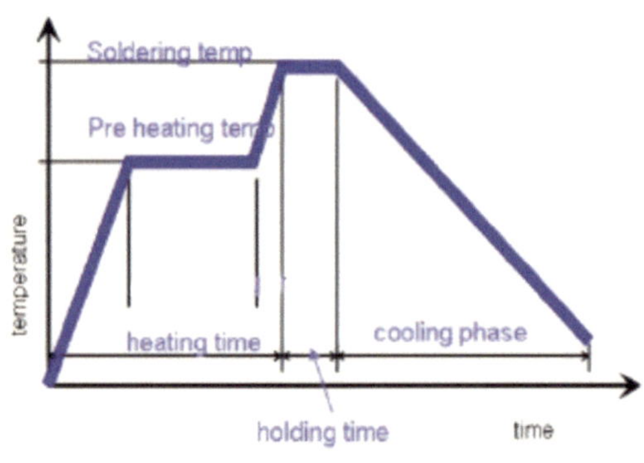

Figure 2. the soldering profile

Pyrometer used as a temperature measuring tool continually monitors the surface temperature of the joint being soldered every 10ms during the process for optimal process control [16]. Moreover, the use of pyrometer speeds up the soldering program generation process.

1.4. Types of laser soldering processes

There are three different laser soldering processes: single spot, simultaneous and Mask soldering [19, 20]:

Single spot soldering: This method of soldering is particularly applicable for multi-dimensional configurations, tight packages and heat-sensitive components or substrates. In this process, a single contact is quickly and precisely soldered by the laser spot, offering high flexibility, quality and reproducibility. A point laser beam controlled by software moves to each programmed point. Here the solder points are heated with individually matched beam diameter and laser power profile.

Simultaneous soldering: This concept allows fast soldering of individual components or complete connection arrays. Several contact points are radiated and soldered simultaneously and no relative movement between the laser beam and the workpiece is necessary. Here line-shaped laser beams are used. it allows short cycle times with high output volume.

Mask soldering: Mask soldering is applicable for very sensitive substrates or for the soldering of fine contacts and structure. In this process, a mask is placed between the collimated laser beam which is mostly line shaped and the target. The mask covers all parts which should not be exposed to the laser radiation. The laser beam only passes through the openings in the mask only where a contact to be soldered lies underneath which ensures locally confined heating. The areas between the contact points are protected and no damage is done to the substrate. It is fast and precise.

2. Types of lasers used in soldering

Three main types of lasers have been found suitable for soldering process. They are carbon dioxide laser (gas laser), Nd:YAG laser (a solid state laser) and the semiconductor Laser (diode laser).

2.1. Carbondioxide laser

Carbon dioxide laser is a gas laser and it has a wavelength of 10.6μm in the far infrared region [21 - 23]. It is reflected from metal surfaces but is strongly absorbed by flux and the heated flux then in turn transfers heat to the solder and metal in the joint [22]. The active medium in a carbon dioxide laser is a mixture of carbon dioxide, nitrogen and helium. Carbon dioxide gas is the lasing gas while nitrogen molecules help excite the CO_2 molecules and increase the efficiency of the light generation processes. The helium plays a dual role in assisting heat transfer from excited nitrogen molecules to the co_2 molecules and also helps to maintain the population inversion by getting co_2 molecules to drop from the lower laser level to the ground level [24]. The carbon dioxide laser has an efficiency of up to 20% and is cooled by pumping the gas through the heat exchanger.

2.1.1. The pump source

The active medium requires external pump source in order for lasing to occur. The active medium contains carbon dioxide, nitrogen and helium. The choice of pumping depends on the type of laser medium and since the active medium is in gas state, the best excitation is by electrical discharge of the gas. An electric discharge passing through the laser gas excites the co_2 laser. The energy of the accelerated electrons are transferred by collisions to the nitrogen molecules and then to co_2 molecules. Thus the co_2 molecules are raised to the excited state

2.1.2. Resonator

The design of the resonator has a very significant impact on the quality and the spatial power distribution of the emitted laser beam [25]. The carbon dioxide laser normally has a totally reflective rear mirror and a partly reflective output mirror. The cavity mirrors are often made of metal, with output coupling through a hole in the mirror rather than through a partly transmissive coating [24]. They are made with cavity optics reflective throughout the 9 to 11μm range in order to extract as much energy as possible from the co_2 laser cavity.

2.1.3. Drawbacks of carbon dioxide laser for soldering

There are some technological and economical reasons which prevent the use of carbon dioxide lasers as a laser source for soldering applications [6, 8, 21 - 26].

1. The energy at 10.6μm is strongly reflected by metals but undergoes about more than 90% absorption by the organic materials used for making flux, printed circuit boards and other substrates materials used for the soldering application and soldering alloys (e.g. tin-lead) have a reflective about 74% at the same wavelength. As a result, the risk of burning the circuit board is high by primary or diffused scattered laser radiation.
2. It has a high operating cost due to their electrical inefficiency (the amount of input electrical energy converted into useful laser light),
3. The co_2 laser relies on a constant gas supply and so it has to be maintained
4. The dimension of this laser source is large and the light produced by this co_2 laser cannot be delivered by an optical fiber.

2.2. Nd: YAG laser

This is a solid state laser. It uses Yttrium aluminum garnet doped with neodymium as a lasing substance. The neodymium is an impurity that takes the place of some yttrium atoms of roughly the same size. It emits at a wavelength of 1.06μm which is located in the near infrared region. Neodymium is an excellent lasing material as it produces the highest level of powers than any other doping element [25]. The small size of Nd:YAG rods and the optical properties of neodymium atoms limit the amount of energy that can be stored in a typical rod to about half a joule. Beam energies of 10-20 watts are normally used for soldering. Solder absorbs the thermal radiation of 1μm well which means that the light from the Nd:YAG has a high heating efficiency [23].

2.2.1. The pump source

The Nd:YAG being a solid state laser, has a solid crystal, and it uses light energy as the pump source. The thermal and optical properties of Nd:YAG allows it to be pumped continuously with an arc lamp or series of flash lamp pulses. Flash lamp pump is common for pulsed solid state lasers while continuous –wave solid state lasers can be pumped with an electric arc lamp. Energy pumping selectively energizes the Nd ions that subsequently lead to a cascading effect and stimulated emission of light. This light is bright enough to sustain population inversion in some solid state laser materials that are capable of continuous laser operation. The maximum average power from an Nd:YAG laser can exceed 1000W, although most operate at a much lower powers.

2.2.2. Resonator

The most commonly used resonator design is composed of two spherical or flat mirrors facing each other and the beam propagation properties are determined by the curvature of the reflective mirrors as well as the distance these mirrors are apart [27]. The two ends of the medium which is a solid crystal is silver coated, one slightly and the other heavily and so acts as a resonant cavity. As the laser absorbs the pump energy, the laser rod heats up. If the frequency of the pump energy exceeds the thermal relaxation time of the crystal, the temperature of the crystal increases. This induces temperature gradients in the laser rod crystal that give rise to thermal lensing, whereby the crystal acts as a lens to diffract the laser-which reduces power [25].

2.2.3. Drawbacks of Nd:YAG laser for laser soldering

1. Relative replacement of the light bulbs is required.

2. There is uneven energy intensity due to a decrease in the overall amount of energy delivered by the laser [21].

Nd:YAG laser emits at a wavelength of 1.06μm at near infrared and fortunately the near infrared spectrum is less reflective on metallic surfaces and less absorbent on organic materials thereby making Nd:YAG a more desirable candidate for laser soldering when compared to co_2 laser [21]. The shorter emission also enables flexible and cheaper beam shaping and guidance [26]. The light produced by this laser can be guided using fiber optics

2.3. Diode lasers

This is just like a diode, i.e. a PN-junction which is operated in forward bias mode. They are sometimes called a semiconductor diode offering wavelength ranging from 790nm to 980nm. A population inversion is generated in the active region by injecting electrons. By recombining electrons and holes in the active region laser operation occurs above a characteristic threshold current density. The wavelength of the laser radiation can be tuned within certain limits by varying the Al-doping of the $Al_xGa_{1-x}As$ semiconductor, and so influencing the width of the bandgap [28]. They are mainly based on gallium arsenide (GaAs) and gallium aluminum arsenide (GaAlAs).

In this diode, electrons are injected and they combined with holes. Some of their excess energy is release as photons which interact with more electrons and thereby producing more photons in a kind of self sustaining process called resonance. This repeated conversion of incoming electrons into photons is actually similar to the process of stimulated emission that occurs in a conventional gas-based laser.

The basic type which uses two semiconductor layers is very inefficient. Better designs have several layers which increase the power by combining more charge carriers and trapping more light inside the active layer which serves as a waveguide. The light is confined inside the waveguide where it is reflected and amplified until it exits through one end. The power of a single laser diode is low and many of it can be grouped to form a single high powered laser. When compared with co_2 and Nd:YAG lasers, diode laser offers many technological advances in laser soldering [21] and this includes

1. Due to its shorter wavelength, it has higher absorbance in metals and lower absorbance in organic materials usually used for making PCBs and other substrates.
2. It gives manufacturers a flexible and powerful tool to solve heating problems.
3. The energy density distribution of the laser beam generated remains uniform through the spot regardless of its size.
4. It is maintenance free.
5. High powered laser diode is very compact
6. The maximum conversion efficiency of transforming the input electrical efficiency of about 59%, which translates into a total electrical efficiency of about 40% for a high power diode laser system and this actually caused the operating cost to be low.

2.3.1. Pump source

The source of pumping is by optical excitation. If a photon with more energy band gap strikes a semiconductor, it can raise an electron from the valence band to the conduction band, creating an electro-hole pair, a pair of current carriers. This effect can be used to detect light by generating an electric current proportional to the amount of illumination. The current passing through the semiconductor laser produces a population inversion in the junction which is incidentally the active layer. Inside the junction, the composition of the semiconductor changes. Population inversion in this case would be a large concentration of free electrons and holes in the same region.

2.3.2. Resonator

In conventional laser system, a laser beam is produced by pumping the light emitted from the atoms severally between two mirrors. In a laser diode, an equivalent process occurs when the photon bounces back and forth in the Junction (the active layer) between the N-type and P-type semiconductor. This type of resonator is known as Fabry –Perot resonant cavity. The two parallel polished planes are used to produce multiple reflections in the cavity in order to achieve a very high intensity of laser beam. The amplified laser light eventually emerges from the polished end of the cavity.

http://dx.doi.org/10.5772/51527

3. Diode laser soldering

The diode laser is a semiconductor device that directly converts electrical energy into laser light. Mostly, higher power diode laser output in the near infrared, at either 808nm or 980nm. The size of diode laser is small and this makes them easier to be integrated into workstations. They produce their waste in a relatively small physical area and as a result, can be cooled with a small volume of circulating water and a chiller.

In diode laser soldering, a focused laser beam provides fast and controlled heating of the solder alloy to reflow the solder and form stable joints. Lasers with output power in the range of 30W-80W are sufficient for soldering applications [29].

A typical diode laser soldering system consists of a laser/control unit and a fiber cable is used to deliver the laser spot to any desired location. The laser/control unit provides diode laser temperature and current control and the output can be pulsed using internal and external system interfaces and this makes them easily adaptable to automation.

Higher power diode laser soldering has been adopted by some electronics and telecommunication equipment manufacturers because of the following reasons [2, 29];

1. It provides temporal and spatial process control and this control extends to both the location and the metallurgy of connections, resulting in optimized joints for thermally sensitive components.
2. Automated solutions for complex applications – the complex three dimensional (3-D) circuit geometries with thermally sensitive or high value components. Difficult – to -reach locations and fine –pitch quad flat packs
3. Easily tailored to specific joint design and consistent high quality joints.

3.1. The process of a diode laser soldering

The laser beam is generated from the laser diode and is focused precisely on the solder joint with the use of an optical system. The required temperature at the solder joint is generated by absorption. This method is suitable for both reflow soldering with paste and soldering with solder wire.

For selective reflow soldering, the soldering paste is dispensed first, then warmed slowly and the solder joint is preheated. Finally the soldering paste is melted totally, a meniscus is then formed at the solder joint and the contact is covered by the solder completely.

For soldering with solder wire, the process occurs in three steps [30]

Step 1: Pre-heating

The laser is switched on and the solder wire enters the range of the laser beam. The direct radiating heats the wire up close to melting temperature. If the laser beam hits the PCB outside the solder pad the surface of the PCB may get burned.

Step 2: solderwire feeding

This process is important for an accurate soldering process. The solder wire is fed during the solder feeding time at a specific speed. The solderwire with temperature close to melting temperature hits the solder spot and melts at the preheated solder pad and the preheated pin. If the solder pad and pin are not warm enough, the wire gets burred or bends. The solder wire feeder is driven by a DC-Motor and an encoder is used to control the feeding rate. The feeding length and feeding rate are programmable.

Step 3: Hold time

The solder wire having melted, can spread evenly to build the typical shape of a solder spot. Liquid solder has a significant lower absorption rate. It acts like a mirror. In this phase of the soldering process the laser beam can be partially deflected and can damage possible surrounding components or the plastic housing.

The beams of many single diodes must be bundled to the focus point in order to have the energy of approximately 30 W to a focus point of 0.8 mm diameter the and this is only possible with complex optics (see figure 3). The beam of the laser diodes is at first collimated in one axis by a cylinder lens and is then coupled into the optical fiber.

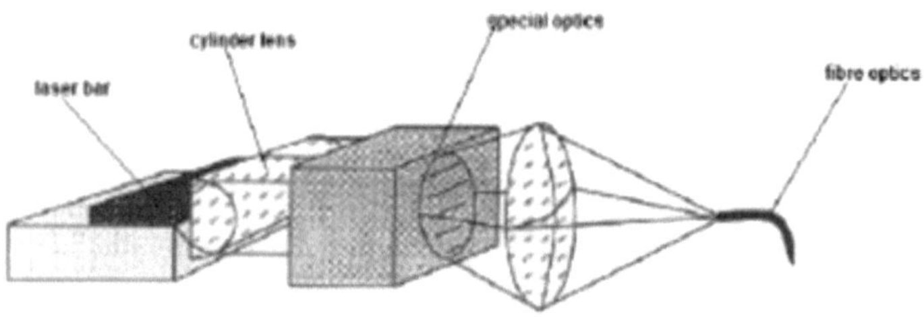

Figure 3. Laser diode with the optics delivery beam. Adapted from [30]

3.2. Soldering with high power diode lasers

Hoult [2] stated that in the microelectronics industry, for soldering tasks to be done, an average laser power of 2 to 80 watts can used. This laser average power depends on solder joint dimensions and the required speed. The soldering tasks are categorized by size into small, medium and large soldering areas [2].

Small, 40 to 100 μm (0.0016 to 0.004 in.) Pads

Typical applications for soldering these small pads are in high-density packaging. A few watts of average power are normally sufficient to solder these joints. Specific spot size and working distance requirements can be addressed by a host of commercially available optical imaging accessories (OIAs).

Medium, 100 to 500 μm (0.004 to 0.02 in.) Pads

In these cases, 25 watts delivered from an 800-μm-diameter optical fiber with an OIA capable of reducing the source size by a factor of 1.8:1 offers some benefits. Dwell time is approximately 1 second per solder joint.

http://dx.doi.org/10.5772/51527

Large, 1 to 3 mm (0.04 to 0.12 in.) Pads

In some cases, it is very essential to scan several joints simultaneously and this can be done by expanding the spot size of the diode laser. This technique is a viable solution for soldering multiple joints on a densely populated printed circuit board (PCB). To increase the scan time, higher average power diode laser systems (up to 80 watts continuous wave) can be employed.

3.3. Laser soldering parameters

Hoult [2] has also identified a list of laser soldering parameters to be considered while developing a soldering process in order to achieve a high quality joint. They include the following

1. Average power: The average power of the laser controls the rate of heat delivered to the joint being soldered. High average power is preferred since it reduces the soldering time, but the excess power causes vaporization and also reduces the quality of the joint.
2. Pulse time/length: Just like average power, pulse time/length controls the amount of energy delivered to the joint by a laser.
3. Pulse duty cycle: This modifies the rate at which heat is delivered to the joint thereby increasing the control of the soldering process. A high duty cycle, because it permits a minimum soldering times, is therefore preferred.
4. Laser power density (intensity): The laser power density (intensity) controls the response of the material to the laser beam and in association with the average power, generally determines the rate of the soldering process.
5. Laser focus position: The accurate placement of the laser focus spot is very critical in order to secure a good quality joint. This is best achieved by using precision x-y positioning tables coupled with a CCD camera and an imaging accessory that allows viewing of the laser beam in real time.

4. Laser soldering as a selective soldering process

Years back, most of the components in all electronic products are through-hole (TH) components but today about 90% of the through –hole components in these electronic products are being replaced by their surface mount counterparts [1]. This may be to the benefits of the surface mount technology (SMT) with regard to through-hole technology (THT) and they include [31]:

- Increased circuit density
- Lower costs in volume application
- The reduction in the size of the board
- The reduction in the size of the components

- Shorter pins
- Shorter interconnections
- Facilitated automation
- Improved electrical performance.

Though many through-hole components are being replaced by their surface mount counterparts, printed circuit boards (PCBs) are still being designed with both technologies. This is because despite the benefits of SMT in relation to THT, the through-hole components will remain in electronic industry for many years because of the [1]:

- Component availability
- Solder joint reliability
- The need for socketability.

The hot air and convection reflow soldering methods are primarily used for SMT components. The hot air is used for low volume application while the convection reflow is used for high volume application. Hand and wave soldering are primarily used for TH components, though a few SMT components can be soldered by these methods.

When confronted with mixed assembly board, there is always a challenge to determine the optimal method that can be used to solder these boards, and then selective soldering becomes the answer.

In selective soldering, only the through-hole components are selectively soldered after the surface mount components have been reflow-soldered. Therefore selective soldering is defined as a process of soldering only through-hole components onto a printed circuit board (PCB) that has surface mount components on the underside.

In a mixed- assembly board, through-hole components can be selectively soldered by three methods [1].

Wave soldering, with or without specially designed fixtures: This can be used for through-hole whenever the board's secondary side contains devices that can go through the wave solder process. The through-hole components must be selectively soldered without impact to the adjacent components already soldered in the reflow oven. Depending on the application, selective wave soldering using text fixtures can be very expensive.

Convection reflow using paste-in-process: This allows reflowing of both the through-hole and surface mount device (SMD) at the same time but the difference and distribution of thermal masses on circuits may lead to different results in terms of the quality for different components. When the component is temperature-sensitive, has too many rows of pins to allow sufficient solder paste deposition, or the board is not designed for the paste-in-hole process to begin with, the paste-in-hole process using convection reflow cannot be used [1].

Hand soldering: This has always been the backbone of electronics assembly for decades. They can be used to complete connections that are not possible using other methods. The compo-

http://dx.doi.org/10.5772/51527

nents that cannot withstand the high temperature involved in the wave soldering process are soldered by hand. This method is not only slow and expensive but unfortunately the quality of the joint being soldered depends on operator's capability.

For cost effective selective soldering of through-hole components when either wave, hand or convection reflow is either not technically possible, desirable or too expensive, laser soldering comes into play. Laser soldering does not need any special fixture and is capable of meeting the requirements of mixed SMT and TH fabrication in terms of the quality, repeatability aswell as the flexibility [1]. The miniaturization and integration of electronic components and the use of temperature-sensitive components used in telecommunication equipments also contributed to the rapid growth of this field.

Laser soldering is a desirable process used to solder areas that are difficult to reach or parts which cannot be soldered by either wave or reflow soldering. These can be parts that are heat/ temperature sensitive and is repeatable [22] or need a special temperature program due to their size. The power control of the laser source and the steadiness of the output power are necessary to guarantee a steady repeatable process [3]. The use of optics developed for the soldering process produced conically shaped beams and this allows it to be moved and focused on a solder joint. The use of laser technology for selective soldering permits precise heating, thereby avoiding the subjection of high thermal stress on the heat-sensitive components at high temperatures.

The benefits of a selective soldering includes but not limited to [33]

1. It allows the user to optimize the solder process down to pin level verses the compromise techniques in flux application.

2. It improves the quality of the solder and reduces the thermal coefficient of expansion issues with minimalistic surface mount components that do not tolerate the thermal shock of wave soldering.

3. Selective soldering is ideal for thermally challenged through-hole components since the flux deposition solder dwell time and peel-off parameters are fully programmable.

4. It is used to solder printed circuit board assembly (PCBAs) with high component density since it can maintain clearances between the through-hole pads and adjacent surface mount pads that are not achievable with masking pallets and permits the soldering of through –hole components on both sides of the PCBAs without restricting the components height.

5. It offers the ability to reduce solder defects, hence improving the first pass yields.

5. Comparison of laser soldering with most of the common soldering methods

Laser soldering is mainly characterized by short duration heating and high intensity radiation which can be focused on a very small spot. There are other types of soldering techniques - Iron soldering, wave soldering, induction soldering etc. They also offer a lot of advantages and disadvantages but when compared to laser soldering, they have the following characteristics;

1. Iron Soldering

– Contact heating

– It requires high maintenance since the tips wear with use and has to be replaced periodically.

– Inexpensive

– Iron's heat spreads beyond the joint area and so the sensitive components in close proximity may be destroyed.

– Regular cleaning of the tip to remove flux and oxide build up

– Long processing/soldering time

2. Induction Soldering

– Soldering from underneath – not above

– It has high energy capacity –Ability to melt high temperature solders

– Complexity of determining appropriate settings based on material conductivity

– Nitrogen atmosphere is necessary.

– Appropriate for large scale productiion

– Flux application is necessary

3. Wave Soldering

– The need for nitrogen atmosphere is necessary

– Low cost

– Separate preheating and flux application is necessary

– Soldering is from underneath

– Difficult to use with diverse package types especially ball types SMD packages and narrow lead pitch SMD packages.

– High thermal stress for SMD but low for THD.

– Appropriate for mass production

4. Infrared Soldering

– High thermal stress.

– Difficult to heat components that are in the shadows.

– Uneven heating (temperature variations) arise in places being soldered due to component shapes.

http://dx.doi.org/10.5772/51527

– Low cost

– Separate preheating and flux application is necessary

5. Convection Reflow Soldering

– Nitrogen atmosphere is necessary

– Separate preheating and high demand for flux activity

– High thermal stress

– Easy heating of components that are in the shadows

– Uniform heating across product is possible i.e. even temperature distribution

– Long processing time.

6. Vapour Phase Soldering

– Separate preheating and inert liquid atmosphere is necessary

– No temperature control system is required

– High thermal stress

– Minimal oxidation and contamination of soldered areas

– Appropriate for mass production.

– Even heating is achieved regardless of the shape of the components

7. Resistance Soldering

– Contact heating

– Heat may not reach the right spots due to temperature variations in contact resistance

6. Industrial applications of laser soldering

Industrial lasers deliver large amounts of heat with great precision and without contact, making them ideal for applications such as soldering. Laser soldering uses the well-focused, highly controlled beam to deliver energy to a desired location for a precisely measured length of time. Laser soldering is applied in the following areas.

6.1. Photovoltaic module manufacturing

In photovoltaic module production, solar cells get interconnected by strings which are then laminated in modules. The solar cells are stressed thermally and mechanically during cell soldering and string handling. As solar cells are currently becoming thinner and thinner and thus more fragile, soldering and handling is becoming more difficult. Due to the trend in the decreasing thickness of the solar cell (<200μm) [34, 35], the demands for gentle production methods to reduce breaking of the silicon wafer during manufacturing and soldering method without any mechanical contact has led to the application of laser soldering in photovoltaic module production.

The string handling can be avoided completely by laser soldering directly on the laminate layers. This method is called In-Laminate-Laser-soldering (ILL). Laser-soldering is a connection method that does not induce mechanical loads to the solar cells and reduces thermal stress on the cell. The laser beam produces mechanically and electrically stable solder joints between the connector and the solar cell in fractions of a second. A high power diode is used because it has all the properties for contacting thin film solar cell [35]. It uses a non contact technology with a precise and locally limited thermal input.

The pyrometer sensor integrated in the processing head and aligned in the optical path of the laser beam is used to control the process.

Figure 4. With In-Laminate Laser Soldering (ILLS) manufactured solar cell module. Adapted from [35].

6.2. Electronic manufacturing

The automation in the telecommunication equipment industry, miniaturization of electronic devices in consumer electronics and other biomedical applications which produces high density microelectronics with fine-pitch leads and small pad diameters has led to the demand for highly controllable selective laser soldering. This is because these devices often have the complex three dimensional (3-D) circuit geometries with thermally sensitive or high value components such as sensors, lenses central processing units etc that cannot be soldered with conventional wave soldering techniques.

Laser soldering has a lot of attributes but the main attributes are short duration heating and high intensity radiation which can be focused on a spot as small as 0.050mm in diameter [31] and this is beneficial to soldering densely packed regions, where local solder joints can be made without affecting the nearby components.

Nd:YAG lasers and carbondioxide lasers have been successfully applied in industrial production but the development of high power diode lasers recently has offered a new laser source for soldering with technological advantages. The absorptivity of laser radiation on metals, generally increases with shorter wavelength and, consequently, diode lasers may lead to a higher process efficiency compared to Nd:YAG lasers and carbon dioxide lasers [21, 36]. The high power diode laser has been adopted by some electronics and telecom equipment manufacturers because it offers high reliability, ease of automation and temporal and spatial process control [2]. This control extends to both the location and the metallurgy of connections, resulting in optimized joints for thermally sensitive components, special substrates and some areas difficult to reach [6]. Selective laser soldering permits the delivery of precise amount of energy to specific locations without causing heat related damage to the surrounding components.

6.3. Automobile application

In automobiles, Laser soldering offers a better and perfect alternative to conventional techniques due to the increasing number of electrical contacts in combination with more complex and miniaturized components [36].

Laser application for soldering permits highly automated production system, providing enormous flexibility for a wide variety of fabrication tasks at a very high processing speed because it is a non-contact technique. The mating parts are joined by a solder and the melting point of the solder is usually lower than that of the components materials. When the solder is melted, it flows into the gap between the parts and bonds with the surface of the workpiece. The thin gap between the components provides the capillary function, thereby drawing the liquid solder into the joint.

The surface of the solder seam is smooth and clean and normally requires no refinishing; they are often used in automobile industry for making body parts such as car roofs or trunk lids.

6.4. Medical application

The desire for minimal invasive surgical joining procedure led to the use of laser soldering in medical applications and this is because for a fast healing process, a successful joining of tissue is very important. Laser tissue soldering is the process of using laser energy to join tissues without sutures and it is based on some soldering material (such as albumin proteins – a biomedical solder) that coagulates on interaction with a laser beam to form joint [9].

Laser tissue soldering has been successfully utilized in bonding different type of tissues, including the cartilage & blood vessels [10], skin [11], liver [13]..

Diode lasers can be used to repair without damaging the surrounding tissues. This technique shortens the surgical times and helps minimized trauma to tissue speeds healing [29].

Laser soldering of biological tissue is a unique and promising technique because of the following reasons [29];

- No damage is done to the surrounding tissues

- Precise targeting
- Highly reliable
- Compact size and portability.

Laser tissue soldering is gaining importance because the laser technology actually fulfills the high quality requirements to perfect hygienic surfaces, dross free and free of any material residues.

7. Conclusion

Laser soldering provides a clean non-contact process which involves transferring energy to the soldering location using a precisely controlled beam. The laser beam is mainly absorbed by the solder leading to a fast and high quality solder joint. The very short heating and cooling times leads to a fine-grained structure of inter metallic bond. Due to its precise and contact-free process, laser soldering produces the best quality of solder joint

As a selective soldering process, it enables the precise amount of energy to be delivered to specific soldering areas, even those areas difficult to reach, without causing collateral heat-related damage. It represents the best method to solder surface mount components onto the printed circuit boards.

Author details

Ezeonu Stella Ogochukwu*

Address all correspondence to: ogooezeonu@yahoo.co.uk

Department of Physics/Industrial Physics, Nnamdi, Azikiwe University, Awka, Nigeria

References

[1] Prasad, R P. (2004). Laser soldering: Technology of Today and Tomorrow. SMT magazine, April 2004, 18 (4) : 14 http://www.smtonline.com/pages/zone.cgi?a=64726,Accessed 2012 April 25.

[2] Hoult, T. (2004). Laser Solutions for Soldering. Circuit Assembly.j, http://www.coherent.com/Downloads/circuitsassembly0204Rev2.pdf.,Accessed 2012 March 30.

[3] Raga, A, & Carlson, K. (2011). Selective Soldering by Laser. http://www.us-tech.com/Relld/.../selective_soldering_by_laser.htm, Accessed 2012 May 15.

[4] Messler R. W & Millard D. L (1994). Laser Soldering: New Light on an Old Joining Process. Welding J , 73(10), 43-48.

[5] Semerad E., Musiejovsky, L., and Nicolis, J. (1993). Laser Soldering of Surface-Mounted devices for high Reliability Applications. J of Material Sci. 28: 5065-5069. Avalilable: www.springerlink.com/index/pdf.Accessed: 2013 March1

[6] Beckett, P. M, Fleming, A. R, & Foster, R. J. Gilbert, J.M & Whitehead D.G ((1995). The Application of Semiconductor Diode Lasers to the Soldering of Electronic Components. Optical and Quantum Electronics Available: www.springerlink.com/index/r140g33338565063.pdf.Accessed 2013 Feb 24, 27, 1303-1311.

[7] Lea, C. (1987). Laser Soldering of Surface Mounted Assemblies. Microelectronics Int. Avaliable: www.emeraldinsight.com/journals.htm?articleid=1666621.Accessed: 2013 March 9, 4(1), 36-42.

[8] Beckett, P. M, & Fleming, A. R. Gilbert, J.M & Whitehead D.G ((1997). Practical Application of Laser Soldering in Electronics Assembly. Proceedings of IEEE Int. Symposium on Ind. Electronics. ISIE'97. Avaliable: ieeexplore.ieee.org/iel4/5230/14218/00651791.pdf. Accessed: 2013 Feb 24, 1, 59-64.

[9] Ion, J. (2005). Laser Processing of Engineering Materials; Principles, Procedures and Industrial Applications. 335 p.

[10] Ott, B, Zueger, B. J, Koestli, K, Erni, D, Schaffner, T, & Banic, A. Weber H. P & Frenz, M ((2002). Dye-Enhanced Laser Soldering of Cartilage and Blood Vessels. Laser Physics.12 (4): 635-640. Accessed 2013 March 5. Available: www.maik.ru/full/lasphys/02/4/lasphys4_02pdf.Accessed 2013 March 3

[11] Khosroshahi, M. E, Nourbakhsh, M. S, Saremi, S, & Tabatabaee, F. (2010). Characterization of Skin Tissue Soldering using Diode Laser and Indocyanine green: In Vitro Studies. Lasers in Med. Sci. Avaliable: link.springer.com/article/10.1007%2Fs10103-009-0679-y. Accessed: 2013 March 3, 25(2), 207-212.

[12] Kirsch, A. J, Duckett, J. W, Snyder, H. M, Canning, D. A, Harshaw, D. W, & Howard, P. Macarak E.J & Zderic S.A ((1997). Skin Flap Closure by Dermal Laser Soldering: A Wound Healing Model Sutureless Hypospadies Repair. Pedia. Urology (Elsevier). Available: www.sciencedirect.com/science/article/pii/S0090429597002781.Accessed: 2013 March 3, 50(2), 263-272.

[13] Wadia, Y, Xie, H, & Kajitani, M. (2000). Liver Repair and Hemorrhage Control by using Laser Soldering of Liquid Albumin in a porcine Model, Lasers Surg. Med., Available: http://www.ncbi.nlm.nih.gov/pubmed/11074508.Accessed : 2013 March 5, 27(4), 319-328.

[14] Xiong, Z, Zou, X, Wang, Y, & Ding, H. (2010). Development and Error Compensation ofLlaser Soldering System. Assembly Automation, Available: http://dx.doi.org/10.1108/01445151011061109. Accessed 2013 Feb 13, 30

[15] Renil S & Hoppert B ((2011). Pyrometer Optimized Laser-based Selective Soldering Process; A Proven Process in an Industrial Manufacturing Environment. Avaliable : www.dilas.com/.../DILAS_Pyrometer%20optimized%20laser-...Accessed: 2013 Feb 22

[16] Cerda, E. (2000). Laser Soldering Applications for RF Shield Rework. Emerging technologies conference Proceedings., http://www.pcb-prototype.net/article/Laser_Soldering.html,Accessed 2012 April 10.

[17] Basic Principle Soldering- Applications-Lasersystems (2012). http://www.leister.com/en/grundprinzip-loeten.html,Accessed 2012 March 12.

[18] Strauss, R. (1998). SMT Soldering Handbook Newnes. 221.

[19] ((2012). http://www.leister.com/en/linienloeten.html,Accessed April 25.

[20] ((2012). http://www.plasticweldingstools.com/techtips/laser_solder_desolder.php,Accessed May 5.

[21] Vivari, J, & Kasma, A. (2007). Laser Solder Reflow: A Process Solution. http://www.nordson.com/.../solder/Nordson-EFD-Laser-Reflow-Soldering/,Accessed 2012 Feb. 20.

[22] Juddy, M, & Brindley, K. (1999). Soldering in Electronics Assembly. Newnes, , 178-179.

[23] Strauss, R. (1998). SMT Soldering Handbook Newnes. , 219-220.

[24] Hecht, J. (1994). Understanding Lasers: An entry Level. IEEE.

[25] Unitek Miyachi Corporation (2003). Laser Welding Fundamentals. http://www.unitekmiyachi.com,Accessed 2012 March 3.

[26] Qin, Yi. (2009). micro manufacturing Engineering & Technology. 187.

[27] Ezeonu, S O. (2006). Introduction to Laser Technology. Rex Charles & Patrick, 27.

[28] Tonshoff, H. K, Berndt, A, Sturmer, M, Golla, D, & Schumacher, J. (1994). Diode Lasers for Material Processing. Journal de Physique, http://dx.doi.org/10.1051/jp4:1994411, Accessed 2012 May 2.

[29] Visotek Inc, (2012). Diode Laser soldering. http://www.visotekinc.com/diode-laser-applications/diode-laser-soldering,Accessed 2012 April 3.

[30] Wolf Produktionssysteme (2012). Laser Soldering with High power Diode laser. http://www.orion-technology.fr/robot_laser/BrochureLaserSoldering.pdf,Accessed 2012 May 10.

[31] Hwang, J S. (2004). Solder technologies for Electronic packaging and Assembly. Haper C A, Electronic Materials and Process Handbook. McGraw-Hill, 5.

[32] Macapagal, R. (2003). Automated selective Soldering. http://www.empf.org/emphfasis/archive/1203automated.htm,Accessed 2012 June 15.

[33] Zarrow, P, & Klenke, B. (2003). Selective Soldering-The future is now: Opimize your Through-hole Soldering Process down to Pin Level. Circuit Assembly j. 14(4):16, Available: http://www.novaenginc.com/articles/columnselective%soldering.pdf,Accessed 2012 June 12.

[34] Horn, W. (2009). Welding and Soldering with High Powered Diode Laser. http://www.dilas.com/gdresources/downloads/.../DILAS_IIW_June09,Accessed 2012 April 12.

[35] Horn, W. (2007). High Power Diode Lasers for Industrial Applications. Laser Technik j, Available: onlinelibrary.wiley.com/doi/10.1002/latj.200790169/pdf , 4(3), 62-65.

[36] Brandner, M, Seibold, G, & Chang, C. Dausinger F & Hugel H ((2000). Soldering with Solid State and Diode Lasers: Energy Coupling, Temperature Rise, Process Window. J of laser Appl. Available: http://dx.doi.org/10.2351/1.1317551. accessed: 2013 March 8, 12(5), 194-199.

Chapter 16

Automatic Optical Inspection of Soldering

Mihály Janóczki, Ákos Becker, László Jakab,
Richárd Gróf and Tibor Takács

Additional information is available at the end of the chapter

http://dx.doi.org/10.5772/51699

1. Introduction

Automatic Optical Inspection (AOI) or Automated Visual Inspection (AVI) is a control process. It evaluates the quality of manufactured products with the help of visual information. Amongst its several uses, one is the inspection of PWB (Printed Wiring Boards) after their assembling sequences i.e. paste printing, component placement and soldering. Nowadays, surface mount technology is the main method of assembly. It can be automated with ease. The increasing widespread use of SMT (Surface Mount Technology) in PWB assembly results in down-scaling of component size, increasing of lead count and component density. Parallel to this the latest manufacturing assembly lines have a very high rate of productivity. Not only is productivity required but a high quality also is expected. The quality requirements for electronic devices have already been standardized, e.g. IPC, ANSI-JSTD standards. Modern machines used in manufacturing lines such as paste printers, component placement machines etc. are capable of producing significantly better results than those required in normal standards specifications. Nowadays, the capability of modern manufacturing machines now reaches 6 σ as usually applied specification norm and 5 σ for more stringent ones. Even so, manufacturing processes are still kept under constant supervision. There still occasions when even a modern assembly line fails to create fully operational devices.

Besides the "classic" electric tests, such as in-circuit-test (ICT) and/or functional tests, there are in-line inspection possibilities: automatic optical inspection and automatic X-ray inspection systems. Because of their capabilities and properties, mostly the AOI systems are used as in-line quality inspection appliances. The main advantage of these systems is their ability to detect failures earlier and not only when the product has been assembled. AOI systems can be used to inspect the quality at each stage of the manufacturing process of the electronic device. Accordingly, there are real financial advantages by using such systems because the sooner a failure can be detected, the smaller the likelihood of refuse device manufacturing. Because of component down-scaling and increasing in density, optical inspection is now only possible with the help of machine vision as opposed to manual inspection.

2. Rise of the AOI systems

Manufacturing electronic devices necessitates the constant controlling and inspection of the product. Previously, ICT was the main appliance used for this purpose. It inspected electronic components (e.g. resistor, capacitor etc.), checked for shorts, opens, resistance, capacitance and other basic quantities. Finally, it checked the proper operation of the whole circuit to show whether the assembly had been correctly fabricated. It operated by using a bed of measure-nails type test fixtures, designed for the current PWB and other specialist test equipment. It had the following disadvantages. As the dimensions of components were shrinking and the emplacement density was increasing, the positioning of the test-points became increasingly difficult. Beds of measure-nails are also relatively expensive and they are PWB-specific. This problem was however, partially solved by using flying-probe ICT systems (but at the expense of speed). Another disadvantage of ICT was that only finished product could be examined. It was able to detect failures but not to prevent them. It is also was not suitable for inspecting the quality of various assembling technologies. A further disadvantage was also in the case of functional testing. Extra measurement procedures had to be developed to ensure the enhancement of the quality of the manufacturing process.

Previously, the quality of solder joints had only been verified by manual visual inspection (MVI). The disadvantage of manual inspection, which at best was subjective, was that the tolerance limits were narrower than used in automated machines. A magnifying glass could help for a while, but as the mounting number of components per unit area exceeded the capabilities of manual testing, this option was already proving to be difficult or not even applicable as described in [1]. Because of the rapid development of digital computing, machine vision and image processing, it was obvious that it was becoming necessary to automate the process with the help of various high-resolution cameras, novel lighting devices [2], illumination techniques [3]-[5] and efficient image processing algorithms. Such state-of-the-art devices and solutions are described in detail in the following books: [6]-[8].

In cases where the manufacturing of large quantities of precise and high quality products takes place, the capabilities of production appliances can only be used effectively if the inspections, after various technological sequences are automated (in-line), are fast and reliable. As a result, the automatic optical inspection or testing appliances has been developed to replace manual inspection. The words, Automated Optical Inspection imply that when used in the manufacturing and assembly of PWBs, the nature of the inspection process itself, using digital machine vision and image processing, will give objective results.

AOI inspects bare and mounted PWBs automatically and uses optical information. It is faster, more accurate and cheaper than manual inspection. In preparing the parameters for such

http://dx.doi.org/10.5772/51699

inspections, parametric test procedures can be used to evaluate the digital image and on the basis of this they classify the inspected PWB, component or solder joint. Automatic Optical Inspection systems offer a reliable, flexible, fast and cost-effective solution when inspecting each step of the manufacturing process. Using AOI systems also has financial advantages. Detailed calculations show this in [1]. Further works give more reasons why AOI should be used. Several economic, efficiency and suitability studies have been undertaken about these systems [9]-[29].

3. Sensors, image capturing methods, structure

In the early 1970s, CCD (Charge Coupled Device) and CMOS (Complementary Metal–Oxide–Semiconductor) sensors were invented. It presented an opportunity to capture digital images that could be processed and evaluated by a computer. Machine vision was born. The subsequent exponential development resulted in an infinite number of these applications. One such development was the automatic optical inspection. Comparison between these sensors is reported in detailed [30]-[36].

Two kinds of methods exist to capture the images: FOI (Field of Interest) based matrix camera and line scan camera. The first captures several images on an optimized course, the second scans the whole surface of PWB. Both have their advantages and disadvantages. Line-scan is the faster method but the design of a proper source of illumination is more difficult or sometimes not possible at all because the position of components themselves affects the efficiency of illumination. If the component is parallel or perpendicular to the scanning line, the captured image could differ. In case of paste inspection, component positioning is out of question, so line-scan is better choice. For components and meniscus inspection, FOI is better. A new FOI generation method is shown in [37].

Basically AOI systems have three main parts: optical unit (illumination, cameras), positioning mechanism, and control system (Fig. 1).

4. Identifying PWB

AOI appliances identify the PWBs with the help of a separate built-in unit i.e. laser-scanner or by using its inbuilt functionality. On this basis, machines can decide what inspection is necessary. According to data contained in a barcode, the AOI system loads the appropriate inspection program. As barcodes (Fig. 2.a) became more widely used, in some cases the amount of data that could be stored in them was too limited and this became a barrier to its applicability. To solve this problem, the so-called 'matrix codes' (Fig. 2.b) were developed. In [30] 22 types of linear barcodes and 48 types of matrix codes are described.

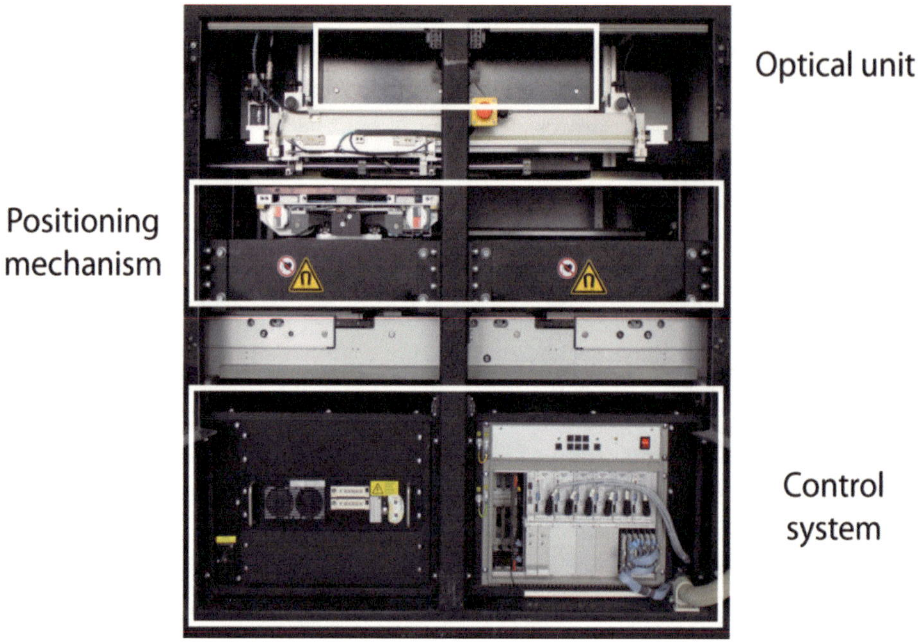

Figure 1. The three main part of an AOI system

(a) (b)

Figure 2. Example for: a) linear barcode b) DMC

5. Inspection of bare PWBs

There are several possibilities, appliances and algorithms when inspecting bare PWBs optically. These are able to inspect the copper wire-patterns on a PWB surface with high precision. Optical inspection gives rapid and reliable results regarding the quality of the PWB. Electrical detection methods, (e.g. ICT, Flying Probes) are slower and more expensive. Bare PWB inspecting AOIs have a special name: Automatic Optical Test (AOT) systems. There are several research and survey papers about this topic [31]-[43] and two manufacturers now have AOT machines [44]-[49]. In Table I a comparison between these appliances is shown.

http://dx.doi.org/10.5772/51699

Manufacturer	Amistar Automation Inc.	Lloyd Doyle Limited				
Model	K5L	doutech	Excalibur	phasor	redline	LD 6000
Board size max. [mm x mm]	510 x 410	610 x 760	610 x 760	760 x 760 (extended: 1000 x 2000)	610 x 760	610 x 760 (extended: 915 x 1525)
Board size min. [mm x mm]	50 x 50	n.a.	n.a.	n.a.	n.a.	n.a.
Board thickness [mm]	0.5 - 3.0	n.a.	n.a.	n.a.	n.a.	n.a.
Board warp [mm]	+ 0.5; - 1.0	n.a.	n.a.	n.a.	n.a.	n.a.
Board clearance top [mm]	50	n.a.	n.a.	n.a.	n.a.	n.a.
Board clearance bottom [mm]	50	n.a.	n.a.	n.a.	n.a.	n.a.
Camera type	CCD	n.a.	n.a.	n.a.	n.a.	n.a.
Illumination	3-stage LED dome lighting (Upper: IR; Middle: WH; Lower: WH)	n.a.	n.a.	n.a.	n.a.	n.a.
View size [mm]	30.4 x 22.8	500 x 635	500 x 610	500 x 635 (extended: 915 x 1980)	500 x 635 (extended: 1000 x 2000)	500 x 610 (extended: 865 x 1475)
Resolution [μm]	19	50	50	60	50	n.a.
Inspection time	0.45 sec/screen	20 - 60 sq. m/h	45 sq. m/h	n.a.	20 - 60 sq. m/h	15 sec @ 3/3 mil line/space; 27 sec @ 2/2 mil line/space (for 480 x 600 mm board size)
Applicability	missing components, position shift, rotation error, wrong components, polarity check, bridge, character recognition	functional and cosmetic faults	functional and cosmetic faults	functional and cosmetic faults	functional and cosmetic faults	functional and cosmetic faults

Table 1. Comparison of Automatic Optical Test (AOT) machines

6. Inspections following SMT sequences

In the SMT assembling process there are three main phases where AOI plays an important role; after-paste printing, component placement and soldering. In the case of wave, r selective or partial soldering there are other sequences e.g. glue dispensing, through-hole component insertion etc. But SMT processes are used mainly for inspection in discussions about this method of assembly. Possible locations where AOI can be placed in an SMT line are: the post-paste, post-placement or pre-reflow and post reflow (Fig. 3.).

At each location AOI appliances have a special name. These are: Solder Paste Inspection (SPI, also known as Post-Printing Inspection), Automatic Placement Inspection (API. also known as Post-Placement Inspection) and Post-Soldering Inspection (PSI). The AOI systems able to inspect each manufacturing sequence are called: Universal AOI (UAOI). If there is a possibility that the equipment can inspect the finished product optically, it is then called the Automatic Final Inspection (AFI).

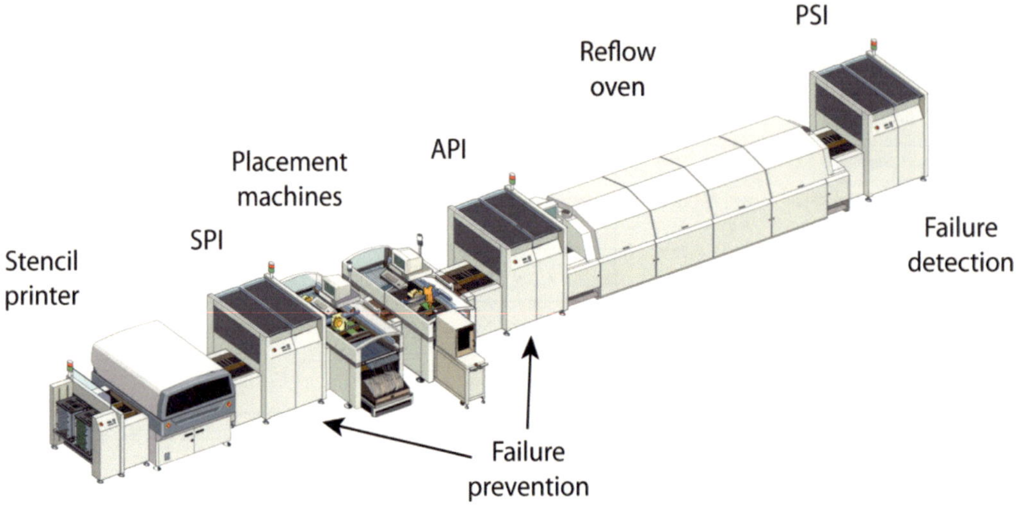

Figure 3. Possible places of AOI systems

6.1. Solder paste inspection

According to PWB assemblers, it is very important that the quality of the print solder paste is inspected because it heavily influences the quality of solder joints. In some papers it has been reported that 52%–71% of SMT defects are related to the printing process [50]-[53]. As failures can be detected much earlier, this obviously results in cost savings. According to some other opinions, inspection of the solder paste is not so relevant: "Contrary to the common, frequently quoted opinion that paste faults represent the primary percentage or 70% of all faults in the printed circuit assembly process, this detailed analysis shows that those faults amounted to only 8.3%." [54].

Special AOI machines are able to inspect the quality of print of the solder paste. It is an important option because in case of failure, the product can be repaired with minimum cost and without scrap loss. The size of the print in the 3 dimensions examined (latitude, longitude, altitude) must fall within the limits specified. To measure these parameters, so-called SPI (Solder Paste Inspection) machines have been developed. These machines are able to inspect only one step i.e. paste printing, but they are cheaper than universal AOI machines. As the control of solder paste presence is one of the easier tasks, then only the width, length and position needs to be inspected and so several failures can be detected such as bridges [55]. This can be solved using image capturing (usually with the help of line scan cameras) and subsequent evaluation.

But to measure volume as well the paste thickness is equally as important. Comparison between 2D and 3D solder paste inspections are reported in [56] and [57]. There are several possibilities to enable the measurement of paste volume optically: laser scanner [58]-[63]; projected sinusoidal fringe pattern as in [64]; the development of this technique for solder paste geometry measurement in [65], [66] and some special methods shown in [67]-[69]. Nine manufacturers offer SPI systems [70]-[82]. Several different solutions have been developed in these appliances as can be found in the scientific literature, described above. Comparison between the different methods is shown in Table 2.

http://dx.doi.org/10.5772/51699

Manufacturer	Model	Board size max. [mm x mm]	Board size min. [mm x mm]	Maximum board weight [kg]	Maximum inspection area [mm]	Maximum pad size in field of view [mm]	Typical inspection speed @ high speed [sq. cm/sec]	Typical inspection speed @ high resolution [sq. cm/sec]	Typical inspection speed @ unload and fiducial find [sec]	X and Y pixel size @ high speed [μm]	X and Y pixel size @ high resolution [μm]	Paste height range [μm]	Height resolution [μm]	Measurement types
CyberOptics	SE 300 Ultra	508 x 508	101 x 40	3	508 x 503	5 x 10	29.0	16.0	3 - 4	40	20	50 - 610	0.13	height, area, volume, registration, bridge detection
CyberOptics	SE 500	510 x 510	50 x 50	3 - 5	508 x 503	15 x 15	80.0	50.0	4 - 5	30	15	50 - 500	0.20	height, area, volume, registration, bridge detection
CyberOptics	SE 500-X	810 x 610	100 x 100	10	810 x 605	15 x 15	80.0	50.0	4 - 5	30	15	50 - 500	0.20	height, area, volume, registration, bridge detection
Koh Young Technology	aSPIre	n.a.	n.a.	n.a.	n.a.	n.a.	41.0	n.a.	n.a.	10	20	20 - 400	0.37	height, area, volume, offset, bridge detection, shape deformity
Koh Young Technology	KY-8030	n.a.	n.a.	n.a.	n.a.	n.a.	15.0	19.7	n.a.	20	20	20 - 400	0.37	height, area, volume, offset, bridge detection, shape deformity
Marantz	Power Spector	510 x 460	50 x 50	n.a.	n.a.	n.a.	80.0	80.0	n.a.	n.a.	n.a.	max. 600	n.a.	height, area, volume, offset, bridge detection, shape deformity
Omron	VT-RNS-P	510 x 460	50 x 50	n.a.	n.a.	n.a.	n.a.	n.a.	n.a.	10; 15; 20	10; 15; 20	n.a.	n.a.	presence of solder, insufficient solder, excessive solder, solder shifting, grazing, bridging, spreading
Omron - CKD	VP5000L	510 x 460	50 x 50	n.a.	n.a.	n.a.	n.a.	n.a.	n.a.	12	12	n.a.	n.a.	average height, volume, excessive deposition, insufficient solder, smearing, misalignment, bridging
Orpro Vision	Symbian P36	508 x 540	n.a.	n.a.	n.a.	n.a.	60	60	4 - 6	20	20	50 - 300	5.00	n.a.
Saki	BF-SPIder-M	250 x 330	50 x 60	n.a.	n.a.	n.a.	n.a.	n.a.	3	12	12	max. 450	0.40	height, area, volume, shift, shape, spread, bridge
Saki	BF-SPIder-L	460 x 500	50 x 60	n.a.	n.a.	n.a.	n.a.	n.a.	3 - 5	12	12	max. 450	0.40	height, area, volume, shift, shape, spread, bridge
ScanCAD	ScanINSPECT SPI	457 x 508	50 x 50	n.a.	419 x 508	n.a.	n.a.	n.a.	n.a.	n.a.	n.a.	n.a.	n.a.	n.a.
TRI Innovation	TR7006/L/LL-20	330 x 280	50 x 50	3	n.a.	n.a.	98.8	24.7	n.a.	20	20	40 - 600	n.a.	n.a.
TRI Innovation	TR7006/L/LL-16	510 x 460	50 x 50	5	n.a.	n.a.	63.2	15.8	n.a.	16	16	32 - 480	n.a.	n.a.
TRI Innovation	TR7006/L/LL-12	660 x 610	50 x 50	10	n.a.	n.a.	35.5	8.8	n.a.	12	12	24 - 360	n.a.	n.a.
TRI Innovation	TR7066-20	510 x 460	50 x 50	3	n.a.	n.a.	153.6	23.0	n.a.	20	20	40 - 600	n.a.	n.a.
TRI Innovation	TR7066-16	510 x 460	50 x 50	3	n.a.	n.a.	122.9	18.4	n.a.	16	16	32 - 480	n.a.	n.a.
TRI Innovation	TR7066-12	510 x 460	50 x 50	3	n.a.	n.a.	92.2	13.8	n.a.	12	12	24 - 360	n.a.	n.a.
Viscom	S3088-II QS	450 x 350	n.a.	n.a.	n.a.	n.a.	110	110	n.a.	22	22	n.a.	n.a.	n.a.
Vi Technology	3D-SPI	510 x 460	n.a.	n.a.	n.a.	n.a.	20.0	20.0	n.a.	n.a.	n.a.	n.a.	n.a.	height, area, volume, bridge, shape, position

Table 2.

Manufacturer	Model	Conveyor speed [mm/sec]	Conveyor height [mm]	Board warp	Volume repeatability on a certification target	Volume repeatability on a circuit board	Gage R&R	Measurement types
CyberOptics	SE 300 Ultra	150 - 450	889 - 990	<2% of PCB diagonal or 6.35mm total	n.a.	n.a.	n.a.	height, area, volume, registration, bridge detection
CyberOptics	SE 500	150 - 450	810 - 970	<2% of PCB diagonal or 6.35mm total	< 1%; 3	<3%; 3	<< 10%; 6	height, area, volume, registration, bridge detection
CyberOptics	SE 500-X	150 - 450	810 - 970	<2% of PCB diagonal or 6.35mm total	< 1%; 3	<3%; 3	<< 10%; 6	height, area, volume, registration, bridge detection
Koh Young Technology	aSPIre	n.a.	830 - 970	± 5.0 mm	< 1%; 3	<3%; 3	<< 10%; 6	height, area, volume, offset, bridge detection, shape deformity
Koh Young Technology	KY-8030	n.a.	870 - 970	± 3.5 mm	< 1%; 3	<3%; 3	<< 10%; 6	height, area, volume, offset, bridge detection, shape deformity
Marantz	Power Spector	n.a.	830 - 970	± 5.0 mm	n.a.	n.a.	n.a.	height, area, volume, offset, bridge detection, shape deformity
Omron	VT-RNS-P	n.a.	n.a.	n.a.	n.a.	n.a.	n.a.	presence of solder, insufficient solder, excessive solder, solder shifting, grazing, bridging, spreading
Omron - CKD	VP5000L	n.a.	n.a.	n.a.	n.a.	n.a.	n.a.	average height, volume, excessive deposition, insufficient solder, smearing, misalignment, bridging
Orpro Vision	Symbion P36	n.a.	870 - 930	+ 3.0 mm; - 6.0 mm	n.a.	n.a.	< 10%	n.a.
Saki	BF-SPIder-M	n.a.	max. 900	n.a.	< 1%; 3	n.a.	< 10%	height, area, volume, shift, shape, spread, bridge
Saki	BF-SPIder-L	n.a.	max. 900	n.a.	< 1%; 3	n.a.	< 10%	height, area, volume, shift, shape, spread, bridge
ScanCAD	ScanINSPECT SPI	n.a.	n.a.	n.a.	n.a.	n.a.	n.a.	n.a.
TRI Innovation	TR7006/L/LL-20	n.a.	n.a.	n.a.	n.a.	n.a.	n.a.	n.a.
TRI Innovation	TR7006/L/LL-16	n.a.	n.a.	n.a.	n.a.	n.a.	n.a.	n.a.
TRI Innovation	TR7006/L/LL-12	n.a.	n.a.	n.a.	n.a.	n.a.	n.a.	n.a.
TRI Innovation	TR7066-20	n.a.	n.a.	n.a.	n.a.	n.a.	n.a.	n.a.
TRI Innovation	TR7066-16	n.a.	n.a.	n.a.	n.a.	n.a.	n.a.	n.a.
TRI Innovation	TR7066-12	n.a.	n.a.	n.a.	n.a.	n.a.	n.a.	n.a.
Viscom	S3088-II QS	n.a.	850 - 960	n.a.	n.a.	n.a.	n.a.	n.a.
Vi Technology	3D-SPI	n.a.	max. 950	± 3.5 mm	n.a.	n.a.	< 10%	height, area, volume, bridge, shape, position

Table 3. Comparison of Solder Paste Inspection (SPI) machines

http://dx.doi.org/10.5772/51699

6.2 Automatic placement inspection

Inspection of the PWB after component placement is the next possibility. With this method possible placement failures can be detected and some defective paste printing phenomena as well. If there is a sign or mark on the component, it can be read and identified with the help of modern image processing algorithms even if it has more than one different-looking label type. APIs (Table III) are able to measure most parameters of components objectively e.g. X-Y shift, rotation, polarity, labels, size etc. [83]-[90]. Four manufacturers have this special appliance in stock. [91]-[94].

Manufacturer	BeamWorks	Landrex	Omron	Viscom
Model	Inspector cpv	Optima II 7301 Express	VT-RNS-Z	S3054QV
Field of view	12 x 9 mm @ 15 μm; 48 x 36 mm @ 73 μm	10 x 10 mm; 15 x 15 mm	n.a.	1280 x 1024 pixel
Pixel size [μm]	15; 73	n.a.	10; 15; 20	10; 22
Depth of field (max. component height for inspection) [mm]	10; 15	n.a.	n.a.	n.a.
Number of cameras	1	1 vertical, 4 angled	1	1
Lighting method	oblique ring, white LED light	n.a.	ring-shaped RGB LED	n.a.
Board size max. @ single board operation [mm x mm]	508 x 406	609 x 558	510 x 460	443 x 406
Board size max. @ dual board operation [mm x mm]	none	none	none	370 x 406
Board size min. [mm]	40 x 28	51 x 76	50 x 50	n.a.
Board thickness [mm]	0.8 - 3.2	n.a.	n.a.	n.a.
Conveyor height [mm]	n.a.	n.a.	n.a.	850 - 960
Board edge clearance [mm]	4	n.a.	n.a.	3
Board edge clearance top [mm]	25; 37	63	20	35
Board edge clearance bottom [mm]	25; 37	63	50	50
Inspection speed [sq. cm/sec]	2 @ 12 x 9 mm field of view	n.a.	250 ms/screen @ 10 sq. mm field of view	20 - 30
Applicability	missing component, wrong component, polarity check, offset, skew	missing components, misoriented components, extra components, component placement, tombstoned and bill boarded components, lifted leads, insufficient solder and excess solder, wrong part, through hole pins	presence of solder, component shifting, polarity error, missing components, wrong components, solder balls, skewing, bridging, foreign objects	n.a.

Table 4. Comparison of Automatic Placement Inspection (API) machines

6.3 Post soldering inspection

Most manufacturers agree from a strategic point of view, that optical inspection after soldering has been completed should not be missed out. At the very least, the defective products must be eliminated because many failures are generated during soldering: "Forty-nine percent of the true faults were detectable only after soldering. These consisted of component and soldering faults. Forty-eight percent of the optically recognizable faults could not be recognized electrically." [54].

In consequence, this is the most important part of the AOI inspection process. Most scientific papers are preoccupied with this subject [95]-[123]. The quality of the solder joint (and the soldering process) can be inspected with the methods described in this section. The quality of the solder joints is determined from geometric and optical properties of the solder meniscus. These parameters determine the reflection properties of the meniscus which is formed from the liquid alloy during the soldering process. After cooling, the meniscus becomes solid and reflects illumination which means that we can evaluate it (Fig.5, Fig. 6). From these reflection patterns and with the help of image processing algorithms we are able to determine the quality of the solder joints.

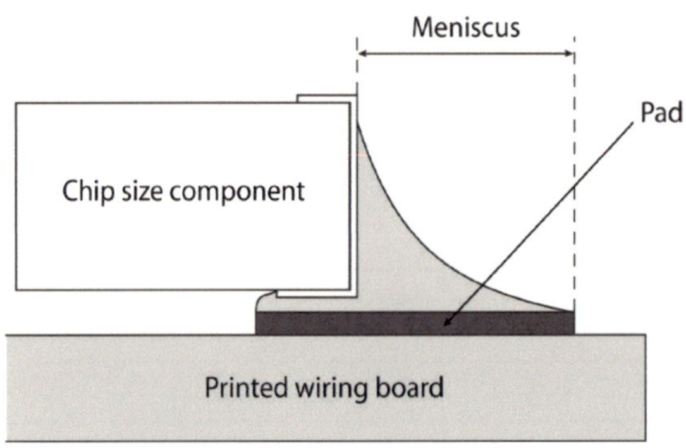

Figure 4. Schematic of the meniscus

Figure 5. Reflection pattern on meniscus model with white ring illumination

Figure 6. Reflection pattern on meniscus model with RGB ring illumination

Using reflection patterns is the basis of all papers that have been published in this field of study. There are two solutions: gray-scale or colour inspection. Supplier and appliances are shown in Table IV [124]-[126].

One interesting area is wave soldering. It needs different types of algorithms because of the circular solder shape and the pin. Some solutions for this kind of inspection are reported in [127]-[129]. A summary of possible failures and appliances that can detect them are shown in Table V.

6.4. Combined appliances

There are systems that are able to inspect more sequences. These are combined systems, namely API&PSI [140]-[146] (Table VI).

And the all-in-one machines are the UAOI systems, detailed: SPI&API&PSI [147]-[163] (Table VII).

Manufacturer	TRI Innovation	Viscom	
Model	TR7530	S3016	S3054QC
Field of view (orthogonal camera) [pixel]	1024 x 768	2752 x 2048 @ 55 x 43 mm	672 x 512
Pixel size (orthogonal camera) [μm]	10; 15; 20	22; 10	22; 10
Number of cameras (orthogonal)	1	4	1, 2 or 4
Resolution (angled view camera) [μm/pixel]	n.a.	15	n.a.
Number of cameras (angled view)	n.a.	4	n.a.
Illumination	ultra-low angle, multi-segment, RGB LED lighting	n.a.	n.a.
Inspection speed	72 sq. cm/sec @ 20 μm; 40 sq. cm/sec @ 10 μm; 18 sq. cm/sec @ 10 μm	typical connector with 100 pins 15 sec	typical connector with 100 pins 15 sec
Board size max. [mm]	400 x 300	430 x 406	443 x 406
Board size min. [mm]	50 x 50	n.a.	n.a.
Board weight max. [kg]	3	n.a.	n.a.
Conveyor height [mm]	n.a.	850 - 960	850 - 960
Board edge clearance [mm]	3	3	3
Board clearance top [mm]	25	50	15
Board clearance bottom [mm]	40	35	50
Applicability	missing component, tombstone, billboard, polarity, skew, marking, defective, insufficient and excess solder, bridge, trough hole pin, lifted lead, golden finger, scratch, blur	selective and special solder joints	selective and special solder joints

Table 5. Comparison of Post Soldering Inspection (PSI) machines

	UAOI					
	SPI	API	PSI		API	PSI
	paste inspection	pre reflow component in paste	post reflow	adhesive inspection	pre wave component in adhesive	post wave
missing paste	✓	✓				
completeness or/and volume of solder paste	✓	✓ whether it is not covered by component				
misalignment	✓					
paste bridge	✓	✓				
smeared solder paste	✓	✓ whether it is not covered by component				
contamination	✓	✓ whether it is not covered by component	✓		✓	✓
missing component		✓	✓		✓	✓
component position (x-y shift, rotation, face lift)		✓	✓		✓	✓
polarity		✓	✓		✓	✓
damaged component		✓	✓		✓	✓
unsoldered component			✓			✓
insufficient solder joint			✓			✓
solder bridge			✓			✓
lifted lead			✓			✓
tombstone			✓			✓
missing adhesive				✓		
smeared adhesive on pad				✓		
missing pin end						✓
insufficient pin solder						✓
dewetting						✓
PWB registration	✓	✓	✓	✓	✓	✓

Table 6. Possible failures and appliances

Manufacturer	Model	Camera movement	Board movement	Component inspection	Printing/paste inspection	Distinction principles	Distinction parameters	Camera type	Camera field of view/resolution	Lens	Illumination
Machine Vision Products	Supra E	n.a.	n.a.	n.a.	n.a.	n.a.	n.a.	5MP color camera	7 - 17 μm	n.a.	programmable variable LED strobe lighting, proprietary multi-color illumination
Marantz	iSpector HDL 350L	X + Y direction	stationary during inspection	presence, polarity, offset, correctness, soldering	offset, smearing, bridges, uniformity	synthetic imaging, spectral analysis, grayscale limits	brightness, hue, saturation via filters	UXGA CCD digital camera with CameraLink	32 x 24 mm @ 20 μm; 40 x 30 mm @ 25 μm; 16 x 12 mm @ 10 μm	telecentric lens with built in prism for DOAL lighting	omnidirectional triple LED rings; side, main, line sourced DOAL (diffused on axis lighting (coaxial))
Marantz	iSpector HDL 650L	X + Y direction	stationary during inspection	presence, polarity, offset, correctness, soldering	offset, smearing, bridges, uniformity	synthetic imaging, spectral analysis, grayscale limits	brightness, hue, saturation via filters	UXGA CCD digital camera with CameraLink	32 x 24 mm @ 20 μm; 40 x 30 mm @ 25 μm; 16 x 12 mm @ 10 μm	telecentric lens with built in prism for DOAL lighting	omnidirectional triple LED rings; side, main, line sourced DOAL (diffused on axis lighting (coaxial))
Marantz	iSpector HML 350L	X + Y direction	stationary during inspection	presence, polarity, offset, correctness, soldering	offset, smearing, bridges, uniformity	synthetic imaging, spectral analysis, grayscale limits	brightness, hue, saturation via filters	UXGA CCD digital camera with CameraLink	32 x 24 mm @ 20 μm; 40 x 30 mm @ 25 μm; 16 x 12 mm @ 10 μm	high resolution telecentric	omnidirectional 4-angle LED: RGB-DOAL (coaxial)
Marantz	iSpector HML 650L	X + Y direction	stationary during inspection	presence, polarity, offset, correctness, soldering	offset, smearing, bridges, uniformity	synthetic imaging, spectral analysis, grayscale limits	brightness, hue, saturation via filters	UXGA CCD digital camera with CameraLink	32 x 24 mm @ 20 μm; 40 x 30 mm @ 25 μm; 16 x 12 mm @ 10 μm	high resolution telecentric	omnidirectional 4-angle LED: RGB-DOAL (coaxial)
Orpro Vision	Symbion S36	stationary during inspection	X + Y + Z direction	presence and absence of components, placement accuracy and polarity, optical character recognition	insufficient solder, tombstone, billboard, coplanarity, lifted leads, shorts	n.a.	n.a.	4MP XGA high resolution top and 4 side cameras with symmetric image acquisition and color capability	n.a.	n.a.	axial, direct, diffuse and RGB multi-color illumination
TRI Innovation	7500	n.a.	n.a.	missing, tombstone, billboard, polarity, shift	insufficient solder, excess solder, bridge	n.a.	n.a.	1 top view XGA 3CCD camera @ 1024 x 768 pixel, 4 angle view XGA mono camera & 1024 x 768 pixel	10 μm; 15 μm; 20 μm; 25 μm	n.a.	multi segment, multi angle LED, RGB+W
TRI Innovation	7500L	n.a.	n.a.	missing, tombstone, billboard, polarity, shift	insufficient solder, excess solder, bridge	n.a.	n.a.	1 top view XGA 3CCD camera @ 1024 x 768 pixel, 4 angle view XGA mono camera & 1024 x 768 pixel	10 μm; 15 μm; 20 μm; 25 μm	n.a.	multi segment, multi angle LED, RGB+W
TRI Innovation	7550	n.a.	n.a.	missing, tombstone, billboard, polarity, skew, marking, defective	insufficient solder, excess solder, bridge, trough hole pins, lifted lead, golden finger scratch, blur	n.a.	n.a.	1 top view XGA 3CCD camera @ 1024 x 768 pixel, 4 angle view XGA mono camera & 1024 x 768 pixel	10 μm; 15 μm; 20 μm; 25 μm	n.a.	multi segment, multi angle LED, RGB (coaxial lighting optional)
Sony	SI-V200	n.a.	n.a.	inaccurate mounting, reversed, polarity	missing, solder, bridging, solder quantity	n.a.	n.a.	2MP color CCD camera	24.8 x 18.6 mm @ 15.5 μm; 17.6 x 13.2 mm @ 11 μm	n.a.	high intensity white LED
Saki	BF-Tristar	n.a.	n.a.	presence and absence of components, misalignment, tombstone, reverse, polarity, bridge, foreign material, lifted lead, lifted chip	absence, insufficient solder, fillet defect	n.a.	n.a.	line color CCD camera	10 μm	n.a.	LED lighting system

Table 7.

Manufacturer	Model	Minimum inspection component size	Position accuracy	Board clearance top [mm]	Board clearance bottom [mm]	Board size max. [mm x mm]	Board size min. [mm x mm]	Movement speed [mm/sec]	Inspection capacity typical	Conveyor speed [mm/sec]	Conveyor height [mm]	Board clamping	Board thickness [mm]
Machine Vision Products	Supra E	0201 and 01005 (0.4 x 0.2 mm)	0.5 μm	n.a.	n.a.	508 x 508	n.a.	n.a.	90 sq. cm/sec	n.a.	n.a.	n.a.	n.a.
Marantz	iSpector HDL 350L	01005 (0.4 x 0.2 mm @ 10 μm)	pixel related feedback loop	40	35, 55	350 x 250	50 x 50	720	1500 cps/min	10 - 500	830 - 950	ruler blade, top & edge clamping, sensor stopper	0.6 - 4.0
Marantz	iSpector HDL 650L	01005 (0.4 x 0.2 mm @ 10 μm)	pixel related feedback loop	40	35, 55	650 x 550	50 x 50	720	1500 cps/min	10 - 500	830 - 950	ruler blade, top & edge clamping, sensor stopper	0.6 - 4.0
Marantz	iSpector HML 350L	01005 (0.4 x 0.2 mm @ 10 μm)	pixel related feedback loop	30	35, 55	350 x 250	50 x 50	720	1500 cps/min	10 - 500	n.a.	ruler blade, top & edge clamping, sensor stopper	0.6 - 4.0
Marantz	iSpector HML 650L	01005 (0.4 x 0.2 mm @ 10 μm)	pixel related feedback loop	30	35, 55	650 x 550	50 x 50	720	1500 cps/min	10 - 500	n.a.	ruler blade, top & edge clamping, sensor stopper	0.6 - 4.0
Orpro Vision	Symbion S36	01005 (0.4 x 0.2 mm) down to 0.3 mm pitch	n.a.	90	90	550 x 470	n.a.	n.a.	40 sq. cm/sec	n.a.	n.a.	n.a.	n.a.
TRI Innovation	7500	n.a.	n.a.	50	50	510 x 460	n.a.	n.a.	110 sq. cm/sec @ 25 μm; 72 sq. cm/sec @ 20 μm; 40 sq. cm/sec @ 15 μm; 18 sq. cm/sec @ 10 μm;	n.a.	n.a.	n.a.	max. 4.0
TRI Innovation	7500L	n.a.	n.a.	50	50	660 x 610	n.a.	n.a.	110 sq. cm/sec @ 25 μm; 72 sq. cm/sec @ 20 μm; 40 sq. cm/sec @ 15 μm; 18 sq. cm/sec @ 10 μm;	n.a.	n.a.	n.a.	max. 4.0
TRI Innovation	7550	n.a.	n.a.	40	40	540 x 460	50 x 50	n.a.	110 sq. cm/sec @ 25 μm; 72 sq. cm/sec @ 20 μm; 40 sq. cm/sec @ 15 μm; 18 sq. cm/sec @ 10 μm;	n.a.	n.a.	n.a.	0.6 - 4.0
Sony	SI-V200	0402 @ high resolution, 0603 @ normal resolution	n.a.	20	40	460 x 510	40 x 50	n.a.	0.27 sec/frame	n.a.	n.a.	n.a.	0.4 - 2.0
Saki	BF-Tristar	n.a.	n.a.	30	30	250 x 330	50 x 70	n.a.	20 sec / 250 x 350 mm	n.a.	900	n.a.	0.6 - 2.5

Table 8. Comparison of combined (API&PSI) machines

Manufacturer	Model	Pre-reflow inspection	Post-reflow inspection	Camera type	Orthogonal camera field of view [mm]	Orthogonal camera resolution [µm]	Number of orthogonal cameras	Angular camera resolution [µm]	Number of angular cameras	Illumination
Agilent Technologies	Medalist SJ50 Series 3	missing, offset, 2D paste, skewed, polarity, bridging, wrong part, traceability	missing, offset, skewed, polarity, billboard, tombstone, lifted/bent leads, excess/insufficient solder, bridging, wrong part, traceability	4 megapixel digital camera	44.7 x 32.8	19; 12	n.a.	n.a.	n.a.	multiple color, multiple angle, multiple segment ED lighting head, auto-calibration
Agilent Technologies	Medalist SJ50 Series 3 XL	missing, offset, 2D paste, skewed, polarity, bridging, wrong part, traceability	missing, offset, skewed, polarity, billboard, tombstone, lifted/bent leads, excess/insufficient solder, bridging, wrong part, traceability	4 megapixel digital camera	44.7 x 32.8	19; 12	n.a.	n.a.	n.a.	multiple color, multiple angle, multiple segment ED lighting head, auto-calibration
Agilent Technologies	Medalist sj5000	missing, offset, 2D paste, skewed, polarity, bridging, billboard, wrong part , extra part, traceability	missing, offset, skewed, polarity, billboard, tombstone, lifted/bent leads, excess/insufficient solder, bridging, wrong part, traceability	4 megapixel digital camera	44.7 x 32.8	21; 12	n.a.	n.a.	n.a.	multiple color, multiple angle, multiple segment ED lighting head, auto-calibration
Amistar Automation Inc.	K2	position shift, blur, solder area, bridge	missing components, position shift, rotation error, wrong components, polarity check, bridge, character recognition, solder quantity, lifted leads	CCD	30.4 x 22.8	19	n.a.	n.a.	n.a.	auto-adjust 3-stage LED dome lighting (Upper: IR; Middle: WH; Lower: WH)
Amistar Automation Inc.	K2L	position shift, blur, solder area, bridge	missing components, position shift, rotation error, wrong components, polarity check, bridge, character recognition, solder quantity, lifted leads	CCD	30.4 x 22.8	19	n.a.	n.a.	n.a.	auto-adjust 3-stage LED dome lighting (Upper: IR; Middle: WH; Lower: WH)
CyberOptics	Flex HR 8	missing, polarity, billboard, flipped, wrong part, gross body and lead damage, gold-finger contamination,	missing, polarity, billboard, flipped, wrong part, gross body and lead damage, gold-finger contamination, tombstone, solder bridge, opens, lifted leads, wettability, excess/insufficient solder, debris	5 megapixel color CMOS camera	n.a.	17	5	n.a.	n.a.	fluorescent white light
CyberOptics	Flex HR 12	missing, polarity, billboard, flipped, wrong part, gross body and lead damage, gold-finger contamination,	missing, polarity, billboard, flipped, wrong part, gross body and lead damage, gold-finger contamination, tombstone, solder bridge, opens, lifted leads, wettability, excess/insufficient solder, debris	5 megapixel color CMOS camera	n.a.	17	8	n.a.	n.a.	fluorescent white light
CyberOptics	Flex HR 11	missing, polarity, billboard, flipped, wrong part, gross body and lead damage, gold-finger contamination,	missing, polarity, billboard, flipped, wrong part, gross body and lead damage, gold-finger contamination, tombstone, solder bridge, opens, lifted leads, wettability, excess/insufficient solder, debris	5 megapixel color CMOS camera	n.a.	17	11	n.a.	n.a.	fluorescent white light
Machine Vision Products	Ultra IV	n.a.	n.a.	n.a.	n.a.	8 - 17	n.a.	n.a.	n.a.	programmable variable LED strobe lighting, proprietary multi-color illumination
Mirtec	MV-7	n.a.	n.a.	1.3, 2 or 4 megapixel digital color camera	14.0 x 10.5 to 37.2 x 37.2	9.8 - 18.2	n.a.	n.a.	4	n.a.
Mirtec	MV-7L	n.a.	n.a.	1.3, 2 or 4 megapixel digital color camera	14.0 x 10.5 to 37.2 x 37.2	9.8 - 18.2	n.a.	n.a.	4	n.a.
Mirtec	MV-7xi	n.a.	n.a.	1.3, 2 or 4 megapixel digital color camera	14.0 x 10.5 to 37.2 x 37.2	9.8 - 18.2	n.a.	n.a.	4	n.a.
Mirtec	MV-7U	n.a.	n.a.	1.3, 2 or 4 megapixel digital color camera	14.0 x 10.5 to 37.2 x 37.2	9.8 - 18.2	n.a.	n.a.	4	n.a.
Omron	VT-RNS-S	presence of solder, wrong components, missing components, bridging, component shifting, lead bending	presence of solder, wrong components, missing components, bridging, tombstone, component shifting, fillet, wettability, lead bending, adhesive, solder balls	3-CCD camera	n.a.	10; 15; 20	n.a.	n.a.	n.a.	ring-shaped LED (red, green, blue)
Omron	VT-WIN II	presence/absence, skewed, shifted, wrong component, un-inserted, upside-down/backward, polarity, lead bent	presence/absence of solder, excessive solder, insufficient solder, blow holes, wettability, bridges, solder balls, skewed, shifted, wrong component, polarity, lead bent	triple element CCD camera	n.a.	10; 13; 15; 20; 25; 30; 35; 50	n.a.	n.a.	n.a.	3 ring-shaped LED arrays with automatic brightness control
Saki	BF-Frontier	presence/absence, misalignment, polarity, bridge	presence/absence, tombstone, reverse, polarity, bridge, foreign material, absence of solder, insufficient solder, lifted lead, lifted chip, fillet	line color CCD camera	n.a.	18	n.a.	n.a.	n.a.	LED lighting system
Saki	BF-Planet-X	presence/absence, misalignment, polarity, bridge	presence/absence, tombstone, reverse, polarity, bridge, foreign material, absence of solder, insufficient solder, lifted lead, lifted chip, fillet	line color CCD camera		10	n.a.	n.a.	n.a.	LED lighting system
Viscom	S3088-III	n.a.	n.a.	megapixel	57.6 x 43.5	23.4; 11.7	4	16.1; 8.05	4; 8	n.a.
Viscom	S3088-II	n.a.	n.a.	megapixel	57.6 x 43.5	23.4; 11.7	4	16.1; 8.05	4; 8	n.a.
Viscom	S6056-ST1	n.a.	n.a.	megapixel	57.6 x 43.5	23.4; 11.7	4	16.1; 8.05	4; 8	n.a.
Viscom	S6056-DS1W	n.a.	n.a.	megapixel	57.6 x 43.5	23.4; 11.7	4	16.1; 8.05	4; 8	n.a.
Viscom	S6056-DS2W	n.a.	n.a.	megapixel	57.6 x 43.5	23.4; 11.7	4	16.1; 8.05	4; 8	n.a.
Vi Technology	3K Series	n.a.	n.a.	1620 x 1220 pixel; 2352 x 1728 pixel	42.1 x 31.7; 61.1 x 44.9	8 - 12	n.a.	n.a.	n.a.	i-LITE (red, green, blue); axial and peripheral
Vi Technology	5K Series	n.a.	n.a.	1620 x 1220 pixel; 2352 x 1728 pixel	42.1 x 31.7; 61.1 x 44.9	8 - 12	n.a.	n.a.	n.a.	i-LITE (red, green, blue); axial and peripheral
Vi Technology	7K Series	n.a.	n.a.	1620 x 1220 pixel; 2352 x 1728 pixel	42.1 x 31.7; 61.1 x 44.9	8 - 12	n.a.	n.a.	n.a.	i-LITE (red, green, blue); axial and peripheral
Vi Technology	Vi-5000	n.a.	n.a.	1360 x 1040 pixel	44.5 x 33.6	12	n.a.	n.a.	n.a.	amber, green, blue
Vi Technology	Vi-5000-2	n.a.	n.a.	1600 x 1152 pixel	41.6 x 29.9	8	n.a.	n.a.	n.a.	green, white, blue, axial and peripheral
Vi Technology	Vi-5000-3	n.a.	n.a.	2048 x 2048 pixel	53.2 x 53.2	8	n.a.	n.a.	n.a.	green, white, blue
YES Tech	YTV-F1	position, missing, wrong, polarity, skew	polarity, skew, tombstone, bent lead, lifted, bridging, open solder, insufficient, short, solder balls	Multiple Thin Camera megapixel color top-down viewing camera @ 1280 x 1024 pixel	n.a.	25; 12	n.a.	n.a.	n.a.	LED top light, proprietary bi-color multiangle LED lighting
YES Tech	YTV-F1S	position, missing, wrong, polarity, skew	polarity, skew, tombstone, bent lead, lifted, bridging, open solder, insufficient, short, solder balls	Multiple Thin Camera megapixel color top-down and 4 side viewing camera @ 1280 x 1024 pixel	n.a.	25; 12	n.a.	n.a.	n.a.	LED top light, proprietary bi-color multiangle LED lighting
YES Tech	YTV-M1	position, missing, wrong, polarity, skew	polarity, skew, tombstone, bent lead, lifted, bridging, open solder, insufficient, short, solder balls	YESTech 3 Megapixel Thin Camera top-down viewing camera and telecentric lens @ 2048 x 1536 pixel	n.a.	25; 12	n.a.	n.a.	n.a.	proprietary Fusion Lighting

Table 9.

http://dx.doi.org/10.5772/51699

Manufacturer	Model	Inspection capacity typical	Board size max. [mm x mm]	Board size min. [mm x mm]	Board warp [mm]	Board clearance top [mm]	Board clearance bottom [mm]	Board thickness [mm]	Board weight max. [kg]	Dual lane capable	Conveyor height [mm]
Agilent Technologies	Medalist SJ50 Series 3	41 sq. cm/sec @ pre-reflow; 32 sq. cm/sec @ post reflow	510 x 510	50 x 50	n.a.	n.a.	n.a.	0.5 - 4.0	3	Yes	n.a.
Agilent Technologies	Medalist SJ50 Series 3 XL	41 sq. cm/sec @ pre-reflow; 32 sq. cm/sec @ post reflow	620 x 620	75 x 50	n.a.	n.a.	n.a.	1.5 - 15	13	n.a.	n.a.
Agilent Technologies	Medalist sj5000	41 sq. cm/sec @ pre-reflow; 32 sq. cm/sec @ post reflow	510 x 510	50 x 50	n.a.	n.a.	n.a.	0.5 - 4.0	3	Yes	n.a.
Amistar Automation Inc.	K2	0.25 sec/screen	330 x 250	50 x 50	+ 0.5; - 1.0	28	25	0.5 - 2.0	n.a.	n.a.	n.a.
Amistar Automation Inc.	K2L	0.25 sec/screen	485 x 410	50 x 50	+ 0.5; - 1.0	28	25	0.5 - 2.0	n.a.	n.a.	n.a.
CyberOptics	Flex HR 8	50 sq. cm/sec	203 x 508	110 x 63	± 0.7	32	3	n.a.	n.a.	n.a.	813 - 965
CyberOptics	Flex HR 12	50 sq. cm/sec	305 x 508	110 x 63	± 0.7	32	3	n.a.	n.a.	n.a.	813 - 965
CyberOptics	Flex HR 11	50 sq. cm/sec	457 x 508	110 x 63	± 0.7	32	3	n.a.	n.a.	n.a.	813 - 965
Machine Vision Products	Ultra IV	90 sq. cm/sec	500 x 546	n.a.	n.a.	n.a.	n.a.	n.a.	n.a.	n.a.	n.a.
Mirtec	MV-7	4.94 sq. mm/sec	350 x 250	50 x 50	n.a.	25 - 45	50.8	n.a.	n.a.	n.a.	n.a.
Mirtec	MV-7L	4.94 sq. mm/sec	500 x 400	50 x 50	n.a.	25 - 45	50.8	n.a.	n.a.	n.a.	n.a.
Mirtec	MV-7xi	4.94 sq. mm/sec	510 x 460	50 x 50	n.a.	25 - 45	50.8	n.a.	n.a.	n.a.	n.a.
Mirtec	MV-7U	4.94 sq. mm/sec	660 x 510	50 x 50	n.a.	25 - 45	50.8	n.a.	n.a.	n.a.	n.a.
Omron	VT-RNS-S	0.25 sec/screen @ 10 sq. mm field of view	510 x 460	50 x 50	n.a.	20 - 40	40 - 50	n.a.	n.a.	n.a.	n.a.
Omron	VT-WIN II	0.40 sec/screen	460 x 510	50 x 50	n.a.	50	50	0.3 - 4.0	n.a.	n.a.	n.a.
Saki	BF-Frontier	24 sec/screen	460 x 500	50 x 60	± 0.2	40	40	0.6 - 2.5	n.a.	n.a.	max. 900
Saki	BF-Planet-X	23 sec/screen	250 x 330	50 x 60	n.a.	20	30	0.6 - 2.5	n.a.	n.a.	max. 900
Viscom	S3088-III	20 - 40 sq. cm/sec	508 x 508	n.a.	n.a.	35	40	n.a.	n.a.	n.a.	850 - 960
Viscom	S3088-II	20 - 40 sq. cm/sec	450 x 350	n.a.	n.a.	35	40	n.a.	n.a.	n.a.	850 - 960
Viscom	S6056-ST1	20 - 40 sq. cm/sec	457 x 356	n.a.	n.a.	35	60	n.a.	n.a.	n.a.	830 - 960
Viscom	S6056-DS1W	20 - 40 sq. cm/sec	457 x 356	n.a.	n.a.	35	60	n.a.	n.a.	n.a.	830 - 960
Viscom	S6056-DS2W	40 - 80 sq. cm/sec	457 x 356	n.a.	n.a.	35	60	n.a.	n.a.	n.a.	830 - 960
Vi Technology	3K Series	4 - 20 ms	458 x 406	50 x 50	n.a.	34	34	0.7 - 4.0	3	Yes	n.a.
Vi Technology	5K Series	4 - 20 ms	533 x 533	50 x 50	n.a.	34	60	0.5 - 4.0	3	n.a.	n.a.
Vi Technology	7K Series	4 - 20 ms	533 x 610	50 x 50	n.a.	34	60	0.5 - 4.0	3	Yes	n.a.
Vi Technology	Vi-5000	4 - 20 ms	508 x 458	50 x 50	n.a.	34	40	0.7 - 5.0	7	n.a.	n.a.
Vi Technology	Vi-5000-2	4 - 20 ms	508 x 458	50 x 50	n.a.	34	40	0.7 - 5.0	7	n.a.	n.a.
Vi Technology	Vi-5000-3	4 - 20 ms	508 x 458	50 x 50	n.a.	34	40	0.7 - 5.0	7	n.a.	n.a.
YES Tech	YTV-F1	35 sq. cm/sec	560 x 510	n.a.	n.a.	50	50	n.a.	n.a.	n.a.	n.a.
YES Tech	YTV-F1S	35 sq. cm/sec	560 x 510	n.a.	n.a.	50	50	n.a.	n.a.	n.a.	n.a.
YES Tech	YTV-M1	35 sq. cm/sec	350 x 250	50 x 50	n.a.	25	50	n.a.	n.a.	n.a.	max. 950

Table 10. Comparison of Universal Automatic Optical Inspection (UAOI) machines

Assuming that the component is fully operational, these systems practically are able to prove that the whole circuit board is working correctly thus replacing the ICT. However, because they are usually connected to SPC (Statistical Process Control) servers, they can also provide much information about the SMT process itself and provides help as to how to improve it.

But of course there are disadvantages to using AOI systems. They are not able to inspect hidden failures such as soldered BGA (Ball Grid Array) bumps and usually the parameters of inspection algorithms cannot be adjusted perfectly. So from time to time they do not detect real failures which are called 'slip-through failures'. These are the most significant malfunctions during the operation of AOI systems because in these cases, they fail to do what they were programmed for. So the number of slip-throughs must be zero and - if they arise - close investigation is necessary to prevent and eliminate them. However if this occurs repeatedly, then the appropriate parts would seem to be defective. These are the pseudo-failures which can reduce productivity so their number should as close to zero as possible. [164] ALSO indicates some other image processing problems. The problems of AOI systems will be described in more detail later in the chapter.

Another disadvantage is that they are usually in the 'bottle-neck' of the manufacturing production line because they are not able to inspect the whole circuit board as fast as the line can produce them. Therefore, the practice is usually to place more machines behind each other to enable inspections to take place in parallel. Of course, this also has financial implications which should be taken into consideration.

7. Special AOI solutions — Inspection of lead-free solder joints, flexible substrates, wire bonding and semiconductors

According to RoHS and WEEE directives, lead-free solder alloys have to be used in commercial electronics. This has presented a new challenge for AOI systems because of the differing optical properties of lead-free alloy. Some solutions are shown in the following studies [166]-[173]. AOI has several further application possibilities in electronic device manufacturing e.g. semiconductor and wire-bonding inspection. These appliances need extremely high-resolution cameras to detect defects in the μm scale. Another interesting area is flexible substrate inspection. Some of these special inspections are described in [174]-[179].

7.1. Differences between lead-based and lead-free solder alloys

Solders that contain lead are available with a tin content of between 5% and 70%. The composition of the most commonly used lead solder is 63/37 Sn/Pb; this was the main type used in electronics manufacturing until strict controls were imposed on its use for environmental reasons. The homogeneity of the solder meniscus that formed was beneficial in that the melting point of eutectic solder really is manifested as a single point on the phase diagram; in other words the molten alloy solidifies at a specific temperature, rather than within a broader temperature range. The solidified alloy can be broken down into tiny lead and tin phases of almost 100% purity, without intermetallic layers.

In the case of non-eutectic solders, the crystallisation begins around cores of differing composition and crystal structure, and at differing temperatures, so that during the accretion of the individual cores the composition of the residual melt also changes. Due to this, in the case of lead-free, non-eutectic solder alloys, certain phases solidify earlier, and these solid cores do not form a completely mirror-like, smooth surface on the face of the solder meniscus (and naturally, they also cause differences in the volume of the material).

Lead-free solders usually contain tin, silver and copper. Compared to lead-based solders they have several negative properties: they are more expensive, their melting point is higher, and they give rise to problems that do not occur when soldering with lead (the phenomenon of whisker formation has still not been fully explored). Because their surface differs from that of

http://dx.doi.org/10.5772/51699

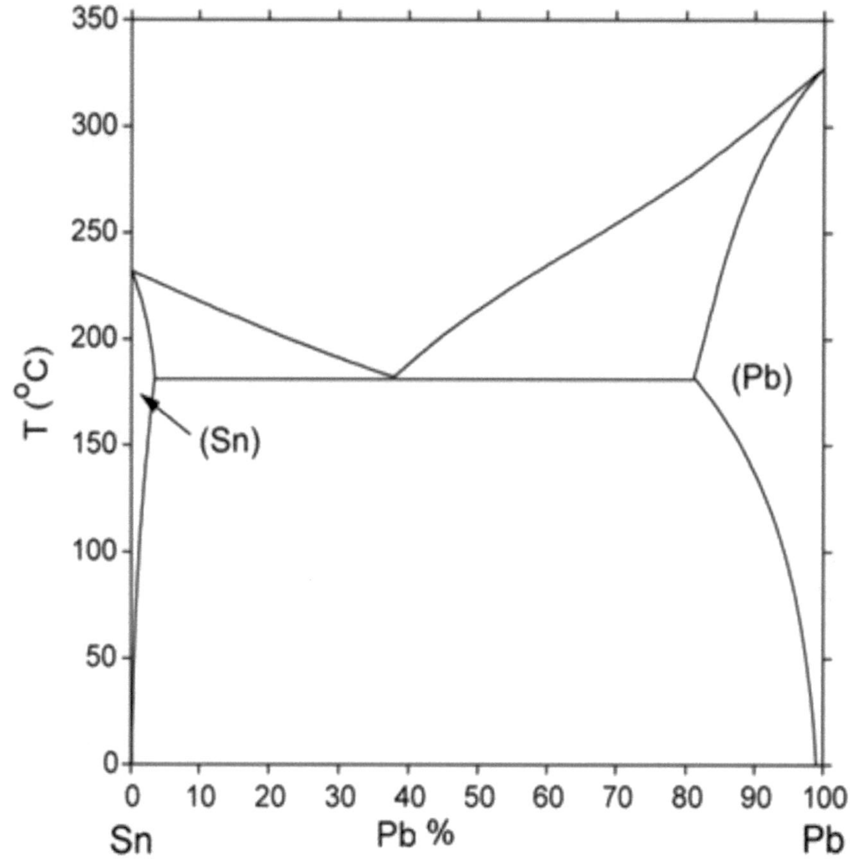

Figure 7. Tin-lead phase diagram

lead-based solder alloys – which are much more even and mirror-like – they reflect light differently, so different procedures may be used to verify the presence and quality of the solder menisci.

In the case of tin-lead solders, the solidification of the melt begins around the cores that are solid at melting point, and the individual solid phases grow at a virtually consistent rate as the two elements separate from the melt. This is how the volumes that are rich in lead and tin become a smooth-surfaced alloy consisting of lead and tin patches, typical of eutectic solder, that can easily differentiated on the cross-section.

Lead-free solders do not usually form eutectic alloys, and exist in many variants with different compositions. Tin is usually alloyed with copper and silver, but there are also alloys containing, for example, bismuth and indium.

In the case of the non-eutectic alloys (the vast majority of lead-free alloys), however, one of the phases begins to solidify earlier, and the alloying metal concentration of this phase will be smaller than that of the melt. This means that the composition of the remaining part of the alloy, which is still in a liquid state, continues to change until the eutectic composition is

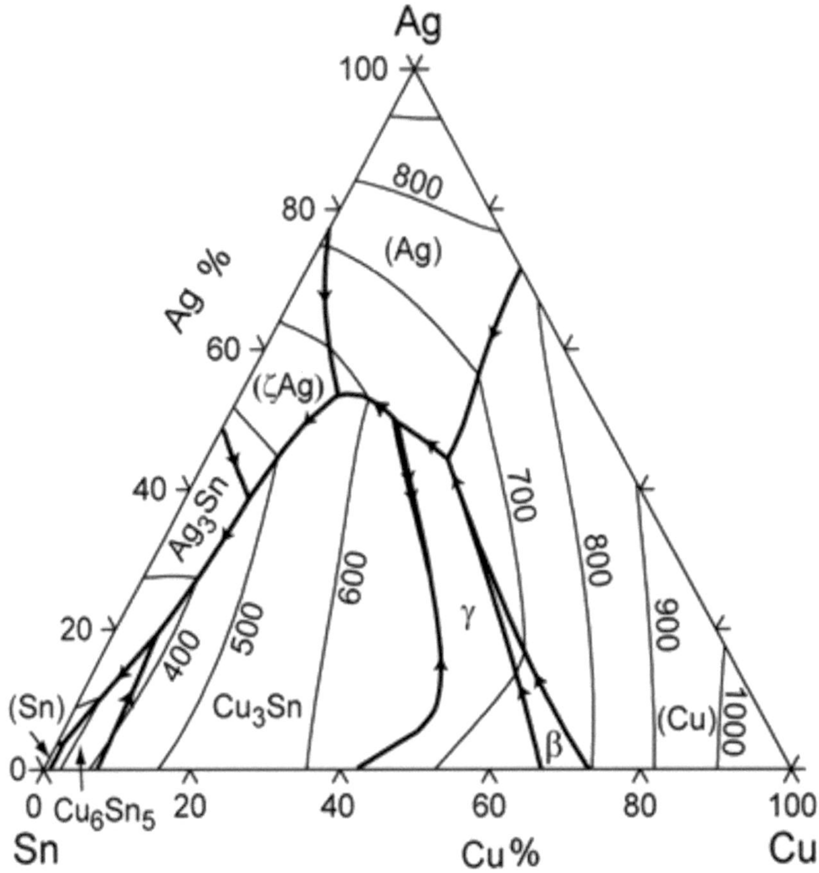

Figure 8. Tin-copper-silver phase diagram

achieved, when it cools and solidifies. As a result of this, the microstructure that is created has a greater surface roughness than in the previous case: as the eutectic melt ebbs away, the intermetallic crystals that were the first to solidify create a more irregular surface. This surface scatters light much more than the smoother, more even surface of the tin-lead solder; in other words the proportion of diffuse reflection will be greater than that of specular reflection.

An attempt to measure the two solders with AOI equipment using the same settings will probably result in several errors, because after the necessary image conversion procedures the images made by the equipment will differ. For this reason, it would clearly be useful to calibrate the AOI equipment specifically for the different solders.

In what follows we present a series of images of tin-lead eutectic and lead-free Sn-Ag-Cu solder alloys made using a scanning electron microscope (SEM). This instrument is not suitable for measuring the surface roughness, but it does provide an accurate, high-resolution image of the examined surfaces and of the two solder alloys with differing composition and surface roughness, showing the differences in height and material with spectacular contrast.

http://dx.doi.org/10.5772/51699

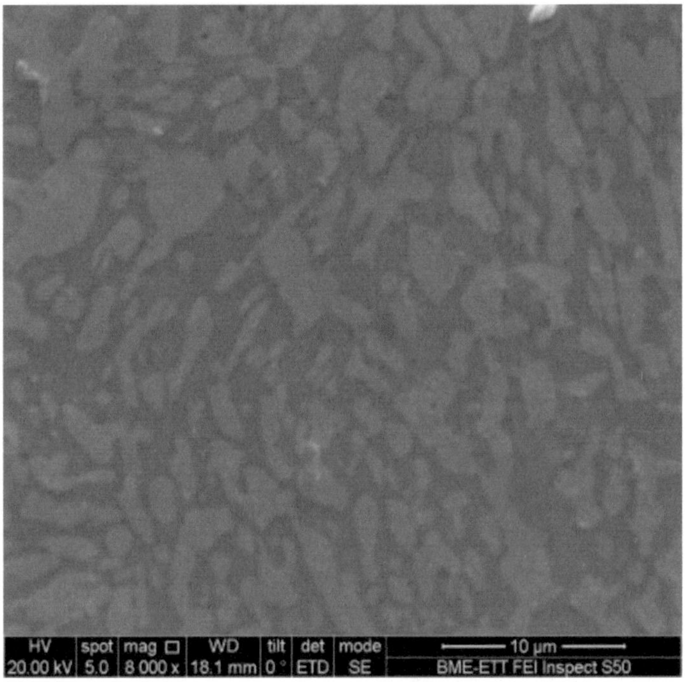

Figure 9. Electron microscope image of the surface of a tin-lead solder meniscus

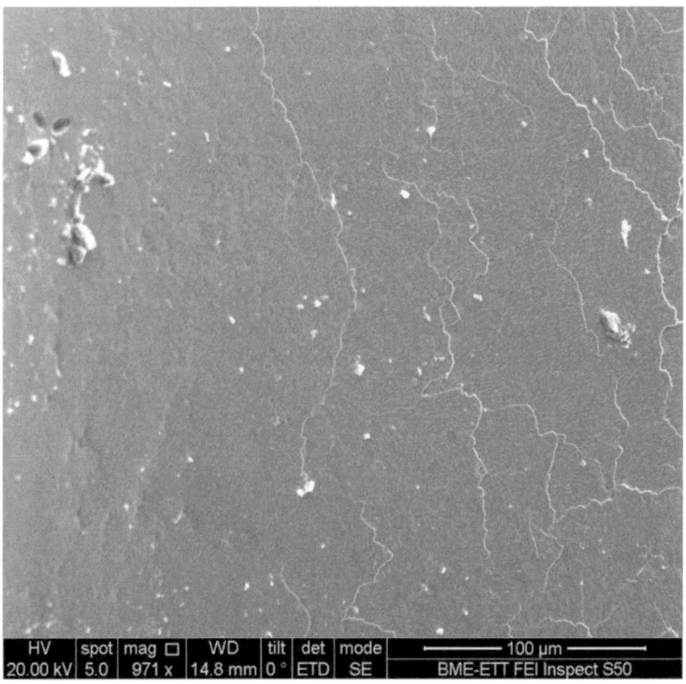

Figure 10. SEM image of the surface of a lead-based solder meniscus

An important question is precisely what the roughness of the pattern formed on the surface of lead-free solder alloys depends on, and how "reliably" predictable the process of its formation is.

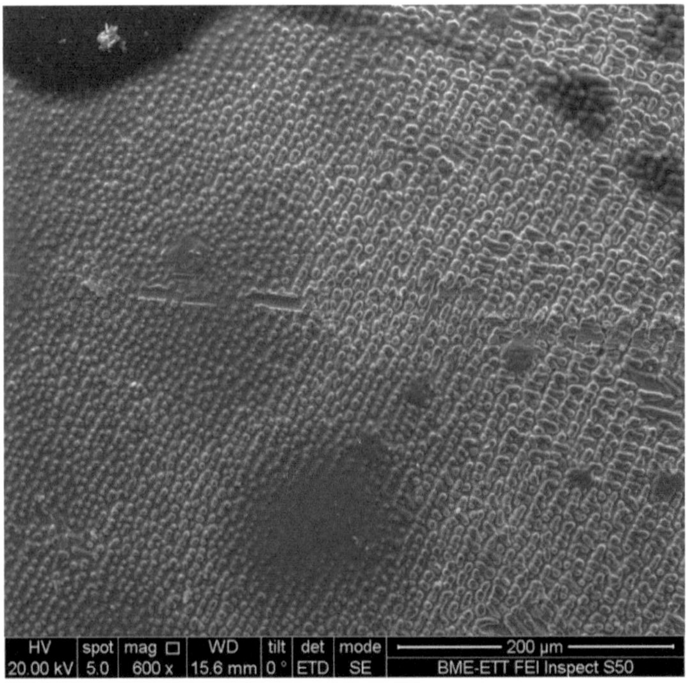

Figure 11. SEM image of the surface of a lead-free solder meniscus

In the above image the two types of solder can be clearly differentiated due to the rougher surface that is typical of the lead-free alloy. In places the surface looks quite similar to the one assumed by the microfacet model, which simplifies reality for the purpose of mathematical manageability; in other words, small semi-spherical formations can be observed side by side with each other. On other parts of the picture, however, areas with no unevenness are also visible; and we have taken the electron microscope image of an area that gives a good illustration of a particular property of lead-free, non-eutectic solder alloys (in this case a tin-copper-silver alloy), namely that due to the unevenness of the surface it reflects the light more diffusely (in other words, it scatters the light more) than the smoother surface of a eutectic solder. In the applied Cook-Torrance model, the roughness of the surface is described by a single parameter, which describes the surface in average terms.

The above picture shows an SEM image of a cross-section in which the tin (light) and lead (dark) phases of the eutectic alloy are clearly differentiated.

The above image was made at a lower magnification (400x rather than 1500x), but the phase boundaries can still be made out, and the smooth meniscus surface typical of lead-based solder

http://dx.doi.org/10.5772/51699

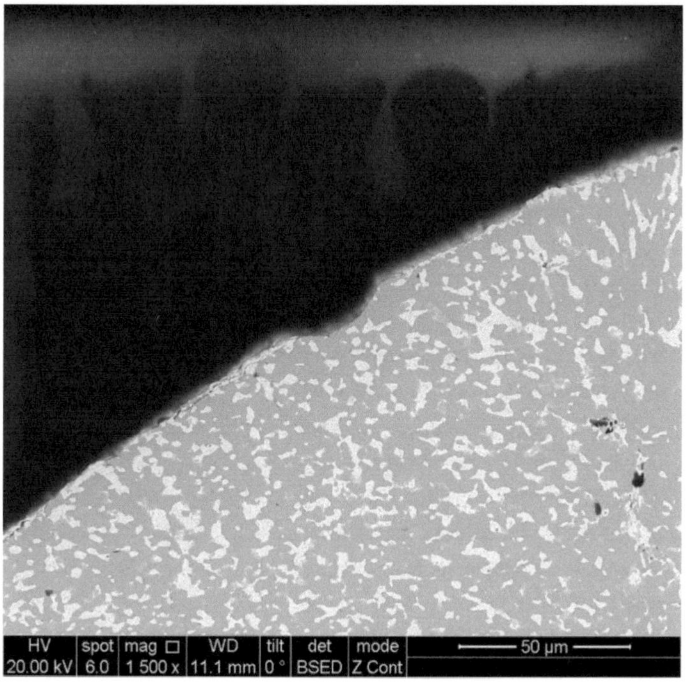

Figure 12. SEM image of a cross-section of a lead-based solder meniscus

Figure 13. SEM image of a cross-section of a lead-based solder meniscus

alloys is even more visible. In the following SEM images the rough surface typical of lead-free solder alloys can be observed.

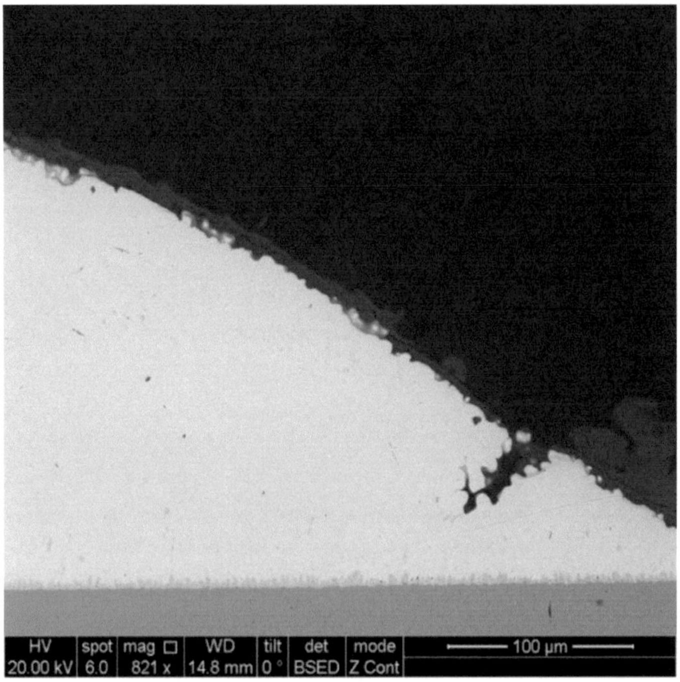

Figure 14. SEM image of a cross-section of a lead-free solder meniscus

The fracture in the solder meniscus seen in the above image is probably due to a contaminant, but within the fracture is a particularly clear example of just how uneven a surface can be formed by the lead-free solder alloy.

The surface roughness of the lead-free solder meniscus is visibly greater than that of the tin-lead solder. Taking the scale bar as a guide we can also estimate that the size of the uneven protrusions that increase the surface roughness, in terms of both their breadth and height, is in the order of 10 μm. It is also worth noting that the simplification of the microfacet model described by the Cook-Torrance model is clearly visible, as a visual inspection reveals that the surface is not closely similar to the surface made up of tiny flat plates that is assumed by the microfacet model. This simplification, however, is more than made up for by the model's simplicity and general ease of use.

7.2. Measuring the surface roughness

To measure the surface roughness we used a Tencor *Alpha Step* 500 surface profilometer. Based on the 10 measurements of each solder, made on the lead-based (Heraeus F816 Sn63-90 B30) and lead-free (Senju Ecosolder M705-GRN360-K1-V) joints, the two solders yielded the following values:

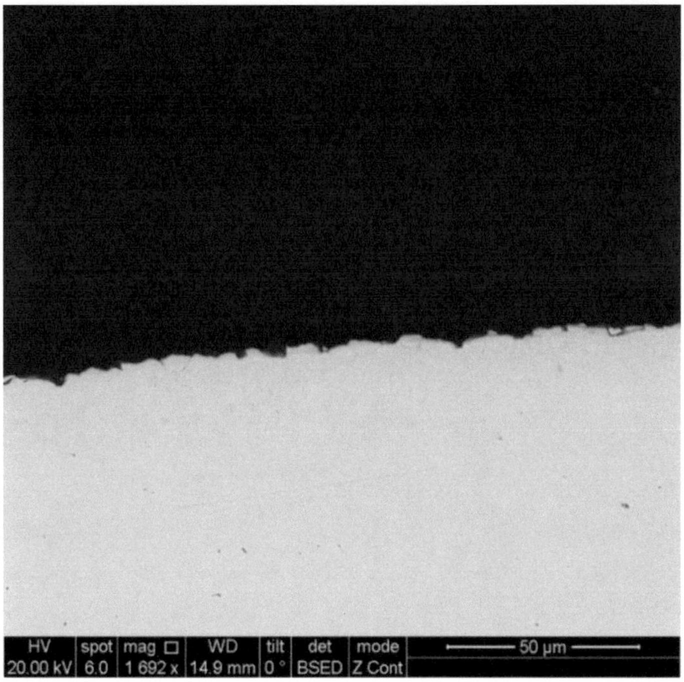

Figure 15. SEM image of a cross-section of a lead-free solder meniscus

Type of measurement result	Lead-based solder paste	Lead-free solder paste
Measured surface roughness (RMS)	0.03092	0.0986
Distribution of measured RMS value	0.004451	0.054626

Table 11. Measured surface roughness values

At first glance the measured surface roughness values appear realistic; the surface roughness of the lead-free solder turned out to be approximately three times that of the lead-based solder. The distribution of the roughness values for the lead-based solder was below 1%, which is satisfactory because the divergence between the shape of the actual solder and that modelled by the computer showed a greater error (a few percent), and because greater fluctuation than this can be expected to result from the differing heat profiles, printed circuit boards or solder-handling requirements of real production lines. The distribution of the surface roughness values for the lead-free solder was over 5%, which is due to diversity of the size and shape of the surface protrusions that appear with this type of solder, which the Cook-Torrance model handles using statistical simplification, by assuming the surface to be of a consistent roughness.

7.3. Simulation created with the computer model

We checked the measured surface roughness values by comparing the images made using optical microscopes with the computer-generated graphic representations. The Surface

Evolver software uses finite element analysis to calculate the surface profile at certain points on the surface. In areas with a greater radius of curvature, where the energies are closer to each other in terms of magnitude (in other words none are dominant in comparison to any others) the software uses more measurement points, that is a denser grid, for displaying the graphic representation.

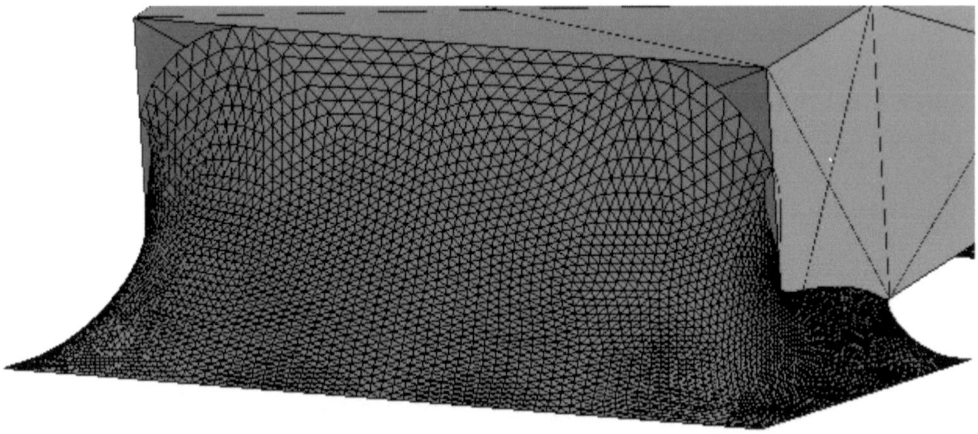

Figure 16. Example of a graphic representation generated using the Surface Evolver software

Of the models that use a formula based on the Bi-directional Reflectance Distribution Function (BRDF), which is based on a physical approach, the most widely used is the Cook-Torrance model, which has surface roughness as one of its input parameters and is also capable of handling Fresnel distribution. During our simulation we used this, in a Direct3D environment, so when generating the rendered graphics we were able to use the measured roughness values as input parameters.

The majority of optical microscopes – including the Olympus BX51 microscope used by me at the department – are capable of operating in bright field (BF) and dark field (DF) imaging mode.

In bright field imaging, both the incident and reflected light fall almost perpendicularly onto the sample, naturally through a focusing lens. Dark field microscopes, on the other hand, collect beams of light that arrive not perpendicularly but from the side, from below a given angle, through a lens, in the direction of the observer; in other words the beams of light travel in the opposite direction but along the same path as would the beams of light that enter perpendicularly but are diffracted, not reflected.

Dark field microscopy gives a good resolution and microscopes with this capability are usually more expensive, but they are eminently suitable for the detection of phase boundaries or the examination of surface irregularities highlighted by the side illumination. In the case of metals, in which the proportion of diffuse components is smaller and the incident light is reflected much more in accordance with the principle of optical reflection, bright field microscopy results in a darker image and in the case of observation along the z axis (from above), as is typical of microscopes, only the surfaces that are parallel to the horizontal are illuminated. We also modelled both of these different types of illumination using the Direct3D software.

What follows is a comparison of the images made using the optical microscope and the graphic representations rendered with Direct3D that most closely replicated the actual light and surface conditions. Where not indicated separately, the soldered joint (at the SMT resistors) is illuminated with scattered light.

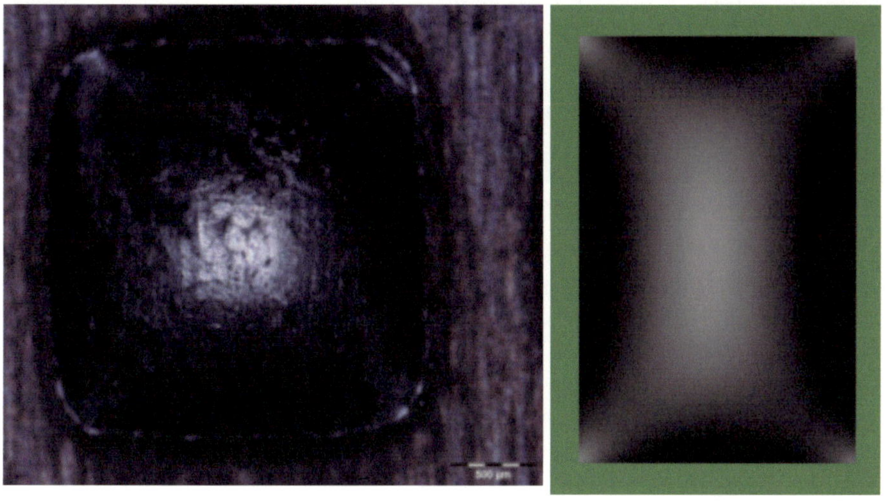

Figure 17. Photograph and graphic representation of empty solder pad covered in lead-free solder (BF imaging)

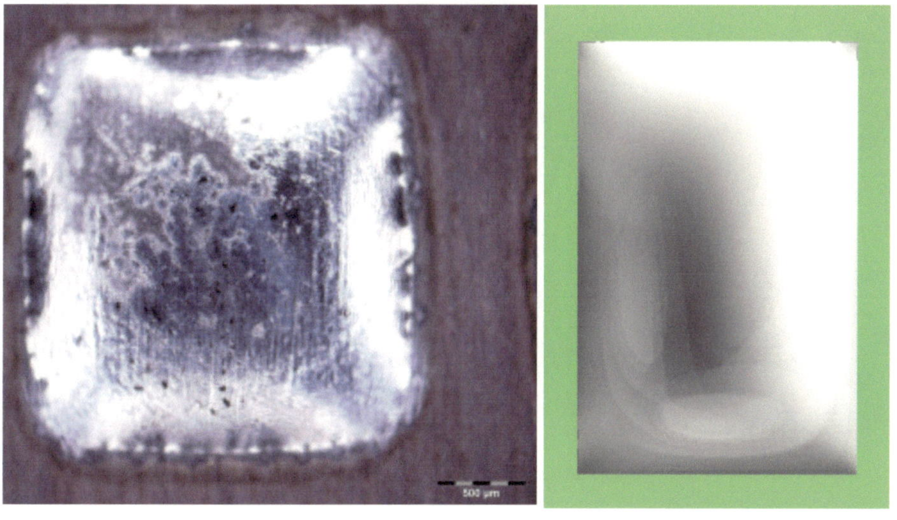

Figure 18. Photograph and graphic representation of empty solder pad covered in lead-free solder (DF imaging)

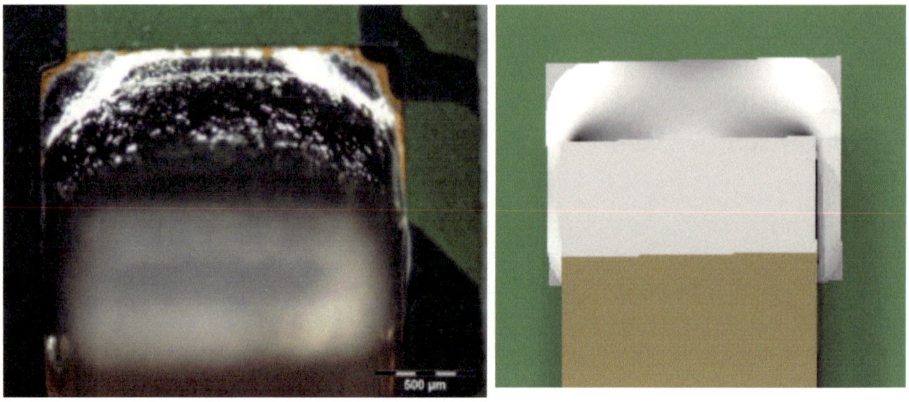

Figure 19. Photograph and graphic representation of SMT joint made with lead-free solder

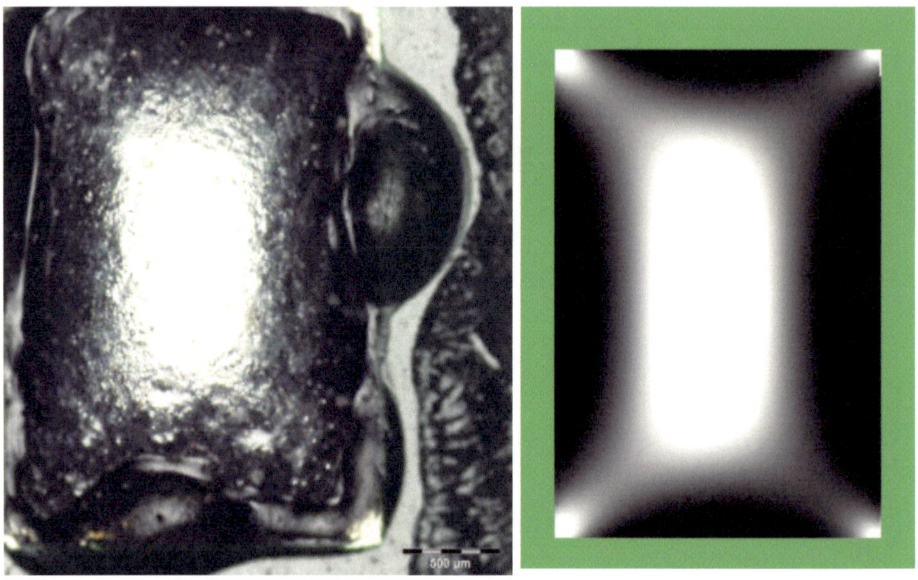

Figure 20. Photograph and graphic representation of empty solder pad covered in lead-based solder (BF imaging)

http://dx.doi.org/10.5772/51699

Figure 21. Photograph and graphic representation of empty solder pad covered in lead-based solder (DF imaging)

Figure 22. Photographs (above: BF, below: DF) and graphic representation of SMT joint made with lead-based solder

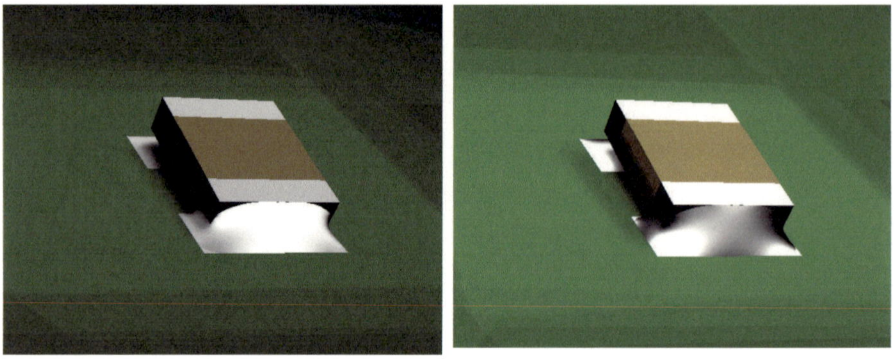

Figure 23. Photograph and graphic representation of SMT joint made with lead-free (left) and lead-based (right) solder

8. Detailed analysis of AOI systems

As can be seen in the previous section, AOI machines handle several tasks. Much literature is dedicated to the intelligence of these systems, but from a technical point of view we can also examine other aspects. A large number of these AOIs work on high-mix-high-volume SMT lines where the most important key factors are the inspection duration and quality. The attributes of this system relate to the following sections:

1. actuating parts (drives and axes)
2. image acquisition system (sensors, optics, illumination)
3. software processing part

They work in close relationship to each other, so the speed of each has to be in sync. There are three well-defined mechanical constructions for an AOI system:

- without special moving parts / drives inside
- with PWB positioning table
- with camera-module actuating unit

The simplest case is when the working-process of the system does not include special positioning steps. The PWB is positioned/placed in "one step" into the field of camera system, an image is acquired and the PWB is then taken out for the next process. This could be of great benefit because the machine does not need to synchronize any movements during the image acquisition process. The speed affecting factor can thus be ignored. This is used typically in Automatic Final Inspection (AFI) systems. This does not mean that the system has to only contain one camera. For more complex applications the number of cameras can be increased. More cameras mean more complex image transformation and manipulation tasks so it follows that these systems are only capable of use when looking at pre-defined areas.

http://dx.doi.org/10.5772/51699

In case of larger inspection areas, the systems are mounted with special drives which can move the camera system or the inspected part. The movements of these drives have to be synchronized with the process of illumination and image acquisition. When the system contains small number of cameras and the illumination devices are also built-in, then the module itself should be the moving section. When there are even more cameras, each with its separate illumination (matrix arrangement), then the PWB should lie on a positioning table.

Two big groups of drive systems are commonly used for this purpose. The first is the conventional electromechanical drive. It is used for some 2D paste inspection machines. Here the velocity of the camera can be constant, while in most cases it contains line-CCD sensor. The other type of motion system is the linear drive which is more accurate and faster and therefore in more frequent use.

The directional route of the moving part highly depends on a second factor, that of speed and the properties of image acquisition system. Here also, three main parts can be singled out:

- optics / lenses
- camera / sensor type
- illumination module / lighting source

The system has to get the necessary amount of information and resolution from even the smallest components. In the SMT field this means zooming down to a 10μm pixel resolution. To ensure the constant magnification at all points of the entire Field of View (FOV) the use of telecentric optics is essential. This criterion enables the system to make the required size-measurements. On an image seen through traditional lenses, the apparent shape of components changes with the distance from the centre of the FOV, therefore sometimes making shape recognition a hard task.

But it is not just the permanency of magnification that is important, so too is the need to select the correct level. On one hand, the larger detection area of the image sensor can help solve this task, but it also increases the computational resources needed. On the other hand, higher magnification levels give a better resolution but at the expense of reducing the field of view. The best scenario is if the system is capable of optional magnification. Generally, a relative large FOV, between 10-25 cm2, could be used and only in certain cases should dedicated Field of Interests (FOI) should be zoomed out.

In most AOI applications, the LED based lighting is used for illumination purposes. But independent of the type of illumination source used, the amount of illumination should be only as much as is required. The optimum depends on the application. For example, a 2D paste or a through-hole-technology (THT) components solder-joint inspection system needs only just a small amount of illumination. As the number of failure types / inspection tasks increase so too the number of illumination modes also increase. The programmable illumination module is a good tool to develop lighting requirements for dedicated purposes, but it also carries the risk of inhomogeneous and reduced FOV.

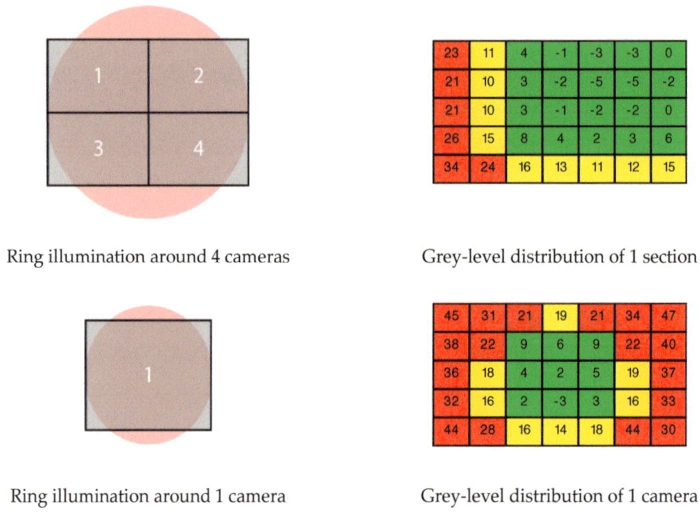

Figure 24. Problem of inhomogeneous grey-level by ring-illumination types

Fig. 24 illustrates two types of camera-illumination systems. The first system contains 4 cameras, the second only 1 camera. Both have LED-ring illumination modules. The grey-level distribution maps shown above have been measured with the same type of illumination and grey-reference flat. The green areas indicate the valuable field of the camera. This example clearly points out the importance of the homogeneity. Of course this phenomenon is also present when the illumination system is multi-coloured.

Most optical inspection / control appliance decisions are based on image-processing methods that have been set experientially. The stress is on the word "experientially". Most of the AOI machines make some kind of template matching. These sample-templates can be colour or greyscale, stand from parts/windows or form a complete pattern. The machine can be 'self-learning or directed by means of an "external trainer". Due to the fact that the overall reliability of these machines is not 100%, the defined limits between good and bad classified patterns are not strict. In some cases it could be that just two pixels differ between the data provided. If the phenomenon which the system needs to detect is not so unambiguous, then it should search for another method to make the gap wider between the 2 classes.

9. Software questions

One of the most wide-spread criticisms against the principles and methods of automated optical inspection systems stems from a very interesting paradox. As we have mentioned earlier, the introduction of AOI devices in the manufacturing lines was a result of the growth in manufacturing process complexity. These inspection and control devices have to fulfill certain reliability criteria which need to be validated. But unfortunately, these validation

http://dx.doi.org/10.5772/51699

processes can only be used to a limited degree because of the high-complex manufacturing process and the equally complex and highly varied appearance of devices under test (DUTs).

This contradiction invokes the conclusion that accuracy and reliability of AOI system depends very much on the competence and working quality of the engineers and operators, the correct management of the setting up and controlling the inspection devices. In reality, this sets out several very serious challenges to experts. The quality inspection algorithms have many parameters – in some cases several hundred – (image processing, region of interest, threshold parameters etc.). Their setup requires experience, intuition and inspiration from the process engineers themselves.

In addition, during parameter tuning, the engineers need to solve the following contradiction, where the difference between images showing correct and faulty components is often only a few pixels which need to be detected by the AOI devices (Fig. 25). In the case of incorrect parameter settings these small signals can disappear and the system classify a bad component as good ("slip-through"). Certainly this false classification is totally intolerable in quality inspection processes; therefore it is necessary to aim for the complete elimination of this possibility by fine tuning the algorithm's parameters. Unfortunately because of this, engineers can easily set the algorithm to be too strict, meaning also that some correct components will be dropped out during the inspection process. Although these "false calls" (also known as pseudo failure) do not cause catastrophic consequences nevertheless they are the source of a very serious problem. Namely, in this instance, the human operators performing the re-inspection of components considered "faulty" can easily get used to the repeated mistakes of the AOI system. Therefore they can eventually take the inspection device's decisions out of consideration even where there is a cases of real errors. This implies that the reliability of the inspection device itself would be in doubt; the fact of which would result in one of the biggest catastrophic effects on AOI systems. In addition, it seems insignificant but it is important to note that many bad classifications slow the manufacturing process, decrease productivity and increase the product overall production costs. To avoid false calls, process-engineers need to reduce the strictness of the inspection parameters which – as we have mentioned earlier – is inconsistent with principle used by the parameter settings preventing the slip-through.

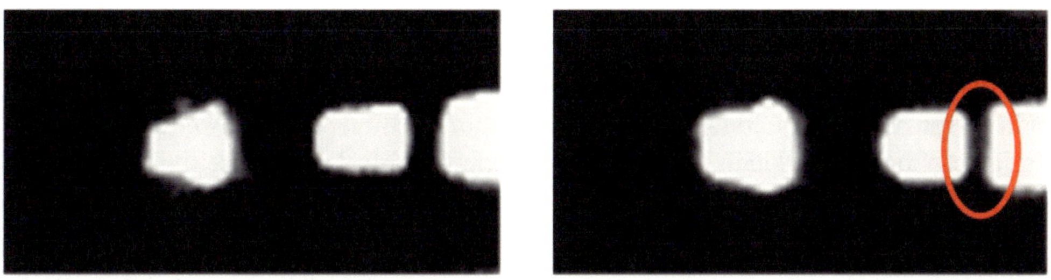

Figure 25. An example for the tiny differences between the images containing correct and faulty components

In addition, AOI engineers need to cope with several other difficulties a major one of which is that the production process changes continuously e.g. the settings of devices on the manufacturing line have to be modified, and this needs to be followed also by modifications to the AOI devices. Therefore the need to monitor the inspection algorithms and adapt to different parameters is a serious challenge to the process engineers.

Furthermore, it is necessary to satisfy some practical requirements when selecting and adjusting the inspection algorithms. Usually, electronic factories manufacture more products in parallel in which several similar or identical components can be located. If all the components were to be inspected with a separate AOI algorithm, the code management, version tracking and fixing etc. would be impossible. Therefore, engineers often use only one inspection method for similar mountings to achieve simpler AOI algorithm version management. Unfortunately, this strategy cannot always be used successfully because of the very heterogeneous appearance of the same components. Fig. 26. shows an image sequence of the C0805 capacitor which illustrates the enormous differences between images taken of similar components.

In this varied environment it is very hard to develop an inspection method which results in highly reliable classification of each type of image for the same component. In addition, a parameter setting process that reduces the number of bad classifications in case of one component influences not only the selected manufacturing line but has an effect on the whole factory. Therefore it can happen that whilst a parameter optimization process reduces the number of bad classifications in the first part of the factory, it increases them on other manufacturing lines. This paradox is one of the reasons why the AOI macro optimization process is a very long and "Sisyphean" task of AOI process engineers.

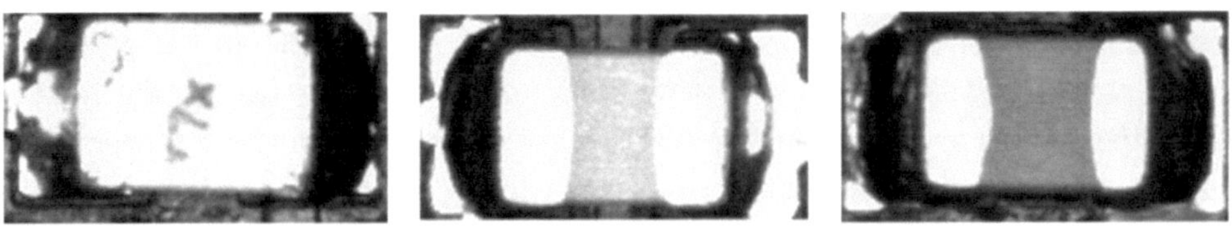

Figure 26. Differences between the appearances of similar components (capacitor C0805)

A very interesting and important question is the optimization of classification thresholds. One of the most important requirements of an inspection system is high-level robustness, but this condition can hardly be guaranteed if the classification decision (namely whether a component gets "faulty" or "good" label) is dependent on only one pixel. Therefore the quality results close to the decision threshold need to be classified in a separate group ("limit error") and it is necessary to apply a different strategy to them. It follows that AOI experts – apart from the fact that they need to solve the optimization paradox mentioned earlier – have to strive to find such an algorithm parameter setting where during the classification, the number of components classified near the decision threshold are as few as possible. Efficiency of AOI appliances

http://dx.doi.org/10.5772/51699

can be significantly improved with the help of macro optimization. In the first task, the pseudo rate was reduced while slip-throughs remains zero (Table 12, Fig. 27)

Before optimization (30 days testing period)			
Inspected components (solder joints) [pieces]	Detected failures [pieces]		***Pseudo rate [ppm]***
	Real failures [pieces]	Pseudo failures [pieces]	
347 130 (694 260)	4 676		***13 459***
	4	4 672	

After optimization (30 days testing period)			
Inspected components (solder joints) [pieces]	Detected failures [pieces]		***Pseudo rate [ppm]***
	Real failures [pieces]	Pseudo failures [pieces]	
223 006 (446 012)	52		***224***
	2	50	

Table 12. Results of macro optimization

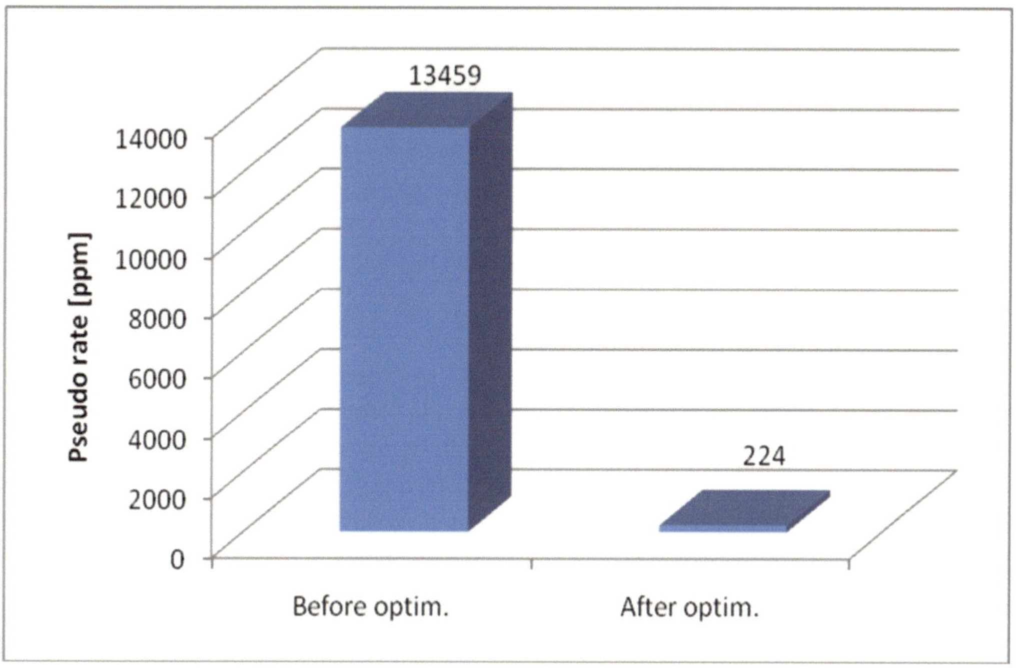

Figure 27. Pseudo reduction

Secondly parallel pseudo and slip-through reduction were carried out (Table 13, Fig. 28, Fig. 29).

Before optimization (30 days testing period)			
Inspected components (solder joints) [pieces]	Detected failures [pieces]		***Pseudo rate [ppm]***
	Real failures (***Quasi-tombstone***) [pieces]	Pseudo failures [pieces]	
2 423 334 (4 846 668)	68 205		***27 995***
	364 (***0***)	67 841	

After optimization (30 days testing period)			
Inspected components (solder joints) [pieces]	Detected failures [pieces]		***Pseudo rate [ppm]***
	Real failures (***Quasi-tombstone***) [pieces]	Pseudo failures [pieces]	
4 655 392 (9 310 784)	58 654		***12 560***
	627 (***62***)	58 027	

Table 13. Results of macro optimization

Another very serious question is about the parameter optimization process, namely how can the AOI engineers validate the new parameter values determined by the optimization process? Certainly a correction of a bad classification cannot be validated only by examination of the specified image, but it is necessary to check several other instances. Therefore, to execute a reliable validation process, the engineers have to collect a large image database ("image base") covering all cases as they occur in the best possible way. Unfortunately, creating a good and usable image base is a long and sometimes impossible task because of several – often contradicting – criteria. A manual image collection by the engineers is very time-consuming and in case of automatic systems (like AOIs) there is only a limited possibility because of the high number and varied type of data. Automatic methods are faster but during the collection, some falsely classified images can be put in the image base which makes the parameter optimization impossible. For example, if an image containing a faulty component is placed into the "good" part of the image base, the optimization process will try to adjust to the parameters that the AOI algorithm has classified the image as "good". As a result, the optimized macro cannot recognize this specified error which can indicate slippages causing the greatest type of inspection catastrophe.

The number of stored images is also a very important factor. If the image base contains too many images, the resources (processor, hard drive, network etc.) become overloaded and the optimization process can only be executed slowly. On the other hand in case of a small image base the algorithm validation is neither reliable nor accurate enough.

http://dx.doi.org/10.5772/51699

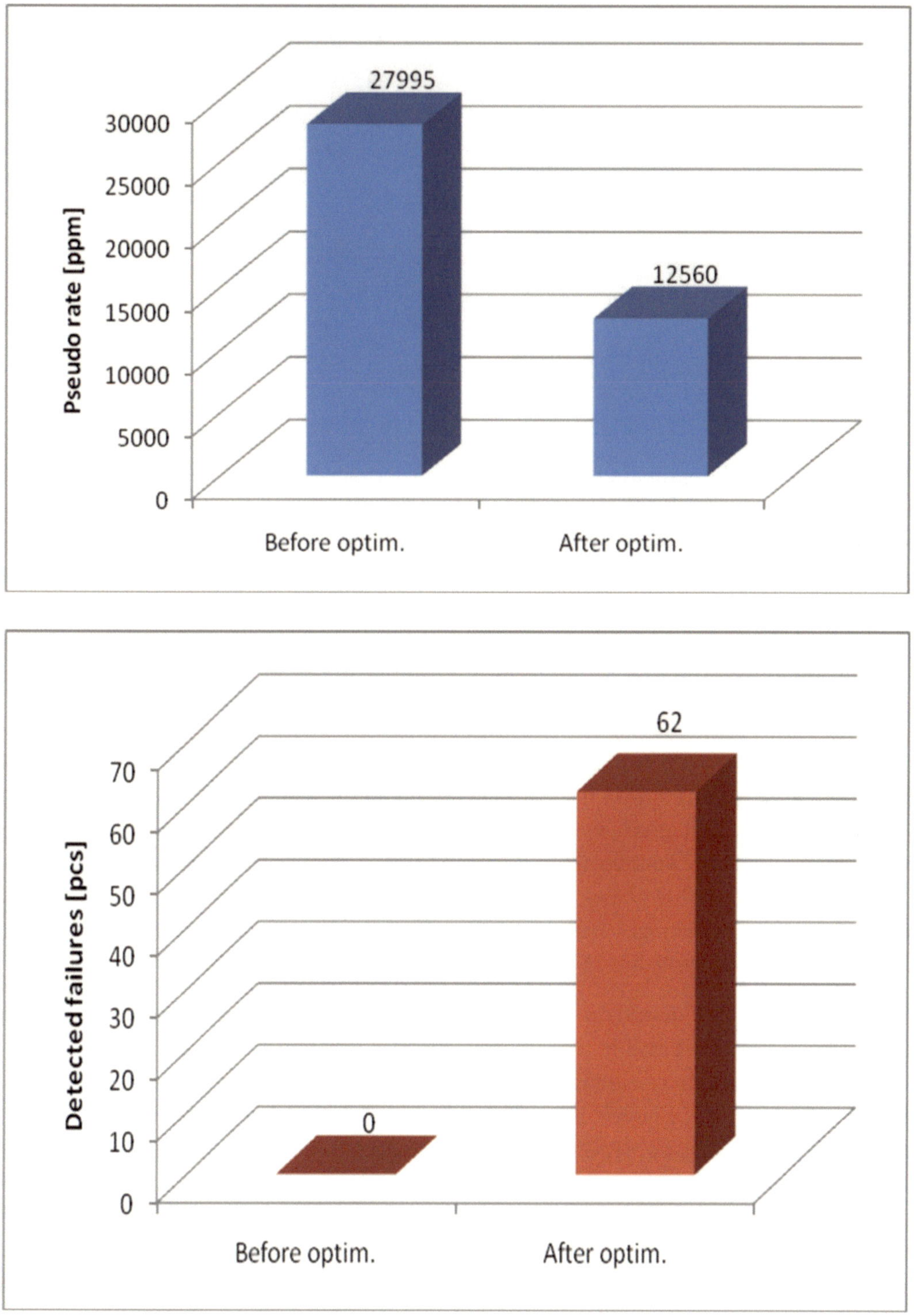

Figure 28. Pseudo and slip-through reduction

Figure 29. Quasi-tombstone

If we suppose that the optimal size of image base is determined (and which cannot be exceeded) and relevant images are collected resulting in a reasonably good image base. At this point another question arises: how can the engineers update the database with new images? It is very hard to determine which images from the new image-set need to be stored and which images need to be deleted from the current image base. There are several criteria – such as the date created, the number of similar images etc. – which can be used as the basis of the updating decision but a precise numerical factor which shows the usefulness of pictures in the database is much more difficult to determine.

The aspects and concepts mentioned in this section have shown that the usage and perfect operation of automated optical inspection system requires human control and supervision. Although the devices' algorithms are able to execute fast, accurate, efficient, reliable, "assiduous" and continuous inspection (they appear to be much more suitable than human operators as a consequence!) without being fed sufficient intelligence they cannot adapt immediately and independently to changes in manufacturing. Therefore the quality inspection process can hinder the increased spreading of autonomous electronic manufacture.

Several researches and developments are focusing on the problem to redeem the status of the human operators' work and to provide help for AOI engineers. Very interesting research directions are in automatic algorithm parameter optimization methods. The AOI devices on the manufacturing line monitor the quality of the algorithms (number of false calls and slip-through, if possible) and on occasions they adjust the parameters using the image base to create a better, higher quality algorithm. The engineers only need to take care of special cases like changing the lighting or creating new inspection methods. Although the automatic parameter optimization methods do not have to satisfy high real-time criteria, it is important to determine the optimized parameter values in a relatively short time. It is easy to verify that even in the case of having some dozen parameters; the analysis of all parameter-combinations takes a very long time (years) therefore heuristic search methods have to be used to solve the optimization problem.

http://dx.doi.org/10.5772/51699

Certainly the automatic optimization methods also need to collect the relevant images autonomously to create the reference image base. This work sets serious challenges for optimization processes because of the problems and difficulties mentioned earlier.

As a summary, we can establish that AOI systems offer a powerful solution for a complex problem by means of simple principles, but the analysis of details can reveal several problems, difficulties and contradictions. Finding a solution for them is an essential condition for the automated optical inspection systems in the future.

10. 3D Inspection

But the analyze-development is just one route for improving the AOI process. The other is the "extended" optical inspection system with measuring capability. The pioneers of this property are 3D SPI machines. In the last few years, a wide variety of these machines have been developed. The inspection in this application - checking the SMT printing process - means the 3 dimensional measurements of solder paste bumps. These bumps are shaped like cylinders or cubes so that the geometries and surfaces are relative simple. This fact makes the 3D optical techniques a viable option. Several measurement techniques are used for this process, some of these are shown in Fig. 30.

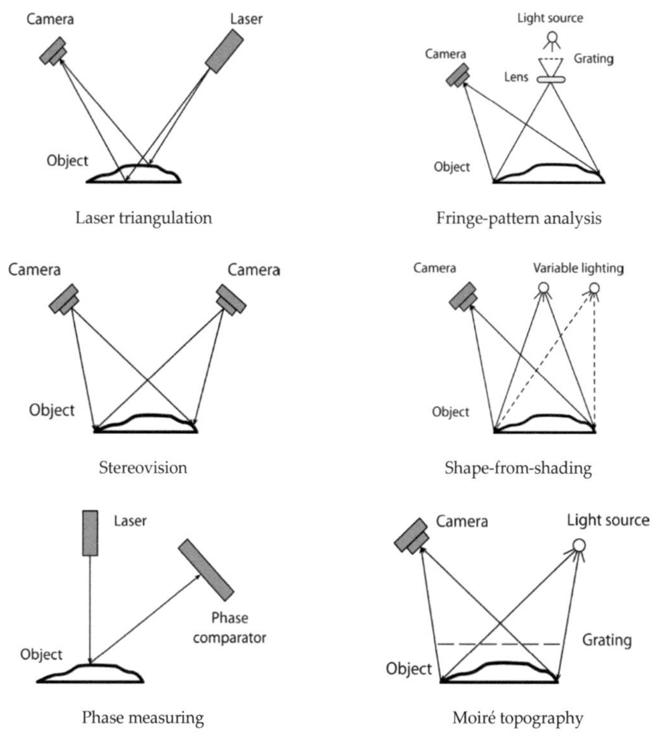

Figure 30. optical measurement techniques

The same is true for inspecting component presence, but for solder-joint detection these technologies are in their infancy at present. The shape of different components' solder-joints is complex and the specular surface also makes the task even more difficult. There have been a number of research efforts, optical 3D shape measurement technologies, based on several technologies as shown in Fig. 30. Some of these researches can be found in the following studies [181-198]. Also some companies are in the development phase such as Koh-Young Technology [180]. So the evaluation of these geometries is as yet more difficult, but with the development of optical metrology there will be more AOI machines with measuring capability.

11. Further developments, the future of AOI

AOI systems are following the worldwide trends i.e. multi-task integration, adaptivity, speed, etc. There are already appliances that integrate optical inspection with repair functions: Ersa's AOI+R solution or optical and X-ray inspection together. Some suppliers have AOI+AXI or Viscom's AOXI (simultaneous inspection). Another possible area of development is the inspection speed. Faster image capturing (with larger FOV, faster camera positioning etc), parallel inspection of two PWBs are some possible ways for this to be done.

The other important area is adaptivity. Mainly adaptive illumination is the future of AOI systems. It would help to drastically reduce pseudo-failures rates and eliminate slip-through failures.

A third area is image processing. 3D inspection, neural networks, fuzzy systems, intelligent algorithms which will help to increase the efficiency and reliability of these systems.

12. Conclusion

Inspection systems are widely used to determine the quality of electronics modules after assembly sequences. Nowadays this is usually the automatic, non-contact and non-destructive process, such as automatic optical inspection (AOI), supplemented with automatic X-ray inspection (AXI) if necessary. These appliances inspect the ready or the incomplete printed wiring boards to determine the quality of it's given property in any technological sequence, such as paste printing, component placement or soldering. The rapid development of electronics module assembly manufacturing requiring parallel development of test procedures. The automatic optical inspection is potential multi-disciplinary research area, because from image acquiring, (illumination, the detection of the reflected light etc.) through image processing, to the evaluation each area can be optimized to reach to goal, that the qualification of the inspected object in the field of interest (FOI) by the used appliance, matches the specifications as stated. Most manufacturers agree that, from a strategic point of view, the optical inspection after soldering should not be ignored. As a consequence, this is the most important part of an AOI inspection. The quality of solder joints is determined from geometric and optical properties of the solder meniscus. These parameters determine the reflection properties of the meniscus. The meniscus forms from the liquid alloy during the soldering process. After cooling, the meniscus becomes solid and reflects illumination which means that we can classify them. From these reflection patterns and with the help of image processing algorithms we are able to determine the quality of the solder joints. As described above, the correct source of illumination is essential. There are several different kinds of approach: white or RGB; directed or diffuse; ring or hemisphere.

This survey gives state of the art review of current automated optical inspection systems in the electronic device manufacturing industry. The aim of the chapter is to give an overview about the development phases, operating mechanisms, advantages and disadvantages of AOI appliances, their technical parameters, field of usage, capabilities and possible trends for further developments.

Author details

Mihály Janóczki[1*], Ákos Becker[2], László Jakab[3], Richárd Gróf[4] and Tibor Takács[5]

*Address all correspondence to: mihaly.janoczki@eu.agc.com

1 AGC Glass Hungary Ltd., H-2851 Környe, Hungary

2 DENSION Audio Systems Ltd., Budapest, Hungary

3 Department of Electronics Technology, Budapest University of Technology and Economics, Budapest, Hungary

4 Epcos AG, Heidenheim, Deutschland

5 Department of Control Engineering and Information Technology, Budapest University of Technology and Economics, Budapest, Hungary

References

[1] Matthew T. Holzmann, Automatic Optical Inspection Of Circuit Assemblies In a High Mix/Low Volume Environment, Christopher Associates, Inc. Santa Ana, California USA

[2] Miran Burmen, Franjo Pernuš and Boštjan Likar, LED Light Sources: A Survey of Quality-Affecting Factors and Methods for Their Assessment, Measurement and Science Technology Vol. 19, No.12, 2008, 122002 (15pp)

[3] Sheng-Lin Lu, Xian-Min Zhang, Yong-Cong Kuang, Optimized Design of an AOI Illuminator, Proceedings of the 2007 International Conference on Wavelet Analysis and Pattern Recognition, Beijing, China, 2-4 Nov. 2007, pp. 924-928

[4] Yuji Takagi, Seiji Hata, Susumu Hibi, Visual Inspection Machine for Solder Joints Using Tiered Illumination, SPIE Machine Vision Systems Integration in Industry, Vol. 1386, 1990, pp. 21-29

[5] Y.J. Roh, D.Y. Lee, M.Y.Kim, H.S. Cho, A Visual Inspection System with Flexible Illumination and Auto-focusing, Proceedings of SPIE, Vol. 4902, 2002, pp. 463-475

[6] Alexander Hornberg, Handbook of Machine Vision, Wiley-VCH Verlag GmbH Co KGaA, Weinheim, 2006

[7] E.R.Davies, Machine Vision: Theory, Algorithms, Practicalities, Elsevier, 2005

[8] Kjell J. Gasvik, Optical Metrology 3rd Edition, John Wiley & Sons Ltd., ISBN: 0-470-84300-4, 2002

[9] Shree K. Nayar, Arthur C. Sanderson, Lee E. Weiss, David A. Simon, Specular Surface Inspection Using Structured Highlight and Gaussian Images, IEEE Transaction on Robotics and Automation, Vol. 6, No. 2, April 1990

[10] Peter Conlon, AOI A Strategy for Closing the Loop, Surface Mount Technology, April 2006, pp. 24-29

[11] Pamela R. Lipson, Imagen and Landrex Technologies, AOI Systems Simulate Human Brain, Test & Measurement World, February 2007, pp. 35-42

[12] Herbert Tietze, Jens Kokott (GOEPEL electronic GmbH), Application of AOI Systems in Backplane Manufacturing

[13] Titus T. Suck (Orbotech), Controlling the Process: Post-Reflow AOI (Automated Optical Inspection) to Ascertain Machine and Process Capability

[14] Don Miller (YesTech), Exploring AOI and X-ray http://www.dataweek.co.za/news.aspx?pklNewsId=31727&pklCategoryID=49, (accessed 12 July 2009)

[15] Mark J. Norris, Advances in Automatic Optical Inspection, Gray Scale Correlation vs. Vectoral Imaging, Journal of Surface Mount Technology, Vol. 15, January 2002

[16] Christopher C. Yang, Michael M. Marefat, Frank W. Ciarallo, Error Analysis and Planning Accuracy for Dimensional Measurement in Active Vision Inspection, IEEE Transactions on Robotics and Automation, Vol. 14, No. 3, June 1998

[17] Jens Kokott, The capability of modern AOI systems, Global SMT & Packaging, November/December 2006, pp. 16-17

[18] Matthew Holzmann, AOI in a High-Mix/Low-Volume Environment, Circuits Assembly, June 2004, pp. 30-35

http://dx.doi.org/10.5772/51699

[19] Keith Fairchild, Evaluating ROI of AXI vs. AOI, Circuits Assembly, October 2006, pp. 20-25

[20] "Realising Expectations of AOI http://www.dataweek.co.za/news.aspx?pklNewsId=30534&pklCategoryID=4914, (accessed 03 July 2009)

[21] David Doyle, Challenges for Second Generation Automated Optical Inspection (AOI) Solutions

[22] K C Fan, C Hsu, Strategic Planning of Developing Automatic Optical Inspection (AOI) Technologies in Taiwan, 7th International Symposium on Measurement Technology and Intelligent Instruments, Journal of Physics, Conference Series 13, 2005, pp. 394-397

[23] Peter Krippner, T&M, Zero-defect IC Inspection Strategy With AOI, 2006 Electronics Manufacturing, March 2006, http://www.emasiamag.com/currentIssue.asp?d=1&m=3&y=2006# (accessed 12 July 2009)

[24] Mukul Shirvaikar, Trends in Automated Visual Inspection, Real-Time Image Processing, Springer, 2006, 1:41–43

[25] Bob Ries, New Advances in AOI Technologies, Surface Mount Technology, January 2001, pp. 62-66

[26] Duncan Nicol, The Key (or Start Button) to Success, EPP Europe, November 2008, pp. 47-49

[27] Douglas W. Raymond, Dominic F. Haigh, Why Automated Optical Inspection, International Test Conference, 1997, pp. 1033

[28] Pamela Lipson, To Be or Not to Be in Color:A 10 Year Study of the Benefits and Pitfalls of Including Color Information in AOI Systems, Proceedings of the IPC APEX Technical Conference EXPO, 2009

[29] Pamela Lipson, Lyle Sherwood, The landscape of PCB Technology is Changing Rapidly. How Will AOI Testing Keep Up?, Proceedings of the IPC APEX Technical Conference EXPO, 2009

[30] Teledyne,DALSA,http://www.teledynedalsa.com/public/dc/documents/Image_Sensor_Architecture_Whitepaper_Digital_Cinema_00218-00_03-70.pdf (accessed 12 July 2009)

[31] Teledyne DALSA Applications Set Imager Choices, Dalsa Aplication Note, http://www.dalsa.com/public/corp/applications_set_imager_choices.pdf (accessed 12 July 2009)

[32] Dave Litwiller, CMOS vs. CCD: Maturing Techologies, Maturing Markets, Reprint from the 2005 August issue of Photonics Spectra, Laurin Publishing

[33] Stuart A. Taylor, CCD and CMOS Imaging Array Technologies: Technology Review, Technical Report EPC-1998-106, Xerox Research Centre Europe http://www.research.microsoft.com/pubs/80353/CCD.pdf (accessed 12 July 2009)

[34] Sharma, A Look at CCD Sensors…, What Digital Camera Magazine; November 1997, pp. 54-56

[35] News Item, CMOS to Signal end of Line for CCD?, What Digital Camera Magazine; June 1997

[36] Dunn, James F.; "A New Digital Camera Startup Busts Price/Performance Standards with CMOS Sensor, Advanced Imaging; January 1997

[37] Young Y. Cha, J.H. Oh, A Field-of-View Generation Algorithm Using Neural Network, Mechatronics Vol. 11, 2001, pp. 731-744

[38] http://en.wikipedia.org/wiki/Barcode#Matrix_.282D.29_barcodes (accessed 12 July 2009)

[39] Ja H. Koo, Suk I. Yoo, A Structural Matching for Two-Dimensional Visual Pattern Inspection, IEEE International Conference on Systems, Man and Cybernetics, Vol. 5, 1998, pp. 4429-4434

[40] Wen-Yen Wu, Mao-Jim J. Wang, Chih-Ming Liu, Automated Inspection of Printed Circuit Boards Through machine Vision, Computers in Industry, Vol. 28, 1996, pp. 103-111

[41] H. Rau, C.-H.Wu, Automatic Optical Inspection for Detecting Defects on Printed Circuit Board Inner Layers, The International Journal of Advanced Manufacturing Technology, Vol. 25, 2005, pp. 940-946

[42] Zuwairie Ibrahim, Zulfakar Aspar, Syed Abdul Rahman Al-Attas, Musa Mohd Mokji, Corse Resolution Defect Localization Algorithm for an Automated Visual Printed Circuit Board Inspection, 28th Annual Conference of the IECON 02, Vol. 4, 2002, pp. 2629-2634

[43] Fikret Ercal, Filiz Bunyak, Hao Feng, Context-Sensitive Filtering in RLE for PCB Inspection, Conference of Intelligent Systems in Design and Manufacturing, Vol. 3517, 1998, pp. 286-293

[44] M Moganti, F Ercal, CH Dagli, S Tsunekawa, Automatic PCB Inspection Algorithms: A Survey, Computer Vision and Image Understanding, Vol. 63, No. 2, March 1996, pp. 287-313

[45] Péter Szolgay, Katalin Tömördi, Optical Detection of Breaks and Short Circuits on the Layouts of Printed Circuit Boards Using CNN, Cellular Neural Networks and their Applications, Fourth IEEE International Workshop, 24-26 Jun. 1996, pp 87-92, DOI: 10.1109/CNNA.1996.566498

[46] Ji-joong Hong, Kyung-ja Park, Kyung-gu Kim, Parallel Processing Machine Vision System for Bare PCB Inspection, IEEE, 1998, pp. 1346-1350

[47] Madhav Moganti, Fikret Ercal, Cihan H. Dagli, Shou Tsunekawa, Automatic PCB Inspection Algorithms: A Survey, Computer Vision and Image Understanding, Vol. 63, No. 2, March, 1996, pp. 287-313

[48] Timót Hidvégi, Péter Szolgay, Some New Analogic CNN Algorithms for PCB Quality Control, International Journal of Circuit Theory and Applications Archive, Vol. 30, Issue 2-3, March 2002, pp. 231-245

[49] Fabiana R. Leta, Flávio F. Feliciano, Flavius P. R. Martins, Computer Vision System for Printed Circuit Board Inspection, ABCM Symposium Series in Mechatronics, Vol. 3, 2008, pp.623-632

[50] Noor Khafifah Khalid, Zuwairie Ibrahim, and Mohamad Sshukri Zainal Abidin, An Algorithm To Group Defects On Printed Circuit Board For Automated Visual Inspection, IJSSST, Vol. 9, No. 2, May 2008, pp. 1-10

[51] Nam Hyeong Kim, Jae Young Pyun, Kang Sun Choi, Byeong Doo, Choi, SungJea Ko, Real-Time Inspection System for Printed Circuit Boards, Lecture Notes in Computer Science, Vol. 2781, 2003, pp. 458-465

[52] Amistar Automation Inc.: K5/K5L Desktop Board Inspector, Data Sheet, Updated November, 2007

[53] Lloyd Doyle Limited duotech, Data Sheet, Updated Septembe, 2005

[54] Lloyd Doyle Limited Excalibur 5000 X2, Data Sheet, Updated September, 2005

[55] Lloyd Doyle Limited LD 6000, Data Sheet

[56] Lloyd Doyle Limited phasor, Data Sheet, Updated September, 2005

[57] Lloyd Doyle Limited redline, Data Sheet, Updated September, 2005

[58] I. Fidan, R. P. Kraft, L. E. Ruff, S. J. Derby, Designed Experiments to Investigate the Solder Joint Quality Output of a Prototype Automated Surface Mount Replacement System, IEEE Transactions on Components, Packaging, Manufacturing Technology, Vol. 21, No. 3, Jul. 1998, pp. 172-181

[59] J. Pan, G. L. Tonkay, R. H. Storer, R. M. Sallade, D. J. Leandri, Critical Variables of Solder Paste Stencil Printing for Micro-BGA and Fine Pitch QFP, Proceedings on the 24th IEEE/CPMT International Electronics Manufacturing Technology Symposium, 1999, pp. 94-101

[60] D. He, N. N. Ekere, M. A. Currie, The Behavior of Solder Pastes in Stencil Printing With Vibrating Squeegee, IEEE Transactions on Components, Packaging, Manufacturing Technology, Vol. 21, No. 4, Oct. 1998, pp. 317-324

[61] S. C. Richard, The Complete Solder Paste Printing Processes, Surface Mount Technology, Vol. 13, 1999, pp. 6-8

[62] Peter Krippner, Detlef Beer, AOI Testing Position in Comparison, Circuits Assembly, April 2004, pp. 26-32

[63] David P. Prince, Bridge Detection in the Solder Paste Print Process, 1st January 2006, http://www.speedlinetech.com/docs/Bridge-Detection-Solder-Paste.pdf (accessed 12 July 2009)

[64] K. Fauber, S. Johnson, 2D Versus 3D Solder Paste Inspection http://www.agilent.com/see/aoi 2003 (accessed 12 July 2009)

[65] Rita Mohanty, Vatsal Shah, Paul Haugen, Laura Holte, Solder Paste Inspection Technologies: 2D-3D Correlation, Proceedings of the APEX Conference, April 2008

[66] R. R. J. Lathrop, Solder Paste Print Qualification Using Laser Triangulation, IEEE Transactions on Components, Packaging, Manufacturing Technology, Vol. 20, 1997, pp.174-182

[67] Okura, M. Kanai, S. Ogata, T. Takei, H. Takakusagi, Optimization of Solder Paste Printability With Laser Inspection Technique, Proceedings ot the IEEE/CPMT International Electronics Manufacturing Technology Symposium, 1997, pp. 361-365

[68] E. H. Rideout, Lowering Test Costs With 3-D Solder-Joint Inspection, Test Measurement World, Vol. 10, 1990, pp. 744-1657

[69] Y. K. Ryu, H. S. Cho, New Optical Measuring System for Solder Joint Inspection, Optics and Lasers in Engineering, Vol. 26, No. 6, April 1997, pp. 487-514

[70] H. Tsukahara, Y. Nishiyama, F. Takahashi, T. Fuse, M. Ando, T. Nishino, High-Speed 3-D Inspection System For Solder Bumps, Proceedings of SPIE, Vol. 2597, 1995, pp. 168-177

[71] J. L. Horijon, W. D. Amstel, F. C. Couweleer, W. C. Ligthart, Optical System of an Industrial 3-D Laser Scanner for Solder Paste Inspection, Proceedings of SPIE, Vol. 2599, 1995, pp. 162-170

[72] T. Xian, X. Su, Area Modulation Grating for Sinusoidal Structure Illumination on Phase-Measuring Profilometry, Applied Optics, Vol. 40, 2001, pp.1201-1206

[73] Hsu-Nan Yen, Du-Ming Tsai, Jun-Yi Yang, Full-Field 3-D Measurement of Solder Pastes Using LCD-Based Phase Shifting Techniques, IEEE Transactions on Electronics Packaging, Manufacturing, Vol. 29, No. 1, January 2006, pp. 50-57

[74] XinYu Wu, WingKwong Chung, Hang Tong, A New Solder Paste Inspection Device: Design and Algorithm, IEEE International Conference on Robotics and Automation. Roma, Italy, 10-14 April 2007, pp. 680-685

[75] Fang-Chung Yang, Chung-Hsien Kulo, Jein-Jong Wing, Ching-Kun Yang, Reconstructing the 3D Solder Paste Surface Model Using Image Processing and Artificial

Neural Network, IEEE International Conference on Systems, Man and Cybernetics, Vol. 3, 2004, pp. 3051-3056

[76] Feng Zhang, AEWMA Control Chart for Monitoring Variability Sources of Solder Joints Quality, IEEE Transactions on Components and Packaging Technologies, Vol. 29, No. 1, March 2006, pp. 80-88

[77] Xinyu Wu, Wing Kwong Chung, Jun Cheng, Hang Tong, Yangsheng Xu, A Parallel-Structure Solder Paste Inspection System, IEEE/ASME Transactions on Mechatronics, Vol. 14, No. 5, October 2009, pp. 590-597

[78] CyberOptics SE 300 Ultra, Data Sheet, 8009067, Rev E 1/08

[79] CyberOptics SE 500 Solder Paste Inspection System, Data Sheet, 8013863, Rev A 3/09

[80] Koh Young Technology aSPIre, Data Sheet, aSPIre-B-08-2008-E

[81] Koh Young Technology KY-8030 Series, Data Sheet, KY-8030-B-08-2008-E

[82] Marantz Power Spector SPI Series, 5D Solder Paste Inspection System, Data Sheet, Printed October, 2009, Updated November, 2009

[83] Omron VT-RNS-P, VT-RNS Series, Automated Optical Inspection System, Data Sheet, Printed in Japan, 0807-5M (1004) (D)

[84] Omron CKD VP5000L, Solder Printing Inspection Machine, Data Sheet, Printed July, 2007

[85] Saki BF-SPIder, 3D In-Line Solder Paste Inspection System, Data Sheet, Printed August 2009, PM01DCF01-08.5E

[86] TRI Innovation TR7006 Series, 3D Solder Paste Inspection System, Data Sheet, Updated November, 2008, C-7006-EN-0810

[87] ScanCAD ScanINSPECT SPI, Solder Paste Inspection, Data Sheet, Updated January, 2007

[88] TRI Innovation TR7066 Series, 3D Solder Paste Inspection System, Data Sheet, Updated November, 2009, C-7066-EN-0909

[89] Viscom S3088-II QS, Data Sheet, Updated February, 2009, Viscom_SYS_S3088-II QS_EN09020001

[90] ViTechnology 3D-SPI, Data Sheet, 3D-SPI EN Rev 0507-2009

[91] Michael E. Zervakis, Stefanos K. Goumas, George A. Rovithakis, A Bayesian Framework for Multilead SMD Post-Placement Quality Inspection, IEEE Transactions on Systems, Man and Cybernetics - Part B, Vol. 34, No. 1, February 2004, pp. 440-453

[92] Gao Hongxia, Hu Yueming, Liu Haiming, Fang Xiaosheng, A Fast Method for Detecting and Locating BGA Based on Twice Grading and Linking Technique, 22nd

IEEE International Symposium on Intelligent Control Part of IEEE Multi-conference on Systems and Control Singapore, 1-3 October 2007, pp. 375-378

[93] Bernard C. Jiang, Szu-Lang Tasi, Chien-Chih Wang, Machine Vision-Based Gray Relational Theory Applied to IC Marking Inspection, IEEE Transactions on Semiconductor Manufacturing, Vol. 15, No. 4, November 2002

[94] John Weisgerber, Doreen Tan, Pre-Reflow, Inline, 3-D Inspection, Circuits Assembly, November 2003, pp. 34-36

[95] Chun-Ho Wu, Da-Zhi Wang, Andrew Ip, Ding-Wei Wang, Ching-Yuen Chan, Hong-Feng Wan, A Particle Swarm Optimization Approach for Components Placement Inspection on Printed Circuit Boards, Journal of Intelligent Manufacturing, 2008, DOI 10.1007/s10845-008-0140-2

[96] L. Shih-Chieh, C. Chih-Hsien, S. Chia-Hsin, A Development of Visual Inspection System for Surface Mounted Devices on Printed Circuit Board, Proceedings of the 33rd Annual Conference of the IEEE Industrial Electronics Society (IECON), Taipei, Taiwan, 2007, pp. 2440-2445

[97] A. J. Crispin, V. Rankov, Automated Inspection of PCB Components Using a Genetic Algorithm Template-Matching Approach, The International Journal of Advanced Manufacturing Technology, Vol. 35, No. 3-4, December 2007, pp. 293-300

[98] E. Guerra, J.R.Villalobos, A Three-Dimensional Automated Visual Inspection System for SMT Assembly, Computer and Industrial Engineering, Vol. 40, 2001, pp. 175-190

[99] BeamWorks Inspector cpv, An In-Line AOI System from BeamWorks, Data Sheet, Updated April, 2002

[100] Landrex Optima II 7301 Express, Data Sheet, Updated July, 2006

[101] Omron VT-RNS-Z, VT-RNS Series, Automated Optical Inspection System, Data Sheet, Printed in Japan, 0807-5M (1004) (D)

[102] Viscom S3054QV, Data Sheet, Updated May, 2006

[103] Hongwei Xie, Yongcong Kuang, Xianmin Zhang, A High Speed AOI Algorithm for Chip Component Based on Image Difference, International Conference on Information and Automation, Zhuhai/Macau, China, 22-25 June, 2009, pp. 969-974

[104] G. Acciani, G. Brunetti, G. Fornarelli, A Multiple Neural Network System to Classify Solder Joints on Integrated Circuits, International Journal of Computational Intelligence Research, Vol. 2, No. 4, 2006, pp. 337-348

[105] Y. K. Ryu, H. S. Cho, A Neural Network Approach to Extended Gaussian Image Based Solder Joint Inspection, Mechatronics, Vol. 7, No. 2, 1997, pp. 159-184

[106] Fan-Hui Kong, A New Method for Locating Solder Joint Based On Rough Set, Proceedings of the Sixth International Conference on Machine Learning and Cybernetics, Hong Kong, 19-22 August 2007, pp. 3678-3681

[107] Fan-Hui Kong, A New Method of Inspection Based on Shape from Shading, Congress on Image and Signal Processing, 2008, pp. 291-294

[108] Giuseppe Acciani, Gioacchino Brunetti, Girolamo Fornarelli, Application of Neural Networks in Optical Inspection and Classification of Solder Joints, Surface Mount Technology, IEEE Transasctions on Industrial Informatics, Vol. 2, No. 3, 2006, pp. 200-209

[109] N.S.S. Mar, C. Fookes, P.K.D.V. Yarlagadda, Design of Automatic Vision-Based Inspection System for Solder Joint Segmentation, Journal of Achievements in Materials and Manufacturing Engineering Vol. 34. Issue 2, 2009, pp. 145-151

[110] Da-Zhi Wang, Chun-Ho Wu, Andrew Ip, Ching-Yuen Chan, Ding-Wei Wang, Fast Multi-Template Matching Using a Particle Swarm Optimization Algorithm for PCB Inspection, Lecture Notes in Computer, No. 4974, 2008, pp. 365-370

[111] Hugo C. Garcia, J. René Villalobos, George C. Runger, An Automated Feature Selection Method for Visual Inspection Systems, IEEE Transactions on Automation Science and Engineering, Vol. 3, No. 4, 2006, pp. 394-406

[112] Y. K. Ryu, H. S. Cho, New Optical Measuring System for Solder Joint Inspection, Optics and Lasers in Engineering, Vol. 26, 1997, pp. 487-514

[113] Huibin Zhao, Jun Cheng Jianxun Jin, NI Vision Based Automatic Optical Inspection (AOI) for Surface Mount Devices: Devices and Method, Proceedings of 2009 IEEE International Conference on Applied Superconductivity and Electromagnetic Devices Chengdu, China, 25-27 September 2009, pp. 356-360

[114] Horng-Hai Loh, Ming-Sing Lu, Printed Circuit Board Inspection Using Image Analysis, IEEE Transactions on Industry Applications, Vol. 35, No. 2, March/April 1999, pp. 673-677

[115] Shao-Nung Chiu, Ming-Hwei Perng, Reflection-Area-Based Feature Descriptor for Solder Joint Inspection, Machine Vision and Applications, Vol. 18, 2007, pp. 95–106

[116] Li Ni, Pan Kai-Lin, Li Peng, Research on Solder Joint Intelligent Optical Inspection Analysis, 2008 International Conference on Electronic Packaging Technology & High Density Packaging (ICEPT-HDP), 28-31 July 2008, pp. 1-4

[117] Kuk Won Ko, Hyung Suck Cho, Solder Joints Inspection Using a Neural Network and Fuzzy Rule-Based Classification Method, IEEE Transactions On Electronics Packaging Manufacturing, Vol. 23, No. 2, April 2000, pp. 93-103

[118] Andy Yates, Steven Brown, Jim Hauss, Paul Hudec, Inspection Strategies for 0201 Components, CyberOptics Corporation 2002, First Published: SMTA International, Chicago 2002

[119] T. Y. Ong, Z. Samad, M. M. Ratnam, Solder Joint Inspection With Multi-Angle Imaging and an Artificial Neural Network, The International Journal of Advanced Manufacturing Technology, Vol. 38, No. 5-6, August 2008, pp. 455-462

[120] Y. Ousten, S. Mejdi, A. Fenech, J.Y. Deletage, L. Bechou, M.G. Perichaud, Y. Danto, The Use of Impedance Spectroscopy, SEM and SAM Imaging for Early Detection of Failure in SMT Assemblies, Microelectronics Reliability, Vol. 38, Issue 10, October 1998, pp. 1539-1545

[121] T.H.Kim, T.H.Cho, Y.S.Moon, S.H.Park, Visual Inspection System for the Classification of Solder Joints, Pattern Recognition, Vol. 32, No. 4, 1999, pp. 565-575

[122] T. Hiroi, K. Yoshimura, T. Ninomiya, T. Hamada, Y. Nakagawa, Development of Solder Joint Inspection Method Using Air Stimulation Speckle Vibration Detection Method and Fluorescence Detection Method, IAPR Workshop on Machine Vision Applications, Tokyo, 7-9 December 1992, pp. 429-434

[123] R. Vanzetti, A. C. Traub, A. A. Richard, Automated Laser Inspection of Solder Joints, ISTFA, 1981, pp.85-96

[124] Z.S. Lee, R.C. Lo, Application of Vision Image Cooperated With Multi-Light Sources o Recognition of Solder Joints For PCB, TAAI, Artificial Intelligence and Applications, 2002, pp. 425-430

[125] B. C. Jiang, C.C. Wang, Y.N. Hsu, Machine Vision and Background Remover-Based Approach for PCB Solder Joints Inspection, International Journal of Production Research, Vol. 45, No. 2, 2007, pp. 451-464

[126] J.H. Kim, H.S. Cho, S. Kim, Pattern Classification of Solder Joint Images Using a Correlation Neural Network, Engineering Applications of Artificial Intelligence, Vol. 9, No. 6, 1996, pp. 655-669

[127] T.H. Kim, T.H. Cho, Y.S. Moon, S.H. Park, Visual Inspection System For The Classification of Solder Joints, Pattern Recognition Vol. 324, 1999, pp. 565-575

[128] J.H. Kim, H.S. Cho, Neural Network-Based Inspection of Solder Joints Using a Circular Illumination, Image and Vision Computing, Vol. 13, No. 6, 1995, pp. 479-490

[129] K.W. Ko, H.S. Cho, Solder Joints Inspection Using a Neural Network and Fuzzy Rule-Based Classification Method, IEEE Transactions on Electronics Packaging Manufacturing, Vol. 23, No. 2, 2000, pp. 93-103

[130] T.S. Yun, K.J. Sim, H.J. Kim, Support Vector Machine-Based Inspection of Solder Joints Using Circular Illumination, Electronics Letters Vol. 36, No. 11, 2000, pp. 949-951

[131] D.W. Capson, S.K. Eng, A Tiered-Color Illumination Approach for Machine Inspection of Solder Joints, IEEE Transactions on Pattern Analysis and Machine Intelligence Vo. 10, No. 3, 1988, pp. 387-393

[132] J. H. Kim, H. S. Cho, S. Kim, Pattern Classification of Solder Joint Images Using a Correlation Neural Network, Engineering Applications of Artificial Intelligence, Vol. 9, No. 6, 1996, pp. 655-669

[133] Tae-Hyeon Kim, Tai-Hoon Cho, Young Shik Moon, Sung Han Park, Visual Inspection System for the Classification of Solder Joints, Pattern Recognition Vol. 32, 1999, pp. 565-575

[134] TRI Innovation 7530 Series, Automated Optical Inspection System, Data Sheet, Updated September, 2009, C-7530-EN-0909

[135] Viscom S3016, Data Sheet, Updated Marc, 2007, VISCOM_SYS_S3016_EN06100001

[136] Viscom S3054QC, Data Sheet, Updated June, 2007, VISCOM_SYS_S3054QC_EN06050002

[137] S. Jagannathan, Automatic Inspection of Wave Soldered Joints Using Neural Networks, Journal of Manufacturing Systems, Vol. 16, Issue 6, 1997, pp. 389-398

[138] K. Sundaraj, Homogeneous Pin-Through-Hole Component Inspection Using Fringe Tracking, WSEAS Transactions on Signal Processing, Vol. 4, Issue 7, July 2008, pp. 419-429

[139] F. Wu, X. Zhang, Y. Kuan, Z. Z. He, An AOI Algorithm for PCB Based on Feature Extraction Inspection, Proceedings of the 7th World Congress on Intelligent Control and Automation, Chongqing, China, 25-27 June 2008, pp. 240-247

[140] Machine Vision Products AutoInspector Series - Supra E, Data Sheet, Updated May, 2009

[141] Marantz iSpector HDL Series, In-Line Automatic Optical Inspection Systems, Data Sheet, Printed September, 2009

[142] Marantz iSpector HML Series, In-Line Automatic Optical Inspection Systems, Data Sheet, Printed September, 2009

[143] Saki BF-Tristar, Simultaneous High Speed Inspection for Both Sides of PCB Automated Optical Inspection System, Data Sheet, Printed August, 2009, SJ08DCF01-02.5E

[144] Sony SI-V200, PWB Visual Inspection Machine, Data Sheet, Updated May, 2007, 043E-0705-05-01

[145] TRI Innovation TR7500 Series, Automated Optical Inspection System, Data Sheet, Updated May, 2009, C-7500-EN-0905

[146] TRI Innovation TR7550 Series, Automated Optical Inspection System, Data Sheet, Updated October, 2009, C-7550-EN-0910

[147] Agilent Medalist SJ50 Series 3, Automated Optical Inspection (AOI) and Measurement, Data Sheet, Printed in the USA September, 2006, Updated December 26, 2006 5989-5518EN

[148] Agilent Medalist sj5000, Automated Optical Inspection Solution, Data Sheet, Printed in the USA April, 2008, 5989-7547EN

[149] Amistar Auomation Inc. K2/K2L Optical Circuit Card Inspector, Data Sheet, Updated November, 2008

[150] CyberOptics Flex HR AOI System, Data Sheet, 8013862, Rev A 3/09

[151] Machine Vision Products AutoInspector Series – Ultra IV, Data Sheet, Updated Marc, 2009

[152] Omron VT-RNS-S, VT-RNS Series, Automated Optical Inspection System, Data Sheet, Printed in Japan, 0807-5M (1004) (D)

[153] Omron VT-WIN II, Printed Circuit Board Inspection System, Data Sheet, Printed in the USA 2002

[154] Saki BF-Frontier, Inline High Resolution, High Speed Automated Optical Inspection System, Data Sheet, Printed August, 2009, SJ16DCF01-02.5E

[155] Saki BF-Planet-X, Inline High Resolution, High Speed Automated Optical Inspection System, Data Sheet, Printed August, 2009, SJ06DCF01-02.5E

[156] Viscom S3088-III, Data Sheet, Updated October, 2009, Viscom_SYS_S3088-III_EN09100001

[157] Viscom S3088-II, Data Sheet, Updated August, 2008, Viscom_SYS_S3088-II_EN08080003

[158] Viscom S6056, Data Sheet, Updated July, 2009, Viscom_SYS_S6056_EN09070007

[159] ViTechnology 3K Series, AOI/AOM Systems, Data Sheet, 3K Series EN Rev 04 07-2009

[160] ViTechnology 5K Series, AOI/AOM Systems, Data Sheet, 5K Series EN Rev 03 08-2009

[161] ViTechnology 7K Series, AOI/AOM Systems, Data Sheet, 7K Series CH Rev 02 08-2009

[162] ViTechnology Vi-5000 Series, AOM Systems, Data Sheet, Vi-5000 Series EN Rev 03 01-2008

[163] YES Tech YTV F1 Series, Automated PCB Inspection, Data Sheet, Updated November, 2005

[164] YES Tech YTV M1 Series, Automated PCB Inspection, Data Sheet, Updated Marc, 2009

[165] Egidijus Paliulis, Raimondas Zemblys, Gintautas Daunys, Image Analysis Problems in AOI Systems, Information Technology and Control, Vol. 37, No. 3, 2008, pp. 220-226

[166] Stig Oresjo, Thorsten Niermeyer, Stacy Johnson, Putting Pb-Free to the Test, Circuits Assembly, May. 2005, pp. 42

[167] Detlef Beer, AOI In The Lead-Free Age, OnBoard Technology, June 2005, pp. 36-38

[168] Detlef Beer, Lead-Free: AOI in High-Volume Production Assemblies, Surface Mount Technology, January 2006, pp. 40-41

[169] Thorsten Niermeyer, Controlling Pb-Free Processes Through AOI, Circuits Assembly, October 2004, pp. 40-41

[170] Paul R. Groome, Lead-Free, PCB Test and Inspection, Surface Mount Technology, November 2005, pp. 36-39

[171] Shu Peng, Sam Wong TS, Test Implications of Lead-Free Implementation in a High-Volume Manufacturing Environment, Proceedings of the International Test Conference (ITC), 8-8 November 2005, pp. 620-627

[172] Paul R. Groome, Lead-Free: PCB Test and Inspection, Surface Mount Technology, 1 November 2005, pp. 36-39

[173] Jim Fishburn, Advances in Lead-Free Soldering and Automatic Inspection, Surface Mount Technology, November 2002, pp. 46-47

[174] Du-Ming Tsai, Cheng-Hsiang Yang, A Quantile–Quantile Plot Based Pattern Matching for Defect Detection, Pattern Recognition Letters, No. 26, 2005, pp. 1948-1962

[175] Ahmed Nabil Belbachir, Mario Lera, Alessandra Fanni, Augusto Montisci, An Automatic Optical Inspection System for the Diagnosis of Printed Circuits Based on Neural Networks, Proceedings of the Industry Applications Conference, 2005, pp. 680-684

[176] Immanuel Edinbarough, Roberto Balderas, Subhash Bose, A Vision and Robot Based On-Line Inspection Monitoring System for Electronic Manufacturing, Computers in Industry, No. 56, 2005, pp. 986-996

[177] Du-Ming Tsai, Chien-Ta Lin, Fast Normalized Cross Correlation for Defect Detection, Pattern Recognition Letters No. 24, 2003, pp. 2625-2631

[178] Chiu-Hui Chen, Chun-Chieh Wang, Chun-Yu Lin, Yu-Sen Shih, Chung-Fan Tu, Realization of Defect Automatic Inspection System for Flexible Printed Circuit (FPC), Proceedings of the 35th International MATADOR Conference, 2007, pp. 225-228

[179] Sean P. Cunningham, Scott MacKinnon, Statistical Methods for Visual Defect Metrology, IEEE Transactions on Semiconductor Manufacturing, Vol. 11, No. 1, February 1998, pp. 48-53

[180] Dr. Kwangli Koh (Koh-Young), Eliminating False Calls With 3D AOI Technology, EPP Europe, Issue 4, 2009, pp. 40-43

[181] Wei-Hung Su, Kebin Shi, Zhiwen Liu, Bo Wang, Karl Reichard, Shizhuo Yin, A Large-Depth-of-Field Projected Fringe Profilometry Using Supercontinuum Light Illumination, Optics Express, Vol. 13, No. 3, February 2007, pp. 1025-1032

[182] Günther Wernicke, Matthias Dürr, Hartmut Gruber, Andreas Hermerschmidt, Sven Krüger, Andreas Langner, High Resolution Optical Reconstruction of Digital Holograms, Fringe 2005, Session 4, 2005 pp. 480-487

[183] Liu Xiao-Li, Li A-Meng, Zhao Xiao-Bo, Gao Peng-Dong, Tian Jin-Dong, Peng Xiang, Model-Based Optical Metrology and Visualization of 3-D Complex Objects, Optoelectronics Letters, Vol. 3, No.2, 15 Mar. 2007, pp. 115-118

[184] Marek Wegiel, Malgorzata Kujawinska, Fast 3D Shape Measurement System Based on Color Structure Light Projection, Fringe 2005, Session 3, 2005, pp. 450-453

[185] Simon Davis, Lead-Free Inspection And Qualification With 3D AOI, OnBoard Technology November 2005, pp. 54-57

[186] Zulki Khan, A Primer on AOI and AOT, Circuits Assembly, September 2006, pp. 38-41

[187] Dongwon Shin, Thomas R. Kurfess, Three-Dimensional Metrology of Surface Extracted From a Cloud of Easured Points Using a New Point-to-Surface Assignment Method: An Application to PCB-Mounted Solder Pastes, Precision Engineering Vol. 28, Issue 3, July 2004, pp. 302-313

[188] Tae-Hyeon Kim, Tai-Hoon Cho, Young Shik Moon, Sung Han Park, An Automated Visual Inspection of Solder Joints Using 2D and 3D Features, Proceedings on the 3rd IEEE Workshop on Applications of Computer Vision, 1996. pp. 110-115

[189] Deokhwa Hong, Hyunki Lee, Min Young Kim, Hyungsuck Cho, Jeon Il Moon, Sensor Fusion of Phase Measuring Profilometry and Stereo Vision for Three-Dimensional Inspection of Electronic Components Assembled on Printed Circuit Boards, Applied Optics, Vol. 48, No. 21, 20 July 2009, pp. 4158-4169

[190] S. S. Wong, K. L. Chan, 3D Object Model Reconstruction From Image Sequence Based on Photometric Consistency in Volume Space, Pattern Analysis and Application, November 2009, DOI 10.1007/s10044-009-0173

[191] Deokhwa Hong, Heechan Park, Hyungsuck Cho: Design of a Multi-Screen Deflectometer for Shape Measurement of Solder Joints on a PCB, Proceedings of the IEEE International Symposium on Industrial Electronics (ISIE), 2009, pp 127-132

[192] Jiquan Ma, Solder Joint's Surface Recovery Based on Linear Hybrid Shape-From-Shading, Proceedings of the Second Asia-Pacific Conference on Computational Intelligence and Industrial Applications (PACIIA), 2009, pp. 245-249

[193] Akira Kusano, Takashi Watanabe, Takuma Funahashi, Takayuki Fujiwara, Hiroyasu Koshimizu, 3D Inspection of Electronic Devices by Means of Stereo Method on Single Camera Environment, Proceedings of the Industrial Electronics (IECON), 2008, pp. 3391-3396

[194] Sheng Liu, Dathan Erdahl, I. Charles Ume, Achyuta Achari, Juergen Gamalski, A Novel Approach for Flip Chip Solder Joint Quality Inspection: Laser Ultrasound and Interferometric System, IEEE Transactions On Components and Packaging Technologies, Vol. 24, No. 4, December 2001, pp 616-624

[195] Grantham K.H. Pang, Ming-Hei Chu, Automated Optical Inspection of Solder Paste based on 2.5D Visual Images, Proceedings of the IEEE International Conference on Mechatronics and Automation, 2009, pp 982-987

[196] Atsushi Teramoto, Takayuki Murakoshi, Masatoshi Tsuzaka, Hiroshi Fujit, Automated Solder Inspection Technique for BGA-Mounted Substrates by Means of Oblique Computed Tomography, IEEE Transactions on Electronics Packaging Manufacturing, Vol. 30, No. 4, October 2007, pp. 285-292

[197] Jun Cheng, Chi-Kit Ronald Chung, Edmund Y. Lam, Kenneth S. M. Fung, Fan Wang, W. H. Leung, Structured-Light Based Sensing Using a Single Fixed Fringe Grating: Fringe Boundary Detection and 3-D Reconstruction, IEEE Transactions on Electronics Packaging Manufacturing, Vol. 31, No. 1, January 2008, pp. 19-31

[198] Yunxia Gao, Jun Wang, Testing Failure of Solder-Joints by ESPI on Board-Level Surface Mount Devices, Proceeding of the International Conference on Electronic Packaging Technology & High Density Packaging (ICEPT-HDP), 2009, pp. 1256-1259

Section 4

Modeling

Computational Methods for Creation Materials with Required Composition and Structure

Konstantin Borodianskiy and Michael Zinigrad

Additional information is available at the end of the chapter

http://dx.doi.org/10.5772/54794

1. Introduction

One of the most important and complicated problems in modern industry is to obtaining materials with the required chemical composition, structure and mechanical and physical properties. Solving this problem involves great deal of time and expense, and the results obtained might be far from the optimal solution.

The development of computer technology and its accessibility have made it possible to solve problems for which there were previously unknown solutions or these methods were so tedious that they proved to be unsuitable for practical application.

There are some works where models were developed to predict the chemical composition [1, 2] and the structure [3-20] of the required deposit metal during high temperature processes, such as welding, joining, and build-up processes. These complicated models include the physical and chemical parameters of solid, liquid and gas phases, phase transition parameters, hydrodynamics' parameters, *etc.*

It is also necessary to be aware of the material's phase structure, which has a significant influence on its final properties. Predicting the phase-structure composition of a metal has been the subject of numerous papers [21-27] that included graphical representations of the phase-structure composition of the metal as a function of its chemical composition, as well as computational methods for determining the percentage of its phase.

Such a method is poorly suited to complex systems and processes described by systems of equations. In the case of mathematical modeling, the process is studied on a mathematical model using a computer, and not on a physical object. The input parameters of the mathematical model are fed into the computer, and the computer supplies the output parameters calculated in the process. The first stage in the mathematical modeling of physicochemical systems is generally the construction of thermodynamic models. This stage is very important both for ascertaining the fundamental possibility of the combined occurrence of particular chemical processes and for listing the most important thermodynamic characteristics.

In recent years mathematical modeling has been applied not only to the investigation of theoretical aspects of physicochemical processes, but also to the analysis of real technologies.

The areas of the prediction and optimization of the composition and properties of materials obtained in different technological processes are especially promising. Some of the results were obtained from the modeling of the process of the formation of a weld pool, from modeling of weld metal transformations. Important results were obtained from the studies of the physical and chemical parameters of high temperature processes, such as welding or casting, and development of kinetic model of alloy transfer. By determining the chemical composition of the weld metal the kinetic model has been developed [3-9]. Based on this model, we described the transfer of alloying elements between the slag, which is the residue left on a weld from the flux consists mostly of mixed metal oxides, sulfides and nitrides, and the metal during arc welding. The model takes into consideration the practical weld process parameters such as voltage, current, travel speed, and weld preparation geometry, and it was experimentally tested.

2. Structural composition

Structural approach is only a method development tool, *i.e.*, a means to structurize a problem, to establish connections and the order of priorities, to structurize data, *etc.*, using structural analysis.

A brief review of the major stages of structural analysis of the welding materials design problems is presented below:

- Determining the composition (structure) of the design object subject domain.
- Establishing functional relationship between the design object and the subject domain elements (direct and reverse connections).
- Establishing connections between the subject domain and the design tool (expert system).
- Determining the operational algorithm structure and the subject domain representation method for the design tool.
- Design stages structurization. Setting priorities.
- Establishing functional relationship between the design stages, tools and design object.
- Determining input and output parameters of the design tool.

This is schematically shown in Fig 1.

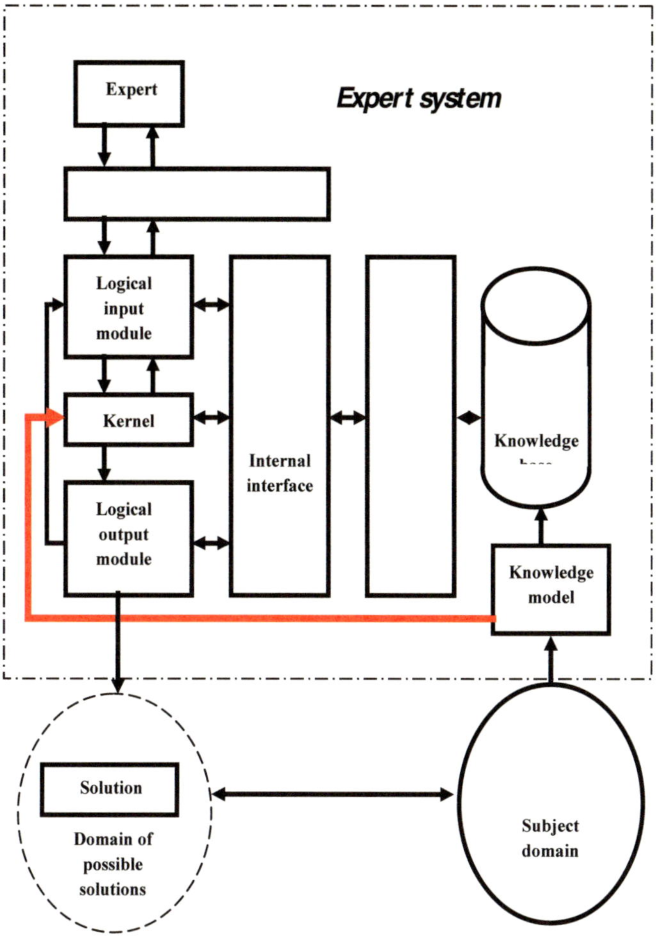

Figure 1. Simplified flow-chart of the expert system.

3. Mathematical modeling of phase interaction in real technological process

A detailed analysis of a welded joint and its interaction with the environment are presented in the followed part. It was shown that an effective method for developing a new welding material involves solving the inverse problem of finding the formula of the material as a function of the service characteristics of the weld metal. The most important problems for the new methodology in the area of determining the electrode formula of a new welding material include devising a model of the required structure of the weld metal under service conditions and calculating the primary structure and chemical composition of the weld deposit. The chemical composition of the weld metal is determined by the initial chemical composition of the welding material and the base metal and by the nature of the physiochemical processes accompanying the interaction between the molten metal and slag.

Prediction of the chemical composition of a weld deposit and, consequently, determine its mechanical properties, is based on a kinetic analysis of the simultaneous diffusion-controlled reactions that occur between the molten metal and slag [5]. The mutual influence of the reactions and the diffusion of all the reactants in the metal and slag are also taken into account.

The analysis of the kinetics and mechanism of individually occurring reactions does not present any special difficulties at the present time and that, as a rule, its results faithfully describe the real processes. A kinetic analysis of the interaction of multi-component metallic and slag melts with consideration of the mutual influence of reactions taking place in parallel is considerably more complicated. The theoretical basis of the method consists of two assumptions:

- under diffusion-controlled conditions the concentration ratio at the phase boundary for each reaction is close to the equilibrium value;
- the rate of transfer of the reactants to the phase boundary or away from it is proportional to the difference between their concentrations in the bulk and on the boundary of the metallic and oxide melts.

The oxidation of elements in a metallic melt can be represented by the reaction.

$$\frac{n}{m}[E_i]+(\mathrm{FeO})\rightarrow\frac{1}{m}(E_{i_n}\mathrm{O}_m)+\mathrm{Fe} \tag{1}$$

where E_i denotes the elements dissolved in the molten metal (Mn, Si, W, Mo, V, *etc.*), and $E_{in}O_m$ denotes the oxides in the molten slag.

A calculation of the rates of reactions of type (1) for each element does not present any difficulties. However, a separate analysis of each reaction does not correspond to the real industrial processes occurring in the weld pool. The mutual influence of both the components of the interacting molten phases and the heterogeneous reactions taking place in these complex systems must be considered. Within the approach developed, the rate v_{Ei} of mass transfer of any element (mol/cm²s) for reactions of type (1) for all the metal components with consideration of their mutual influence are defined by the expression:

$$V_{Ei}=\frac{x^m-K_{Ei}^m\times\dfrac{(E_{in}O_m)}{[Ei]}}{\dfrac{x^m}{v_{Ei}^{\lim}}+\dfrac{K_{Ei}^m\times(E_{in}O_m)}{[Ei]\times v_{E_{in}O_m}^{\lim}}}, \tag{2}$$

at n=1.

$$V_{Ei}=v_{Ei}^{\lim}\left[1+b_i\times\frac{v_{Ei}^{\lim}}{4v_{Ei_2O_m}^{\lim}}-\sqrt{\left(1+b_i\times\frac{v_{Ei}^{\lim}}{4v_{Ei_2O_m}^{\lim}}\right)^2-1+b_i}\right], \tag{3}$$

where b is defined by the equation:

$$b_i = \frac{K_{Ei}^m (Ei_2 O_m)}{x^m [E_i]^2}, \tag{4}$$

at n=2.

Here K_{Ei} is the equilibrium constant of reaction (1) for the *i-th* component of the molten metal, and n and m are stoichiometric coefficients, x is the ratio between the concentration of iron oxide in the slag and the concentration of iron in the molten metal on the boundary between the interacting phases:

$$x = \frac{(FeO)'}{[Fe]'}, \tag{5}$$

$v^{lim}{}_{Ei}$ and $v^{lim}{}_{EinOm}$ are the limiting diffusion fluxes of the components (j) of the molten metal or slag phases, [Ei] and $(E_{in}O_m)$ are the initial concentrations (wt.%) of the elements and oxides in the molten phases, respectively calculated by:

$$v_j^{\lim} = \beta \times D_j^{1/2} \times C_j, \tag{6}$$

where β is the mass-transfer coefficient (cm/s), D_j is a diffusion coefficient (cm^2/s), and C_j is a reagent's concentration at the phase boundary (mol/cm^3).

The rate V_{FeO} of mass transfer of iron monoxide (mol/cm^2s) which is the second reagent in reactions of type (1) defined by the expression:

$$V_{FeO} = \frac{\frac{(FeO)}{[Fe]} - x}{\frac{x}{v_{Fe}^{\lim}} + \frac{(FeO)}{[Fe] \times v_{FeO}^{\lim}}}. \tag{7}$$

It follows from the stoichiometry of the reaction (1) that:

$$V_{FeO} = \sum_1^k V_i = \Sigma \frac{m}{n} V_{Ei}. \tag{8}$$

Having substituted V_{FeO} and V_{Ei} in expression (8), we will have an equation with one unknown $-x$. Having found x from (2) and (3) we can find V_{Ei}.

The scheme of the analyzed technological process of fusion welding process is presented in Fig. 2.

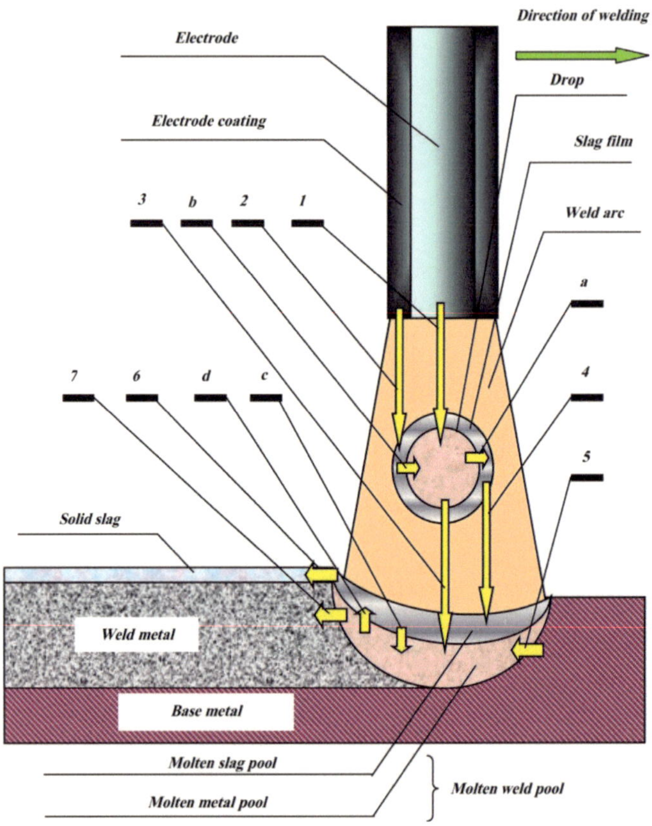

Figure 2. Scheme of the fusion welding process.

On the scheme, figures denote the direction of material transfer, and letters denote the interaction of phases:

- melting of the electrode bare and formation of a drop (1);
- melting of the electrode coating and formation of slag film over the drop (2);
- transfer of the drop metal (which has reacted with slag film at the stage of transfer) to the metal pool (3);
- transfer of the slag film (which has reacted with the drop metal at the stage of transfer) to the slag pool (4);
- melting of base metal (5);
- crystallization of the slag pool (6);
- crystallization of the metal pool (7);
- a, b—redox reaction at slag–metal boundary in a welding drop;
- c, d—redox reaction at slag–metal boundary in a welding pool.

The final composition of the drop in general case is determined by the concentrations in each of the powdered components in the flux of the flux-cored wire or in the electrode coating $[Ei]^{l}_{pd}$ and, accordingly, by the melting rates of these components v^{l}_{pd}, the concentration of each element in the metal sheath or in the electrode bare $[Ei]_b$ and its melting rate v_b, as well as by the rates of passage of the elements through the interface V_{Ei} of the drop and the slag film on its surface, which can be calculated in accordance with the methods described above. The values of v^{l}_{pd} and v_b are found from empirical relations as functions of the technological parameters of the process. Thus, the concentration of the *i-th* element in the drop at any moment in time s, can be calculated from the equation:

$$[Ei]^{\tau} = \frac{v_b \cdot [Ei]_b \cdot d\tau + \sum_{l=1}^{L} v^{l}_{pd} \cdot [Ei]^{l}_{pd} \cdot d\tau + 100 \cdot M_{Ei} \cdot A_d \cdot V_{Ei} \cdot d\tau}{m^{\tau}_{d}}, \tag{9}$$

where A_d is the surface area of the liquid drop, ^{l} labels the type of powder, L is the number of types, M_{Ei} molar mass of the *i-th* element, and m^{τ}_{d} is the mass of the metal drop at the time s.

The final drop composition thus calculated $[Ei]_d$ is used to calculate the concentration of the *i-th* element in the weld pool at any time s. The composition of the pool and therefore the composition of the weld metal are determined by the concentrations of the elements in the liquid drop $[Ei]_d$ and accordingly by the rate of descent of the drops into the liquid pool v_d, the concentration of each element in the base metal $[Ei]_{bm}$, the melting rate of the base metal v_{bm} and, as in the case of the drop stage, by the rate of passage of each element through the interface between the metal and slag pools.

In accordance with the foregoing, the expression for calculating the final composition of element *i* in the crystallized metal can be written in the form:

$$[Ei]^{\tau} = \frac{v_d \cdot [Ei]_d \cdot d\tau + v_{bm} \cdot [Ei]_{bm} \cdot d\tau + 100 \cdot M_{Ei} \cdot A_p \cdot \int_0^{\tau} V_{Ei}\, d\tau}{m^{\tau}_{p}} \tag{10}$$

where A_p is the interfacial interaction area between the metal and the slag, and m^{τ}_{p} is the mass of the weld pool at the time s.

Thus, the proposed method can be used to find the chemical composition of the molten metal in the weld pool, *i.e.*, of the metal in a welded joint. This chemical composition is the starting point for determining the quantitative and qualitative composition of the phases of the weld deposit.

4. Physicochemical analysis of primary crystallization and carbide formation

The subsequent transformations of the molten metal are associated with the primary and secondary crystallization processes, *i.e.*, the phase transformations in the multi-component alloy.

After determining the chemical composition of the *i-th* component in the solid phase at the crystallization time τ, we should determine its distribution between the austenite and the carbide phases that have formed at the primary crystallization process.

Let us use the chemical composition of the liquid molten metal in the weld pool as a starting point for examining the primary crystallization process. As we know from the theory of welding processes, crystallization of the weld pool proceeds under highly non-equilibrium conditions in the absence of convective stirring of the metal in the "tail" of the weld pool, *i.e.*, at the crystallization front. Therefore, the process of distributing the components between the liquid and solid phases is controlled only by diffusion. Another important factor that determines the distribution of the components is the concentration buildup occurring at the crystallization front. These factors produce concentration-induced supercooling, which, together, with thermal supercooling, is responsible for the cyclic character of weld pool crystallization and the chemical nonuniformity of the crystallized weld metal. At any moment during crystallization of the weld pool, the amount of the *i-th* component that has passed from the liquid phase into the solid phase can be defined as:

$$Ei^s = Ei^0[1-(1-K_{eff})\exp(-\frac{L^t v_{cryst}}{D_i^l})] \tag{11}$$

where Ei^s is the concentration of the *i-th* component in the solid phase at the crystallization time τ, Ei^0 is the initial mean concentration of the *i-th* component in the molten phase, K_{eff} is the effective distribution coefficient, L^t is the distance from the crystallization starting point (the length of the crystallite at the crystallization time τ), v_{cryst} is the crystallization rate, and Di^l is the diffusion coefficient of the *i-th* component in the molten phase.

After determining the concentration of the *i-th* component in the solid phase at the crystallization time *t*, we still cannot determine its distribution between the austenite and the carbide phases that have formed at the crystallization process. The factors that influence carbide formation can be divided into two groups:

physicochemical factors, which directly determine the nature of the carbide-formation process.

technological factors, which indirectly influence the carbide-formation process by altering the physicochemical factors parameters.

In our work, the principles governing carbide formation in an alloyed iron-carbon weld deposit were formulated on the basis of a detailed physicochemical analysis of the formation of primary carbides as compounds of carbon with *d* metals according to the quantum-chemical theories of the electronic structure of *d* metals and primary carbides. The carbide forming reaction can be described as follows:

$$xEi + y\mathrm{C} \rightarrow (Ei)_x\mathrm{C}_y. \tag{12}$$

According to these principles, the amount of carbon that is used to form the carbide of the *i-th* metal is proportional to the atomic radius of the metal (R_i) and is inversely proportional to the number of electrons in the d sublevel of the metal (d_i). We introduce the concept of the absolute Carbide Forming Ability (CFA) tendency of the *i-th* d metal (Θ_i) as the ratio:

$$\Theta_i = \frac{R_i}{d_i}. \tag{13}$$

It follows from an analysis of (12) that the carbide-forming tendency increases along the series consisting of: Fe, Mn, Cr, Mo, W, Nb, V, Ta, Ti, Zr, and Hf, in good agreement with the results in [24, 25]. The distribution of the alloying elements and carbon between the liquid and the solid phases is given by (11). Diffusionless decomposition of the supersaturated solid solution to austenite and carbide phases occurs during crystallization. The amount of carbon bound by any carbide-forming element is determined by the stoichiometry of the compound (Me_xC_y) and can be found from the following expression:

$$E_{Ci}^{(c)} = Ei^{(c)} \frac{y A_C}{x A_i}, \tag{14}$$

where x and y are stoichiometric coefficients, A_C and A_i are the atomic weights of carbon and the carbide-forming element, respectively, and $E_i^{(c)}$ is the concentration of the carbide-forming element in the carbide phase. For primary carbides, the value of x is always equal to 1, and y takes values from 0.4 to 1.0, depending on the homogeneity region of the respective carbide. It is logical to assume that only the portion of the alloying elements and carbon that cannot be dissolved in austenite at the respective temperature is used for carbide formation:

$$E_C^{(c)t_k} = E_C^{(s)t_k} - E_C^{(\lim)t_k} \tag{15}$$

where $E_i^{(c)tk}$ is the concentration of carbon that is not dissolved in austenite, $E_i^{(s)tk}$ is the carbon concentration given by (13) at the crystallization time, and $E_i^{(lim)tk}$ is the solubility limit of carbon in austenite at the respective crystallization temperature at the time t_k. The distribution of carbon between the carbide phases and the alloy will be proportional to the relative carbide-forming tendency of the respective transition element

$$\frac{\Theta_i}{\sum_{i=1}^{l} \Theta_i}$$

and its concentration in the alloy a_i. It is now clear that the proportionality factor for the *i*-th carbide-forming element is:

$$\eta_i = \frac{1}{2}\left(\frac{a_i}{100} + \frac{\Theta_i}{\sum_{i=1}^{n} \Theta_i} \right). \tag{16}$$

Then the concentration of the *i-th* carbide-forming element bound in the corresponding carbide phase at the time t_{kt} can be defined as (wt.%):

$$Ei_i^{(c)t_k} = \eta_i \cdot E_C^{(c)t_k} \frac{xA_i}{yA_C}, \tag{17}$$

and the concentration of the *i-th* carbide-forming element dissolved in austenite at the time t_k determined by (wt.%):

$$Ei_i^{(a)t_k} = Ei_i^{(s)t_k} - Ei_i^{(c)t_k}. \tag{18}$$

The concentration of carbides formed at the time t_k (wt.%) is the sum of the carbon concentration and the total concentration of the carbide-forming elements that have participated in carbide formation:

$$Q_k^{t_k} = E_C^{(c)t_k} + \sum_{i=1}^{l} Ei_i^{(c)t_k}. \tag{19}$$

Then the austenite content (wt.%) is:

$$S^{(a)} = 100\% - Q^I \tag{20}$$

The mean concentrations (wt.%) of carbon and the alloying elements in the austenite phase can be found, respectively, as:

$$E_C^{(a)} = \frac{\sum_{k=1}^{z} E_C^{t_k}}{zS^{(a)}} 100\%, \tag{21}$$

$$Ei_i^{(a)} = \frac{\sum_{k=1}^{z} Ei_i^{t_k}}{zS^{(a)}} 100\%. \tag{22}$$

http://dx.doi.org/10.5772/54794

Thus, at the end of primary crystallization, we know the mean chemical composition of the austenite phase, as well as the quantitative and qualitative composition of the carbide phases in different zones of the formed metal. Equations (11) and (13)-(22) comprise a phenomenological model of the primary non-equilibrium crystallization of the weld pool and the formation of the weld metal. At the end of primary crystallization, we have a weld deposit of complex phase and structural composition that consists of primary carbides and of austenite phases.

5. Physicochemical analysis of secondary crystallization

Secondary crystallization is accompanied by diffusion-controlled evening of the composition of the crystallized metal to the composition specified by expressions (21) and (22), and partial coagulation of the primary carbides along their grain boundaries during cooling. When the temperature for the limiting solubility of carbon and the alloying elements in austenite is reached, the isothermal decomposition of austenite occurs, and the distribution of carbon between the carbide phases is proportional to the CFA of the respective transition element.

The evolution of the system in this stage could be predicted theoretically on the basis of the corresponding phase diagrams. However, the construction of such phase diagrams is extremely difficult for the multi-component alloys under consideration, in which the concentrations of the alloying elements can vary from several percent to tens of percent by weight. Taking into account the features of the crystallization of a weld metal noted above, we should be able to predict its phase constitution on the basis of pseudo-binary phase diagrams on the level of a qualitative estimate. There will still be a probability of a high degree of deviation from reality. This is because equilibrium phase diagrams do not take into account the real nature of the crystallization of a weld metal and the effects of the thermal-straining cycle during high temperature processes, such as welding, as well as the cyclic nature of primary crystallization, which results in chemical non-uniformity of the crystallizing metal. More than 70 years, a similar problem has been solved for certain types of molten metals in a weld pool using the phenomenological Schaeffler constitution diagram [21]. The Schaeffler diagram is a real empirical diagram that is constructed for the weld metal in the initial state after welding for ordinary averaged manual arc welding regimes are shown in Fig. 3.

Schaeffler diagram indicates a real metal microstructure formed after secondary crystallization for high alloy steel welds but can be applied for a high variety of high temperature processes in steel making processes. The microstructure composition consists of the ferrite, austenite and martensite phases can be calculated by the following equations which are called equivalents. Chromium equivalent is calculated using the weight percentage of ferrite stabilizing elements:

$$Cr_{eq} = \%Cr + \%Mo + 1.5 \cdot \%Si + 0.5 \cdot \%Nb + 0.8 \cdot \%V + 4 \cdot \%Ti, \tag{23}$$

and nickel equivalent is calculated using the weight percentage of austenite stabilizing elements:

$$Ni_{eq} = \%Ni + 30 \cdot \%C + 0.5 \cdot \%Mn + 1.6 \cdot Al + 19 \cdot N + 0.3 \cdot Cu. \tag{24}$$

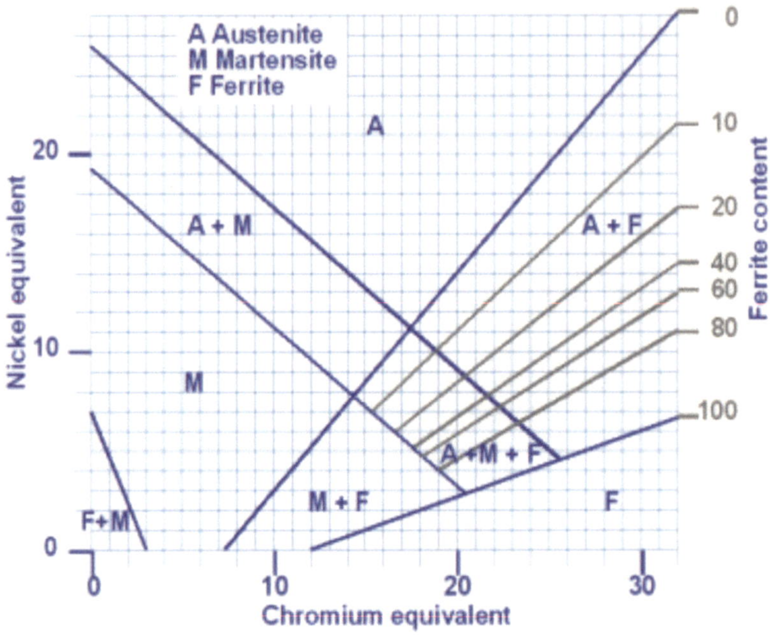

Figure 3. Schaeffler diagram.

However, when a metal contains a considerable amount of carbon (more than 0.12 wt. %) and ferrite-forming elements, are also carbides (NbC, TiC, VC, *etc.*) form in the metal microstructure. The possibility of their formation must be taken into account because the equivalent values of chromium and nickel can deviate significantly in this case from the values calculated using equations (22) and (23) proposed by Schaeffler without consideration of the formation of carbides phases.

The decomposition of austenite begins at 1100-1200 K and is accompanied by the precipitation of secondary carbides, which form mainly with chromium and iron (carbides with the general formulas Me_3C, $Me_{23}C_6$, Me_7C_3, and Me_6C). Empirical relations for predicting the type of carbide formed were determined using literature data [28, 29] and the results of our own research 4-7, 9] on the basis of the ratio between the atomic concentrations of the carbide-forming element and carbon in austenite and the parameters of the thermal-straining cycle during welding. In analogy to (17), we can write:

$$Ei^{(c)} = \sum_{j=1}^{k} w_j \cdot \eta_i \cdot E_C^{(d)} \frac{xA_i}{yA_C}. \tag{25}$$

Here w_j is the fraction of carbon in the *j-th* carbide phase relative to the total amount of carbon used to form carbides of the *i-th* alloying element, $E_C^{(d)}$ is the concentration of carbon in the austenite decomposition products, and η_i is the coefficient defined by (16).

http://dx.doi.org/10.5772/54794

Then the concentrations (wt.%) of the carbide-forming element dissolved in the matrix are given by the expression:

$$Ei^{(b)} = Ei^{(a)} - Ei^{(c)}, \tag{26}$$

and of the carbon dissolved in the matrix are given by the expression:

$$E_C^{(b)} = E_C^{(a)} - \sum_{i=1}^{n}\sum_{j=1}^{k} w_j \cdot \eta_i \cdot Ei^{(c)} \frac{yA_C}{xA_i} \tag{27}$$

The concentration (wt.%) of the carbide phases formed as a result of secondary crystallization is:

$$Q^{II} = E_C^{(a)} - E_C^{(b)} + \sum_{i=1}^{n} Ei^{(c)}, \tag{28}$$

and the total concentration (wt.%) of the carbide phases in the weld metal is:

$$Q_{Hd} = Q^{I} + Q^{II}. \tag{29}$$

The concentration (wt.%) of the matrix in the weld deposit is determined from the expression:

$$S^{(b)} = 100\% - Q_{Hd}. \tag{30}$$

The concentrations (wt.%) of carbon and in the matrix is:

$$E_C^{b} = \frac{E_C^{(b)}}{S^{(b)}} \cdot 100\%, \tag{31}$$

and of the alloying elements in the matrix is:

$$Ei^{b} = \frac{Ei^{(b)}}{S^{(b)}} 100\% \tag{32}$$

Equations (24)-(31) comprise a phenomenological model of the secondary crystallization process in the weld metal which enables us to predict the phase constitution of the weld deposit.

6. Technological experiments using modeling approach

The task of the following step is developing of a new flux-cored wire for forming of build-up layer with shock-abrasion resistance properties. The inverse problem of flux-cored wire computation has been solved (using mentioned model) which can provide us with the required chemical composition of build-up metal and as the result it can give us the required mechanical properties. These required properties are achieved by austenite-martensite matrix structure with 10 wt. % of carbides uniformly distributed in it.

Austenite has a FCC structure allows holding a high proportion of carbon in its solution. In our case austenite is used for shock resistance thanks to its impact energy absorbance ability. Martensite has a BCT structure where the carbon atoms constitute a supersaturated solid solution and as a result it has the hardest and strongest properties. Therefore martensite serves as abrasion resistant according to its mechanical properties. The stable carbide phase brings the better toughness and additional abrasion resistance and also ensures uniform distribution of the hardness properties. Under intensive impact loading, some amount of metastable austenite absorbs part of the impact energy and transforms into addition martensite phase.

The mathematical model permits prediction of the composition of the weld metal as a function of the compositions of the starting materials and the technological parameters of the welding process. Prediction of the microstructure of the weld metal is based on computer simulation of a Schaeffler diagram and the process of carbide formation in steels.

A cold-rolled ribbon (1008 steel) was filled with a powder mixture calculated using the mentioned model. The main alloying elements, in final wire, were: graphite, ferrotitanium, chromium and nickel powders. From Hume-Rothery rules it is known that the crystal structures of the solute and the solvent must be the same. Here the mentioned alloying elements should be dissolved in FCC structure (austenite phase). It is also known that the size difference between solute and solvent must be $< \sim 15\%$.

Austenite and carbide are the only phases crystallize during primary crystallization process. Chromium and nickel dissolves well in the austenite formed matrix. However, titanium, because of its high difference in atomic radii as compared to γ-iron, and because of its different lattice structure, poorly dissolves in austenite. Some amount of non dissolved in austenite chromium and titanium forms carbides.

By the end of the secondary crystallization, the stable carbide phases will be stay and residual amount of the alloying metals will be dissolved in the metal matrix.

The required properties for shock-abrasion resistance were the input, and the output are the needed alloying elements and their wt.% of the flux and wt.% of the final wire. The output is presented in Table 1 which presents chemical composition of the base metal (A516), wire band (cold-rolled ribbon, 1008 steel) and alloying elements the band was filled with.

Component	Density (g/cm³)	Core composition (quantity in 100 kg of FCE, kg)	Relation in the dray mixture of the flux (%)
FeCrC	3.48	11.17	41.86
FeTiC	3.15	8.13	30.49
Ni powder	2.96	4.47	16.74
CaF_2	1.39	2.91	10.92

	Composition of the base materials and the build-up layers, wt%							
	C	Si	Mn	Cr	Mo	Ti	Fe	Ni
A 516	0.28	0.30	1.00	0.00	0.00	0.00	98.35	0.00
Electrode bend	0.08	0.03	0.50	0.12	0.10	0.01	98.82	0.25
Required weld	1.00	0.60	0.80	5.50	0.00	2.00	86.57	3.50
Layer 1	0.96	0.48	0.54	5.30	0.01	2.00	86.32	3.48
Layer 2	1.11	0.53	0.42	6.71	0.13	2.54	83.13	4.41
Layer 3	1.19	0.54	0.39	7.09	0.13	2.68	82.28	4.65

Table 1. Flux cored wire for shock-abrasion resistance output.

Build-up samples were produced using flux cored wire with diameter 1.7 mm by process which was performed by welding machine Kemppi FU 30, PS 3500. The samples were prepared by 3 layers build-up metal. The technological parameters of the process were:

- Current: 250A.
- Voltage: 35V.
- Feed speed: 180 m/h.
- Travel speed: 30 m/h.
- Polarity: Reverse.

The samples were tested on home-made shock-abrasion resistance measuring device. The samples were subjects of the mechanic impact loading with the simultaneous continuous sea sand strew (85 g/min) of the 60 Mesh.

A 12.5 mm thickness sheet of low carbon steel, A516, consisting of 73% ferrite and 27% pearlite phase microstructure, was used as the base metal. The obtained surface (3-rd layer build-up metal) consists from martensite and austenite mixed phase microstructure and carbide stable phase, investigated by Scanning Electron Microscopy (SEM) and shown on Fig. 4.

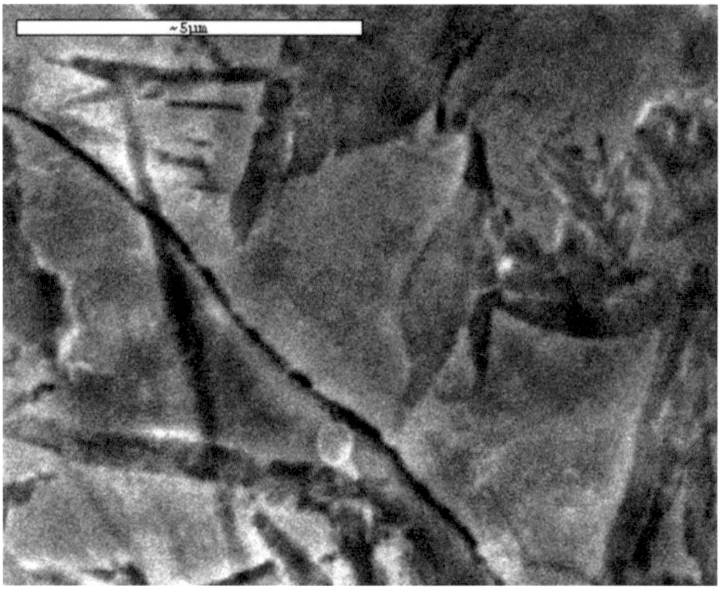

Figure 4. SEM image of surface build-up metal microstructure.

Chemical analysis of the matrix was made with Energy Dispersive X-ray Analysis (EDS) and the received results are shown in Table 2 and compared with the calculated, using the described model, results.

Element	EDS results (wt.%)	Calculated results (wt.%)
C	0.97	1.19
Si	0.49	0.54
Mn	0.40	0.39
Cr	7.26	7.09
Fe	84.01	82.28
Ni	3.77	4.65
Ti	3.10	2.68

Table 2. Chemical composition of the weld obtained by EDS analysis and model calculation.

A little differences as seen in Table 2, caused as the result of technique limitation. The detected chemical elements must be rounded to 100% and this limitation doesn't take into account some chemical inclusions that usually found in steels.

As we see from the presented results, the calculated and the real chemical composition results are closed, that emphasize the correctness of the calculations performed by the model.

Hardness tests were made by Rockwell Hardness Tester. The hardness obtained on the 3-rd layer of build-up metal was 56 HRC as compare to the hardness of the base A516 metal - 90 HRB.

The difference between base metal and build up-metal for shock-abrasion resistance results obtained by shock-abrasion tests using special home-made device are presented in the plot shown in Fig. 5. The results of tested specimens presented as the plot of the weight loss per shocked area as a function of time:

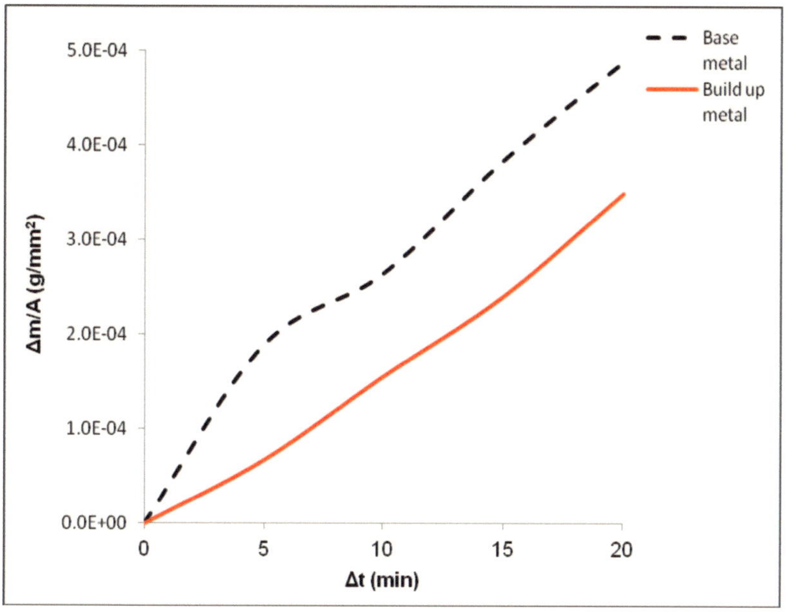

Figure 5. Shock-abrasion test results as a function of time.

It is seen from the plot that the build-up metal has an improved shock-abrasion resistance. It is found that efficiency increased by 29%, what is in a good agreement with the declared aim of the work.

7. Conclusions

A phenomenological model for predicting the chemical composition and structure of the non-equilibrium primary and secondary crystallization which takes into account the carbide phase formation has been developed. This allows the realization of the quantitative prediction of the metal structure and its mechanical properties.

The difficulty which was solved in the purposed model is a kinetic analysis of the interaction of multi-component metallic and slag melts with consideration of the mutual influence of reactions taking place in parallel.

Full-scale testing and investigation have been performed. The obtained results have been analyzed and treated to verify the equivalence of the model. The model was tested for solving the real technological problem, which is developing a new flux-cored wire for forming a build-up layer with shock-abrasion resistance properties. The experimental results confirm the adequacy of the calculations using this model.

Author details

Konstantin Borodianskiy and Michael Zinigrad*

*Address all correspondence to: zinigrad@ariel.ac.il

Laboratory for Metal and Ceramic Coatings and Nanotechnology, Ariel University Center of Samaria, Ariel, Israel

References

[1] Kettner, U. R. The Thermodynamic Modelling of Multi-component Phase Equilibria. Journal of the Minerals, Metals and Materials Society (1997)., 49(12), 14-19.

[2] Ansara, I, Sundman, B, & Willemin, P. Thermodynamic Modelling of Ordered Phases in the Ni-Al System. Acta Metallurgica (1988)., 36(4), 977-982.

[3] Zinigrad, M, Mazurovsky, V, & Borodianskiy, K. Physico-chemical and Mathematical Modeling of Phase Interaction Taking Place during Fusion Welding Processes. Material Science and Engineering Technology (2005)., 36(10), 489-496.

[4] Zinigrad, M. Computational Methods for Development of New Welding Materials. Computational Material Science (2006)., 37(4), 417-424.

[5] Borodianskiy, K, Mazurovsky, V, Zinigrad, M, & Gedanken, A. Creation of Shock-Abrasion Resistance Build-up Metal Using a Physicochemical Model of High-Temperature Processes. Israel Journal of Chemistry (2007).

[6] Borodianskiy, K, Mazurovsky, V, Gedanken, A, & Zinigrad, M. Developing a Required Structure of Metals Using Computational Methods. Material Science A (2008).

[7] Mazurovsky, V, Zinigrad, M, Leontiev, L, Lisin, V, & Borodianskiy, K. Quantitative Estimation of Carbide-Forming Abilities of d-elements: proceedings of the Jubilee Scientific Conference on Physical Chemistry and Technology in Metallurgy, (2005). Yekaterinburg, Russia, , 89-98.

[8] Zinigrad, M, Mazurovsky, V, & Borodianskiy, K. Physico-Chemical and Mathematical Modeling of Phase Interaction taking place during Fusion Welding Processes: proceedings of the 1-st International conference on Diffusion in Solids and Liquids, (2005). Aveiro, Portugal , 2, 841-850.

[9] Boronenkov, V, Zinigrad, M, Leontiev, L, Pastukhov, E, Shalimov, M, & Shanchurov, S. Phase Interaction in the Metal-Oxide Melts-Gas System. Springer, (2012).

[10] Elmer, J, Palmer, T, Zhang, W, & Wood, B. DebRoy T. Kinetic Modeling of Phase Transformations Occuring in the HAZ of C-Mn Steel Welds Based on Direct Observation. Acta Materialia (2003). , 51(12), 3333-3349.

[11] Zhang, W, & Elmer, J. DebRoy T. Kinetic of Ferrite to Austenite Transformation During Welding of 1005 Steel. Scripta Materialia (2002). , 46(10), 753-757.

[12] Takahashi, M, & Bhadeshia, H. K. D. H. Model for Transition from Upper to Lower Bainite. Material Science and Technology (1990). , 6(7), 592-603.

[13] Babu, S, Specht, E, David, S, Karapetrova, E, Zschack, P, Peet, M, & Bhadeshia, H. K. D. H. In-situ Observation of Lattice Parameter Fluctuations in Austenite and Transforamtion to Bainite. Metallurgical and Materials Transaction A (2005). , 36(12), 3281-3289.

[14] Babu, S, David, S, Vitek, J, & Mundra, K. DebRoy T. Development of Macro- and Microstructures of Carbon-Manganese Low Alloy Steel Welds: Inclusion Formation. Material Science and Technology (1995). , 11(2), 186-199.

[15] Grong, Ø. Metallurgical Modelling of Welding. The Institute of Materials Publishing, nd ed., London, UK, (1997). , 2.

[16] Skrzypszak, T, & Wegrzyn-skrzypczak, E. Mathematical and Numerical Model of Solidification Process of Pure Metals. International Journal of Heat and Mass Transfer (2012). , 55-4276.

[17] Assadi, H. A Phase-Field Model for Non-Equilibrium Solidification of Intermetallics. Acta Materialia (2007). , 55-5225.

[18] Serajzadeh, S. A Mathematical Model for Prediction of Austenite Phase Transformation. Materials Letters (2004). , 58-1597.

[19] Krauss, G. Steels Processing, Structure, and Performance. ASM International, Ohio, USA, (2006).

[20] Davis, J. R. Alloying Understanding the Basics. ASM International, Ohio, USA, (2003).

[21] Schaeffler, A. L. Constitution Diagram for Stainless Steel Weld Metal. Metal Progress (1949). b., 56(11), 680-680.

[22] Beres, L. Proposed Modification of Schaeffler Diagram for Chrome Equivalent and Carbon for More Accurate Prediction of Martensite Content. Welding Journal (1998). , 77(7), 273-276.

[23] Kotecki, D. J, & Siewert, T. A. Constitution Diagram for Stainless Steel Weld Metals. Welding Journal (1992). , 71(5), 171-178.

[24] Takalo, T, Suutala, N, & Moisio, T. Austenitic Solidification Mode in Austenitic Stainless Steels Welds. Metallurgical and Materials Transaction A (1979). , 10(8), 1173-1181.

[25] Takalo, T, Suutala, N, & Moisio, T. Ferritic-Austenitic Solidification Mode in Austenitic Stainless Steel Welds. Metallurgical and Materials Transaction A (1980). , 11(5), 717-725.

[26] Chen, S. R, Davies, H. A, & Rainforth, W. M. Austenite Phase Formation in Rapidly Solidified Fe-Cr-Mn-C Steels. Acta Materialia (1999). , 47(18), 4555-4569.

[27] Johnson, E, Gråbæk, L, & Johansen, A. Sarholt Kristensen L. Microstructure of Rapidly Solidified Steel. Mater. Sci. Eng. (1988). , 98-301.

[28] Oyama, S. T. Chemistry of Transition Metal Carbides and Nitrides. Chapman & Hall, Glasgow, (1996).

[29] Grossman, J, Mizel, A, Côté, M, Cohen, M, & Louie, S. Transition Metals and Their Carbides and Nitrides: Trends in Electronic and Structural Properties. Physical review B. (1999). , 60(9), 6343-6347.

[30] Suetin, D. V, Shein, I. R, & Ivanovskii, A. L. Stractural, Electronic and Magnetic Properties of η Carbides (Fe_3W_3C, Fe_6W_6C, Co_3W_3C and Co_6W_6C) from first principles calculation. Physica B (2009). , 404-3544.

[31] Feng, J, Xiao, B, Zhou, R, Jiang, Y. H, & Cen, Q. H. Calculation and Simulation for the Mechanical Properties of Carbides and Borides in Cast Iron. Procedia Engineering (2012). , 31-676.

Chapter 18

New Approaches to Modeling Elastic Media

G. J. Martin, J. Hussan and M. McKubre-Jordens

Additional information is available at the end of the chapter

http://dx.doi.org/10.5772/54771

1. Introduction

In this chapter we suggest new models for the study of deformations of elastic media through the minimization of distortion functionals. This provides a "holistic" approach to this problem and a potential mathematical underpinning of a number of phenomena which are actually observed. The functionals we study are conformally invariant and give measures of the local anisotropic deformation of the material which can be tuned by varying p-norms and weights. That these functions are conformally invariant offers the opportunity to address multiscale problems in an integrative manner - there is no natural scale. When applied to the problem of deforming elastic bodies through stretching we see phenomena such as tearing naturally arising through bad delocalisation of energy. More formally, we find that deformations either exist within certain ranges or fail to exist outside these ranges – a material can only be stretched so far. We dub this the Nitsche phenomenon after a similar phenomenon was observed in connection with minimal surfaces. We see that for these deformation problems the ranges where deformations can be proven to exist depend on invariants picked up from the profile of the material or its density depending on how one interprets the weight. These invariants are akin to conformal invariants such as moduli for complex analysis. Outside the range where we can prove existence we further show how to prove minimisers do not exist. Thus this gives a complete dichotomy depending on certain structural invariants. Further, when properly parameterised by these invariants, existence of otherwise in the borderline cases depends on more subtle invariants such as structural profiles and their degree of regularity. Our results primarily pertain to two dimensional elasticity, but apply in three dimensions in certain circumstances for particular models.

2. Basic notions

Consider an elastic body $\Omega \subset \mathbb{R}^n$. We study deformations $f : \Omega \to \Omega' \subset \mathbb{R}^n$ of Ω of Sobolev class $f \in W^{1,1}_{loc}(\Omega)$ consisting of functions with first derivatives in $L^1_{loc}(\Omega)$. The principle of interpenetrability of matter allows us to focus on the case that f is a homeomorphism. We will also assume that the mapping f has finite distortion. The basics of the theory of mappings of finite distortion were laid down in [3, 16], but see [15, Chapter 6] for a more general discussion. Finite distortion means that the distortion function $\mathbb{K}(x,f)$, defined by the formula

$$\mathbb{K}(x,f) = \frac{\|Df(x)\|^n}{J(x,f)}, \qquad x \in \Omega \tag{1}$$

is finite almost everywhere. Here $Df(x)$ is the Jacobian matrix and $J(x,f)$ its Jacobian determinant. In order to get a viable theory some additional regularity must be placed on $\mathbb{K}(z,f)$. The theory of quasiconformal mappings pertains to the case $\mathbb{K}(x,f) \in L^\infty$. Most recent developments assume $\mathbb{K}(x,f)$ is exponentially integrable (or close to it) and hence a close connection with BMO functions. The theory then exploits the H^1-BMO duality (since we know a Jacobian is in H^1) when considering the distortion inequality

$$\|Df(x)\|^n \leq \mathbb{K}(x,f)J(x,f)$$

so the right hand side here might be given meaning as an integrable function.

Note that the distortion functional is conformally invariant. If $\lambda \in \mathbb{R}_+$, then

$$\mathbb{K}(z,\lambda f) = \mathbb{K}(z,f) \tag{2}$$

Moreover, for sufficiently regular mappings ($W^{1,n}_{loc}$ and $\mathbb{K} \in L^\infty_{loc}$ suffices) we have the multiplicative formula

$$\mathbb{K}(z,f\circ g) \leq \mathbb{K}(z,g)\,\mathbb{K}(g(z),f) \tag{3}$$

for $g : \Omega \to g(\Omega)$ and $f : g(\Omega) \to \mathbb{R}^n$, and in fact if $\mathbb{K}(z,g) = 1$, then $\mathbb{K}(z,f\circ g) = \mathbb{K}(g(z),f)$.

In two dimensions, the equation $\mathbb{K}(z,g) = 1$ linearises to the Cauchy-Riemann equations and $W^{1,1}_{loc}$ solutions will be conformal mappings. In higher dimensions there is the remarkable rigidity theorem of Liouville from 1850 (much improved over the years) that a $W^{1,n}_{loc}(\Omega)$ solution of the equation $\mathbb{K}(z,f) = 1$ is the restriction to Ω of a C^∞ Möbius transformation of $\overline{\mathbb{R}}^n = \mathbb{R}^n \cup \{\infty\}$. A Möbius transformation is a finite composition of reflections, translations and dilations.

2.1. Deformations

The general theory posits that deformations rearrange an object (our domain Ω) so as to minimise a *stored energy functional*. Such functionals are motivated by various physical principals (for instance the principle of least action) and the general theory seeks to

http://dx.doi.org/10.5772/54771

understand the existence, uniqueness and regularity as well as the topological structures of minimisers.

A typical functional might look like

$$f \mapsto \mathcal{E}(f) = \iint_{\Omega} \|Df(x)\|^p + \frac{1}{J^q(x,f)}\, dx \tag{4}$$

For instance one obtains the Dirichlet energy functional with the choice $p = 2$, $q = 0$ and the minimisers are well known to be the harmonic mappings. When $q > 0$ we are penalising the compression of the body as the Jacobian determinant $J(z,f)$ measures the local change in volume. These (and many other) functionals are often set up so that the direct method has a chance of working and some *a priori* modulus of continuity on a minimising sequence can be found. In this way the direct method (examining the limit of a minimising sequence) gives you something to hold in your hand and the problem becomes one of regularity. The associated Euler-Lagrange equations are highly nonlinear in all but the most elementary cases and can seldom be solved. Indeed the solution is usually obtained through the variational process and regularity.

Here we wish to study functionals of the form

$$f \mapsto \mathcal{E}(f) = \iint_{\Omega} \Big(\frac{\|\nabla f(x)\|^n}{J(x,f)}\Big)^p\, dx \quad \Big(= \iint_{\Omega} \mathbb{K}^p(x,f)\, dx\Big)$$

or more generally

$$\mathcal{E}_{\Phi}(f) = \iint_{\Omega} \Phi(\mathbb{K}(x,f))\, \lambda(x)\, dx \tag{5}$$

where here $\Phi(t)$ is convex (eg. $t \mapsto t^n$, $n \geq 1$) and $\lambda(x)$ is a positive weight describing properties (density or profile) of the material body Ω. We call the value of the functional $\mathcal{E}_{\Phi}(f)$ the Φ - conformal energy. It is a measure of the *average local anisotropic stretching* of the material and is clearly conformally invariant. In studying such functionals we find that the direct method in the calculus of variations has little chance of working. For instance

- there are no obvious a priori estimates,
- (provably) no modulus of continuity estimates and
- (provably) no compactness theorems for the classes of mappings one might obtain in a minimising sequence.

However our study is aided by the chance that geometric methods and conformal invariants interplay and these invariants can be used to identify a minima and establish existence or otherwise of a minimum.

2.2. Non-convex case

We take a moment to remark on the hypothesis that Φ is convex. In fact if Φ is a function of sublinear growth minimisers almost never exist. From [5, Theorem 5.3], we note

Lemma 1. *Let $\phi(t)$ be a positive strictly increasing function of sublinear growth:*

$$\lim_{t\to\infty} \frac{\phi(t)}{t} = 0$$

Let $B = D(z_0, r)$ be a round disk and suppose that $f_0 : B \to \mathbb{C}$ is a homeomorphism of finite distortion with $\iint_B \phi\left(\mathbb{K}(z, f_0)\right) < \infty$. Then there is a sequence of mappings of finite distortion $f_n : B \to f_0(B)$ with $f_n(\zeta) = f_0(\zeta)$ near ∂B and with

- $\mathbb{K}(z, f_n) \to 1$ *uniformly on compact subsets of B, and*
- *as $n \to \infty$,*

$$\iint_B \phi\left(\mathbb{K}(z, f_n)\right) \to \iint_B \phi(1)$$

Hence:

Corollary 1. *Let $\phi(t)$ be a positive strictly increasing function of sublinear growth, let Ω be a domain and let $\lambda(z) \in L^\infty(\Omega)$ be a positive weight. Suppose*

$$\iint_\Omega \phi\left(\mathbb{K}(z, f_0)\right)\, dz < \infty.$$

Then

$$\min_{\mathcal{F}} \iint_\Omega \phi\left(\mathbb{K}(z, f_0)\right) \lambda(z)\, dz = \phi(1) \iint_\Omega \lambda(z)\, dz$$

with equality achieved by a mapping of finite distortion if and only if the boundary values of f_0 are shared by a conformal mapping. Here $\mathcal{F}$ consists of homeomorphisms of finite distortion f with $f|_{\partial\Omega} = f_0$.

2.3. Applications and the problem

As we have noted, our functionals measure the local anisotropic stretching of an object, and it is to the physical situations where this appears to be the dominant mechanism of deformation of a body that our models should have most applicability. In particular the modeling of films (eg soap films) and the elastic stretching of tissues and muscles are being currently considered [11]. For instance one achieves a deformation of a 2D-cellular structure of soap films in the following manner: consider the cellular structure of soap bubbles - finite in extent - trapped between two close parallel panes of glass. The steady state with bubbles of uniform size is closely approximated by a regular 2D hexagonal tessellation. Now vary

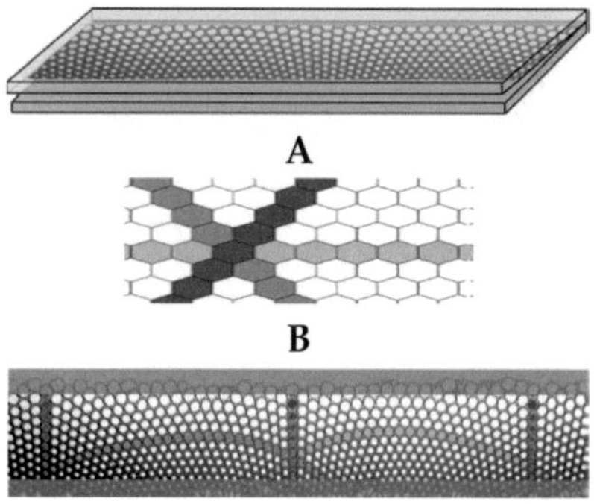

Figure 1. Bubbles trapped between two glass planes. Reproduced from Figure 1 and 3 from [6].

the angle between the panes. The new configuration is expected to be achieved by a "nearly" conformal deformation (that is $\mathbb{K} \approx 1$ - we cannot have $\mathbb{K} = 1$ because of rigidity unless we are in quite exceptional circumstances). This is because the equations for minimising the interfacial (film) energy give, locally, a trivalent equal angle (of $2\pi/3$) hexagonal tiling. The initial configuration also exhibits $2\pi/3$ symmetry. Scale invariance and pressure/volume considerations suggest the free boundary "free energy minimisation" problem may be solved by an angle preserving deformation - that is, a conformal mapping.

Next, in another more constricted regime, one can use microfabrication techniques to create accordion-like honeycomb microstructure to yield porous, elastomeric three-dimensional scaffolds with controllable stiffness and anisotropy. It has been demonstrated that the neonatal rat cardiac cell based tissue, cultured using these honeycomb scaffolds, closely matched the mechanical properties of rat ventricular myocardium. In this case there is a relatively stiff interface which is being deformed *at the boundary* and quite different effects are modeled.

Our aim would be to identify a first attempt at a "holistic" mathematical formalism which might encompass theoretical analysis of structures which appear at both ends of this spectrum. Of course, the energy functional characterising tissues is rather more complex and we should only expect a fairly coarse analogy – however, near a steady state most symmetries and integral invariants are conserved and we will be able to assert existence and regularity of minimisers of our models in reasonable physical situations.

We seek to minimize the energy

$$\inf_{f \in \mathcal{F}} \mathcal{E}_\Phi(f)$$

where our class of mappings $\mathcal{F}$ consists of all mappings $f : \Omega \to \tilde{\Omega}$ of finite distortion, perhaps with prescribed the boundary values,

$$f|\partial\Omega = f_0 : \partial\Omega \to \partial\tilde{\Omega}$$

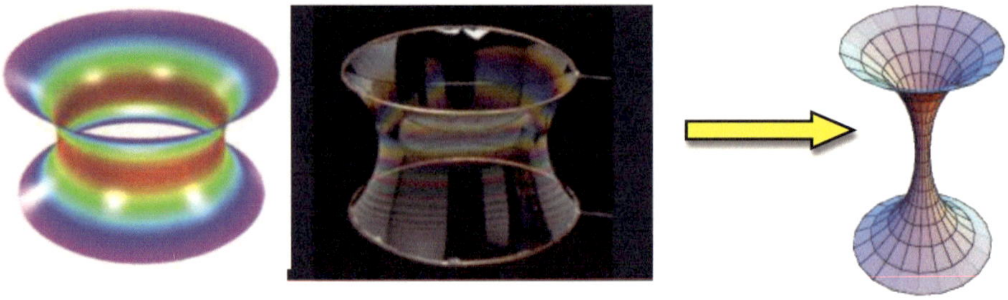

Figure 2. Sequence of catenoids degenerating - mathematical models of soap films

3. Modeling elastic media : Examples

We want to model the degeneration of elastic media under deformations. For instance, consider deforming the soap film between to parallel rings by moving the rings further apart. Classically the solution (shape of the *minimal* surface) is described by the catenoid $\{r = \cosh(z)\}$. Thus we should see a family of constant negative curvature rotationally symmetric surfaces as illustrated Figure 2.

Of course what actually happens is that when the rings are sufficiently far apart other forces com into play and material properties of the film can no longer support a solutions - the film "pops". In fact just before the film pops, the surface looks as indicated.

There are a couple of things we want to note from this example. First, we can actually arbitrarily compress the catenoid - the problem comes when we try to stretch it too far. Second, as we see from this illustration, it appears negative curvature is lost and that the surface has become very flat (curvature = 0) just before the solution ceases to exist. We will see both these phenomena in our model.

As another example to model we would like to consider what happens when we stretch a material with defects. Here is an experiment you can try at home. Take two rubber bands of varying widths $T_1 > T_2$, Depending on the thickness these can be stretched a certain distance d_i before breaking. However if one introduces a slit cut in T_1 so that the width at the end of the slit is T_2 when the first band is stretched it will fail before stretching to d_2 - there is a bad localization of "energy" at the cut. If one makes a V shaped cut, then the band stretches further before failing, but still not as far as the band of with T_2. Yet if one makes a very smooth cut, the band will stretch as far. The moral here is that it is not the thickness of the

http://dx.doi.org/10.5772/54771

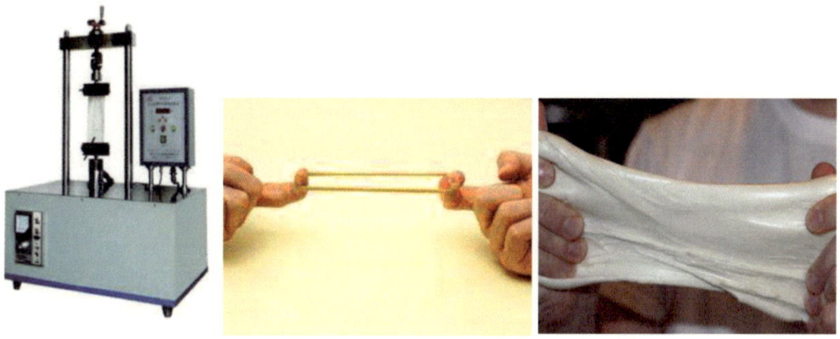

Figure 3. Stretching of highly elastic media.

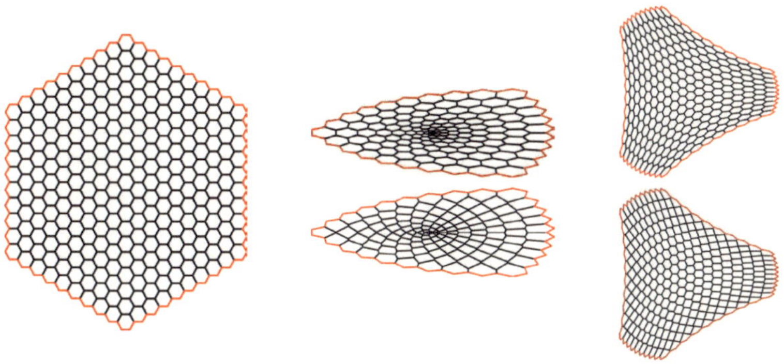

Figure 4. Deformations of cellular structures. Bottom figures minimise the mean distortion of the boundary deformation which is then relaxed to minimise interfacial energy (thus preserving angles) giving top figures.

material but the profile which limits how far it can be deformed. This is what we will see in our models as well.

Images from our computational investigations in Figure 4 show the deformations of elastic cellular structures close to the L^1-case. Here we minimize the conformal energy of all possible extensions of the boundary values (to give the lower configurations). We then minimise the interfacial tension along the cellular walls, leading to the $2\pi/3$ symmetry at vertices.

4. Geometric interpretations of distortion

Our functionals measure average distortion and the distortion of deformation at a point has a couple of important geometric interpretation which we now discuss. We recall $\mathbb{K}(x,f) = \frac{\|Df(x)\|^2}{J(x,f)}$. If we put $\mu(z) = f_{\bar{z}}(z)/f_z(z)$ (known as the Beltrami coefficient of the mapping f), then we compute

$$\mathbb{K}(z,f) = \frac{|f_z|^2 + |f_{\bar{z}}|^2}{|f_z|^2 - |f_{\bar{z}}|^2} = \frac{1+|\mu|^2}{1-|\mu|^2}$$

4.1. Linear distortion

If we define

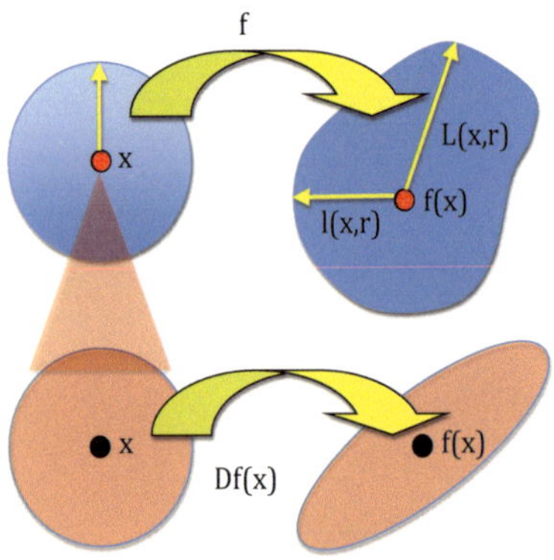

$$L(x,r) = \max_{|x-y|=r} |f(x) - f(y)|$$

$$\ell(x,r) = \max_{|x-y|=r} |f(x) - f(y)|$$

Then

$$\mathbb{K}(x,f) = \limsup_{r \to 0} \frac{1}{2}\left(\frac{L(x,r)}{\ell(x,r)} + \frac{\ell(x,r)}{L(x,r)}\right)$$

A point needs to be made here about the use of $L/\ell + \ell/L$ instead of just L/ℓ say. The point is simply that the quantities L and ℓ are clearly related to the singular values of Df and while these might vary smoothly, as they cross the maximum will not necessarily vary smoothly and the differentiability of the distortion is lost. Thus techniques available through the calculus of variations (looking for Euler Lagrange equations and so forth) can no longer be used.

4.2. Angular distortion

The distortion function $\mathbb{K}(z,f)$ also controls the infinitesimal distortion of angles as explained in [11]. Indeed one may define the maximum and minimum angular distortion for a linear map by consideration of the distortion of the unit circle $\{e^{i\theta} : 0 \leq \theta < 2\pi\}$. If the mapping is

$$z \mapsto \alpha(z + \mu\overline{z}), \qquad \alpha \in \mathbb{C} \setminus \{0\}, \quad |\mu| < 1$$

then the distortions are

$$\Theta_{max} = \max_{\theta \in [0,2\pi]} \left| \frac{d}{d\theta}\left(e^{i\theta} + \mu e^{-i\theta}\right) \right|, \qquad \Theta_{min} = \min_{\theta \in [0,2\pi]} \left| \frac{d}{d\theta}\left(e^{i\theta} + \mu e^{-i\theta}\right) \right|. \tag{6}$$

We calculate that

$$\left| \frac{d}{d\theta}\left(e^{i\theta} + \mu e^{-i\theta}\right) \right|^2 = 1 + |\mu|^2 - 2\Re e(\mu e^{-2i\theta}). \tag{7}$$

Accordingly, we may apply these observations to the differential of a Sobolev mapping via the approximation $f(z) \approx f(z_0) + f_z(z_0)(z - z_0) + f_{\overline{z}}(z_0)(\overline{z} - \overline{z}_0)$ to get

http://dx.doi.org/10.5772/54771

$$\Theta_{max} = 1 + |\mu|, \ \Theta_{min} = 1 - |\mu|.$$

So the angular distortion of a mapping f is

$$\frac{\Theta_{max}(z)}{\Theta_{min}(z)} + \frac{\Theta_{min}(z)}{\Theta_{max}(z)} = 2\frac{1+|\mu|^2}{1-|\mu|^2} = 2\mathbb{K}(z,f) \tag{8}$$

From both of these calculations one can immediately see that the distortion measures the anisotropic properties of the deformation. Further the connections between preserving angles of interfaces (and hence conformal mappings) can be brought out by minimising the deformations of angles at interfaces.

5. Connections with harmonic mappings

The first surprising observation that we want to point out is a remarkable connection with harmonic mappings that motivated some of the computational invesigation above. After a change of variables $f = g^{-1}$

$$\begin{aligned}\int_{\mathbb{A}} \mathbb{K}(z,g)\,\lambda &= \int_{\mathbb{A}'} \frac{\|Dg(f)\|^2}{J_g(f)}\,\lambda(f) J_f = \int_{\mathbb{A}'} \frac{\|Dg(f)\|^2}{J_g(f) J_f}\, J_f^2\,\lambda(f) \\ &= \int_{\mathbb{A}'} \|(Df)^{-1} J_f\|^2\,\lambda(f) = \int_{\mathbb{A}'} \|Df\|^2 \lambda(f)\end{aligned}$$

We must note that there are considerable technical difficulties in ensuring enough regularity here for these mappings and their inverses so that we can make this change of variables, this is discussed in [9].

Stationary points of the functional

$$f \mapsto \int_{\mathbb{A}'} \|Df\|^2 \lambda(f)$$

are the maps which are harmonic with respect to the metric $ds = \lambda(z)|dz|$ and satisfy the nonlinear harmonic map equation

$$\Delta g + (\log \lambda)_{\bar{w}}(g)\, g_z\, g_{\bar{z}} = 0$$

widely studied in geometry, topology and analysis - but unfortunately non-linear in a bad way since the solution is tangled up with the coefficients (the term $\lambda(f)$). We expect that the approach through minimisers of conformal energy functionals might yield a new approach to these problems.

6. Boundary value problems in two-dimensions

Here we give an account of the problem of minimising distortion functions with $\mathbb{K} \in L^1$ for the boundary value problem on the unit disk in the complex plane (see [5] and [18]).

6.1. L^1-Boundary value problem for the disk

Minimise among homeomorphisms of finite distortion $f : \mathbb{D} \to \mathbb{D}$, subject to the boundary values $f|\partial\mathbb{D} = f_0$, the quantity

$$\iint_{\mathbb{D}} \mathbb{K}(z, f)dz$$

It was then shown that there exists a minimiser if and only if

$$E = \int_{\mathbb{S}} \int_{\mathbb{S}} \log |f_0(\zeta) - f_0(\eta)| \, d\zeta \, d\bar{\eta} < \infty$$

In fact the minimum value attained for the problem is E ! When it exists, the minimiser is a smooth diffeomorphism which is quasiconformal if and only if the boundary values f_0 are bilipschitz. What this means is that the minimum has bounded distortion $\mathbb{K}(z, f) \in L^\infty(\mathbb{D})$ if and only if we have, for some $1 \leq L < \infty$, the bilipschitz estimate on the boundary values

$$\frac{1}{L} |\eta - \zeta| \leq |f_0(\eta) - f_0(\zeta)| \leq L |\eta - \zeta| \tag{9}$$

This is surprising in that the minimiser will most often have unbounded distortion, a quite unexpected phenomenon. This occurs for instance if there is no uniform lower bound on $|f_0'(\zeta)|$.

Even more remarkable is that there is a moduli structure to the minimisers. If f is a solution to the minimisation problem, then the Beltrami coefficient

$$\mu = f_{\bar{z}} / f_z \tag{10}$$

itself satisfies the (nonlinear) Beltrami equation

$$\mu_z = \bar{\mu}\mu_{\bar{z}}$$

This implies a number of interesting consequences. For instance the distortion satisfies a maximum principle - the maximal distortion occurs on the boundary [4].

http://dx.doi.org/10.5772/54771

6.2. Teichmüller problems

There is another novel phenomena to observe here. Suppose we try and deform the interior of an object so as to minimise mean distortion, yet try and keep the boundary values fixed (say the identity). In the literature this is known as a "Teichmüller problem", see [18]. Let us discuss two borderline cases. The classical L^∞ problem, where there is always a solution, and the L^1 problem, where there is never a solution ($r \neq 0$).

Problem. For $0 \leq r < 1$, let

$$\mathcal{M}_T^\infty(r) = \|\mathbb{K}(z,f)\|_{L^\infty(\mathbb{D})}, \qquad \text{and} \quad \mathcal{M}_T^1(r) = \inf\left\{\frac{1}{\pi}\iint_{\mathbb{D}} \mathbb{K}(z,f)\, dz\right\}, \tag{11}$$

where the infimum is taken over all mappings $f : \mathbb{D} \to \mathbb{D}$ of finite distortion such that f has a homeomorphic extension to $\overline{\mathbb{D}}$ and

- $f|\partial\mathbb{D} \to \partial\mathbb{D}$ is the identity mapping,
- $f(0) = r$.

6.2.1. *The L^∞ result*

Let $\mathcal{K}$ be the complete elliptic integral of the first kind,

$$\mathcal{K}(r) = \int_0^1 \frac{dx}{\sqrt{(1-x^2)(1-r^2x^2)}}.$$

Set

$$\mu(r) = \frac{\pi}{2}\frac{\mathcal{K}(\sqrt{1-r^2})}{\mathcal{K}(r)} \tag{12}$$

We note the estimate

$$\mu(x) = \log\left(\frac{4}{x} - x - \delta(x)\right) \tag{13}$$

where $x^3/4 < \delta(x) < 2x^3$ given in [17, pp 62].

Teichmüller's theorem states

Theorem 1.

$$\mathcal{M}_T^{\infty}(r) = \frac{1}{2}\left(\coth^2\left(\frac{\mu(r)}{2}\right) + \tanh^2\left(\frac{\mu(r)}{2}\right)\right) \tag{14}$$

and their is a mapping realising this minimum. If we let $d = \rho_{\mathbb{D}}(0, r) = \log\frac{1+r}{1-r}$ *(the hyperbolic distance between 0 and r) we have the asymptotics*

- $\mathcal{M}_T^{\infty}(r) \approx \frac{8}{\pi^4}\, d^2,$ *as* $d \to \infty$.
- $\mathcal{M}_T^{\infty}(r) \approx 1 + \frac{d^2}{4},$ *as* $d \to 0$.

6.2.2. The L^1 result

Theorem 2. *The minimal mean distortion function* $\mathcal{M}_T^1(r)$ *has the following asymptotics*

- *As* $r \to 1$

$$\frac{2}{\pi^2}\log\frac{1+r}{1-r} + \frac{17\log 2}{4\pi^2} \le \mathcal{M}_T(r) \le \frac{2}{\pi^2}\log\frac{1+r}{1-r} + \frac{4}{3} - \frac{7+8\log 2}{2\pi^2}$$

  up to an $O(1-r)$ *term.*

- *As* $r \to 0$

$$\mathcal{M}_T(r) \le 1 + \left(\frac{20 - 8\log(2)}{\pi^2} - \frac{7}{6}\right) r^2 + O(r^4)$$

The minimum value $\mathcal{M}_T^1(r)$ *is never attained for a homeomorphism of finite distortion.*

In terms f the hyperbolic distance $d = \rho_{\mathbb{D}}(0, r)$ we have

- $\mathcal{M}_T^1(r) \approx \frac{2}{\pi^2}\, d,$ as $d \to \infty$.
- $\mathcal{M}_T^1(r) \le 1 + \left(\frac{5-2\log(2)}{\pi^2} - \frac{7}{24}\right) d^2,$ as $d \to 0$.

Notice the constant here $\frac{5-2\log(2)}{\pi^2} - \frac{7}{24} = 0.07447 < \frac{1}{4}$ - the constant for the maximal distortion.

7. Nitsche phenomena

We now want to discuss the results of our paper [19]. A harmonic mapping defined on a domain $\Omega \subset \mathbb{C}$ is a mapping $h = h_1 + ih_2$ and each $h_i : \Omega \to \mathbb{R}$ is harmonic with respect to the Euclidean metric, $\Delta h_i = 0$. Thus conformal mappings are certainly harmonic.

In 1962 Nitsche was able to show, in connection with problems concerning minimal surfaces, that that there was no harmonic homeomorphism $f : \mathbb{A}_1 \to \mathbb{A}_2$ between annuli $\mathbb{A}_1$ and $\mathbb{A}_2$ if $\mathbb{A}_1$ is too thick relative to $\mathbb{A}_2$. That is $Mod(\mathbb{A}_1) \gg Mod(\mathbb{A}_2)$.

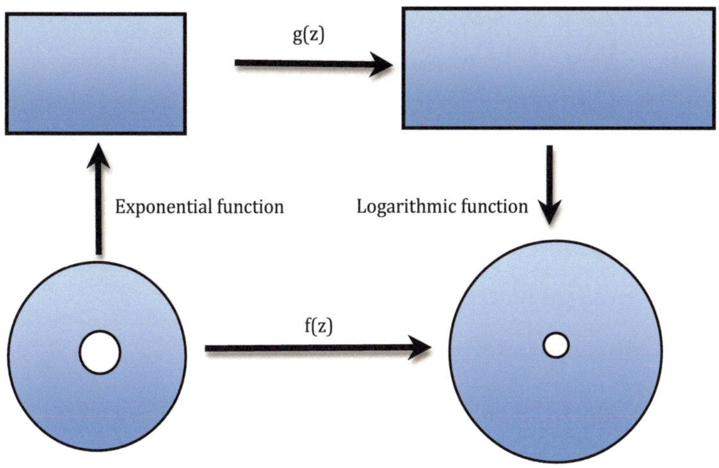

Figure 5. The equivalence between the Grötzsch problem and the Nitsche type problem is effected by conformal change of coordinates

Here the modulus of an annulus $\mathbb{A} = \{z : r < |z| < R\}$ is defined to be $Mod(A) = \log \frac{R}{r}$.

Nitsche's argument was basically by compactness (using a Harnack inequality), but he did conjecture the sharp relationship between moduli necessary for existence.

We call this interesting phenomena, the nonexistence of expected minimisers outside a range of moduli, the **Nitsche phenomenon** and we now explore this in considerably more generality and seek applications in material science.

8. Distortion functionals & Grötzsch problem

In two dimensions the problems of stretching the catenoid or a strip of material have a common formulation as the exp and log functions are conformal.

We can therefore make the following calculation:

$$\mathbb{K}(z,g) = \mathbb{K}\Big(z, \frac{1}{2\pi}\log\big(f(e^{2\pi z})\big)\Big) = \mathbb{K}(z, f(e^{2\pi z})),$$

$$\iint_{\mathbb{Q}} \Phi(\mathbb{K}(z,g))\lambda(z)\ |dz|^2 = \iint_{\mathbb{Q}} \Phi(\mathbb{K}(z, f(e^{2\pi z}))\lambda(z)\ |dz|^2$$

$$= 4\pi^2 \iint_{\mathbb{A}} \Phi\big(\mathbb{K}(w,f)\big)\ \lambda(z)e^{-4\pi\Re e(z)}\ |dw|^2.$$

Thus, basically all that has changed is the weight function (in a well defined way). However, this equivalence is only precise should the minimiser have enough symmetry that it can be lifted from the problem of mapping rectangles to that of annuli. This leads us to the Grötzsch Problem.

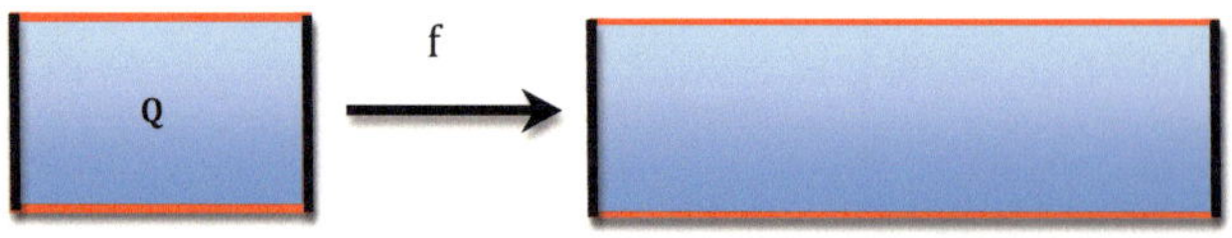

Figure 6. The Grötzsch problem: we seek a miminiser of the conformal energy at (15) and which is a homeomorphism and which respects the boundary as indicated

8.1. The Grötzsch problem

Our conformally invariant functional is

$$\mathcal{E}(f) = \iint_Q \mathbf{\Phi}(\mathbb{K}(z,f))\,\eta(z)\,dz \tag{15}$$

and our problem is to minimize this functional subject to f being a (sufficiently regular) map $Q \to Q'$ preserving the edges.

In order to study this problem we need some basic facts:

- Convexity:

$$(X, Y, J) \to \frac{|X|^2 + |Y|^2}{J}$$

  is convex on $\mathbb{C} \times \mathbb{C} \times \mathbb{R}$; the graph of the function lies above its tangent plane.

- Natural extrema : The mapping between rectangles of the form

$$f_0(z) = u(x) + iy,$$

  is a natural candiate for an extremum if the weigth depends only on x.

- For such a map we have $(f_0)_x = u_x$ and $(f_0)_y = i$. Therefore, if we set $\omega(x) = 1/u_x(x)$, a real valued function, we have the identity

$$|\omega(x)f_x + if_y|^2 \geqslant 0 \tag{16}$$

  with equality holding only for f_0.

We now expand (16) to see

$$\begin{aligned} 0 \leqslant |\omega(x)f_x + if_y|^2 &= (\omega(x)f_x + if_y)(\omega(x)\overline{f_x} - i\overline{f_y}) \\ &= \omega^2(x)|f_x|^2 + |f_y|^2 - 2\Im m(\omega(x)f_y\overline{f_x}) \\ \omega^2(x)|f_x|^2 + |f_y|^2 &\geqslant 2\omega(x)\,\Im m(f_y\overline{f_x}) = 2\omega(x)\,J(z,f). \end{aligned}$$

http://dx.doi.org/10.5772/54771

So

$$\omega^2(x)|f_x|^2 + |f_y|^2 \geqslant 2\omega(x)J(z,f)$$

with equality if and only if $f = f_0$. This gives us estimates on the distortion function (writing $J = J(z,f)$),

$$\mathbb{K}(z,f) \geqslant (1-\omega^{-2}(x))\frac{|f_y|^2}{J} + 2\omega^{-1}(x)$$

Thus, for a general mapping f:

$$\mathbb{K}(z,f) - \mathbb{K}(z,f_0) \geqslant (1-\omega^{-2}(x))\left[\frac{|f_y|^2}{J} - \frac{|(f_0)_y|^2}{J_0}\right]$$

We now want to apply the convexity inequality. If $\Phi : \mathbb{R} \to \mathbb{R}$ is convex, then its graph lies above any tangent line:

$$\Phi(\mathbb{K}) - \Phi(\mathbb{K}_0) \geqslant \Phi'(\mathbb{K}_0)(\mathbb{K} - \mathbb{K}_0).$$

Hence

$$\begin{aligned}&\Phi(\mathbb{K}(z,f)) - \Phi(\mathbb{K}(z,f_0))\\ &\geqslant (1-\omega^{-2}(x))\Phi'(\mathbb{K}_0) \times \left[2\,\Re e\Big(\frac{\overline{(f_0)_y}}{J_0}\,(f_y - (f_0)_y)\Big) - \frac{|(f_0)_y|^2}{J_0^2}(J - J_0)\right]\end{aligned}$$

Now $(f_0)_y = i$ and $(f_0)_x = 1/\omega(x) = J_0$ so

$$\begin{aligned}&\Phi(\mathbb{K}(z,f)) - \Phi(\mathbb{K}(z,f_0))\\ &\geqslant (1-\omega^2(x))\Phi'(\mathbb{K}_0)\left[2\,\Re e(f_x - (f_0)_x) - (J - J_0)\right]\end{aligned}$$

Now we multiply by $\lambda(x)$ and integrate. Choose f_0 so that

$$\lambda(x)(1-\omega^2(x))\Phi'(\mathbb{K}_0) = \alpha \neq 0 \tag{17}$$

for a real constant α.

This equation is

$$\lambda(x)\Big(1-\frac{1}{u_x^2}\Big)\Phi'\Big(u_x+\frac{1}{u_x}\Big)=\alpha$$

Finally, consideration of the boundary values gives

$$\iint_{Q_1}\Re e\big(f_x-(f_0)_x\big)\,|dz|^2=\int_0^1\Big[\int_0^\ell \Re e\big(f_x-(f_0)_x\big)dx\Big]\,dy=0.$$

and for an arbitrary Sobolev mapping

$$\iint_Q J\,|dz|^2\leqslant|Q'|=\iint_Q J_0\,|dz|^2$$

Putting the discussion above together gives us the following theorem.

Theorem 3. *Let $\lambda(x)>0$ be a positive weight, Φ convex and $u:[0,\ell]\to[0,L]$, $u(0)=0$, $u(\ell)=L$ a solution to the ordinary differential equation*

$$\lambda(x)\left(1-\frac{1}{u_x^2(x)}\right)\Phi'\left(u_x(x)+\frac{1}{u_x(x)}\right)=\alpha\in\mathbb{R}\tag{18}$$

Set $f_0(z)=u(x)+iy$. If f is a mapping of finite distortion mapping Q to Q' and satisfying our edge condition, then

$$\iint_Q\Phi(\mathbb{K}(z,f))\lambda(x)\,|dz|^2\geqslant\iint_Q\Phi(\mathbb{K}(z,f_0))\lambda(x)\,|dz|^2.$$

Equality holds if and only if $f=f_0$.

Actually following the construction above, outside the range providing existence by the theorem, one can identify an obvious degenerate minimiser f_0 for which strict inequality still holds. Once we can find a minimising sequence converging to this one can prove more. Namely

Theorem 4. *If (18) can not be solved for f_0 (that is there is no good choice of α), then there cannot be a minimiser in the class of homeomorphisms of finite distortion.*

9. Applications

These theorems motivate us to study the ordinary differential equation

$$\lambda(x)\left(1-\frac{1}{u_x^2(x)}\right)\Phi'\left(u_x(x)+\frac{1}{u_x(x)}\right)=\alpha$$

Actually the transformation from the Nitsche type problem to the Grötzsch problem yields a significantly simpler equation to study – in fact it's not really an ODE at all since the only term here is u_x. This is one of the reasons why we transformed to the Grötzsch problem.

9.1. Sharp bounds for Nitsche conjecture

In the situation of Nitsche's original conjecture we have $\lambda(x) = e^{4\pi x}$ as $\eta(w) = 1$ and $\Phi(t) = t$.

$$1 - \frac{1}{u_x^2(x)} = \alpha e^{-4\pi x}, \qquad u_x(x) = \frac{1}{\sqrt{1-\alpha e^{-4\pi x}}}$$

$$u(x) = \int \frac{e^{2\pi x}\, dx}{\sqrt{e^{4\pi x}-\alpha}} = \frac{1}{2\pi}\int \frac{dt}{\sqrt{t^2-\alpha}}, \qquad t = e^{2\pi x}.$$

So

$$u(x) = \frac{1}{2\pi}\log\left(\frac{e^{2\pi x}+\sqrt{e^{4\pi x}-\alpha}}{1+\sqrt{1-\alpha}}\right), \qquad \alpha \neq 0$$

We must choose α to solve $u(\ell) = L$

$$L = \frac{1}{2\pi}\log\left(\frac{e^{2\pi\ell}+\sqrt{e^{4\pi\ell}-\alpha}}{1+\sqrt{1-\alpha}}\right) \tag{19}$$

As $\alpha \to -\infty$ we can make the RHS of (19) arbitrarily small. Thus there is always a minimiser if $L \leqslant \ell$. Next, if $\alpha > 0$.

$$u_x(x) = \frac{1}{\sqrt{1-\alpha e^{-4\pi x}}}$$

requires $\alpha < 1$ so that

$$L < \frac{1}{2\pi}\log\left(e^{2\pi\ell}+\sqrt{e^{4\pi\ell}-1}\right)$$

and when unwound, this is precisely Nitsche's conjecture.

Most recently Iwaniec, Kovalev and Onninen have completed Nitsche's problem by showing that within this range there are not even local minima - that is, there are no harmonic mappings whatsoever [12].

9.2. More general weights $\lambda(x)$

For more general weights $\lambda(x)$, we put $\lambda_0 = \min_{x\in[0,\ell]} \lambda(x)$. The solution is dominated by the one with the choice $\alpha = \lambda_0$ and the issue is to decide whether

$$\int_0^\ell \frac{1}{\sqrt{\lambda(x) - \lambda_0}}\, dx < \infty.$$

If this integral is finite, then we will observe Nitsche type phenomena; non-existence of minima outside a range of moduli.

9.3. Observing Nitsche phenomena

Without going into excessively fine details, convergence of

$$\int \frac{1}{\sqrt{\lambda(x) - \alpha}}$$

will require that

$$\lambda(x) \approx \lambda_0 + x^{2s}, \qquad s < 1$$

near the minimum λ_0 of $\lambda(x)$. A few simple calculations will reveal

- **Existence I.** if λ is a smooth positive weight and $\lambda'(x) = 0$ at its minimum (which may well occur at the endpoints), then we can always solve the deformation problem.
- **Existence II.** if $\lambda(x)$ is constant, $u_x(x)$ is constant and therefore f will be a linear mapping.
- **Non Existence I.** if $\lambda'(x) \neq 0$ at a minimum (ie minimum at the boundary), then the Nitsche phenomenon occurs.
- **Non Existence II.** if $\lambda'(x) = 0$ at a minimum, but λ is not twice differentiable, then we see the Nitsche phenomenon, (actually $C^{1,1}$ is the cutoff).

Suppose the weight function $\lambda(x)$ is viewed as a thickness profile of the material. Then "cuts" gives a little more insight to Nitsche-type phenomenon. Find α to determine how far the final stretch can be;

This illustrates the sort of thing we were talking about before when we were discussing the stretching of rubber bands. Some precise calculations are indicated in Figure 7.

Next, should the convex function Φ have *unbounded derivative*, then there is *always a minimiser*.

$$\left(1 - \frac{1}{u_x^2(x)}\right) \Phi'\left(u_x(x) + \frac{1}{u_x(x)}\right) = \frac{\alpha}{\lambda(x)}$$

This is simply an application of the maximum principle. If $\alpha \searrow 0$, then $u_x \searrow 1$ and $\alpha \nearrow \infty$ gives $u_x \nearrow \infty$, so the intermediate value theorem provides us with a solution.

http://dx.doi.org/10.5772/54771

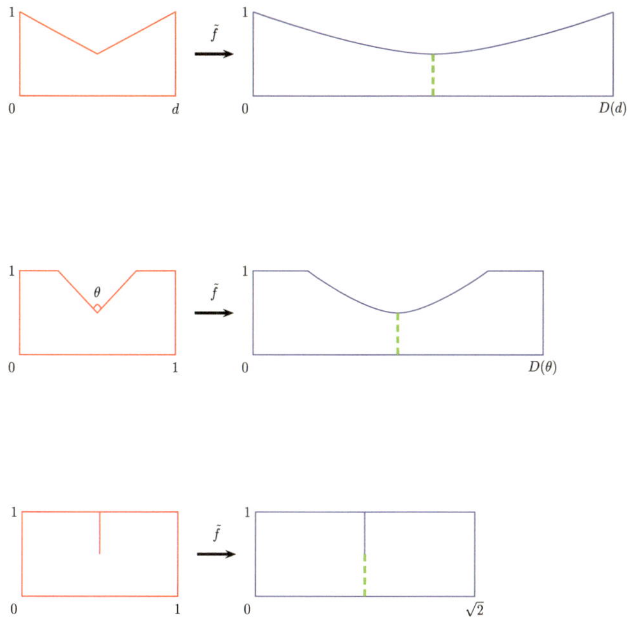

Figure 7. Calculations indicating the maximal stretching deformations of an elastic three dimensional material with given profile. Notice smoothness of the profile at the critical stretching. Beyond this maximal value no homeomorphic solution of finite conformal energy exists

In particular we do not see the Nitsche phenomenon for the L^p–norms of mean distortion: we can always minimise

$$\iint_Q \mathbb{K}(z,f)^p dz, \qquad p > 1$$

While for $p < 1$ there is never a minimiser (unless it is the identity).

10. Conclusions

Here we have suggested new models for the study of deformations of elastic media through the minimisation of distortion functionals which seem to provide a "holistic" approach to this problem and a potential mathematical underpinning of a number of phenomena which are actually observed.

We have developed the theory from a mathematical perspective using conformally invariant functionals measuring the local anisotropic deformation of the material. We expect that there are good physical reasons for the efficacy and validation of these models but we have not considered that in this article. However close reading of related models [6, 8, 21] suggest that these are relatively good models and the relationship between these minimisation problems and conformal mappings evidenced in soap film models and elsewhere is borne out in mathematical calculations in explained in [11]. Basically minimising conformal invariant functionals is "equivalent" to minimising the angular distortion at interfaces when applied to cellular structures (so length scales are implicit). These interfacial angles are often prescribed by the physics involved and therefore themselves are invariant - hence the effectiveness of conformal mappings in free boundary problems as they preserve angles.

The use of conformally invariant functionals in materials science offers the opportunity to address multiscale problems in an integrative manner - particularly when the effects they model dominate. Our models provide a possible mathematical explanation (and perhaps a predictive model when fully understood) for phenomena such as tearing and an analysis of how the process occurs. More formally, we discussed the general Nitsche phenomenon – deformations either exist within certain ranges or fail to exist outside these ranges – a material can only be stretched so far.

Our results primarily pertain to two dimensional elasticity, but in identifying the function $\lambda(z)$ as a profile and not a weight we obtain three dimensions in certain circumstances. There remains the obvious problem of developing this research into higher dimensions. Some related work in higher dimensions can be found in [13, 14] but not for the sorts of conformally invariant functionals we propose to consider. Presently we have only partial results and these will be communicated elsewhere.

Author details

G. J. Martin[1,⋆], J. Hussan[2] and M. McKubre-Jordens[3]

⋆ Address all correspondence to: g.j.martin@massey.ac.nz

1 Massey University, Auckland, New Zealand
2 The University of Auckland, Auckland, New Zealand
3 Canterbury University, Christchurch, New Zealand

References

[1] K. Astala, T. Iwaniec and G.J. Martin, *Elliptic partial differential equations and quasiconformal mappings in the plane*, Princeton Mathematical Series, 48. Princeton University Press, Princeton, NJ, 2009

[2] K. Astala, T. Iwaniec and G.J. Martin, *Deformations of annuli with smallest mean distortion*, Arch. Rational Mech. Anal., 195, (2010), 899–921.

[3] K. Astala, T. Iwaniec, P. Koskela and G.J. Martin, *Mappings of BMO–bounded distortion II*, Math. Annalen, 317, (2000), 703–726.

[4] K. Astala, T. Iwaniec and G.J. Martin, *Bi-Lipschitz homeomorphisms of the circle and non-linear Beltrami equations*, On the notions of solution to nonlinear elliptic problems: results and developments, 105Ð115, Quad. Mat., 23, Dept. Math., Seconda Univ. Napoli, Caserta, 2008

[5] K. Astala, T. Iwaniec , G.J. Martin and J. Onninen, *Extremal mappings of finite distortion*, Proc. London Math. Soc., 91, (2005), 655Ð702.

[6] W. Drenckhan, D. Weaire, and S.J. Cox, *The demonstration of conformal maps with two-dimensional foams*, Eur. J. Phys., 2004, 25, 429-438

[7] P. L. Duren, *Harmonic mappings in a plane*, Cambridge University Press, 2004.

[8] G. C. Engelmayr Jr., M. Cheng, C. J. Bettinger, J. T. Borenstein, R. Langer, and L. E. Freed, *Accordion-like honeycombs for tissue engineering of cardiac anisotropy*, Nature Mat. 7 (2008), no. 12, 1003–1010.

[9] S. Hencl and P. Koskela,*Regularity of the inverse of a planar Sobolev homeomorphism*, Arch. Ration. Mech. Anal., 180, (2006), 75Ð95.

[10] J. Heinonen, T. Kilpeläinen, and O. Martio, *Nonlinear potential theory of degenerate elliptic equations*, Oxford Mathematical Monographs, The Clarendon Press Oxford University Press, New York, 1993, Oxford Science Publications.

[11] J. Hussan and G.J. Martin, *Non-linear models for the deformation of cellular structures*, to appear.

[12] T. Iwaniec , L. Kovalev and J. Onninen, *The Nitsche conjecture*, J. Amer. Math. Soc., 24, (2011), 345–373.

[13] T. Iwaniec and J. Onninen, *Hyperelastic deformations of smallest total energy*, Arch. Ration. Mech. Anal., 194, (2009), 927–986.

[14] T. Iwaniec and J. Onninen, *Deformations of finite conformal energy: existence and removability of singularities* Proc. Lond. Math. Soc., 100, (2010), 1–23.

[15] T. Iwaniec and G.J. Martin, *Geometric Function Theory and Nonlinear Analysis*, Oxford University Press, 2001.

[16] T. Iwaniec, P. Koskela and G.J. Martin, *Mappings of BMO-bounded distortion and Beltrami type operators*, Dedicated to the memory of Tom Wolff., J. Anal. Math. 88, (2002), 337–381.

[17] O. Lehto and K. Virtanen, *Quasiconformal mappings in the plane*, Springer-Verlag, 1973.

[18] G. Martin, *The Teichm§ller problem for mean distortion*, Ann. Acad. Sci. Fenn. Math., 34, (2009), 233–247.

[19] G. Martin and M. McKubre-Jordens, *Deformations with smallest weighted L^p average distortion and Nitsche-type phenomena*, J. Lond. Math. Soc., 85, (2012), 282–300

[20] R Schoen and S-T. Yau, *Lectures on harmonic maps*, Conference Proceedings and Lecture Notes in Geometry and Topology, II, International Press, Cambridge, MA, 1997.

[21] D. Weaire and N. Rivier, *Soap, cells and statistics - random patterns in two dimensions*, Contemp. Phys. 50 (2009), no. 1, 199–239.

Section 5

Biological Aspects of Materials

Chapter 19

Bamboo Based Biocomposites Material, Design and Applications

S. Siti Suhaily, H.P.S. Abdul Khalil,
W.O. Wan Nadirah and M. Jawaid

Additional information is available at the end of the chapter

http://dx.doi.org/10.5772/56057

1. Introduction

Bamboo or *Bambusa* in botanical has 7-10 subfamilies of genres and there are 1575 difference species ranging from the type of wood to bamboo herb. However, each particular species of bamboo has different properties and qualities [1]. Bamboo is easily accessible globally, 64% of bamboo plantation, as can be seen in Figure 1, originated from Southeast Asia, 33% grown in South America, and the rest comes from Africa and Oceania [2]. Bamboo productions dated back to thousands of years ago and thus they are rich with traditional elements. Bamboo naturally, suitable for varieties of uses and benefits. Bamboo often used as materials for constructions or used as the raw materials for the production of paper sheet, they are also used to control erosion and also for embellishments. Therefore, bamboo plant is sometimes regarded by some people as having positive features towards life such as properity, peace and mercy [3]. Recently, issues relating to environmental threatened the life cycle of the environment globally due to the countries using various types of materials that are not biodegradable by industrial sectors globally. It has becoming a serious matter since it is closely related to the Product Lifecycle Phase resulted from extraction or deposition of waste materials that are not disposed properly [4-5]. Increment of logging activities for variety of purposes has resulted to the failing of absorbtion of carbon dioxide emission by the forest of which large amount of CO_2 are released into the atmosphere trapping the heat withing the atmosphere (green house effect) and causing the global warming.

Bamboo as the great potential to be used as solid wood substitute materials, especially in the manufacturing, design, and construction usage. Bamboo properties of being light-weight and high-strength has attracted researchers to investigate and explore, especially

Figure 1. Bamboo plantations in China [10].

in the field of bio-composite bamboo and is acknowledged as one of the green-technology that is fully responsible for eco-products on the environment [6]. Agricultural biomass solid wood made from bamboo have been identified by many researchers as the largest source of natural fiber and cellulose fibre biocomposite, which are provided at minimal cost and will bring a new evolution into production chain and manufacturing world [7]. Bamboo uniqueness are recognized as the source of raw materials that can be processed and shaped into the form of a number of commodities such as veneer, strips, lemon grass and fibre, and also it gives a new dimension, particularly in terms of its value of diversity in the production of bio-composite products. Advancement in science and technology, has led the materials used in manufacturing industries using raw materials from agricultural biomass to replace the use of solid wood and other non-biodegradable materials to improve manufacturing productivity and availability. High elasticity and strength of bamboo are suitable for the construction industry, and bamboo has proven to serve as a foundation structure [8-9]. The creation of bio-composite fibre board is also used in wall construction and are potentially to contribute of making cost effective home possible. Use of bio-composite material is seen increasingly high and the use bamboo as an alternative can be seen in productions such as furniture, automotive and other related productions. The natural colours of bamboo is unique compared to solid wood and other materials. In fact, the effect of the texture and tie on the outer skin of bamboo has the exotic value and at the same time creates a unique identity in the design, particularly furniture.

2. Bamboo biocomposites

2.1. Classification and development of biocomposites

The long-term global impact of furniture production has forced researchers to find solutions to various problems via research and development [11], and this search has given birth to the idea of using bamboo based biocomposite materials. The bamboo based biocomposite industry is important for improving both the quality of manufacturing and production as well as research and development [12-14]. Examples of some of the biocomposite materials that have proven their quality on the international market are medium density fiberboard, plywood and bamboo veneer each of which have been widely used in manufacturing furniture and other products. Distinctive properties of bamboo fiber reinforced biocomposite natural increase and flexural tensile strength, ductility and greater resistance to cracking and larger than a better impact strength and toughness of the composite [15]. All these properties are not accessible in other types of wood-based materials.

2.2. Conventional biocomposites

- Chipboard and Flakeboard

Bamboo chipboard is formed of bamboo shavings as elementary units, which are dried, mixed with certain amount of adhesive and waterproof agent, spread, shaped and hot-pressed at a proper temperature with proper pressure. Shavings are made of small-sized bamboo culm and bamboo wastes. As negative effects of green and yellow matter on adhesion are weakened after shaving, the adhering quality of bamboo chipboard is high. The supply of raw material for making bamboo chipboard is abundant and its production is an effective way to raise utilization ratio of bamboo resources, as can be seen in Figure 2. Bamboo chipboard is produced using water-soluble phenol resin, such a product has higher water tolerance, higher modulus of rupture and modulus of elasticity, and lower moisture expansion in thickness (compared with wood chipboard). Bamboo chipboard can be used as a kind of material for engineering construction. At present, it is mainly used for making ordinary concrete forms.

For the sake of improving utilization ratio of bamboo resources the stems of small diameter and of less known species, stem tops and all bamboo processing residue are used to make bamboo chipboard. The manufacturing process is designed following the technology of wood particleboard; rolling, cutting, chipping, re-drying, gluing, spreading and hot-pressing. The supply of raw material for making bamboo chipboard is abundant. All small bamboo stems of less known species and residue of bamboo cutting on groove land can be used for production. The utilization ratio of raw material for chipboard production is high, from 1.3 ton of raw material 1 m^3 of chipboard can be produced [16]. The technology and equipment for bamboo chipboard production are similar to those of wood particleboard. It is recommended to develop bamboo chipboard for improving the utilization ratio of raw material and the economic performance of enterprise. Bamboo chipboard manufactured with phenol formaldehyde resin is of comparatively high strength and MOE, low expansion rate of water absorbing. In case of need the products can be strengthened by adding bamboo curtain or bamboo mat to the surface. Such products have broad prospect.

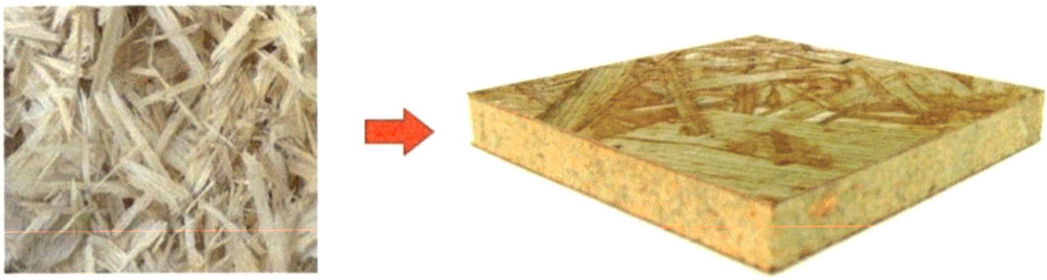

Figure 2. Bamboo flakeboard made from bamboo flake

- Plywood and Laminated

Plywood has been introduced in its application in 1865, since the plywood manufacturing sector began to rapidly developing era, focusing on making buildings and making the walls of the first aircraft using plywood [17]. Instead of plywood, plybamboo is now being used for wall paneling, floor tiles; bamboo pulp for paper making, briquettes for fuel [18]. Plybamboo is a special category in the wide variety of bamboo-based panels. Figure 3 shows plybamboo produced from layered of bamboo veneers with certain desired thickness. Thick strips have higher rigidity; they can hardly be deformed to fill up the blank space between strips even under high pressure therefore leads to the formation of lower the Modulus of Rupture (MOR) and adhering strength. Previously, wood is used to make bottom boards over a long period of time. However, plybamboo was now identified new alternative of make bottom boards. This is because plybamboo is a high quality and have great length which meets following requirements viz low weight, high rigidity, proper friction coefficient (to keep cargo and passengers from sliding) and doesn't rust. Besides, the manufacturing process of plybamboo was found is less laborious and consumes fewer adhesive than other types of composites. The strength, wear ability and rigidity of plybamboo are higher than those of ordinary plywood, thus, plybamboo has a wide prospect in automotive, building industries and engineering construction as well [19].

Due to bamboo's natural hollow tube shape, it is not possible to connect bamboo members with existing standard connections. As a result, it has been of interest to make bamboo available in shapes more suitable to current structural applications. This interest led to the development of Laminated Bamboo Lumber (LBL), which is usually produced as a board of rectangular cross-section [20]. Generally speaking, LBL is produced by flattening bamboo culms and gluing them in stacks to form a laminated composite. The aiming of this research is to examine a new low-technology approach for the fabrication of LBL in an effort to assess the feasibility of using this approach to produce an LBL product that is suitable for use in structural applications. Mechanical properties of bamboo based laminates need to be investigated thoroughly so that the full potential of bamboo as a functionally graded composite could be utilized. This publication reports the mechanical properties evaluation of 5-layered bamboo epoxy laminates [21]. Therefore, the purpose of the present research was to manufacture five types of laminated bamboo flooring (LBF) made from moso bamboo (P. edulis) laminae and investigate their physical (dimensional stability) and mechanical properties (bending properties) by ultrasonic wave techniques and a static bending method [22].

Figure 3. Plybamboo from bamboo veneer

- Medium Density Fibreboard

Medium Density Fiberboard (MDF) is a dry-formed panel product of lignocellulosic fiber mixture of certain synthetic resin such as urea formaldehyde resin (UF), phenol formaldehyde resin (PF) or isocyanate binder [23]. MDF was used commercially in 1970 with the advancement of technologies and materials at that time. However, MDF belongs to the type of wood that is not durable and do not require a very high resistance such as tables, rack, storage and others. In a certain period of time, the MDF can change shape, especially when exposed to water and the weight is too heavy. Presently, the majority of MDF producers in Malaysia are using RW as their major raw material. In order to find alternative of woods due to the arising illegal logging, renewable sources; bamboo fibres is used to produce agro-based MDF. Since bamboo itself has 1250 species, hence each bamboo fibres used in manufacturing MDF is expected not the same. Until now, researchers still in the middle of trying new species of bamboo for examples bamboo *Phyllostachys pubescens* [24-25] and *Dendrocalamus asper* [26] in producing MDF.

Extensive and ongoing research of MDF exhibited with the manufacturing overlaid bamboo fibres board panels using stylus method [27]. This research quantifies the surface roughness of the panel to have better overlaying of the substrate [28] was aiming to evaluate the influence of fibre morphology, slenderness ratios and fibres mixing combinations on the mechanical and physical properties of agro-based MDF using bamboo and bagasse fibres, as shown in Figure 4. It was observed that bamboo fibres had better mechanical performances and were more slender fibres in comparison with bagasse fibres. It appears that manufacturing MDF from bamboo which is non-wood species would provide profitable and marketable panel products in Thailand. Therefore, such panels are not only environmentally friendly but also alternative ways convert under-utilized species into substrate panel products for furniture manufacture.

Figure 4. Bamboo Medium Density Fibreboard (MDF) from bamboo fibre.

- Hybrid Biocomposite

The incorporation of several different types of fibres into a single matrix has led to the development of hybrid biocomposites. Recently, bamboo fibres was also gaining attention to be hybridized with more corrosion-resistant synthetic fibres (glass, carbon or aramid fibres) in order to tailor the composites properties according to the desired structure under consideration. Since synthetic fibres degrade at a much slower rate or does not degrade at all, inclusion with natural fibres may lead to green environmental balances with improvement in performances. Hybrid bamboo-glass fibres composites exhibit enhancement in terms of stiffness, strength and moisture resistance properties. Meanwhile, durability of bamboo-glass fibres composites under environmental aging was improved compared to pure composites [29-30]. Capability of bamboo to produce seven types of shapes encompasses silver, stripes, laths, veneer, particles, strands until bamboo fibres gives a huge impact in creating valuable hybrid biocomposites based on bamboo itself for various applications. In India, continuous ongoing research generates new hybrid bamboo mat veneer composites (BMVC) made from bamboo mats in combination with wood veneer [31]. In BMVC, wood veneer was placed in between bamboo mat. Results revealed presence of woven bamboo mats, BMVC has different mechanical properties along and across the length of the board thereby the properties are comparable to the plywood structure. Instead of bamboo mat, hybridization of bamboo curtain and bamboo mat with wood veneer was limited panels made in China for mainly used in rail coaches. Bamboo mat were also further utilized and commercialized by incorporate with bamboo particleboard for other applications, as can be seen in Figure 5.

Besides, new hybrid biocomposites product mades from bamboo strips and wood veneer bonded with PF resin were also developed. A symmetrical structure with flat and smooth

Figure 5. Crushed bamboo stripes laminated with empty fruit fiber.

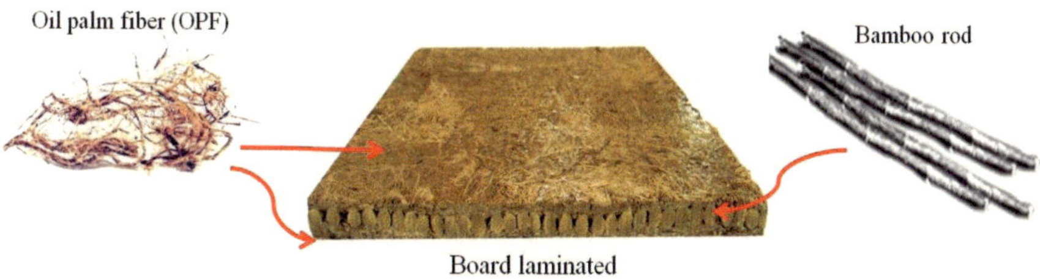

Figure 6. Samples of hybrid biocomposites board. Oil palm fiber laminated with bamboo rod.

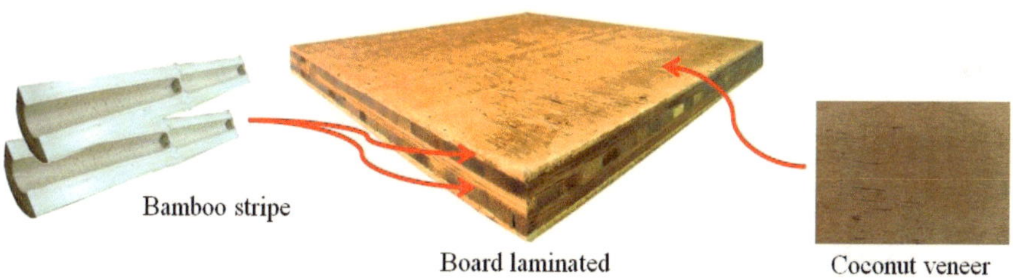

Figure 7. Coconut veneer laminated with bamboo stripe.

surface results from the combination between bamboo strips, bamboo particle and wood veneer plays important role as new material used for concrete formwork and side board of trucks. On the other hand, hybridization between bamboo and other natural fibres were also become a new approach in bamboo development progress. For example, as shown in Figure 6, bamboo rod was stack together with OPF fibres, coconut veneer and bamboo stripe as shown in Figure 7, respectively in order to produce high performances composites and gives variety in design and applications as well.

2.3. Advanced Polymer Biocomposite

- Thermoplastic Based Bamboo Composites

The most common reinforcement of bamboo fibres used today is thermoplastic polypropylene matrices [30]. Apart from various types of bamboo form, bamboo strips have higher cohesive strength than extacted bamboo fibres. For this reason, bamboo strips was reinforced with non-woven polypropylene aiming to produce ultra-light weight unconsolidated composites [32]. Non-woven web allow us to reinforced materials in their native form [6-8] and utilize the unique properties of the reinforcing materials. It was found, bamboo strips-polypropylene (BS-PP) composites has better properties including high flexural, high acoustical properties and good sound dampening that makes them suitable and ideal raw material to replace fibres glass currently used for automotive headliner substrates. Several components can be manufactured using biocomposites such as door insert, trunk liners, pillar trims, parcel shelves and load floors for automotive and field roofing, walling and profiling for building, as can be seen in Figure 8. Some research articles studied the effect of bamboo charcoal addition in the polyolefin thermoplastic polymer [33]. Bamboo charcoal has innumerable pores in its structure making it an excellent medium for preventing static electricity buildup and absorbing volatile chemicals. Taking into consideration these two advantages, bamboo charcoal was chosen as promising material to enhance the water absorption and electrical conductivity of the polyolefin. In another interesting study, bamboo fibres were undergoing autohydrolysis processing as method for obtaining soluble hemicelluloses-derived products reinforced with polylactic acid (PLA). This composite was made with spent autohydrolysis solids presented a markedly reduced water uptake. SEM of reinforced samples showed a satisfactory compatibility between phases, confirming the potential of composites made up of PLA and bamboo fibres as an environmental friendly alternative to conventional petrochemical thermoplastics.

Figure 8. Profiling (a) and roof (b) made from bamboo composites reinforcement thermoplastic.

- Thermoset Based Bamboo Biocomposites

Potential and interest of bamboo used in thermoset composites as expected has the same trend as thermoplastic composites. Previous research studied effects of bamboo fibres reinforced polyester matrix towards various testings for instance tensile and flexural properties [34], dielectric properties [35] and fracture properties [36]. Besides, influence of moisture absorption during storage and composites manufacture of bamboo fibres reinforced vinyl ester was studied by [37]. In another interesting research, bamboo fibres reinforced epoxy composites was subjected to wear and frictional environment in order to achieve widespread acceptance to be used in many applications [38]. It was claimed that, wear volume was superior when the fibres was orientated anti-parallel to the sliding conterface [39]. In another view, bamboo strips epoxy composites was found to be interesting materials to be applied in marine sector worldwide, [40] have produced bamboo boat hull using vacuum bagging and compression moulding process. Figure 9 shows after undergoing several test, this products was confirmed exhibit excellent mechanical properties including material ageing and resist to the marine environment. Exploitation of bamboo epoxy composites was further applied in manufacturing surfboards. Decks of bamboo surfboards are up to 4 layers of bamboo/epoxy laminate in hi-stress areas over a 60 psi medium density foam. Results indicated bamboo decks tend to not dent from normal use unlike glass boards.

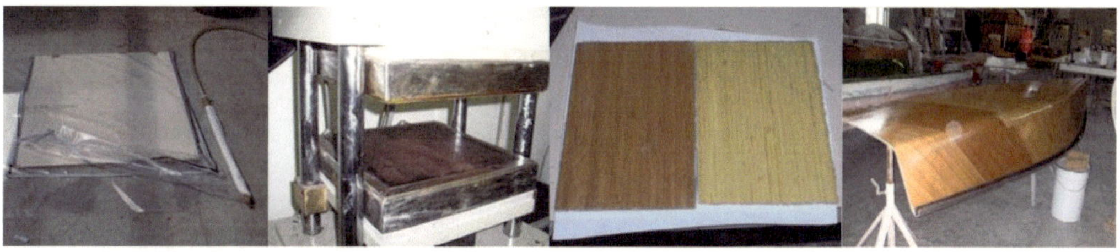

Figure 9. Manufacturing process of bamboo boat hull for water sports activities [40].

- Elastomer Based Biocomposites

Exponential uses of bamboo fibres were expanding into elastomer composites area as new viable alternative filler reinforcement. Short fibres are used in rubber compound due to the considerable processing advantages, improvement in certain mechanical properties and to economic consideration. Addition of short bamboo fibres into elastomer polymer matrix especially natural rubber (polyisoprene) promising great mechanical performances of composites manufactured [41-42]. It was found, bonding agent (silane, phenol formaldehyde and hexamethylenetetramine) plays an important role to obtain good adhesion between fibres and rubber. Results revealed composites properties for instances, hysteresis, fatigue strength, modulus, elongation at failure, creep resistance over particulate filled rubber, hardness, cut, tear and puncture resistance were enhanced. The newest report shows the extraction of cellulose nanowhiskers from bamboo fiber waste were use as a reinforcing phase in natural rubber matrix in producing bio-green nanocomposites [43]. The most excellent starting material for production of nanowhiskers is residue from paper production (bleached pulp fibres). In this study, the processing of cellulose nanocomposites was done via a latex based master batch preparation followed by mill-compounding. It was found to be a viable route to produce rubber based nanocomposites, which can potentially be scaled-up to a commercial scale process.

Applications of elastomer composites included tires, gloves, V belts, hoses and complex shaped mechanical goods. As for tires manufactured, Carbon black has been extensively used for obtaining improved initial modulus and durability [44]. Carbon black mainly used as a reinforcing filler in tires starting from 20th century produced a 10-fold increase in the service life of tires. Apart from various types of natural fibres, bamboo also can be burned in furnace for certain temperature and heat to be synthesized into carbon black formed [45]. Since then, carbon black has remained established the major reinforcing material for use in tires as well as other rubber products. Generally, incorporation of carbon-black comprises about ~30% of most rubber compounds. As people playing more and more attention to environmental protection, therefore utilization of various natural fibres especially bamboo as filler as replacement of burned fossil fuels in natural rubber polymer matrix creates greener tires produced, as shows in Figure 10. In addition, physical and mechanical properties of tires manufactured were enhanced with very satisfactory levels in terms of abrasion resistance and improved a lot of resistance to tread. Thus, exploitation of bamboo was no doubt creates improvement in development of elastomer biocomposites.

Figure 10. Green tire made from bamboo carbon black.

2.4. Inorganic based biocomposites

Inorganic bonded plays important role in the construction industry. Generally, inorganic bonded composites can be formed using three types of inorganic binders consists of gypsum, portland cement, and magnesia cement can be applied for producing shingles, blocks and bricks. In this rapidly developing world, there has been a clear trend toward investigate alternative additions for the manufacture eco-efficient blended cement composites. To meet this satisfaction, utilization of lignocellulosic materials for instance bamboo and oil palm fronds remains an exciting and innovative technology as cement replacement [46]. Figure 11 shows the bamboo cement-boards (BCB) were produced from bamboo flake types *Bambusa Vulgaris* from Malaysia. A bamboo-cement ratio of 1:2:75 and 2% aluminum sulfate alone or in combination with sodium silicate was possible to produce a board which satisfied the strength and dimensional stability requirements of international standards which can be used in a wide range of infrastructure construction applications.

Besides, gypsum bonded particleboards Brazilian giant bamboo (*guadua magna*) has been manufactured by Priscila C. de Araújo 2011 [47]. Results revealed, bamboo cement boards presented higher bending strength and lower moisture content than bamboo gypsum boards. Despites, generally bamboo cement composites and bamboo gypsum composites have superior performances viz higher strength, good weathering resistant ability, good fire resistant and sound insulating as well as containing no synthetic adhesives which will lead to free emission of formaldehyde and other noxious chemicals [48]. Apart from the bamboo structure itself, utilization of bamboo leaf ash as supplementary cementing material for the production of blended cements has been studied by Moises Frias et al 2012. This study has generated the other new possibility in utilizing other side of bamboo structure which is a bamboo leaf in concrete and cement industry. Cement and concrete panels produced can be used for a wide range of applications in the building, housing and other commercial/infra-structure projects for instance wall, partition, roof and pillar materials as well. Further, bamboo as construction materials using bamboo sticks as replacement of steel was studied by Khosrow Ghavami and Mahzuz [49-50].

Figure 11. Inorganic cement board made from inorganic carbon.

3. Biocomposites as potential material in design

Since 1865, the use of agro-based biocomposite material in the manufacturing sector has been introduced and its use has increased consequent to the acceptance of positive users of composite materials [51]. This focusing on sustainable economic stability and protect the corresponding sustainable base resource and environment. Until now, engineer and designers have been succeeded in convincing consumers towards the level of quality and durability of biocomposites material produced through the design. Biocomposites is made of two or more materials combined together to create a new and effective material in terms of quality and production process, based on references relating to the past problems. Activities hybridizing an element of progressive thinking to challenge basic human search for the truth behind the reason for any of the material itself [52]. In the process of biocomposites production, specific characteristics of each original material can be increased or decreased to give the desired effect and this is the reason to why many materials can be found developed through the process of biocomposites, for example, bamboo for product development [53-54]. In the era of science and technology, the use of bamboo has increased its research and development (R&D) to exploit the advantages of bamboo as a fiber substitute other materials because of the advantages of bamboo are sustainable and renewable over time [7, 55]. Many researchers believe that a combination of high-tech applications will be the future standard for the international manufacturing market. Various companies involved in the composites industry began to grow bamboo and compete brings innovative ideas in producing bamboo products from bamboo bicomposites as composite deck, composite bamboo fence, bamboo composite deck tile, bamboo composite bins, bamboo deck accessories [16, 56]. The transformation through creativity and innovation are also on the run, especially in Asian countries such as Japan, China, India, Philippines, Thailand and Malaysia [57-58]. Southeast Asia, the cultivation of the highest quality bamboo has gained attention from many parties, especially the government and the firm of research [52, 59]. Emphasis on innovation is very important because it will not only create jobs but to promote competitiveness [60-61]. Before the design process, the manufacturers able to move to the highest level with the help of four aspects of design, quality, identity, and raw materials. Four aspects are interrelated with each other in the manufacturing industry. Therefore, the cooperation between researchers, engineers, designers and marketing necessary to ensure that products produced in the limelight from customer [62]. Failure of one of the four aspects of the products produced will fail in the market. There are some examples of products that failed in the market not because of poor quality or less attractive design but use less material to meet customer needs [63-64]. In the second stage some product evaluation process will be conducted to identify the level of quality of the products and fulfill the quality standards [65].

Technical assistance, infrastructure improvements to the farm, machinery and factory equipment in place which needed to develop a revolutionary product in the bamboo industry composite. Products such as biomass fuel pellets, particle board, and composite applications are designed with a combination of bamboo fiber to produce strong and durable strength to the conventional wood, steel, and plastic [13]. The physical properties of bamboo are tapered, hollow, have a node at varying distances, easily shaped and not perfectly round bamboo can be a major factor to be alternatives to other sources [16]. Physical property makes bamboo is often chosen as the lead in the design of the building structure as to fit the shape of bent bamboo is not difficult. This is because bamboo can be used in both situations either green bamboo or dried bamboo, because bamboo has determined its shape will remain for long without mindfully stretches [66-67]. Bamboo biocomposite always thought to compete with the strength of steel as well as having the advantage of aesthetic value compared with other materials. Steel production requires the use of fossil fuels is high, therefore, emissions in the steel manufacturing industry increasingly apparent, studied to understand the mechanical behavior of bamboo reinforced concrete members and explain the differences in the structural properties of steel reinforced concrete, reinforced concrete and bamboo. In this chapter, several tests bamboo reinforced concrete beams and columns that run in the laboratory report.

Excellent research to understand the mechanical of behavior bamboo reinforced concrete and explain the differences in the structural properties of steel reinforced concrete and bamboo has been proved by researchers [53]. Composite panels using natural fiber made from bamboo reinforced cement have great potential due to their better strength, dimensional stability and other characteristics compared to panels made from several plantation timbers [31].

Bamboo fiber has an inner impact of natural color and texture is interesting, original, versatile, smooth surface, low cost and sustainable. An example of innovative bamboo research is the design of a Spring Chair based on the elements of swift motion, transforming bamboo's strength and flexibility to produce a reaction from the design, a unique structure as the primary feature of a complete biocomposite material designed by Anthony Marschak [68]. Nowadays, different designers from manufacture company compete each others to create something modern, stylish, beautiful and outstanding product in the market used bamboo biocomposite. The use of bamboo as biocomposites in the design is usefull to create a better experience for the end user, giving more attractive design and allowing efficient manufacturing systems produced as an alternative. Biocomposite bamboo as modern material is a different experiment and does not constraints to the limit of creative thinking [69]. Biocomposite market acceptance of the use of bamboo in furniture manufacturing, automotive, construction and interior decoration is becoming in demand and easily can be found in international furniture fairs and interior design exhibitions every year [54, 69-70]. It is obvious that biocomposite bamboo material has the tendency to tackle resource-efficient challenges, creating virtually no waste when processed properly and at the same time increased the product market, while also promoting the use of sustainable materials [71-72].

4. Commercial applications

In recent years, the use of bamboo has been enhanced to exploit bamboo as a renewable wood fiber. Evolution in theoretical and applied research on bamboo-based products has increased year by year and expanded its use in almost all applications, especially in building, furniture, product, transport, packaging and others. Bamboo composite was accepted in the global market in applications replacing traditional wood interior and exterior products [36]. This proved the strength of bamboo is found 10 times stronger than wood materials [73]. Various positive advantages found in composite products from bamboo as dimensional stability, longevity, weather resistant, high impact resistant, low maintenance, non-toxic, low flame spread, etc. [34]. Table 1 shows the innovative design and application of bamboo fiber biocomposites in various categories.

4.1. Construction applications

Wood has been used as a building material for thousands of years, otherwise the use of bamboo as the main material in construction activities in Southeast Asia have taken place since the era of human civilization began to grow. Community in the early days to know about the benefits of plant bamboo and consider the benefits of life [3]. The use of bamboo in construction design has long proven its excellence by building houses one hundred percent use of natural bamboo. Mechanical properties, durability, suitability as a good absorber of heat and access to source material made it famous and still used until now [74-75]. This is evident based on the tensile strength required in the development of the bamboo bridge before the first world war again. The suspension bridge was first created using bamboo to cross the river and business relations. At that time, the bamboo used classic exterior use only bamboo, which is four times as strong as the interior. Bamboo bridge was built in India, South America, while in Colombia, using a bamboo bridge and cable tension structure created by the tensile strength of up to 3,200 kg/cm2 using Guadua bamboo species [67]. Innovations in bamboo technology offer new opportunities for large-scale construction of this sustainable material. From long-range beam cross laminated laminated bamboo panel, joinery, bamboo has proven to be safe and durable for city buildings and homes and it proved to be used in major cities around the world, from Europe to United States and Southeast Asia [76]. Many architects and designers convince bamboo as the most environmentally friendly in the world. Scientific and technological progress have resulted in hybrid biocomposites from bamboo has the capability to produce various types of reinforced veneer that has a big impact, particularly for construction [6]. Many bamboo transformations were produced by scientists to improve the quality of the various aspects of bamboo, for example in China, bamboo is used in the design of the roof with the tip covered with decorative tiles to protect from rain water, add neatness and aesthetic value to the roof [3]. Various techniques have been developed to produce a strong roof support system. In the Philippines the roof function improved by using the split bamboo roof and produce a soft surface to facilitate the flow of air and water in bamboo [77]. Roof architecture is most suitable as roofing solutions at the time. Design prefabricated truss system has a frame will be covered with bamboo board, lath and plaster to create a waterproof roof and can last up to 15 years with regular maintenance.

The natural beauty of bamboo aesthetic usage has led bamboo to be widely showcased as part of the collections globally. Bamboo has been widely accepted to be more than just the material by the architects and designers but also can be used to decorate and embellish. Bamboo has the same technical performance comparable to the solid wood, concrete, and steel but release smaller carbon footprint [7]. Because of its eco-friendly property, bamboo is often alternatively used as concrete reinforcement. Many studies have been conducted to determine the feasibility of using bamboo to reinforce concrete with flat symmetric structure decisions and smooth surface from a combination of bamboo strips, bamboo and wood veneer particles play an important role as new material is used for concrete formwork [72]. New biocomposite hybrid product made of bamboo strips and wood veneer bonded with PF resin was developed as a result of ongoing research. Prominent architects bamboo, a renowned architect Oscar Hidalgo comes from Chinchina, Colombia make bamboo as the main material in the construction work and thus make bamboo as a symbol of art in every creation. Advantage and uniqueness of unusual bamboo plants, Oscar then has dedicated his whole life to bamboo research. Research on the structural integrity and aesthetic value in bamboo has brought Oscar to Asia, Costa Rica, Brazil, and elsewhere to study this plant and build some experimental structure. He was the first to use a variety of beam culma and uses a unique bolt system as the introduction of concrete in intends to create very strong joints for construction. All in all, Oscar recommend the use of bamboo in housing construction because many of the problems associated with bamboo can be reduced by creating laminated bamboo strips.

In 1942, Oscar has made a study of the use of bamboo laminates in ski pole was commissioned by the government of the United States (U.S.). At that time the bamboo laminate floor tile products applied with a very good quality for heavy floor traffic with soft strips of bamboo from remote culma and can be used safely [67]. Bamboo also excelled as a reinforcement of concrete tested because many studies have been able to determine the possibility of using hybrid bamboo to reinforce concrete in the future [47, 78]. However, ongoing research necessary to solve some other problems of resistance if the water in the bamboo because bamboo can break concrete durability when experiencing the process of expanding and then shrinking.

4.2. Interior design applications

Bamboo biocomposites has excellent impact on creating interior design that has a commercial value of its own way. Biocomposite use in the production of various bamboo products for exterior and interior which have a good demand in the global market. Most users realize the greatness of this material and support the efforts of sustainable for nature in everyday life. This is further enhanced by its excellence as an innovative material to get recognition from various quarters, proving that hybrid bamboo material can overcome other types of materials from various aspects such as physical, mechanical and aesthetic [6]. Nowadays, various types of hybrid bamboo-based products have been produced, from the design of the ceiling, walls, floors, window frames, doors, stairs and up to the home decorative accessories. Bamboo can create special effects, as well as using bamboo joinery can be bent or straightened by heating and clamping. Based on the previous research, a typical wall section created with bamboo stud where distance is determined by the thickness of the bamboo boards which are used in the study [79].

For example, when a board of 1 cm is used, the distance for each stud is 40cm. Bamboo boards are attached, and two layers of plaster are used. Another wall system utilizes the bamboo studs as described above, is by using small pieces of bamboo attached together with 1 1/2-2" nails. Then, the attached bamboos are plastered with a mixture of clay / straw outside the system, this is known to be heavier than the previous example. The prefabricated nature of the bamboo wall panel system which are pre-built on the ground leads to better housing development [80]. Interior design most impact on the industry for interior decoration and architecture is Madrid Barajas International Airport, Spain also it has been recognized the world as a Sustainable Building 2011 [81]. Richard Rogers, designer of the world's most prestigious airports that designs consist of 200,000 m2 ceiling lath gently curved laminated bamboo, the bamboo industry's biggest project in the world. International Airport was built using laminated bamboo laths from all walks of bamboo veneer. Richard Rogers has managed to apply the design process in yield designs with the use of materials and finishes that can create a unique passenger experience, exciting and the atmosphere is peaceful and quiet. Although the simplicity of the concept of architectural features terminal, it still gives comfort to passengers. Therefore, interior design use hybrid bamboo could be emulated by other designers in meeting the demands of the 21st century, so that the designs produced will be efficient, economical and functional.

Category	Year	Inventor/ Designer	Design Name	References	Design
Furniture	2006	Anthony Marschak	Spring Chair	[68]	
	2009	Kenyon Yeh	Jufuku Stoo	[88]	
Automotive	1999	Automotive Manufacturer (Audi, BMW, Peugeot, Volvo, etc.)	Automotive Components (Cloth seats, floor mats, dashboard, door panels, etc.)	[19, 86, 89]	
	2010	Kenneth Cobonpue & Albercht Birkner	Pheonix Bamboo Concept Car	[87, 90]	
Interior Design & Construction	2002	HPP Architects	Parking Garage, Leipzig Zoo, Germany	[76]	

Category	Year	Inventor/ Designer	Design Name	References	Design
	2005	Richard Rogers	Barajas Madrid International Airport, Spain.	[81]	

Table 1. Innovative design and applications from bamboo fiber biocomposites in various categories.

4.3. Furniture applications

The design is a mechanism to display an awareness of the importance of the needs and quality of life through creative and innovative ways. Revenue awareness now, a lot of furniture design in the market focused on the continuity between current needs and environmental concerns to ensure the life cycle assessment as a result of product benefit. Bamboo materials importance to environmental sustainability supported by success in applying design furniture designer to include design elements with environmental relationships to enhance the product in order to gain market attention [82].Many countries have started to establish research and development-based furniture such as the Malaysian government established the agency Forest Research Institute Malaysia (FRIM) for more in-depth research to help the furniture industry because Malaysia is the largest exporter of furniture to more than 100 countries. The Chinese government also provides support and assistance to help establish Chinese Association Ecomaterials materials scientists who research on how to design, produce, reuse, disposal and recycling of materials in an environmentally. Bamboo biocomposite proved by many researchers to have high benefits as an alternative material for the production of furniture and other components. A variety of new furniture designs have been produced using smart materials is based on the proven quality furniture compare to solid wood material. High innovation in bamboo fiber can improve the durability even bent and shaped materials such as solid wood other [66]. Research and development in making advanced bio-based furniture products around the world are able to produce continuous improvement in product innovation success. Initiatives to increase the use of bio-composite value highly praised and encouraged for these materials to reduce environmental impact, improve innovation and advanced technology in the manufacturing process.

Bamboo also has the characteristics of materials and textures are very useful for designers to create a unique design and original, it is imperative that users reacted positively [83]. Now, many examples related to bamboo furniture that can be used as an important consideration in choosing furniture bamboo [55]. There are several designs of amazing furniture use bamboo material simplicity, many furniture designers prove bamboo materials is not only resilient and pliable, but tremendously powerful internal and external. For example young designer

Kenyon Yeh, a designer has produced designs stool named Jufuku from Japanese word means duplication or repetition. Stool Jufuku clearly emphasizes the best quality bamboo to produce designs without parts or fasteners, where each piece of bamboo on Jufuku Stool is made from a single structure shape which is then repeated to complete each form of the object but the result is still beautiful in a simple, minimalist and attractive. The famous designers Anthony Marschak discovered the magic of bamboo while looking for ideas as versatile materials other than solid wood for his design for Spring Chair. Spring Chair is produced from renewable resources has become a sustainable furniture and other luxury furniture comparable. Strength and flexibility of bamboo materials to create natural bending very important in the design and ergonomic nature [68]. Spring Chair bamboo bending technique: made from one continuous sheet, the surface of the first three curves are made to suits the human contour seats. Bamboo biocomposite manage to stand out as a versatile and able to provide a beautiful surface finish, elegant, unique structure and interesting links suited to any modern home. Beauty can be seen in Spring Chair as the pioneer era of the rise of modern bamboo.

4.4. Automotive applications

Natural fiber has experienced rapid growth in the automotive market, especially in Europe and Southeast Asia. Biocomposite based innovation increasing every year in the global research arena because it promises a reasonable cost and performance compared to competing technologies. The first Industrial Revolution progress in transportation with the creation of steam-powered ships and aircraft engines. In 1930, a second industrial revolution is an important era in the manufacture of car compartment using fiber as an alternative to existing materials. Famous automobile inventor, Henry Ford also supports the use of materials from natural fibers start to bring progress in the era of automotive construction. European Union (EU) and the countries of Asia also supported by issuing guidelines in the global automotive manufacturing industry [84]. A study shows that low cost natural fiber bamboo materials are highly potential to be used in automotive parts [84]. Guidelines made in 2006 ordered all automotive manufacturers to produce automotive plastic reinforced using natural fiber. In addition, the European Union (EU) targeting 80% of the vehicle compartment must be reused or recycled and the amount should be increased by 2015 to be 95%. Through previous research has produced various components of the car which has been designed using natural fibers as the main component. Research continues to generate new bamboo mat veneer composite hybrid (BMVC) where bamboo mate was used as the face and back layers of wood veneer and the core layer. In addition, bamboo mat with wood veneer panels used in train carriages. Natural fiber composites with thermoplastic and thermoset matrices have been widely used in the manufacture of door panels, rear seats, headliners, package trays, dashboards, and the interior of the car manufacturers' world [85]. This is supported by many researchers who have proven quality and effectiveness of bamboo as an alternative material in the automotive manufacturing industry.

To make the strip used to laminate, soft bamboo for interior issued by plane, leaving the external hard drive to strip lamination. Natural fibers such as bamboo offers benefits such as reduction in weight, cost, reliance on sources other materials, and has the advantage that can be recycled. However, few studies on technical and mechanical material of bamboo fiber are still being studied by scientists and engineers before getting the confidence to allow large-scale manufacturing, especially to the outside of the car body. In the 21 century, bamboo fiber has become crucial for the development and design (R & D) products [86]. Earlier bamboo has been used to build boats and zeppelins. In aeronautics research, the structure of the kite and early aircraft were built using materials from bamboo fiber because it is lightweight and very strong. The aircraft made entirely of bamboo were built in the Philippines, while the Chinese use in their aircraft during World War II [40, 86]. In 2010, Kenneth Cobonpue designers from the Philippines and Albercht Birkner branded products from Germany managed to create a 'Phoenix', the first car in the world is made of bamboo and natural fiber that can be recycled [87]. Phoenix uniqueness is reflected in his designs made using small bamboo stacked and tied neatly. Use bamboo turned into products with the quality of its own internal and external design makes the Phoenix has a high aesthetic value. Biodegradable materials challenge the notion compact, durable materials in vehicle design. It looks at cheaper and ecological option replaces the shell, on the other hand build and explore the relationship between technology and nature in the bamboo concept car. In fact it can be used as laminated wood, with a bamboo laminate edge is much lighter in weight.

5. Sustainable product and development of bamboo biocomposites

There has identified three key areas that will be noticed in connection with products and sustainable development, ecology, economy and technology is all that should always give priority in life, as shows in Figure 12. All levels will impact specific, largely due to the materials involved in the different stages. This concept can be described as a wise balance between the demands of society's increasing demand for products, the preservation of forest health and diversity of material resources and benefits.

5.1. Ecology capital

In addition to improving environmental quality through the development of a sustainable supply chain resources, and better towards reducing CO_2 and almost zero net greenhouse gas emissions. From previous time, environmental issues is not high on the public agenda, but it is an exciting challenge that can lead to new solutions through the design and prove a wise economic choice. Solid wood product work has been in existence since the era Neolitic [91]. Artisans from many cultures have developed a technique to design and style this way many conventional furniture and rooted in human culture. In addition, problems such as a lack of material resources and the population increases each year is the main reason why designers need to focus their attention on the development of new materials design from bamboo fiber as technology and marks the progress of evolution in style. Product development will help consumers to see the potential of bamboo biocomposites products as part of our culture and heritage, and to enhance the status in our society [16]. From the results of studies carried out in Europe, it can be concluded that the bamboo fiber reaches "CO_2 neutral or better". However, the level is far more excellent if used in bamboo producing countries such as China and India

Figure 12. Three elements to support the sustainable product.

to have a lot of bamboo material resources and ecosystem [92]. Bamboo has roots that spread underground in all directions; land turned into solid and protects us from landslides and earthquakes by heavy rains. This means that the stand of bamboo should be treated as common ground for the public community. Economic trade market incomes only concern about the cost of production, otherwise cost disposable products used completely overlooked. This is one of the main reasons why the serious problem of waste disposal and environmental disruption has been caused recently.

5.2. Economic capital

Bamboo will be considered as one of the most useful resources to maintain sustainable economic development. Successful product development technical or physical demand is insufficient. Factors such as reputation, fashion products, trend, cost and other factor should also be taken into account when developing a sustainable product. An example of addressing this problem is for the exchange of ideas between the designers, engineers, socialists, scientists and marketing experts. Transformation of low-impact materials considering the material is important. Renewable materials, nontoxic materials, and materials that can easily be recycled all the smart choice to shift the perception of beauty designer different reference frames. At the same time, a potential new market in the development of sustainable solutions will be increased globally [93]. The design, which is a practical activity but also part of the culture and research, can make a significant contribution not only to design products and services that require creative community, but also to the development of a more general transformation of the materials industry [94]. To realize sustainable economic development, we need to consider the costs associated with not only the goods but also non-tradable goods such as environmental protection and natural resources. In the case of the total cost of producing bamboo bamboo including the cost of disposable copied and will be cheaper than making chemicals. In addition to bamboo has a better variety of mechanical, anti-bacterial applications and industries that make it an excellent resource for sustainable economic development [95].

5.3. Technology

The history of technology in keeping pace with the development of human culture since Paleolithic times. Through the events in the movement era of human civilization shows some technology is starting created slowly the impact of human knowledge about materials, science and technology. Since the world is faced with many serious problems such as global warming, acid rain, soil erosion, the financial crisis, extinction of flora and fauna habitat and others, which these problems are caused by human behavior-oriented manufacturing profit that could be marketed. Global manufacturers rich with knowledge of high technology need to consider a sustainable technology in each manufacturing process. Over the years, manufacturers already accept green technology at several countries to support the environment. Green technology is important because not only it can increase the profit, but to maintain the ecology, source materials and people will be able to enjoy a peaceful life until next generations. In order to develop a sustainable product, it is important to know the aspects of technology. Sustainability issues have recently become considerations when consumers choose to use green products in everyday life. In other words, the materials and design are very needed as an intermediary with the user in maintaining the quality of life and maintain as it reflects our cultural values [96]. Therefore, other materials needed as an alternative to meet these needs [84]. Therefore, the introduction of new technology in the sciences material is needed to maintain the momentum of the global manufacturing market.

Today, most people recognize that solid wood resources are limited and the progressive consideration must be given to the processing of biomass. Manufacturers need to change the manufacturing process of current technology to build a sustainable world and a strong economy [93]. To be a sustainable ecological community, the values of life we have to change along with the way of life that we see the product from different reference frames. Since bamboo provides a number of specific characteristics, it becomes a typical example to make our communities sustainable and rich thru technology.

6. Conclusion

From this chapter, it is concluded that the new development in innovation bamboo biocomposites from the natural fiber need to be more highlighted. The biodegradability and recyclability of design based material could be the critical problem in the next decade. In addition to the increase of population, the regulation forced manufacturer to use natural fiber in further products. Low cost, environmentally friendly, accessible and easy production of natural fiber composite are attractive benefits for design development and applications. A collaboration between scientist and designer is important to archive the quality materials, produce good design and has implications not only for the companies but also for consumers and the society.

Good design combines the capabilities of a balanced approach in terms of material, commercial design, environment, technology, idealism, and humanitarian concerns will generate benefit product for Life Cycle Assessment without the effect of the ecological system. In other hand, product value can be increased with the use of a material and design that reduces simultaneously the environmental impact and cost effective if manufacturers completely support new material mechanical properties and explore that have the potential to meet the requirements of a new future. The advantages of renewable bamboo fiber and biodegradable at the end of the product cycle including safetyness during handling and processing, and also as a resolution to the problem of resources reduction of other materials. The importance of awareness of the diversity of natural resources such as bamboo can generate new economic resources while protecting natural forests for future generations.

Acknowledgements

The authors highly acknowledge and pay gratitude to the Universiti Sains Malaysia, Penang for providing Research University Individual (RUI) grant 1001/PTEKIND/811195 that made this work possible.

Author details

S. Siti Suhaily[1], H.P.S. Abdul Khalil[1*], W.O. Wan Nadirah[1] and M. Jawaid[2]

*Address all correspondence to: akhalilhps@gmail.com

1 School of Industrial Technology, Universiti Sains Malaysia, Penang, Malaysia

2 Laboratory of Biocomposite Technology, Institute of Tropical Forestry and Forest Products (INTROP), Universiti Putra Malaysia, UPM Serdang, Selangor, Malaysia

References

[1] Mohd, T. B. M, Bhat, I. U. H, Mohmod, A. L, & Aditiawaiti, P. Abdul Khalil H.P.S.,Thermal and FT-IR Characterization of Gigantochloa Levis and Gigantochloa Scortechinii Bamboo, a Naturally Occuring Polymeric Composite. Journal of Polymer and the Environment. (2012).

[2] Bonilla, S, Guarnetti, R. L, Almeida, C. M. V. B, & Giannetti, B. F. Sustainability Assessment of A Giant Bamboo Plantation in Brazil: Exploring The Influence of Labour, Time and Space. Journal of Cleaner Production. (2010)., 18(1), 83-91.

[3] Bar-yosef, O, Eren, M. I, Yuan, J, Cohen, D. J, & Li, Y. Were Bamboo Tools Made in Prehistoric Southeast Asia? An Experimental View From South China. Quaternary International. (2012). , 269(0), 9-21.

[4] Ljungberg, L. Y. Materials Selection and Design for Development of Sustainable Products. Materials & Design. 2005;(2007). , 28(2007), 466-79.

[5] Bovea, M. D, & Vidal, R. Materials Selection for Sustainable Product Design: A Case Study of Wood Based Furniture Eco-Design. Materials & Design. (2004).

[6] Abdul Khalil HP.S., I.U.H. Bhat, M. Jawaid, and A.Z. Hermawan. Bamboo Fibre Reinforced Biocomposites: A Review. Materials & Design. (2012). , 42, 353-68.

[7] Scurlock, J. M. O, Dayton, D. C, & Hames, B. Bamboo: An Overlooked Biomass Resource? Biomass & Bioenergy. 2000;(2000). , 19(2000), 229-44.

[8] Khanam, P. N, & Abdul, H. P. S. Khalil, G. Ramachandra Reddy, and S.V. Naidu. Tensile, Flexural and Chemical Resistance Properties of Sisal Fibre Reinforced Polymer Composites: Effect of Fibre Surface Treatment. Polymer Environment. (2011). , 19, 115-9.

[9] Bhat, I. U. H, Mustafa, M. T, & Mohmod, A. L. and H.P.S. Abdul Khalil. Spectroscopic, Thermal, and Anatomical Characterization of Cultivated Bamboo (Gigantochloa Spp.). Bioresource. (2011). , 6(2), 1752-63.

[10] Akihiro, T, & Eka, A. Application of MDI Binder Towards Environmental Friendly Wood-Based Industry. Seminar on Wood-Based Panel Products; 10-11 July 2001; Kuala Lumpur: Forest Research Institute Malaysia; (2001). , 6.

[11] Liansheng, Y, Maili, S, Tao, W, & Wu, Z. The Pretreatment of Carbon Fibres for 3D C/SiC Composites. Science and Engineering of Composite Materials. (2002). , 10(1), 55-8.

[12] Zhiyong Cai JE.W., editor. Opportunity and Development of Bio-Based Composites. Proceeding of International Workshop on Prefabricated Bamboo Panel Module; (2005). Beijing, China.

[13] John, M. J, & Thomas, S. Biofibres and Biocomposites. Carbohydrate Polymers. (2008). , 71(3), 343-64.

[14] John, M. J, Francis, B, Varughese, K. T, & Thomas, S. Effect of Chemical Modification on Properties of Hybrid Fiber Biocomposites. Composites Part A: Applied Science and Manufacturing. (2008). , 39(2), 352-63.

[15] Qisheng, Z, Shenxue, J, & Yongyu, T. Industrial Utilization on Bamboo (2001).

[16] Shi, S, & Walker, J. Wood-Based Composites: Plywood and Veneer-Based Products. Primary Wood Processing ed. USA (2006). , 391-426.

[17] Sen, H. N. T, & Reddy, J. Application of Sisal, Bamboo, Coir and Jute Natural Composites in Structural Upgradation. International Journal of Innovation, Management and Technology. (2011). June 2011;, 2(3), 186-91.

[18] Makinejad, M. D, Salit, M. S, Ahmad, D, Ali, A, & Abdan, K. A Review of Natural Fibre Composites in Automotive Industry. Research on Natural Fibre Reinforced Polymer Composites. , 2009-28.

[19] Mahdavi, M, Clouston, P. L, Arwade, S. R, & Low-technology, A. Approach Toward Fabrication of Laminated Bamboo Lumber. Construction and Building Materials. (2012). , 29(0), 257-62.

[20] Verma, C. S, & Chariar, V. M. Development of Layered Laminate Bamboo Composite and Their Mechanical Properties. Composites Part B: Engineering. (2012). , 43(3), 1063-9.

[21] Lee, C. H, Chung, M. J, Lin, C. H, & Yang, T. H. Effects of Layered Structure on the Physical and Mechanical Properties of Laminated Moso Bamboo (Phyllosachys Edulis) Flooring. Construction and Building Materials (2012). , 28, 31-5.

[22] Abdul Khalil HP.S., and R. Hashim. Komposit Panel Berasaskan Sumber Kayu. Penang: Penerbit Universiti Sains Malaysia, Pulau Pinang; (2004).

[23] Xiaobo, L. M. Physical, Chemical, and Mechanical Properties of Bamboo and Its Utilization Potential for Fiberboad Manufacturing [Masters Thesis]. Beijing, China: Beijing Forestry University; (2004).

[24] Matsumoto, K, Yamauchi, H, Yamada, M, Taki, K, & Hiroaki, Y. Manufacture and Properties of Fiberboard Made From Moso Bamboo. Journal of The Japan Wood Research Society. (2001). , 47(2), 111-9.

[25] Hiziroglu, S. Bauchongkol, Piyawade, Fueangvivat, Vallayuth, and Soontonbura. Selected Properties of Medium Density Riberboard (MDF) Panels Made From Bamboo and Rice straw. (2007).

[26] Salim, H, Songklod, J, Piyawade, B, & Vallayuth, F. Overlaying Properties of Fiberboard Manufactured From Bamboo and Rice Straw. Industrial Crops and Products (2008). , 28, 107-11.

[27] Lee, S, Shupe, T. F, & Hse, C. Y. Mechanical and Physical Properties of Agro-Based Fiberboard Holz Als Roh-Und Werkstoff. (2006). , 64, 74-9.

[28] Thwe, M. M. and K. Liao Tensile Behaviour of Modified Bamboo-Glass Fibre Reinforced Hybrid Composites. Plastics, Rubber and Composites. (2002). , 31(10), 422-31.

[29] Nayak, S. K, Mohanty, S, & Samal, S. K. Influence of Short Bamboo/Glass Fiber on the Thermal, Dynamic Mechanical and Rheological Properties of Polypropylene Hybrid Composites Materrials Science and Engineering. (2009). , 523, 32-8.

[30] Bansal, A. K, & Zoolagud, S. S. Bamboo Composites: Material of The Furniture. J Bamboo and Rattan. [Journal]. (2002). , 1(2), 119-30.

[31] Shah, H, Reddy, N, & Yang, Y. Ultra-Light-Weight Composites From Bamboo Strips and Polypropylene Web with Exceptional Flexural Properties Composites: Part B. (2012). , 43, 1658-64.

[32] Siriwan, K, & Suthamnoi, W. Physical Properties of Polyolefin/ Bamboo Charcoal Composites Journal of Metals, Materials and Minerals. (2009). , 19(1), 9-15.

[33] Ratna Prasad AV., and K Mohana Rao. Mechanical Properties of Natural Fibre Reinforced Polyester Composites: Jowar, Sisal and Bamboo. Materials & Design. (2011).

[34] Murali Mohan Rao KMohana Mohan Rao K., A.V. Ratna Prasad. Fabrication and Testing of Natural Fibre Composites: Vakka, Sisal, Bamboo and Banana. Materials & Design. (2010). , 31(1), 508-13.

[35] Wong, K. J, Zahi, S, & Low, K. O. C.C. Lim Fracture Characterisation of Short Bamboo Fibre Reinforced Polyester Composites. Materials and Design (2010). , 31, 4147-54.

[36] Hongyan, C, Miao, M, & Ding, X. Influence of Moisture Absorption on the Interfacial Strength of Bamboo/ Vinyl Ester Composites. Composites: Part A (2009). , 40, 2013-9.

[37] Nirmal, U, & Jamilhashim, K. O. Low. Adhesive Wear and Frictional Performance of Bamboo Fibres Reinforced Epoxy Composite. Tribology International (2012). , 47, 122-33.

[38] Tong, J, Arnell, R. D, & Ren, L. Q. Dry Sliding Wear Behaviour of Bamboo. Journal Wear. (1998). , 1998(221), 37-46.

[39] Corradi, S. Composite Boat Hulls with Bamboo Natural Fibers. (2003).

[40] Lsmail, H, Edyham, M. R, & Wirjosentono, B. Dynamic Properties And Swelling Behaviour of Bamboo Filled Natural Rubber Composites : The Effect of Bonding Agent Iranian Polymer Journal. (2001). , 10(6), 377-83.

[41] Ismail, H, Shuhelmy, S, & Edyham, M. R. The Effects of A Silane Coupling Agent on Curing Characteristic and Mechanical Properties of Bamboo Fibre Filled Natural Rubber Composites. European Polymer Journal. (2002). , 38, 39-47.

[42] Visakh, P. M, Kristiina, O, & Mathew, A. P. Crosslinked Natural Rubber Nanocomposites Reinforced with Cellulose Whiskers Isolated From Bamboo Waste: Processing And Mechanical/Thermal Properties. Composites: Part A (2012). , 43, 735-41.

[43] Abdul Khalil HP.S., P. Firoozian, I.O. Bakare, H.M. Akil, and A.M. Noor. Exploring Biomass Based Carbon Black as Filler in Epoxy Composites: Flexural and Thermal Properties. Materials & Design. (2010). , 31(7), 3419-25.

[44] Abdul Khalil HP.S., Noriman N.Z., Ahmad M.N., Nik Fuaad N.A., Rathnam M.M. The Effect of Biological Studies of Polyester Composites Filled Carbon Black and Ac-

tivated Carbon from bamboo (Gigantochloa Scortechinii). Polymer Composites. (2007)., 28(1), 6-14.

[45] Rahim, S, & Narayan, S. Bamboo and Wood Fibre Cement Composites for Sustainable Infrastructure Regeneration. Science & Material Journal. (2006)., 41, 6917-24.

[46] Araujo, P. C. D, Arruda, L. M, Menezzi, C. H. S. D, Teixeira, D. E, & Souza, M. R. Lignocellulosic Composites from Braziian Giant Bamboo (Guadua Magna). Part 2: Properties of Cement and Gypsum Bonded Particleboards. Ciencia Y Technologia. (2011).

[47] Zheng, K, & Chen, X. Potential of Bamboo-Based Panels Serving As Prefabricated Construction Materials (2009).

[48] Mahzuz, H. M. A, Mushtaq, A, Ashrafuzzaman, M, Rejaul, K, & Raju, A. Performance Evaluation of Bamboo With Morter and Concrete. Journal of Engineering and Technology Research (2011)., 3(12), 342-50.

[49] Ghavami, K. Bamboo As Reinforcement In Structural Concrete Elements. Cement & Concrete Composites. (2005).

[50] Suhaily, S. S, Jawaid, M, Abdul, H. P. S, & Khalil, A. R. Mohamed, and F. Ibrahim. A Review of Oil Palm Biocomposites for Furniture Design and Applications: Potential and Challanges. BioResourcescom. (2012). June 2012;, 7(3), 4400-23.

[51] Zuo, T, Wang, T, & Nie, Z. Ecomaterials Research in China. Materials & Design. (2001)., 22, 107-10.

[52] Terai, M, & Minami, K. Fracture Behavior and Mechanical Properties of Bamboo Reinforced Concrete Members. Procedia Engineering. (2011)., 10(0), 2967-72.

[53] Vogtländer, J, Pablo, V. D. L, & Han, B. The Sustainability of Bamboo Products for Local and Western European Applications. LCAs and Land-Use. Journal of Cleaner Production. (2010)., 18(13), 1260-9.

[54] Larsson-brelid, P, & Mepw, M. Westin, and Roger Rowell. Ecobuild A Center for Development of Fully Biobased Material Systems and Furniture Applications. Molecular Crystals and Liquid Crystals. (2010).

[55] Petel, A, Picard, L, Imberty, F, & Raoul, J. Design of An Innovative Road Bridge with Advanced Steel and Concrete. Structural Engineering International: Journal of the International Association for Bridge and Structural Engineering (IABSE). (2010)., 20(2), 166-8.

[56] Aitken, S. C, & An, L. Figured Worlds: Environmental Complexity and Affective Ecologies in Fanjingshan, China. Ecological Modelling. (2012)., 229(0), 5-15.

[57] Proyuth, L, Pillot, D, Lamballe, P, & De Neergaard, A. Evaluation of Bamboo As An Alternative Cropping Strategy in the Northern Central Upland of Vietnam: Above-Ground Carbon Fixing Capacity, Accumulation of Soil Organic Carbon, and Socio-

Economic Aspects. Agriculture, Ecosystems & Environment. (2012). , 149(0), 80-90.

[58] Tae, S, & Shin, S. Current Work and Future Trends for Sustainable Buildings in South Korea. Renewable and Sustainable Energy Reviews. (2009). , 13(8), 1910-21.

[59] Kramer, K. L. Chapter 5- Usable and Sustainable. User Experience in the Age of Sustainability. Boston: Morgan Kaufmann; (2012). , 151-191.

[60] Kubba, S. Chapter 10- Economics of Green Design. LEED Practices, Certification, and Accreditation Handbook. Boston: Butterworth-Heinemann; (2010). , 379-415.

[61] Driver, A, Peralta, C, & Moultrie, J. Exploring How Industrial Designers Can Contribute to Scientific Research. International Journal of Design. (2011). December 20, 2010;, 5(1), 17-28.

[62] Kubba, S. Chapter 1- "Green" and "Sustainability" Defined. Green Construction Project Management and Cost Oversight. Boston: Architectural Press; (2010). , 1-27.

[63] Kubba, S. Chapter 6- Choosing Materials and Products. Green Construction Project Management and Cost Oversight. Boston: Architectural Press; (2010). , 221-266.

[64] Saville, S. Design Secrets: Furniture. 50 Real-Life Projects Uncovered. United States of America: Rockport Publishers; (2008).

[65] Sovacool, B. K, & Valentine, S. V. Bending Bamboo: Restructuring Rural Electrification in Sarawak, Malaysia. Energy for Sustainable Development. (2011). , 15(3), 240-53.

[66] Schroder, S. Guadua Bamboo Costa Rica (2012). cited 2012]; Available from: www.guaduabamboo.com.

[67] Dalcacio, R, & Wiedemann, E. J. Product Design in the Sustainable Era. Germany: TASCHEN; (2010).

[68] KlineCompany. Natural Fiber Composite Market Report. New Jersey (2004).

[69] AnonBamboo Strip Plywood for Automobile Carriage Base Plates: The Forestry Trade Standard of the People's Republic of China: LY , 1055-911992.

[70] Pawlak, J. J. A Sustainable Economy. Bioresource Technology. (2008).

[71] Khairun, A. U. M, Paridah, T. M, Hamdan, H, Sapuan, S. M, & Shuhaimi, B. E. Modification of Plybamboo Through Resin Impregnation. Research on Natural Fibre Reinforced Polymer Composites. (2009).

[72] Xiao, Y, & Ma, J. Fire Simulation Test and Analysis of Laminated Bamboo Frame Building. Construction and Building Materials. (2012). , 34(0), 257-66.

[73] AdnanNormiadilah, Othman, and Noriah. The Relationship Between Plants and The Malay Culture. Procedia- Social and Behavioral Sciences. (2012). , 42(0), 231-41.

[74] Tan, T, Rahbar, N, Allameh, S. M, Kwofie, S, Dissmore, D, & Ghavami, K. Mechanical Properties of Functionally Graded Hierarchical Bamboo Structures. Acta Biomaterialia. (2011). , 7(10), 3796-803.

[75] Hpp, A. Projects/ Transportation/ Industry. Germany (2012). updated 2012; cited 2012 20 October 2012]; Available from: www.hpp.com.

[76] Adams, C. Bamboo Architecture and Construction with Oscar Hidalgo. Designer/ Builder Magazine. (1998). , 1998, 471-4549.

[77] Zhou, A, Huang, D, Li, H, & Su, Y. Hybrid Approach to Determine the Mechanical Parameters of Fibers and Matrixes of Bamboo. Construction and Building Materials. (2012). , 35(0), 191-6.

[78] Anwar, U. M. K, Hiziroglu, S, Hamdan, H, & Abd, M. Latif. Effect of Outdoor Exposure on Some Properties of Resin-Treated Plybamboo. Industrial Crops and Products. (2011). , 33(1), 140-5.

[79] BiswasDaisy, S. Kanti Bose, and M. Mozaffar Hossain. Physical and Mechanical Properties of Urea Formaldehyde-Bonded Particleboard Made From Bamboo Waste. International Journal of Adhesion and Adhesives. (2011). , 31(2), 84-7.

[80] Richard, R. Barajas Madrid International Airport (2005).

[81] Bovea, M. D, & Vidal, R. Increasing Product Value By Intergrating Environment Impact, Cost And Customer Valuation. Resources Conservating & Recycling. (2004).

[82] Zuo, H, Hope, T, Jones, M, & Castle, P. Sensory Interaction with Materials. In: McDonagh D., Hekkert P., J.V. Erp, D. Gyi, editors. Design and Emotion-The Experience of Everyday Things. London: Taylor & Francis Group; (2004). , 223-227.

[83] Holbery, J. Houston, and Dan. Natural Fiber Reinforced Polymer Composites in Automotive Applications. Journal of the Minerals, Metals and Materials Society. (2006). , 58(11), 80-6.

[84] Davoodi, M. M, Sapuan, S. M, Ahmad, D, Aidy, A, Khalina, A, & Jonoobi, M. Concept Selection of Car Bumper Beam with Developed Hybrid Bio-Composite Material. Materials and Design. (2011). , 32(10), 4857-65.

[85] Proemper, E. New Automotive International from Natural Fiber Materials. International AVK-TV Conference for Reinforced Plastics and Thermoset Molding Compounds; Sept Germany (2004). , 28-29.

[86] Kenneth, C, & Albercht, B. Phoenix Bamboo Concept Car. (2010). cited 2012 30 August 2012]; Available from: http://Inhabitat.com.

[87] Bledzki, A. K, Faruk, O, & Sperber, V. E. Cars from Bio-Finers. Macromulecular Materials and Engineering. (2005). , 291, 449-57.

[88] Nishimura, T. Car Components. (2004). Available from:www.bc.bangor.ac.uk/suscomp/assets/pdf/carcomponents.pdf.

[89] Riley, N. The Elements of Design. London: Octopus Publisher Group Ltd.; (2003).

[90] Lugt, P. V. D, Voglander, J. G, Van Der Vegte, J. H, & Brezet, J. C. editors. Life Cycle Assessment and Carbon Sequestration; The Environmental Impact of Industrial Bamboo Products. IXth World Bamboo Congress; (2012). Netherlands.

[91] Ljungberg, L. Y. Materials Selection and Design for Development of Sustainable Products. Materials & Design. (2007). , 28(2007), 466-79.

[92] Lane, J. P, & Flagg, J. L. Translating Three of Knowledge Discovery, Invention, and Innovation (2010). cited 2011 13 August 2012]; 5(9).

[93] Kar, S. P, & Jacobson, M. G. NTFP Income Contribution to Household Economy and Related Socio-Economic Factors: Lessons from Bangladesh. Forest Policy and Economics. (2012). , 14(1), 136-42.

[94] Steffen, D. Product Language as a Reflection on Technical Innovation and Socia-Cultural Change. Design Semantics of Innovation. Germany: Department of Art and Design History, Bergische Universitat Wuppertal, Germany.; (2007). , 7.

Chapter 20

Investigation of Subterranean Termites Nest Material Composition, Structure and Properties

Petr Ptáček, Jiří Brandštetr, František Šoukal and Tomáš Opravil

Additional information is available at the end of the chapter

http://dx.doi.org/10.5772/55145

1. Introduction

Termites (Isoptera[1]) are colourless or white insect sometimes called "white ants[2]" but they are intimate relatives of cockroaches. The organization of termite community definitely belongs to the most complicated in the nature hence termites are described as the eusocial[3] insect. Termites are distributed all over the world from the 47° Northern latitude to the 47° Southern latitude, but they are extremely abundant in the tropical rainforest[4]. Until today there are recognized about 3 000 termite species in the 9 families but two of them became extinct (Fig. 1).

Only species of the genera *Archotermopsis, Hodotermopsis, Zootermopsis* and *Reticulitermes* can be found also in the temperature regions. Only two species[5] are native of the temperature region of Europe. Besides the known kinds of termites an existence of other more than 1000 species is supposed [1-6].

The foundations of thermite taxonomy were established by Holmgren [7-9]. The following studies provided by Snyder [10], Grasse [11] and Emerson [12] define seven families of termites which are today generally accepted. Many of the conventional phylogenetic reconstructions

1 Isoptera is Greek name that means "equal wings" referring to the similar shape and size of four wings.

2 This name is widespread mainly in Australia. However ants belong to the order hymenoptera.

3 Eusocial behaviour is the highest level of social organization with specialized individuals (the developed system of castea). The expression is consisting of Greek words "eu" (real) and "social". Besides Isoptera (termites), the social Hymenoptera insect such as ants, bees and wasps is another example of eusocial insect but the social system of termites is the oldest known.

4 Each square meter of tropical rainforest may contain hundreds of termite individuals.

5*Reticulitermes lucifugus* and *Reticulitermes flavipes*, the cold climate restricts the dissemination of termites into Europe to the Atlantic coast of France and Hamburg in Germany, respectively.

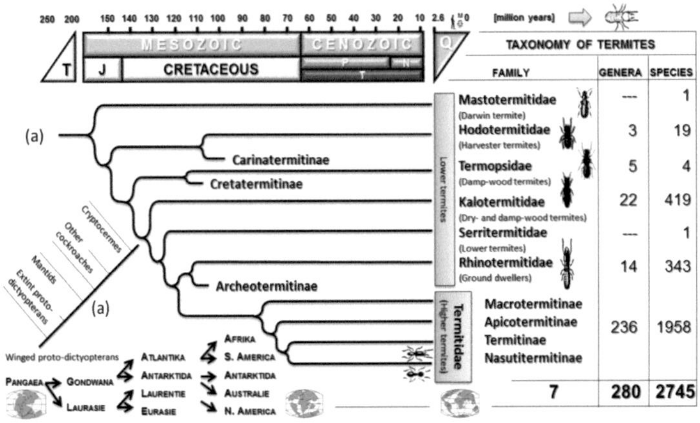

Figure 1. Taxonomy and phylogeny of termites: Triassic (***C***), Jurassic (***J***), Palaeogene (***P***), Neogene (***N***), Tertiary (***T***), Quaternary period (***Q***) and appearing of modern human (***M***).

are based only on the limited subset of characters [13]. For example the hypothesis of Ahmad [14] is based on the worker-imago mandibles, the classifications of Johnson [15] and Noirot [16] use the worker gut and the classification of DeSalle [17] is based on DNA. Others looked at the relationships within as well as between particular families [16, 18, 19]. The hypothesis about morphological phylogenetics of termites is given in the Fig. 1.

The Hodotermitidae are generally considered as the basal group for all other families, except the Mastotermitidae which are the most primitive and closest to the cockroaches. The positions of Kalotermitidae, Termopsidae, Rhinotermitidae and Serritermitidae are less certain. Termopsids living in small colonies inside decaying wood are assumed the most primitive with regard to their caste differentiation and eusociality in general[6]. The family of Termitidae (higher termites) represents about two thirds of all the described termite species [6, 13, 20, 21].

1.1. Evolution of termites and evolution of eusocial behaviour of termites

Molecular [22] and fossil evidence [23] suggests that during late Jurassic termites evolved from subsocial wood-feeding Cryptocercus cockroaches after which a period of rapid radiation and spreading followed [24]. This foundation is in agreement with previous hypothesis of termite origin [25] that was established on the similarity in nesting and feeding habitats as well as the presence of cellulolytic protozoa in their hindguts. The fossil records of termites are known from early Cretaceous deposit [24, 26] although structures from late Triassic and lower Jurassic were described as fossilized termite nests. Those support the hypothesis which explains the worldwide distribution of the social insect including termites by their early Mesozoic origin (Fig. 1), prior to breakup of the Pangaea [27-28].

Comparative studies of the social biology of the spectrum of solitary through eusocial species of bees and wasps elucidated the evolution of complex societies but parallel research on

6 For example relictual Himalayan termite *Archotermopsis wroughtoni*.

http://dx.doi.org/10.5772/55145

termites or ants is impossible hence all living species are eusocial. The two ways are generally accepted for origin of social groups:

- **Subsocial route** when social groups originate from familial units initially composed of parents and offspring;
- **Semisocial route** when social groups are formed by association of individuals of the same or different generation.

Therefore workers and soldiers of termites are specialized juveniles[7] and there is no evidence for exchange of reproductives between different established colonies, termites eusocial behaviour is probably not formed via semisocial route [20].

1.2. Termite social system

Termite eusocial society consists of specialized casts including reproductives[8], sterile workers, soldiers, and immature individuals. The differentiation of individual caste phenotypes is affected by the complex interaction of extrinsic and intrinsic factors. Although experimental data suggest a genetic component to termite caste differentiation, environmental and social condition and are still considered the major trigger that initiates the development of different caste phenotypes [30].

The two types of reproductive species are recognised – primary and supplementary. The primary reproductives are king, queen and fully developed winged adults. Their role is in production of eggs and distribution by colonizing flights. The queen lives up to 25 years and lays about 3000 yellowish-white eggs per day. The eggs are hatch after 50–60 days of incubation. Termite eggs are genetically identical hence the differentiation into each of the different castes is controlled by intrinsic and extrinsic factors that modulate caste-regulatory and developmental gene expression [30, 31]. The lifecycle of termite colony is schematically drawn on Fig. 2(a).

Queen also produces pheromones[9] which have important functions in social regulation of termite society. These substances influence the behaviour and physiology of colony members, for example maintain sterility of colony members or the induction of special worker care. Primer pheromones which cause a psychological change in the receiver are considered the type of extrinsic caste-regulatory factor [30, 32].

The sterile castes, the workers (2 – 15 mm) and the soldiers (up to 20 mm), are wingless and usually lack eyes. Workers (Fig. 2(b)) mature in a year and live up to 3–5 years. Soldiers also mature within a year and live up to 5 years. The colony reaches its maximum size in approximately 4–5 years and it may include 60 000 to 200 000 workers. Workers construct the nest

---

7 Label not fully grown or developed young individuals such as larva or another form before the adult stage is reached.

8 Only the limited numbers of individuals are fertile and fecund in eusocial society while the others are sterile or have reduced reproductive capacity.

9 Pheromones are volatile chemicals (chemical signalling molecules) released by one individual which can have behavioural or physiological effects on other individuals. Using pheromones for communication is common phenomenon for insect.

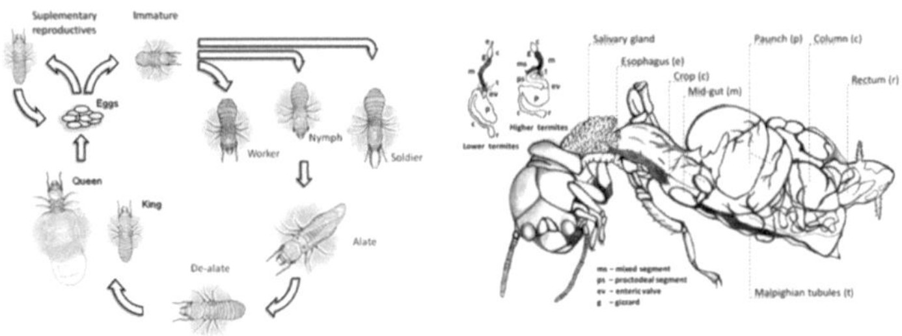

Figure 2. The life cycle of termite colony (b) and segments of a typical termite worker gut (b).

(Fig. 3), distinctive shelter tubes and collect food to feed the young and other members of the colony. Soldier termites guard and defend the colony [3, 31, 35].

The role of the workers in termite defence should not be underestimated. First, they are always the most abundant caste and are fully responsible for the construction and repair of the passive defensive structures. Second, they indeed actively participate in defence [33, 34]. The research on aggression of the termite worker indicates the flexible mechanisms of defence of colony which depend on the social context. The workers show only limited aggression against intruders (ants) if the soldiers are present but they attack the enemy if soldier isn't on the site. [35, 36]. The soldier caste is fully devoted to defensive activities. An impressive variety of defensive adaptations including heavily sclerotized head and the mandibles can be found [37, 38]. As a complement to the mechanical weapons, the chemical defence occurs in soldiers [30, 39, 40].

1.3. Termite nest

The nest built by the termite society refers to the complexity of their social organization, diet, biology and environmental factors on the site but there is the large scale of variations. Some of the termite species create only simple galleries in wood or ground while others build the nest of large dimensions and complexity [5].

From this point of view, the two species can be recognised:

- The wood dwellers;
- The ground dwellers (subterranean termites).

The wood dwelling and eating termites live in the fresh (damp-wood termites) or dead (dry-wood termites) wood in which they also build the nest. The connection of nest with ground is not required. On the contrary the subterranean termites (ground dwellers) live in soil and connection with the ground is necessary for the normal life and breeding of the colony.

The nests of subterranean termites are certainly the most admired natural structures which can reach 6 m of height and 4 m in the diameter of the base but towers built by some of the African Macrotermes species can be even 9 m high. The nest consists of galleries, chambers of different size and shape, system of tunnels and ventilation shafts. The nest of subterranean

http://dx.doi.org/10.5772/55145

termites is supported by extend underground structure that ensures proper temperature and humidity. The tunnels built by termites can be tens of meters long and often reach the level of ground water or the bedrock. The scheme of nest of subterranean termites is shown in Fig. 3.

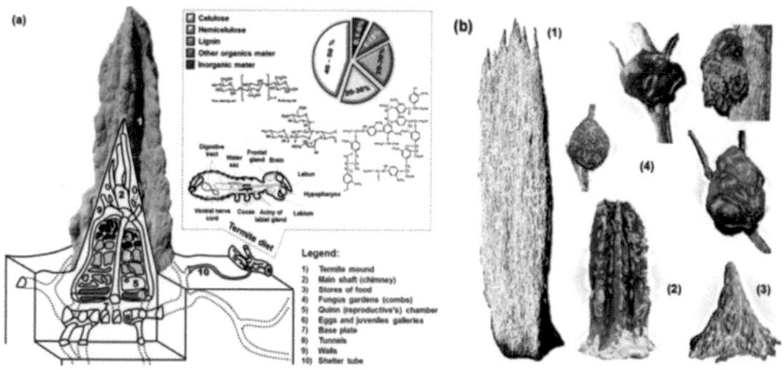

Figure 3. Simplified structure of termite nest with basic building block (a) and examples of ground-dweller (1, 2, 3) and wood-dweller nests (4) of different shape (b).

1.4. The diet of termites

The diet of termites as a group is quite diverse, but it is basically rich in cellulose, hemicellulose and lignin or lignin derivatives (Fig. 3). Termites digest lignocellulosic compounds due to the cooperation of their own enzymes and exogenous enzymes from microorganisms. According to the diet termites are divided into feeding groups i.e. wood feeding, dry-wood feeding, wood and litter feeding, soil feeding, fungus growing and grass feeding termites. The termite gut (Fig. 2(b)) provides a very distinct ecological environment which accommodates and promotes very specialized cellulolytic and hemicellulolytic microorganisms. The association between certain xylophagous termites and their hindgut protozoa is the fascinating and frequently cited example of nutritional symbiosis [5, 13].

On the other hand, Slaytor [41] found cellulose activity in the salivary glands and predominantly in the foregut and midgut of termites (Fig. 2(b)) and cockroaches. These parts are normal site of secretion of digestive enzymes and are devoid or have very limited content of microorganism. Therefore there is no evidence that bacteria are involved in cellulose digestion of termites.

The lower termites produce numerous endogenous enzymes (such as beta-glucosidases, exoglucanases, endoglucanases, chitinases) secreted in particular by the salivary glands and foregut. The higher termite species which do not possess symbiotic microorganism in the gut no longer feed wood. They prefer wood that is partially decayed by fungi or cultivate fungi in elaborate gardens for use as a nutrient source (fungus-growing termites). Fungus-growing termites which are abundant in the African and Asian tropics live in a unique mutualism with basidiomycete fungi of the genus *Termitomyces*.

The symbiotic fungi grow on a special culture within the nest maintained by the termites and called "fungus combs or fungus garden" (Fig. 3) which are made from partly digested foraged plant litter passing rapidly through the termite's gut. The resulting faecal pellets are pressed together to make a comb-like matrix. As the comb matures, mycelium develops and produces conidial nodules, which together with older senescent comb are consumed by workers [5, 42-45].

Since the cellulose is encrusted with the lignin (lignocellulose) in the higher plants, which prevents its digestion, the lignin must be disrupted for the enzyme accessibility. Grassé and Noirot [46] proposed the "lignin degradation hypothesis" that the symbiotic fungi have the ability to degrade lignin, which makes cellulose more easily attacked by the termites' own cellulase. This hypothesis was confirmed by the foundation of Hoyodo at al [43].

1.4.1. Degradation of cellulose

Termites utilize wood as a source of food[10] and feeding experiments with wood show that cellulose and hemicellulose are subjected to degradation with efficiency from 59 to 99 %. Termites consume more than 90% of dry wood in some arid tropical areas. Cellulose is fermented anaerobically by the protozoa or directly by the termite own cellulolytic enzymes to acetate, CO_2 and H_2. Acetate is subsequently absorbed by the termites and used as their major oxidizable energy source [5, 41].

The hydrolysis of cellulose is a complex process that requires the participation of three cellulolytic enzymes at least [5, 47]:

- **Endo-β-(1, 4)-glucanase** (EG, endoglucanase or carboxymethyl cellulase) enzyme (EC 3.2.1.4) that hydrolyses β-1, 4-bonds of cellulose chains;
- **Exo-β-(1, 4)-cellobiohydrolase** enzyme (CBH, EC 3.2.1.91) that releases cellobiose from the non-reducing ends of cellulose;
- **β-glucosidase** enzyme (EC 3.2.1.21) that hydrolyses cellobiose or longer cellulose chains and releases glucose.

1.4.2. Degradation of hemicelluse and other polysacharides

Hemicellulose is digested by termites with high efficiency ranging from 49 to 78 %. Hence Xylans as polysaccharides made from units of xylose are the basic building blocks of hemicellulose. The degradation of beta-1, 4-xylan into xylose by xylanase breaks down hemicellulose chain. The process requires catalysis by enzyme endo-1, 4-β-xylanase (EC 3.2.1.8). In addition, termites can possess different kinds of carbohydrases, such as sucrase, maltase, trehalase and raffinase. In *Neotermes bosei* the chitinase activity could be found, which can be produced by the microbial symbionts. Chitin digestion plays a role during the cannibalism at the times of food shortage [5, 45].

1.4.3. Degradation of lignin

Lignin is a phenolic polymer that makes up 15 to 36% of wood biomass. It serves several functions in the extracellular matrix of plants [48]:

10 Digestion of cellulose is also found in other insect groups such as the Thysanura, Orthoptera, Coleoptera and Hymenoptera.

- Lignin gives the cell walls mechanical support;
- Lignin serves as a barrier against microbial attack;
- Lignin acts as a water impermeable seal for the xylem vessels of the plant vasculature.

Therefore the environment in the termite gut is predominantly anaerobic and the natural anaerobic mechanism of degradation of lignin is not known, the mechanism of degradation is still uncertain [5]. White rot fungi can degrade lignin by oxidative enzymes faster than any other organisms and they are responsible for most of the lignin decomposition in the nature. The degradation of lignin by ligninase (the group of oxidative extracellular fungal enzymes) occurs aerobically inside the fungal cells. The most studied lignin-modifying enzymes of white-rot fungi are [49]:

- 1, 2-bis(3, 4-dimethoxyphenyl)propane-1, 3-diol:hydrogen-peroxide oxidoreductase (LiP, EC 1.11.1.14) often termed as ligninase peroxidase or ligninase I;
- Mn(II):hydrogen-peroxide oxidoreductase (MnP, EC 1.11.1.13) also known as manganese peroxidase or peroxidase-M2.
- Laccases (EC 1.10.3.2), i.e. group of copper-containing oxidase enzymes.

As mentioned above the "lignin degradation hypothesis" assumes that the symbiotic fungi have the ability to degrade lignin, which enables the digestion of the cellulose by the termites' own cellulase [43, 46]. The hypothesis is also supported by the fact that no microorganism breaking down the lignin has been observed in the termite gut yet as well as no specific site for degradation of lignin is known [5].

1.5. The role of termites in the ecosystem

Termites are the dominant invertebrates in the tropical ecosystem where they can make up to 95% of the soil insect biomass. Therefore the termites play the important and irreplaceable role in the ecosystem. The main effect of termites in ecosystem is in mineralization of biomass and humification (enrichment) of soils [42, 50, 51].

In general termites are considered the most important soil ecosystem engineers with influence on [52, 53]:

- Distribution of soil organic matter (SOM);
- Hydraulic properties of soil;
- Erosion;
- Microbial diversity;
- Vegetation diversity and growth.

The termite nests and biomass are characterized as "sinks" as they withdraw large quantities of litter and soil organic matter from the "normal" decomposition pathways. Except for the case of fungus-growing termites, the SOM, Ca^{2+}, Mg^{2+} and K^{+} content is usually higher in termite mounds and in the surrounding soil than in the soil unaffected by termites. The consumption of organic matter by termites is utilized in the production of termite biomass and in the building of nest-constructions. The return input of organic matter and mineral nutrients to the soil environment occurs via faeces, salivary secretions, corpses and predators. Mortality, particularly from ant predation, and mound erosion are the important contributors to the turnover and redistribution of the organic matter and mineral nutrients in the ecosystem.

The role of termites in water infiltration and runoff is closely related to the importance, structure and arrangement of their subterranean biogenic structures, which comprise foraging and storage galleries, feeding chambers and communication channels, in addition to chambered colony centres (in whole underground species) and the foundations of epigeal mounds. The termite biostructures act therefore as a network of horizontal and vertical macropores influencing bulk density, aeration, water infiltration and runoff, then capturing overland flow and determining the hydrological characteristics of watersheds. The termite mounds can also influence water infiltration and nutrient leaching, although this effect likely depends on the age and whether the mounds are inhabited by termites or abandoned.

The erosion of termite mound would take ten years. The process is very rapid in the first years after the abandonment but then the rate subsides gradually until it is more or less negligible due to the stabilization effect of vegetation growth on the remains of the structure. Temporal scale is important: some termite mounds persist in the landscape for periods of a decade or more, and their individual influence on soil profile development during that time may be negligible.

However, over 100s or 1000s of years, the importance of termites in the turnovers of both mineral materials and organic matter in the landscape and in the whole soil profiles may be much greater, such that long-term pedogenesis is mediated by their activity. The abandoned termite nest does not lack purpose hence it can be occupied by other animal, insect or it can provide substrate for the seeds of trees and plants.

Termites modulate the availability of resources for other species, such as soil macro- and microfauna and microorganisms, and create habitats which can be used by a remarkable number of organisms for nesting, roosting or accommodating stages of their life cycles.

The concentration of SOM and nutrients in termite nest structures exaggerates resource patchiness and the structures can be considered as "fertile islands" which are beneficial resources for plants [52, 54-57].

1.6. Termite control

Therefore the termites are the excellent decomposers of cellulose; they become a serious menace to both the plants and the wood structures. Termites cause significant losses to the annual and perennial crops and damage the wooden components in buildings, especially in the semi-arid and subhumid tropics [3]. The methods developed in order to control termites are shown in Fig.4.

The physical barriers are the most popular methods used against subterranean termites to prevent wooden structures from their attack. The barriers are of two types - toxic and nontoxic.

http://dx.doi.org/10.5772/55145

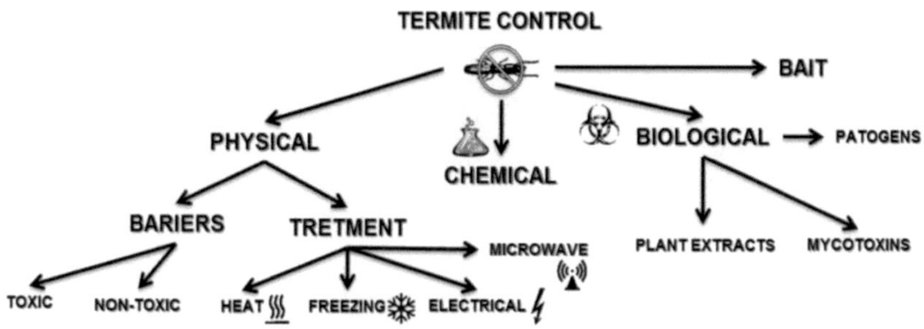

Figure 4. Method developed for termite control.

Toxic physical barriers include the use of chemical termiticides in the soil around the structure. Nontoxic physical barriers are substances (e.g., sand or gravel aggregates, metal mesh or sheeting) which exclude termites because they are impenetrable, thus act as physical/mechanical barriers preventing termite penetration and damage to the building. Other physical methods include heat, freezing, electricity, and microwaves. Chemical control is a successful method of preventing termite attack, but the effects of these chemicals are of concern as they create problems to our health and the environment.

Baiting is the method of termite pest control that is more environmentally friendly as it uses very small amounts of insect toxicants. In bait technology the termite colonies can be eliminated by the use of toxic or nontoxic baits. Bait is a wood or a cellulose matrix favoured by termites that is impregnated with a slow-acting toxic chemical or nontoxic substance such as fungal spores, mycelium (that grows through termite cuticle and utilizes entire termite body) and infective stages of nematodes (corynebacterium which produces toxins lethal to termites). The termite workers are exposed to the lethal dose of desirable food bait inside the bait stations.

Plant-derived natural products and biological control agents are promising replacements of chemical methods for termite control. These techniques utilize insecticidal activity of essential oils, extracts and resins, parasites and various pathogens (bacteria and fungi) [3]. Although many studies are overly optimistic the more than 50 years of the development of biological control of termites using pathogens demonstrated that this technology is not successful [58]. The high carbon dioxide and naphthalene contents and secretions of termites may also inhibit the growth of pathogens in termite nests or galleries, and therefore the termite control with pathogens stays unsuccessful under the field conditions [59].

2. Experimental

2.1. Sample and sample treatment

The nest of subterranean termites (ground dwellers) from Gambia was used for investigation of the material composition and properties in this work (Fig.5).

The four portions (samples) were prepared from termite nest material:

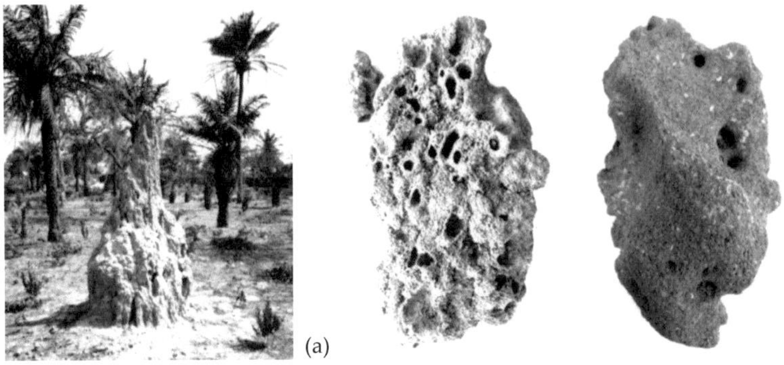

(a) (b)

Figure 5. Termite mound (a) and sample of termite nests analysed in this work (b).

- **Sample A**: untreated material;
- **Sample B**: finely dispersed (milled) material of the nest;
- **Sample C**: sand grains rich fraction ≥63 μm;
- **Sample D**: clay minerals rich fraction ≤63 μm.

The sand grains and clay minerals rich fractions were prepared via careful dispersion of the sample in the porcelain dish using a force insufficient to crush the grains of sand. The ground material was divided into both fractions by sieving.

2.2. Analysis of samples

The composition and properties of the material were investigated by scanning electron microscopy (SEM), x-ray diffraction analysis (XRD), infrared spectroscopy (IR), thermal analysis (TA) techniques including simultaneous termogravimetry and differential thermal analysis (TG-DTA), effluent gas analysis (EGA) and the high-temperature x-ray diffraction analysis (HT-XRD). other tests and analyses (Table 1) including the assessment of humidity, loss on ignition, pH, etc., were performed according to the standards.

2.2.1. Scanning electron microscopy

The scanning electron microscopy was performed with the microscope JEOL JSM-7600F. The samples were coated by the gold layer. The energy dispersive x-ray fluorescence spectroscopy (EDX) was used for the assessment of the sample composition.

2.2.2. X–ray diffraction analysis

The quantitative Rietveld phase analysis of phase composition was performed using x-ray diffractometer (PANanalytical Empyrean) with Cu(K_α) radiation at 40 kV and 40 mA. For the high temperature x-ray diffraction analysis (HT-XRD) the sample was inserted into the heating chamber (Anton-Paar HTK16) and heated on the platinum pane at the heating rate 10 °C•min^{-1} up to temperature 1250 °C.

2.2.3. *Infrared spectroscopy*

The infrared spectra were collected within the mid-IR region by means of the KBr pellets technique using FT-IR spectrometer iS10 (ThermoScientific) using the sample to KBr ratio of 1:100.

2.2.4. *Thermal analysis*

Thermal analysis was carried out using TG-DTA analyser Q600 (TA Instruments) connected to measuring cell (TGA/FT-IR Interface, Thermo Scientific) of FT-IR spectrometer Nicolet iS10 (Thermo Scientific) through heated capillary (200 °C).

2.2.5. *Particle size distribution*

The particle size distribution was determined by the sieve analysis and the particle size analyser Helos (Sympatec). The specific surface area was calculated from the obtained data.

3. Results and discussion

The basic properties of investigated sample from termite nest are listed in Table 1.

Feature		Value	Description
Humidity	[%]	0.5±0.1	Performed on sample B by drying to constant weight at temperature 105 °C.
Loss on ignition	[%]	3.1±0.1	Performed on the dry sample B. Burning of organic matter and dehydroxylation of clay.
Pour density	[$g \cdot cm^{-3}$]	2.4±0.1	Weight of granular materials that related to volume unit.
True density	[$g \cdot cm^{-3}$]	2.6 ±0.1	Volumetric technique. The value corresponds to both the main components (kaolinite and quartz)
pH		6.4±0.1	Sample D in the suspension with re-boiled (dissolved CO_2 free) distilled water.
Zeta potential	[mV]	−17±1	Electrokinetic potential of particles ≤ 63 μm (Sample D).
Conductivity	[$W(m\,K)^{-1}$]	0.7±0.1	Coefficient of thermal conductivity measured on material of Sample A (TCI, C-Therm).

Table 1. The basic properties of the sample of termite nest.

3.1. Granulometry

The particle size distribution was assessed by the sieve analysis for the rough grains of sand (Sample C) and by laser analyser for the fraction of particles ≤ 63 μm (Sample D).

3.1.1. Sample C

The results of the sieve analysis including histogram of log-normal particle size distribution and cumulative particle size distribution curve are shown in Fig. 6.

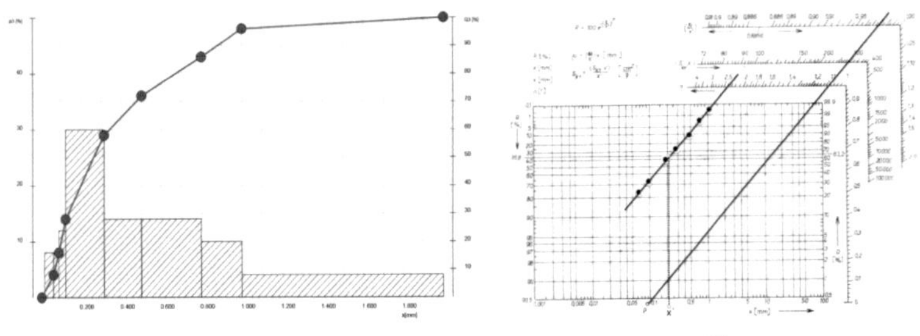

(a) (b)

Figure 6. The sieve analysis of the Sample C (a) and RRSB diagram (b)

Using RRSB diagram[11] the granulometry of the sample is determined by the value of exponent n = 1.08. The value is close to one and that corresponds to the rounded particles of termites applied sand (please referee to Fig. 5). The characteristic size of grain (**x**′) was assessed to 0.2 mm[12]. The specific surface area of 1450 $cm^2 \bullet g^{-1}$ is resulting from these results.

3.1.2. Sample D and B

The particle size distribution in the fraction ≥63 μm is shown in Fig.7 (a). The median of particle size is 22.9±0.3 μm. Combination of the results of both samples leads to Fig.7 (b) hence the material of sample B should be described as sandy clayish (in this case kaolinite) sand.

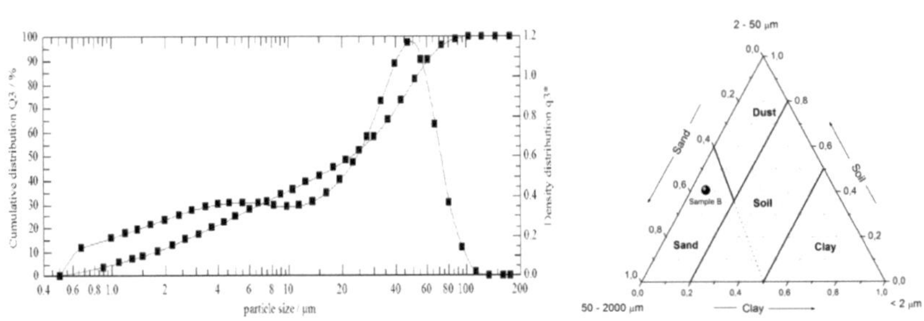

(a) (b)

Figure 7. The granulometry of the Sample D (a) and total granulometric composition (b).

11) The acronym for the Rosin-Rammler-Sperling-Benett diagram.

12) The characteristic value for sand is typically higher (~1.4). The x′ corresponds to R = 36.8 %.

3.2. Electron microscopy

The structure of the termite nest wall is shown in Fig.8. Termites build their nest from the sand grains and use the clays and faeces as the binding material. The parts of plants serve as the reinforcement. The detail shows that smaller grains fill the spaces between the larger and reinforcing plants material. The surface of sand grains is treated by layers of clay (Fig.9). It is obvious that the resulting composite structure of termite nest shows excellent mechanical properties which enable to build structures of several meters of height / structures several meters high.

Figure 8. Termite nest microphotography (Sample A).

While the structure inside of the wall is rough the surface of tunnels is polished by the mixture of clay (Fig.10) and faeces (Fig.11). The clay phase loses plasticity via evaporation of water and formation of new weak (van der Walls) bonds between lamellar particles (Fig.10) which leads to the formation of hard aggregates fixing the silica grains on their positions. It stands to reason that the silica particles as the phase of high Young modulus serve as the reinforcement as well as the filling material. The conceived particle size distribution of grains of "aggregate" (packing density[13] of ~92 %) and mechanical and saliva activation clay phase (mainly kaolinite) form very solid structure that is analogical to the ceramic green body.

Hence the structure of termite nest is tailored well, and it is possible that the role of organic matter (SOM) and faeces [52] often discussed in literature do not increase mechanical proper-

13 Defined as the ratio of pour and true density (Table 1).

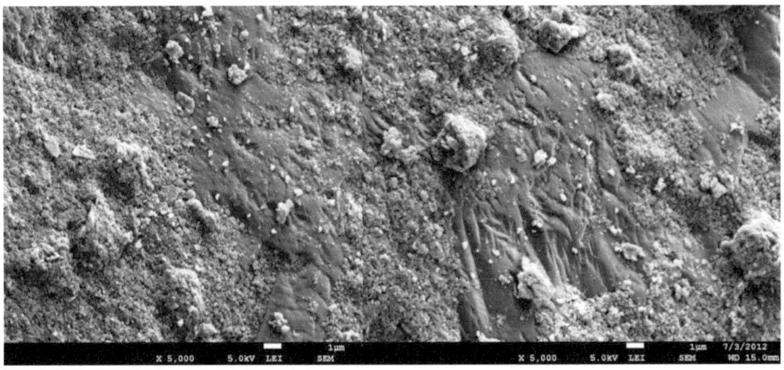

Figure 9. Microphotography of surface of clay layer treated sand grain (Sample C).

ties of material but more probably prevent spreading of pathogenic organism [2] and support fungi inside fungus garden [5] of fungus growing termites.

Figure 10. Clay from termite nest (Sample D).

It is supported by the foundation that the highest concentration of elements such as phosphorus, potassium and nitrogen coming from termite diet is located near the surface of tunnels. Hence the faeces are used for the treatment of surface (and to provide plasticity to shelter tubes of course) but the main part of the weight of mould bearing material does not contain these elements and content of SOM is in common level[14].

http://dx.doi.org/10.5772/55145

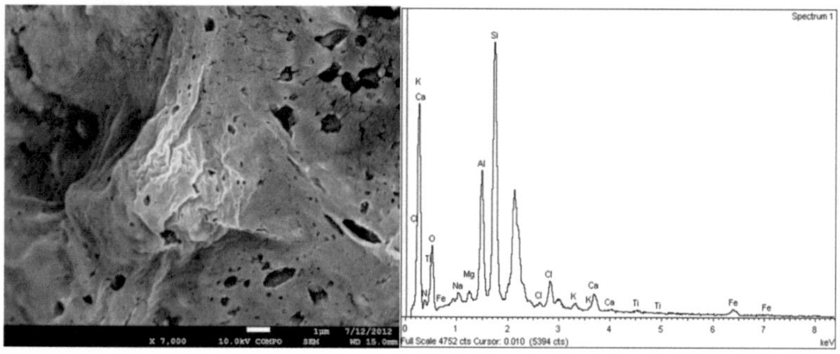

Figure 11. The faeces rich layer covering the surface of tunnel in termite nest.

Therefore the secret of ground dweller termite material engineering lies more probably in the well-tailored structure analogical to the dry ceramic body than in their diet or enzymes in their digestive tract.

Site	Si	Al	Mg	Fe	Ti	Ca	K	Na	N	Cl
Surface	46.5	18.6	1.6	6.9	1.4	6.3	2.0	1.6	12.7	2.5
Inside	67.1	24.4	0.3	4.6	1.1	0.2	0.7	0.7	0	0.9

Table 2. The composition of termite nest material inside and on the surface of wall of tunnel.

The fact that the faeces of herbivore animals increase the plasticity and reduce the drying shrinkage of clays has been known well. This foundation is used by many cultures over the world in the construction of clay houses from prehistoric until today[15]. There is the possibility that the role of faeces rich layer on the surface of galleries and tunnels may indirectly contribute to the strength as it slows down the rate of drying (reduce the drying shrinkage) and weathering. Considering the previous applied comparison to the dry ceramic body the layer can serve as the glaze on the surface but its thickness lower than 1 μm is not enough to reach a significant influence on the mechanical properties of the material.

3.3. X–ray diffraction analysis

The x-ray diffraction patterns of the sample B and D are shown in Fig. 8 (a) and (b), respectively. The kaolin and quartz were identified as the main constituents of the sample. Furthermore small amount of clay mineral halloysite was identified.

14 Various clays used by ceramic industry contain even higher amount of organic matter what can lead to failure known as "black core" which is formed if fired under improper forint regime ("firing curve").

15 For example clay pyramids known as "Temples of the sun" built by Mochica culture, in Europe were clay houses built from the Stone Age through Medieval to industrial revolution (in some countries even later), Massai houses, etc. Nowadays probably more than ½ of human population lives in clay houses.

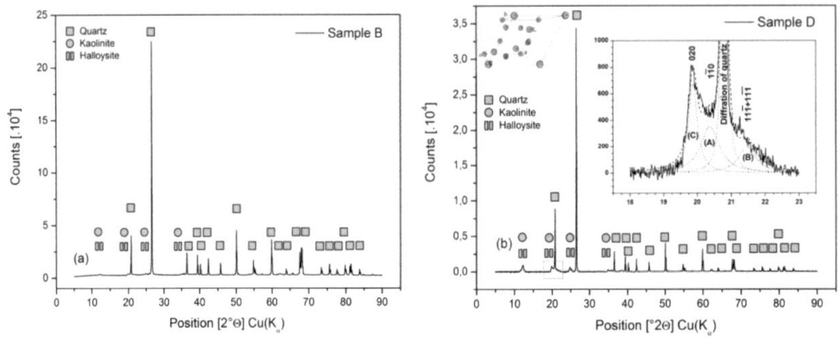

Figure 12. The XRD pattern of Sample B (a) and Sample D (b).

The structure refinement performed on the Rietveld analysis data shows slightly deformed triclinic kaolinite structure with lower parameter b and angle α. Table 1 shows the sample composition according to the results of quantitative Rietveld analysis.

Sample	Quartz [%]	Kaolinite [%]	Halloysite [%]
Sample B	97, 4	2, 6	---
Sample D	70.3	29.5	0, 2

Table 3. Phase composition of the thermite nest material.

In order to estimate the effect of termite treatment or activation of clays the structure order of the kaolinite was evaluated using weighed ratio intensity index[16] (WIRI) that is based on weighed intensity of (02, 11) diffractions and calculated as follows [60]:

$$AGFI = (Ia + Ib)/2Ic \qquad (1)$$

$$WIRI = 1 - \exp\left[\frac{-w_1^{-1}I(1\bar{1}0) + w_2^{-1}I(11\bar{1}) + w_3^{-1}I(1\bar{1}\bar{1})}{w_4^{-1}I(020)}\right] \qquad (2)$$

$$HI = 0.1 + 1.44\, WIRI \qquad (3)$$

where w is half-width of the diffraction line. The results of the XRD pattern fitting procedure are shown in Fig.12. The values of HI, AGFI and WIRI were calculated to be 0.40, 0.92 and 0.21 corresponding to the kaolinite of the poor level of structure order.

16 The WIRI index ranged from zero to one while for other often applied indexes including HI (Hinckley Index) and AGFI (Aparicio-Galán-Ferrell Index) the value higher than one can be reached.

The sequence of transformations of the Sample D during thermal treatment was investigated by high temperature X-ray diffraction analysis (Fig.9) up to temperature 1250 °C.

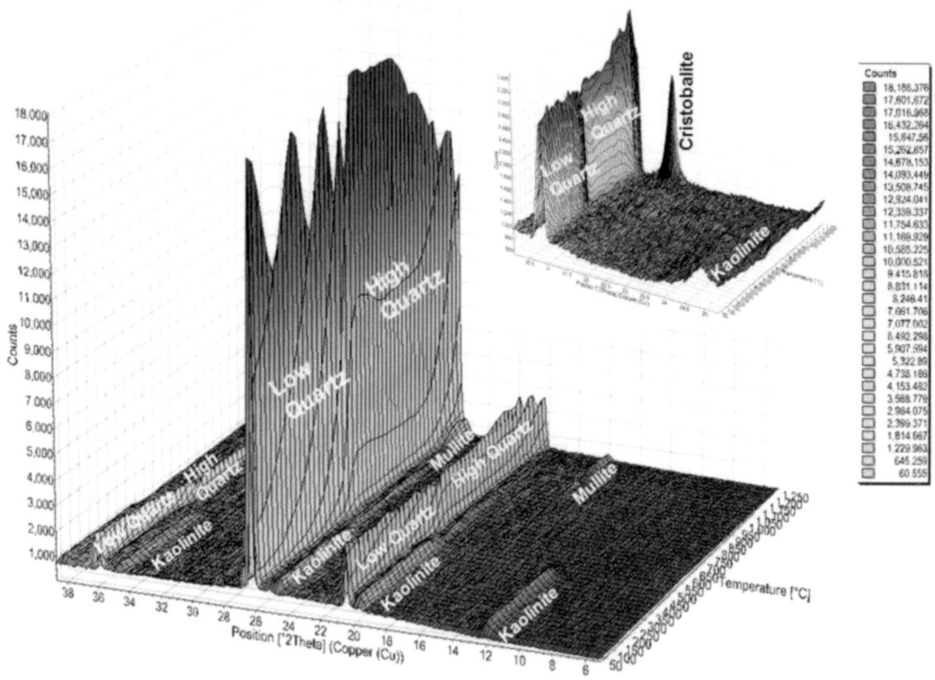

Figure 13. HT-XRD analysis of the Sample D heated to 1250 °C using heating rate 10 °C•min^{-1}.

The following process takes place in the sample:

- Dehydroxylation of kaolinite and formation of roentgen-amorphous metakaolinite (Eq.4) leads to disappearing of diffractions of kaolinite (K) from the HT-XRD plot.
- Transformation of low quartz to high temperature phase leads to the shift of diffraction to lower position. For example the most intensive diffraction line at 26.31 is shifted to 26.23 °2Θ.
- Formation of mullite (Eq.6).
- Formation of cristobalite (Eq.7) from the amorphous silica phase that is formed during the synthesis of cubic Al-Si spinel phase[17] and mullite as well.

3.4. Infrared spectroscopy

The infrared spectrum of the sample C and D is in Fig.14 (a) and (b), respectively. The results show the typical pattern of quartz and kaolinite.

17 While the process shows sharp exothermic effect on DTA curve (please see Fig.15 and discussion) the response in XRD diagram is low with uncertain interpretation in the literature. Some of the works bring evidence on the formation of mullite or γ-Al2O3 but the final consensus has not been reached yet.

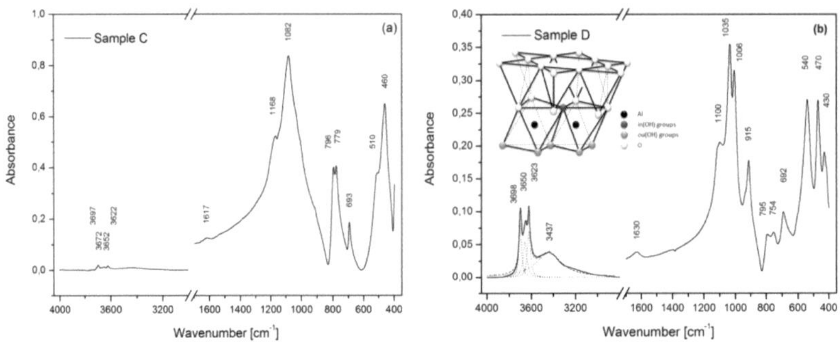

Figure 14. The infrared spectrum of the Sample C (a) and Sample D (b).

The adsorbed water shows broad band of stretching and deformation at 3440 - 3430 and 1630 – 1610 cm^{-1}, respectively. The interpretation of spectra is given in Table 4.

Sample C – Typical spectrum of quartz			Sample D – Typical spectrum of kaolinite		
Wavenumber	**Fragment**	**Note**	**Wavenumber**	**Fragment**	**Note**
3700 - 3620	AlO-H from kaolinite bound to the silica surface.		3698, 3672, 3650, 3623	AlO-H	ν_1(ou), ν_2(ou), ν_3(ou), ν_4(in)
1168	Si-O	LO part of ν_{as} mode	1100	Si-O_{ap}	ν(Si-O) of apical oxygen
1082		TO part of ν_{as} (TO_3) mode	1035, 1006	Si-O-Si	ν_{as}, ν_s in layer of tetrahedra
776, 779		Doublet of bending (TO_2)	935, 915	AlO-H	δ_2(ou), δ_2(in)
693		Perpendicular mode	795, 754	AlO-H	Translation modes
510		(LO_1) rocking mode	692	Si-O	Perpendicular mode
460		(TO_1) rocking motion	540	Al-O	ν(AlO_6) in the gibbsite layer
LO (longitudinal-optic), TO (transverse-optic) , ν_{as} and ν_s (antisymmetric and symmetric stretch), δ (deformation)			470, 430	Si-O	Deformation modes

Table 4. Interpretation of spectrum of samples C and D.

3.5. Thermal analysis

The results of thermal analysis of the sample including the simultaneous TG-DTA (a) and EGA (b) are shown in Fig. 15. The evaporation of water causes the first broad endothermic effect. The SOM mater begins to burn at temperature higher than 200 °C. The formed carbon dioxide is well visible in the plot of EGA. The dehydroxylation of kaolinite to metakaolinite takes place at 571 °C. The EGA is shown by increasing intensity of water vapour in the gas phase. The transformation of low quartz (trigonal) into the high quartz (hexagonal) shows very sharp endothermic peak[18] at temperature 571 °C. The sharp exothermic peak of the formation of spinel phase[20] appears at 975 °C.

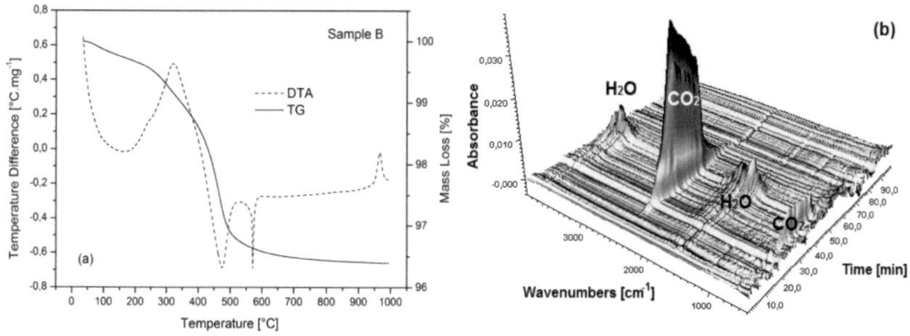

Figure 15. Thermal analysis of the sample B: TG-DTA (a) and EGA (b).

Therefore it is difficult to distinguish between individual steps of TG curve. The changes in the mass of the sample during evaporation of water, burning of the SOM and dehydroxylation were evaluated via the series of isothermal steps (Fig. 16(a)).

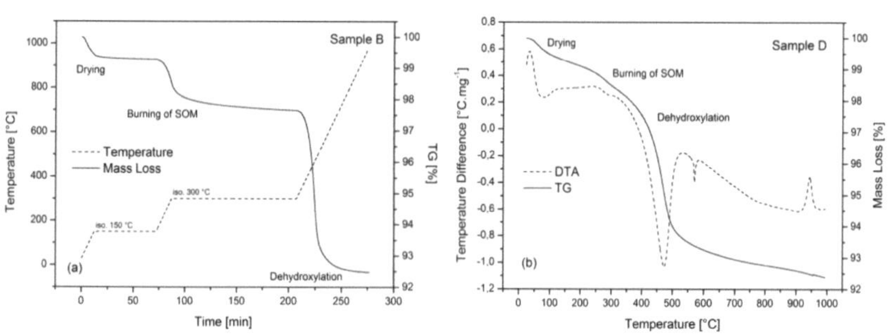

Figure 16. Evaluation of the SOM and kaolinite content in the Sample B (a) and thermal analysis of the sample D (b).

The sequence of thermal transitions for the sample B should be identified in the same way for the sample D (Fig. 16 (b)).

3.6. Thermal transformation of the clay from termite nest

Thermal transformations of kaolinite as the main part of clay phase of the termite nest material include the following processes [61-63]:

- Dehydroxylation of kaolinite into metakaolinite[19];

18) Transformation is independent on heating rate.

$$Al_2O_3 \cdot 2SiO_2 \cdot 2H_2O \xrightarrow{400-700°C} Al_2O_3 \cdot 2SiO_2 + 2H_2O(g) \tag{4}$$

- Formation of the Al-Si spinel phase [20];

$$2(Al_2O_3 \cdot 2SiO_2) \xrightarrow{925-1050°C} 2Al_2O_3 \cdot 3SiO_2 + SiO_2(amorphous) \tag{5}$$

- Formation of the mullite;

$$3(2Al_2O_3 \cdot 3SiO_2) \xrightarrow{\geq 1050°C} 2(3Al_2O_3 \cdot 2SiO_2) + 5SiO_2(amorphous) \tag{6}$$

- Formation of the cristobalite.

$$SiO_2(amorphous) \xrightarrow{\geq 1200°C} SiO_2(cristobalite) \tag{7}$$

The kinetics triplet, i.e. apparent activation energy (***E***$_a$), frequency factor (***A***) and mechanism of the process and thermodynamics of these processes are the objective of numerous works [64-66]. The processes, especially dehydroxylation, are affected by the structural changes in kaolinite. In order to estimate the influence of potential treatment or activation of clay by the termites, the kinetics of the dehydroxylation and formation of spinel phase was investigated via Kissinger kinetic equation [67]:

$$\ln\left[\frac{T_m^2}{\Theta}\right] = \ln\left[\frac{AR}{E_a} n(1-\alpha)^{n-1}\right] - \frac{E_a}{RT_m} = const. - \frac{E_a}{RT_m} \tag{8}$$

The T_m denotes the peak temperature of thermal analysis (DTA, DTG, DSC EGA...), Θ is the heating rate, n is the empirical reaction order (kinetic exponent), α is conversion degree (fractional conversion) and R is the universal gas constant. The constant term is equal to ln(AR/E_a) for the pseudo first-order type of reaction. The plot of $ln[\Theta/T_m^2]$ on reciprocal temperature $1/T_m$ (Kissinger plot) provides the straight line with slope of E_a/R.

The mechanism of the investigated process can be estimated from the shape of TDA peak. The peak parameters of TA curves enable to determine the value of kinetic exponent (n) as follows [68-70]:

$$n = \frac{2.5\, R\, T_m^2}{w_{1/2}\, E_a} \tag{9}$$

19 The course of dehydroxylation is affected by the burning of organic matter (actually there is feedback between both processes) hence The formation of water during pyrolysis of SOM decreases equilibrium constant of the process [71].

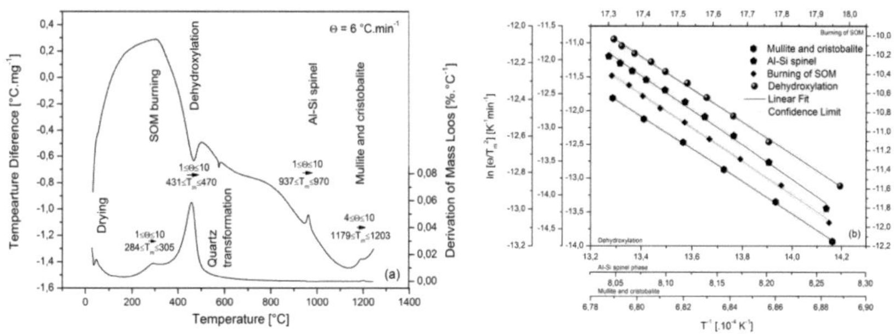

Figure 17. The shift of DTA effect with heating rate (a) and Kissinger plot (b).

The DTA results of sample D and Kissinger plot are shown in Fig. 17 (a) and (b), respectively. The formations of mullite (Eq.4) and cristobalite (Eq.5) show compound inexpressive effect on DTA hence the determined value is related to both processes. The overview of kinetic results including activation energy, frequency factor and the most probable mechanism are listed in Table 5.

Process	E_a [kJ•mol^{-1}]	A [s^{-1}]	*n*	Description
Burning of SOM	203	$3.23 \bullet 10^{16}$	0.80	Process limited by the rate of diffusion.
Dehydroxylation	201	$6.26 \cdot 10^{11}$	1.06	Random nucleation.
Al-Si spinel	852	$7.17 \cdot 10^{33}$	2.52	Diffusion controlled growth; const. or increasing nucleation rate.
Mullite and cristobalite	677	$5.85 \cdot 10^{21}$	3.79	Decreasing nucleation rate.

Table 5. The kinetics of thermal transformations of kaolin from termite nest.

The most frequently published value of the activation energy for dehydroxylation process of kaolinite ranges from 140 to 250 kJ•mol^{-1} and formation of Al-Si spinel shows E_a of 856 kJ•mol^{-1} [72]. The results indicate that the mechanism and kinetics of investigated process are not changed hence there is no supporting evidence for an activation process that would have a significant effect of termites on the structure of clay minerals.

The correlation between kinetic and thermodynamic parameters of the investigated process results from the combination of Arrhenius with Eyring or Wertera and Zenera laws related to the temperature dependence of rate constant (***k***(***T***)) [73]:

$$k(T) = A\exp\left[-\frac{E_a}{RT}\right] = \frac{k_B T}{h}\exp\left[\frac{\Delta S^{\#}}{R}\right]\exp\left[-\frac{\Delta H^{\#}}{RT}\right] = \nu\exp\left[-\frac{\Delta G^{\#}}{RT}\right] = \nu K^{\#} \tag{10}$$

where k_B, h, $\nu = k_B T/h$ and $K^{\#}$ are Boltzmann, Plank constant, vibration frequency and equilibrium constant of formation of activated complex, respectively.

The thermodynamic parameters of activated complex, including free energy ($\Delta G^{\#}$), enthalpy ($\Delta H^{\#}$) and entropy ($\Delta S^{\#}$) of process were calculated using Eyring equations:

$$\Delta H^{\#} = E_a - RT \tag{11}$$

$$\Delta S^{\#} = R\left[\ln\left(\frac{h\,A}{k_B T}\right) - 1\right] \tag{12}$$

$$\Delta G^{\#} = \Delta H^{\#} - T\,\Delta S^{\#} = -RT\,\ln K^{\#} \tag{13}$$

Table 6 shows the overview of average values of thermodynamic results calculated for activated complex according to Eq. 10 – 13 for interval of ΔT according to Fig.17.

Process	$\Delta H^{>\#}$ [kJ·mol^{-1}]	$\Delta S^{>\#}$ [J·(mol·K)$^{-1}$]	$\Delta G^{>\#}$ [kJ·mol^{-1}]	$K^{>\#}$
Burning of SOM	201	63	183	$1.96 \bullet 10^{-17}$
Dehydroxylation	195	-35	220	$2.60 \bullet 10^{-16}$
Al-Si spinel	844	385	475	$8.87 \bullet 10^{-21}$
Mullite and cristobalite	667	152	486	$4.79 \bullet 10^{-18}$

Table 6. Thermodynamics of thermal transformation of kaolin from termite nest.

The negative value of entropy ($\Delta S^{\#} < 0$) indicating formation of more ordered transition state during dehydroxylation is out of the usual findings for thermal decomposition of kaolinite. Since pure well defined sample of kaolin shows mostly $\Delta S^{\#} > 0$, SOM interacting with clay phase are responsible for this behaviour.

The assessed kinetic triplet of combustion of SOM shows that the activation energy of the process is close to the activation energy for dehydroxylation but the burning of SOM proceeds faster due to the higher ***A*** and the lower ***n***. The gas products of burning of SOM (Fig.15(b)) diffuse through the kaolinite aggregates and increase the pressure of water vapour affecting the thermodynamics of dehydroxylation via the equilibrium constant (Eq.13; the $\Delta G^{\#}$ increases due to negative value of $\Delta S^{\#}$)[20]. The temperature dependence of $\Delta G^{\#}$ determined by extrapolation of the values calculated from the experiment is shown in Fig.18 for both processes. From the kinetic point of view which is given by Eq.10, the rate constant of dehydroxylation with temperature increases more slowly than for the combustion process of SOM. In the other words, increasing pressure

20 It must be pointed that combustion of SOM is strongly affected by the content of individual kind of humic substances and the both processes (combustion of SOM and dehydroxylation) show mutual relationship. For example, intensive origin of water vapor slows down diffusion of oxygen, leads to reducing conditions, changing composition of product and slowing down the rate of process [74].

of water vapour slows down the rate of decomposition of activated complex into product and the rate of dehydroxylation process decreases.

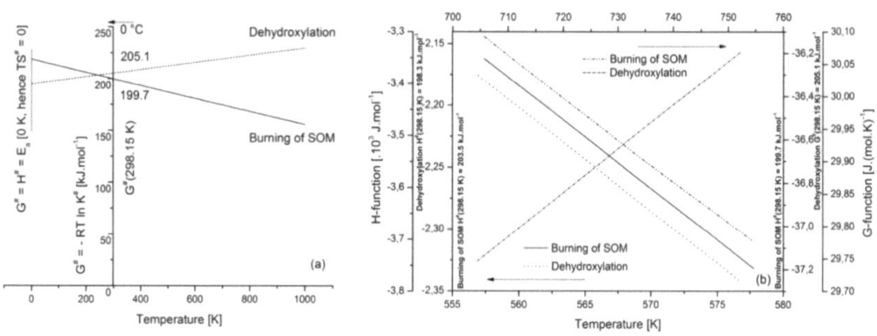

Figure 18. Temperature dependence of $\Delta \boldsymbol{G}^{\#}$ for burning of SOM and dehydroxylation (a) and temperature dependence of $\boldsymbol{G}^{\#}$-function and $\boldsymbol{H}^{\#}$-function.

In Fig. 18(b) the temperature dependence of $\Delta H^{\#}$ and $\Delta G^{\#}$ is expressed in usual form of $G^{\#}$-function and $H^{\#}$-function:

$$G^{\#} - function = \frac{G^{\#}(T) - H^{\#}(298.15\ K)}{T} \tag{14}$$

$$H^{\#} - function = H^{\#}(T) - H^{\#}(298.15\ K) \tag{15}$$

3.7. Properties of artificial termite nest material analogue

The material with the particle size distribution (Fig. 7(b)) and the composition analogical to the termite nest material was prepared from the mixture of sand and kaolin in order to evaluate material properties. The determination of dry bending strength (Fig. 19(a)) and crushing (compressive) strength were performed using 40×40×160 mm test pieces. The microphotography of fracture planes (Fig. 19(b)) shows the grains of sand surrounded by the clay phase like in the termite nest material. The mechanical properties and some other investigated material properties are listed in Table. 7.

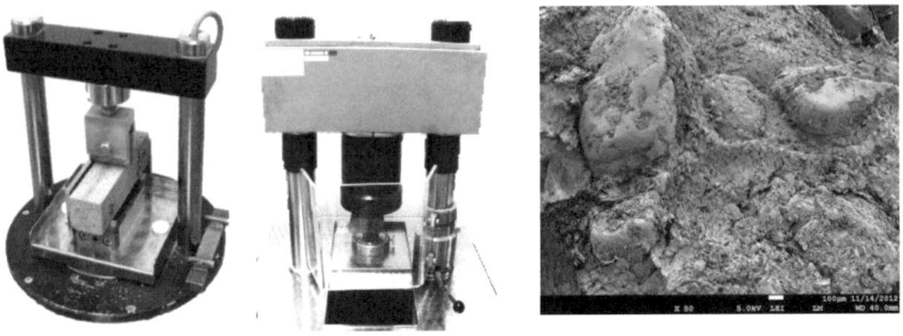

(a) (b) (c)

Figure 19. Examination of bending strength (a) crushing strength (b) and fracture plane (c).

Property		**Value**	**Description**
Pour density	[g•cm^{-3}]	2.3± 0.1	Weight of granular material related to volume unit.
Volume density	[g•cm^{-3}]	1.9 ± 0.1	The weight of volume unit of porous material.
True density	[g•cm^{-3}]	2.6 ± 0.1	Volumetric technique.
Packing density	[%]	88	The ratio of the pour and true density.
Drying shrinkage	[%]	1.6 ± 0.3	Shrinkage of dry material with regard to original dimensions.
Dry bending strength	[MPa]	0.52 ± 0.05	Measured on the dry material using testing hydraulic press with device for testing of bending and crushing strength.
Dry compressive strength	[MPa]	1.23 ±0.02	Measured on the dry material using testing hydraulic press with device for testing of bending and crushing strength.
Thermal conductivity	[W(m K)$^{-1}$]	0.5±0.1	Coefficient of thermal conductivity (TCI, C-Therm).

Table 7. Properties of artificial termite nest material analogue.

The material shows a small drying shrinkage as the consequence of high content of aggregates. The lower value of pour density is reached in comparison to the termite nest material (Table 1). That means a lack of fine particles of sand in the material and gives the reason for measured poor value of compressive strength. The results indicate that an optimal moisture and weighted granulometry is necessary to reach the good material properties while the role of SOM that usually increases plasticity of clay minerals pastes is of peripheral importance for the materials with as high content of aggregates as in the investigated one.

4. Conclusion

Behind all of the presented experimental data there is one significant foundation. The building engineering of soil-dwelling termites shows us that the relationships between the particle size distributions, particle shape, packing density of aggregate grains and the plasticity as well as the amount of binding phase (clays) are sufficient to cover the surface of the aggregates by the continuous thin layer as well as the amount of mixing water has the same relevance for the construction of clay buildings as for the preparation of concrete structures. Therefore the research that finds out and describes the mutual influence of these factors and condition of treatment on the resulting properties of clay based material (similarly as for preparation of concrete mixtures) is required.

Acknowledgements

This work was supported by the project of Ministry of Education, Youth and Sports of the Czech Republic No. CZ.1.05/2.1.00/01.0012 "Centre for Materials Research at FCH BUT" supported by operational program Research and Development for Innovations.

Author details

Petr Ptáček*, Jiří Brandštetr, František Šoukal and Tomáš Opravil

*Address all correspondence to: ptacek@fch.vutbr.cz

Brno University of Technology, Faculty of Chemistry, Centre for Material Research, Purkyňova, Brno, Czech Republic

References

[1] Pearce, M. J, & Waite, B. S. A list of termite genera (Isoptera) with comments on taxonomic changes and distribution. Sociobiology (1994). , 23(3), 247-63.

[2] Chouvenc, T, & Su, N. Yao, Grace JK. Fifty years of attempted biological control of termites- Analysis of failure. Biological Control (2011). , 59(2), 69-82.

[3] Verma, M, Sharma, S, & Prasad, R. Biological alternatives for termite control: A review. International Biodeterioration & Biodegradation (2009). , 63(8), 959-72.

[4] Hartke, T. R, & Baer, B. The mating biology of termites: a comparative review. Animal Behaviour (2011). , 82(5), 927-36.

[5] Varma, A, Kolli, B. K, Paul, J, Saxena, S, & König, H. Lignocellulose degradation by microorganisms from termite hills and termite guts: A survey on the present state of art. FEMS Microbiology Reviews (1994)., 15(1), 9-28.

[6] Emerson, A. E. The biogeography of termites. Bulletin of the American Museum of Natural History (1952)., 99(3), 217-25.

[7] Holmgren, N. T, & Termitenstudien, I. Anatomische Untersuchungen. Kungliga Svenska Vetenskapsakademiens Handlingar: Almqvist u. Wiksel; (1909).

[8] Holmgren, N. T, & Termitenstudien, I. I. Systematik der Termiten. Die Familien Mastotermitidae, Protermitidae und Mesotermitidae. Kungliga Svenska Vetenskapsakademiens Handlingar: Almqvist u. Wiksel; (1911).

[9] Holmgren, N. T. Termitenstudien III. Systematik der Termiten. Die Familien Metatermitidae. Kungliga Svenska Vetenskapsakademiens Handlingar: Almqvist u. Wiksel; (1912).

[10] Snyder, T. E. Catalog of the termites (Isoptera) of the world. Smithsonian Miscellaneous Collections. University of Minnesota: Smithsonian Institution; (1949).

[11] Grassé, P-P. editor. Ordre des Isoptères ou Termites. Traité de Zoologie IX. Paris: Masson; (1949).

[12] Emerson, A. E. A review of the Mastotermitidae (Isoptera), including a new fossil genus from Brazil. American Museum novitates (1965)., 2236, 1-46.

[13] Donovan, S. E, Jones, D. T, Sands, W. A, & Eggleton, P. Morphological phylogenetics of termites (Isoptera). Biological Journal of the Linnean Society (2000). , 70(3), 467-513.

[14] Ahmad, M. The phylogeny of termite genera based on imago-worker mandibles. Bulletin of the American Museum of Natural History (1950)., 1950(95), 2-37.

[15] Johnson, R. A. Configuration of the digestive tube as an aid to identification of worker Termitidae (Isoptrera). Systematic Entomology (1979)., 4(1), 31-38.

[16] Noirot, C. The gut of termites (Isoptera). Comparative anatomy, systematics, phylogeny. I. Lower termites. Annals de la Societé Entomologique de France (1995)., 31(3), 197-226.

[17] Desalle, R, Gatesy, J, Wheeler, W, & Grimaldi, D. DNA sequences from a fossil termite in Oligo-Miocene amber and their phylogenetic implication. Science (1992). , 257, 1933-36.

[18] Krishna, K. A generic revision and phylogenetic study of the family Kalotermitidae (Isoptera). Bulletin of the American Museum of Natural Histrory (1961). , 122(4), 303-408.

[19] Emerson, A. E, & Krishna, K. The termite family Serritermitidae (Isoptera). American Museum Novilates (1975)., 2570, 1-31.

[20] Thorne, B. L. Evolution of eusociality in termites. Annual Review of Ecology and Systematics (1997)., 28, 27-54.

[21] Thorne, B. L, & Carpenter, J. M. Phylogeny of the Dictyoptera. Systematic Entomology (1992)., 17(3), 253-68.

[22] Lo, N, Tokuda, G, Watanabe, H, Rose, H, Slaytor, M, Maekawa, K, Bandi, C, & Noda, H. Evidence from multiple gene sequences indicates that termites evolved from wood-feeding cockroaches. Current Biology (2000)., 10(13), 801-4.

[23] Thorne, B. L, Grimaldi, D. A, & Krishna, K. Early fossil history of the termites. In: Abe T, Bignell D, Higashi M. (Eds.), Termites: Evolution, Sociality, Symbiosis, Ecology. Dordrecht: Kluwer Academic Publishers; (2000).

[24] Ware, J. L, Grimaldi, D. A, & Engel, M. S. The effects of fossil placement and calibration on divergence times and rates: An example from the termites (Insecta: Isoptera) Arthropod Structure & Development (2010).

[25] Cleveland, L. R, Hall, S. K, & Sanders, E. P. Collier J: The wood-feeding roach Cryptocercus, its protozoa, and the symbiosis between protozoa and roach. Memoirs of the American Academy Of Arts And Sciences (1934)., 17, 185-342.

[26] Grimaldi, D. A fossil mantis (Insecta: Mantodea) in Cretaceous amber of New Jersey, with comments on the early history of Dictyoptera. American Museum novitates (1997).

[27] Gay, F. J, & Calaby, J. H. Termites of the Australian region. In; Krishna K Weesner FM (Eds.) Biology of Termites, Vol. II; Academic Press; (1970).

[28] Bordy, E. M, Bumby, A. J, Catuneanu, O, & Eriksson, P. G. Possible trace fossils of putative termite origin in the Lower Jurassic (Karro Supergroup) of South Africa and Lesotho. South African Journal of Science (2009)., 105, 356-62.

[29] Emerson, A. E. Geographic origins and dispersion of termite genera. Fieldiana: Zoology; Chicago: Natural History Museum; (1955).

[30] Tarver, M. R, Schmelz, E. A, & Scharf, M. E. Soldier caste influences on candidate primer pheromone levels and juvenile hormone-dependent caste differentiation in workers of the termite Reticulitermes flavipes. Journal of Insect Physiology (2011)., 57(6), 771-7.

[31] Thompson, G. Termites. Tropical Topics Newsletter Tropical Savanna, Australia; (2000). (64)

[32] Himuro ChYokoi T, Matsuura K. Queen-specific volatile in a higher termite Nasutitermes takasagoensis (Isoptera: Termitidae). Journal of Insect Physiology (2011)., 57(7), 962-5.

[33] Thorne, B. L. Termite-termite interactions: workers as an agonistic caste. Psyche (1982).

[34] Binder, B. F. Intercolonial aggression in the subterranean termite Heterotermes aureus (Isoptera: Rhinotermitidae). Psyche (1988).

[35] Ishikawa, Y, & Miura, T. Hidden aggression in termite workers: plastic defensive behaviour dependent upon social context. Animal Behaviour (2012)., 83(3), 737-45.

[36] Šobotník, J, Jirošová, A, & Hanus, R. Review: Chemical warfare in termites. Journal of Insect Physiology (2010)., 56(9), 1012-21.

[37] Deligne, J, Quennedey, A, & Blum, M. S. The enemies and defense mechanisms of termites. In: Hermann, H.R. (Ed.), Social Insects, vol. II. Academic Press, New York, (1981)., 1981, 1-76.

[38] Prestwich, G. D. Defense mechanisms of termites. Annual Review of Entomology (1984)., 29(1), 201-32.

[39] Haverty, M. I, & Howard, R. W. Production of Soldiers and Maintenance of Soldier Proportions by Laboratory Experimental Groups of Reticulitermes Flavipes (Kollar) and Reticulitermes Virginicus (Banks) (Isoptera: Rhinotermitidae). Insectes Sociaux (1981)., 28(1), 32-9.

[40] Korb, J, Roux, E. A, & Lenz, M. Proximate factors influencing soldier development in the basal termite Cryptotermes secundus (Hill). Insectes Sociaux (2003). , 50(4), 299-303.

[41] Slaytor, M. Cellulose digestion in termites and cockroaches: What role do symbionts play? Comparative Biochemistry and Physiology Part B: Comparative Biochemistry (1992)., 103(4), 775-84.

[42] Buxton, R. D. Termites and the turnover of dead wood in an arid tropical environment. Oecologia (1981)., 51(3), 379-84.

[43] Hyodo, F, Inoue, T, Azuma, J, Tayasu, I, & Abe, I. T. Role of the mutualistic fungus in lignin degradation in the fungus-growing termite Macrotermes gilvus (Isoptera; Macrotermitinae). Soil Biology and Biochemistry (2000)., 32(5), 653-8.

[44] Sieber, R, & Leuthold, R. H. Behavioural elements and their meaning in incipient laboratory colonies of the fungus-growing termite Macrotermes michaelseni (Isoptera: Macrotermitinae). Insectes Sociaux (1981)., 28(4), 371-82.

[45] Mattéotti Ch., Bauwens J, Brasseur C, Tarayre C, Thonart P, Destain J, Francis F, Haubruge E, De Pauw E, Portetelle D, Vandenbol M. Identification and characterization of a new xylanase from Gram-positive bacteria isolated from termite gut (Reticulitermes santonensis). Protein Expression and Purification (2012)., 83(2), 117-127.

[46] Grassé, P, & Noirot, P. C. Le meule des termites champignonnistes et sa signification symbiotique. Annales des Sciences Naturelles. Zoologie et Biologie Animale (1958). , 11, 113-28.

[47] Oppert, C, Klingeman, W. E, Willis, J. D, Oppert, B, & Jurat-fuentes, J. L. Prospecting for cellulolytic activity in insect digestive fluids. Comparative Biochemistry and Physiology Part B: Biochemistry and Molecular Biology (2010). , 155(2), 145-54.

[48] Eriksson, K. EL. Concluding remarks: Where do we stand and where are we going?: Lignin biodegradation and practical utilization. Journal of Biotechnology (1993). , 30(1), 149-58.

[49] Tuomela, M, Vikman, M, Hatakka, A, & Itävaara, M. Biodegradation of lignin in a compost environment: a review. Bioresource Technology (2000). , 72(2), 169-83.

[50] Grimaldi, D, & Engel, M. S. Evolution of the Insects. Cambridge University Press; (2005).

[51] Eggleton, P, & Tayasu, I. Feeding groups, lifetypes and the global ecology of termites. Ecological Research (2001). , 16(5), 941-60.

[52] Jouquet, P, Traoré, S, Choosai, C, Hartmann, C, & Bignell, D. Influence of termites on ecosystem functioning. Ecosystem services provided by termites. European Journal of Soil Biology (2011). , 47(4), 215-22.

[53] Black HIJOkwakol MJN. Agricultural intensification, soil biodiversity and agroecosystem function in the tropics: the role of termites. Applied Soil Ecology (1997). , 6(1), 37-53.

[54] Whitford, W. G, Ludwig, J. A, & Noble, J. C. The importance of subterranean termites in semi-arid ecosystems in south-eastern Australia. Journal of Arid Environments (1992). , 22, 87-91.

[55] Dawes, T. Z. Impacts of habitat disturbance on termites and soil water storage in a tropical Australian savannah. Pedobiologia (2010). , 53(4), 241-6.

[56] Ackerman, I. L, Teixeira, W. G, Riha, S. J, & Lehmann, J. Fernandes ECM. The impact of mound-building termites on surface soil properties in a secondary forest of Central Amazonia. Applied soil ecology (2007). , 37, 267-76.

[57] Mando, A. The impact of termites and mulch on the water balance of crusted Sahelian soil. Soil Technology (1997). , 11(2), 121-38.

[58] Chouvenc, T, Su, N-Y, & Grace, J. K. Fifty years of attempted biological control of termites- Analysis of a failure. Biological Control (2011). , 59(2), 69-82.

[59] Wang, C, & Powell, J. E. Cellulose bait improves the effectiveness of Metarhizium anisopliae as a microbial control of termites (Isoptera: Rhinotermitidae). Biological Control (2004). , 30(2), 523-9.

[60] Chmielová, M, & Weiss, Z. Determination of structural disorder degree using an XRD profile fitting procedure. Application to Czech kaolins. Applied Clay Science (2002).

[61] Chen, C. Y, & Tuan, W. H. The processing of kaolin powder compact. Ceramics International (2001). , 27(7), 795-800.

[62] Chen, Y-F, Wang, M-C. h, & Hon, M-H. Phase transformation and growth of mullite in kaolin ceramics. Journal of the European Ceramic Society (2004). , 24(8), 2389-97.

[63] Frost, R. L, Horváth, E, Makó, É, Kristóf, J, & Rédey, Á. Slow transformation of mechanically dehydroxylated kaolinite to kaolinite-an aged mechanochemically activated formamide-intercalated kaolinite study. Thermochimica Acta (2003).

[64] Balek, V, & Murat, M. The emanation thermal analysis of kaolinite clay minerals. Thermochimica Acta (1996).

[65] Chakraborty, A. K. DTA study of preheated kaolinite in the mullite formation region, Thermochimica Acta (2003).

[66] Heide, K, & Földvari, M. High temperature mass spectrometric gas-release studies of kaolinite $Al_2[Si_2O_5(OH)_4]$ decomposition. Thermochimica Acta (2006).

[67] Kissinger, H. E. Reaction Kinetics in Differential Thermal Analysis. Analytical chemistry (1957). , 29(11), 1702-6.

[68] Friedman, H. L. Kinetics of thermal degradation of char-forming plastics from thermogravimetry. Application to a phenolic plastic. Journal of Polymer Science Part C: Polymer Symposia (1964). , 6(1), 183-5.

[69] Augis, J. A, & Bennett, J. D. Calculation of Avrami parameters for heterogeneous solid-state reactions using a modification of Kissinger method. Journal of Thermal Analysis (1978). , 13(2), 283-92.

[70] Ray, C. S, Yang, Q, Huang, W-H, & Day, D. E. Surface and internal crystallization in glasses as determined by differential thermal analyses. Journal of the American Ceramic Society (1996). , 79(12), 3155-60.

[71] Ptácek, P, Šoukal, F, Opravil, T, Havlica, J, & Brandštetr, J. The kinetic analysis of the thermal decomposition of kaolinite by DTG technique. Powder Technology (2010). , 208(1), 20-5.

[72] Ptácek, P, Šoukal, F, Opravil, T, Nosková, M, Havlica, J, & Brandštetr, J. The kinetics of Al-Si spinel phase crystallization from calcined kaolin. Journal of Solid State Chemistry (2010). , 183(11), 2565-2569.

[73] Straszko, J, Olszak-humienik, M, & Možejko, J. Kinetics of thermal decomposition of $ZnSO_4$ $7H_2O$. Thermochimica Acta (1997).

[74] Ptácek, P, Šoukal, F, Opravil, T, Nosková, M, Havlica, J, & Brandštetr, J. The non-isothermal kinetics analysis of the thermal decomposition of kaolinite by Effluent Gas Analysis technique. Powder Technology (2010). , 2010(2), 272-276.